VOLUME ONE HUNDRED AND TWO

Current Topics in DEVELOPMENTAL BIOLOGY

Gametogenesis

CURRENT TOPICS IN DEVELOPMENTAL BIOLOGY

"A meeting-ground for critical review and discussion of developmental processes"
A.A. Moscona and Alberto Monroy (Volume 1, 1966)

VOLUME ONE HUNDRED AND TWO

CURRENT TOPICS IN DEVELOPMENTAL BIOLOGY

Gametogenesis

Edited by

PAUL M. WASSARMAN
Department of Developmental and Regenerative Biology, Mount Sinai School of Medicine, New York, NY, USA

ELSEVIER

AMSTERDAM • BOSTON • HEIDELBERG • LONDON
NEW YORK • OXFORD • PARIS • SAN DIEGO
SAN FRANCISCO • SINGAPORE • SYDNEY • TOKYO
Academic Press is an imprint of Elsevier

Academic Press is an imprint of Elsevier
525 B Street, Suite 1900, San Diego, CA 92101-4495, USA
225 Wyman Street, Waltham, MA 02451, USA
The Boulevard, Langford Lane, Kidlington, Oxford, OX51GB, UK
32, Jamestown Road, London NW1 7BY, UK
Radarweg 29, PO Box 211, 1000 AE Amsterdam, The Netherlands

First edition 2013

ISBN: 978-0-12-416024-8
ISSN: 0070-2153

Printed and bound by CPI Group (UK) Ltd, Croydon, CR0 4YY
Transferred to Digital Printing, 2013

CONTENTS

CONTRIBUTORS

David F. Albertini
Department of Molecular and Integrative Physiology, Kansas University Medical Center, Kansas City, Missouri, USA

Manfred Alsheimer
Department of Cell and Developmental Biology, Biocenter, University of Würzburg, Würzburg, Germany

Binyamin D. Berkovits
Department of Genetics and Development, Columbia University Medical Center, New York, USA

Robert Blelloch
Department of Urology, Eli and Edythe Broad Center of Regeneration Medicine and Stem Cell Research, Center for Reproductive Sciences, University of California, San Francisco, California, USA

Matthew S. Cook
Department of Urology, Eli and Edythe Broad Center of Regeneration Medicine and Stem Cell Research, Center for Reproductive Sciences, University of California, San Francisco, California, USA

A. Darszon
Departamento de Genética del Desarrollo y Fisiología Molecular, Instituto de Biotecnología, Universidad Nacional Autónoma de México, Cuernavaca, Morelos, Mexico

Janet E. Holt
Center for Reproductive Sciences & School of Biomedical Sciences, University of Newcastle, Callaghan, New South Wales, Australia

Rolf Jessberger
Institute of Physiological Chemistry, Medical Faculty Carl Gustav Carus, Dresden University of Technology, Dresden, Germany

Keith T. Jones
Center for Reproductive Sciences & School of Biomedical Sciences, University of Newcastle, Callaghan, New South Wales, Australia, Centre for Biological Sciences, Southampton University, SO17 1BJ, UK

Martin P. Kracklauer
Department of Physiology, Wayne State University Medical School, Detroit, Michigan, USA

Amber R. Krauchunas
Department of Molecular Biology and Genetics, Cornell University, Ithaca, New York, USA

Simon I.R. Lane
Center for Reproductive Sciences & School of Biomedical Sciences, University of Newcastle, Callaghan, New South Wales, Australia, Centre for Biological Sciences, Southampton University, SO17 1BJ, UK

Xiajun Li
Black Family Stem Cell Institute, Department of Developmental and Regenerative Biology, Department of Oncological Sciences, and Graduate School of Biological Sciences, Mount Sinai School of Medicine, New York, USA

S. Darlene Limback
Department of Molecular and Integrative Physiology, Kansas University Medical Center, Kansas City, Missouri, USA

Jana Link
Department of Cell and Developmental Biology, Biocenter, University of Würzburg, Würzburg, Germany

Eveline S. Litscher
Department of Developmental and Regenerative Biology, Mount Sinai School of Medicine, New York, USA

Huacheng Luo
Key laboratory for the Genetics of Developmental and Neuropsychiatric Disorders (Ministry of Education), Bio-X Institutes, Shanghai Jiao Tong University, and School of Life Sciences and Biotechnology, Shanghai Jiao University, Shanghai, China

Lynda K. McGinnis
Department of Molecular and Integrative Physiology, Kansas University Medical Center, Kansas City, Missouri, USA

François McNicoll
Institute of Physiological Chemistry, Medical Faculty Carl Gustav Carus, Dresden University of Technology, Dresden, Germany

Makoto C. Nagano
Department of Obstetrics and Gynecology, Division of Experimental Medicine, McGill University, Montreal, Quebec, Canada

G. Orta
Departamento de Genética del Desarrollo y Fisiología Molecular, Instituto de Biotecnología, Universidad Nacional Autónoma de México, Cuernavaca, Morelos, Mexico

Sandrine Puverel
Neural Development Section, Mouse Cancer Genetics Program, Center for Cancer Research, NCI, Frederick, Maryland, USA

L. Salkoff
Department of Anatomy and Neurobiology, Washington University School of Medicine, St Louis, Missouri, USA

C.M. Santi
Department of Anatomy and Neurobiology, Washington University School of Medicine, St Louis, Missouri, USA

Michelle Stevense
Institute of Physiological Chemistry, Medical Faculty Carl Gustav Carus, Dresden University of Technology, Dresden, Germany

Lino Tessarollo
Neural Development Section, Mouse Cancer Genetics Program, Center for Cancer Research, NCI, Frederick, Maryland, USA

C.L. Treviño
Departamento de Genética del Desarrollo y Fisiología Molecular, Instituto de Biotecnología, Universidad Nacional Autónoma de México, Cuernavaca, Morelos, Mexico

P.E. Visconti
Department of Veterinary and Animal Science, University of Massachusetts, Amherst Massachusetts, USA

Hu Wang
Key laboratory for the Genetics of Developmental and Neuropsychiatric Disorders (Ministry of Education), Bio-X Institutes, Shanghai Jiao Tong University, China, and School of Life Sciences and Biotechnology, Shanghai Jiao University, Shanghai, China

Paul M. Wassarman
Department of Developmental and Regenerative Biology, Mount Sinai School of Medicine, New York, USA

Mariana F. Wolfner
Department of Molecular Biology and Genetics, Cornell University, Ithaca, New York, USA

Debra J. Wolgemuth
Department of Genetics and Development, Department of Obstetrics and Gynecology, The Institute of Human Nutrition, and The Herbert Irving Comprehensive Cancer Center, Columbia University Medical Center, New York, USA

Ji Wu
Key laboratory for the Genetics of Developmental and Neuropsychiatric Disorders (Ministry of Education), Bio-X Institutes, Shanghai Jiao Tong University, and School of Life Sciences and Biotechnology, Shanghai Jiao University, Shanghai, China

Jonathan R. Yeh
Department of Obstetrics and Gynecology, Division of Experimental Medicine, McGill University, Montreal, Quebec, Canada

David Zarkower
Department of Genetics, Cell Biology, and Development, University of Minnesota, Minneapolis, Minnesota, USA

PREFACE

Reproduction is a fundamental attribute of all living things. Of the 14 million or so species on Earth, approximately 99% of them reproduce sexually. It is gametogenesis, the production of specialized germ cells during oogenesis in females and spermatogenesis in males, that provides the promise of a new individual exhibiting characteristics of the species. Consequently, gametogenesis has been of considerable interest to scientists since as far back as the latter part of the seventeenth century and this interest continues today. A perusal of PubMed indicates that more than 20,000 papers have been published since 1910 under the heading "gametogenesis."

This volume of *Current Topics in Developmental Biology* includes 14 chapters by experienced investigators in the field and highlights several areas of recent research on gametogenesis, from identification of spermatogonial stem cells and regulation of genomic imprinting to identification of molecules involved during maturation and activation of eggs and in fertility.

In Chapter 1, Francois McNicoll, Michelle Stevense, and Rolf Jessberger review the roles of cohesin in sister chromatid cohesion as well as other vital processes during meiosis. At least 18 different cohesin core complexes have been identified during meiosis and at least six core complexes have been identified in somatic cells.

In Chapter 2, Xiajun Li reviews research that suggests "Genomic imprinting is a parental effect established in mammalian germ cells." Mammals have more than 100 imprinted genes that are regulated by a cis-acting imprinting control region, or ICR. The ICR contains a germline-derived differentially methylated region, or DMR, a DNA methylation imprint. Maintenance of the imprint is dependent on two maternal effect genes, *Zfp57* and *PGC7/Stella*, and during development, there is erasure, resetting, restoration, and maintenance of differential methylation.

In Chapter 3, Makoto C. Nagano and Jonathan R. Yeh review research on spermatogonial stem cells, or SSCs. The authors focus on the heterogeneity of primitive spermatogonia and maintenance and exit from the stem cell state. Using a comparative approach, they raise hypotheses for potential mechanisms for SSC fate decisions involving self-renewal and differentiation.

Chapter 4, by Ji Wu, Huacheng Luo, and Hu Wang, reviews current information about the origin, location, characteristics, regulatory mechanisms, and role of mammalian germline stem cells in spermatogenesis and oogenesis.

In Chapter 5, Martin P. Kracklauer, Jana Link, and Manfred Alsheimer review studies on LINC complexes and nucleocytoskeletal connections in gametogenesis. Particular attention is given to mechanisms involved in nuclear shaping and in controlling sperm head formation.

In Chapter 6, Matthew S. Cook and Robert Blelloch review current information about the role of miRNAs, endo-siRNAs, and piRNAs in post-transcriptional regulation during germline development in various organisms, particularly the mouse. Primordial germ cells depend on these small RNAs for their specification, migration, and differentiation into competent gametes.

Chapter 7, by Janet E. Holt, Simon I. R. Lane, and Keith T. Jones, reviews research on "The control of meiotic maturation in mammalian oocytes." The focus is on a kinase, CDK1, whose activity is regulated by the anaphase-promoting complex, APC. The activity of APC is regulated in turn by the spindle assembly checkpoint. As a result of this regulation, CDK1 activity is suppressed during meiotic arrest of oocytes and raised during meiotic maturation.

In Chapter 8, Lynda K. McGinnis, S. Darlene Limback, and David F. Albertini review modalities of intercellular signaling with respect to regulation of oocyte gene expression, metabolism, spatial patterning, and the cell cycle during oogenesis. They consider signaling during assembly of primordial follicles and ovulation, as well as feedback loops that coordinate the timing of oogenesis and folliculogenesis.

In Chapter 9, Paul M. Wassarman and Eveline S. Litscher review research on the biosynthesis of the zona pellucida, ZP, in "Biogenesis of the mouse egg's extracellular coat, the zona pellucida." Expression of mouse ZP genes, synthesis of ZP proteins, and assembly of the proteins into an extracellular coat are restricted to growing oocytes, that is, are gender specific and are essential for normal fertilization of eggs as well as early embryogenesis.

Chapter 10, by Amber R. Krauschunas and Mariana F. Wolfner, reviews "Molecular changes during egg activation," an essential step along the path to creation of an embryo. In particular, the authors review the role of calcium and zinc during egg activation, as well as the macromolecular changes in protein and messenger-RNA composition that accompany activation of eggs.

In Chapter 11, Binyamin D. Berkovits and Debra J. Wolgemuth review "The role of the double bromodomain-containing *BET* genes during mammalian spermatogenesis." These genes encode transcriptional regulators and histone-interacting chromatin remodelers. In particular, the authors discuss the testis-specific gene, *Brdt*, which is essential for proper chromatin organization, regulation of transcription, and cotranscriptional processing during spermatogenesis.

In Chapter 12, David Zarkower reviews research on transcription factors containing a DM domain that regulate gonadal differentiation, gametogenesis, or sex determination in several invertebrate and vertebrate species. One of these factors, DMRT1, is required for germ cell differentiation in males, regulates meiosis in males and females, and is required for establishment and maintenance of male fate in the testis.

Chapter 13, by Sandrine Puverel and Lino Tessarollo, reviews the role of "RanBPM, a scaffolding protein for gametogenesis." RanBPM is a critical factor in establishing fertility in both male and female mice and here the RanBPM sterility phenotype is examined in relation to other genetic mutations affecting gametogenesis and fertility.

Chapter 14, by Celia Santi, Gerardo Orta, Lawrence Salkoff, Pablo Visconti, Alberto Darszon, and Claudia Treviño, reviews research on two ion channels, K^+ and Cl^- channels, that participate in several sperm functions, including motility, capacitation, maturation, and the acrosome reaction.

Finally, I thank all of the authors for their scholarly contributions to this volume and for their considerable efforts to complete their manuscripts on time in order to meet the publisher's deadlines.

Paul M. Wassarman

CHAPTER ONE

Cohesin in Gametogenesis

François McNicoll[1], Michelle Stevense[1], Rolf Jessberger[2]
Institute of Physiological Chemistry, Medical Faculty Carl Gustav Carus, Dresden University of Technology, Dresden, Germany
[1]Equally contributed.
[2]Corresponding author: e-mail address: rolf.jessberger@tu-dresden.de

Contents

Abstract

Sister chromatid cohesion depends on cohesin, a tripartite complex that forms ring structures to hold sister chromatids together in mitosis and meiosis. Meiocytes feature a multiplicity of distinct cohesin proteins and complexes, some meiosis specific, which serve additional functions such as supporting synapsis of two pairs of sister chromatids and determining the loop–axis architecture of prophase I chromosomes. Despite considerable new insights gained in the past few years into the localization and function of some cohesin proteins, and the recent identification of yet another meiosis-specific cohesin subunit, a plethora of open questions remains, which concern not only fundamental germ cell biology but also the consequences of cohesin impairment for human reproductive health.

1. INTRODUCTION

Cohesins are ring-shaped protein complexes whose multiple functions depend mostly on their ability to bring two different DNA molecules or two distant parts of the same DNA molecule into close proximity. Originally discovered for their essential role in sister chromatid cohesion (SCC), they were found to participate in various nuclear processes such as the

Current Topics in Developmental Biology, Volume 102
ISSN 0070-2153
http://dx.doi.org/10.1016/B978-0-12-416024-8.00001-5

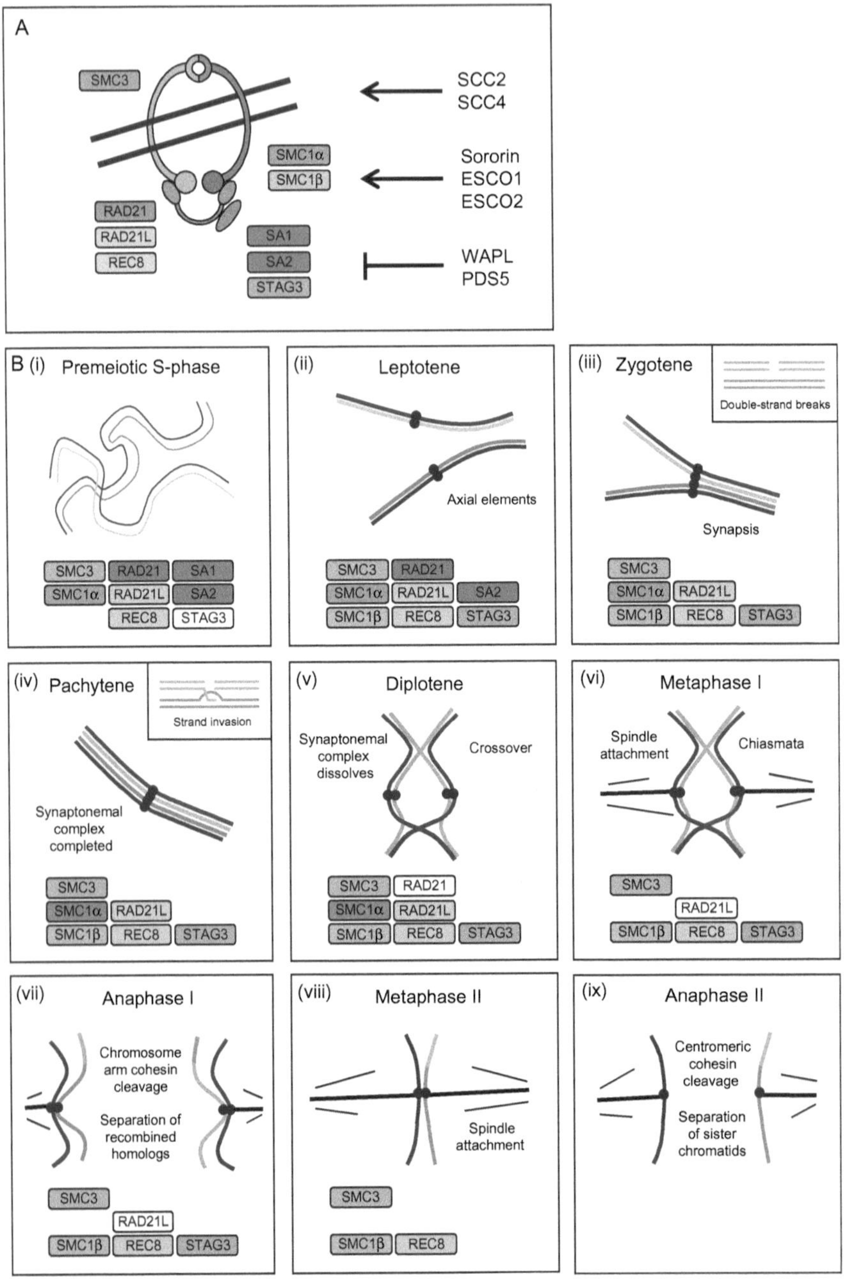
A
SMC3
SMC1α
SMC1β
RAD21
RAD21L
REC8
SA1
SA2
STAG3
SCC2
SCC4
Sororin
ESCO1
ESCO2
WAPL
PDS5
B (i) Premeiotic S-phase
SMC3 RAD21 SA1
SMC1α RAD21L SA2
REC8 STAG3
(ii) Leptotene
Axial elements
SMC3 RAD21
SMC1α RAD21L SA2
SMC1β REC8 STAG3
(iii) Zygotene
Double-strand breaks
Synapsis
SMC3
SMC1α RAD21L
SMC1β REC8 STAG3
(iv) Pachytene
Strand invasion
Synaptonemal complex completed
SMC3
SMC1α RAD21L
SMC1β REC8 STAG3
(v) Diplotene
Synaptonemal complex dissolves
Crossover
SMC3 RAD21
SMC1α RAD21L
SMC1β REC8 STAG3
(vi) Metaphase I
Spindle attachment
Chiasmata
SMC3
RAD21L
SMC1β REC8 STAG3
(vii) Anaphase I
Chromosome arm cohesin cleavage
Separation of recombined homologs
SMC3
RAD21L
SMC1β REC8 STAG3
(viii) Metaphase II
Spindle attachment
SMC3
SMC1β REC8
(ix) Anaphase II
Centromeric cohesin cleavage
Separation of sister chromatids

Figure 1.1 Cohesin in meiosis. (A) Model of the cohesin ring encircling two chromatids. SMC3 (gray) is present in all cohesin complexes. There are two SMC1 genes and proteins: SMC1α (dark blue) and a meiosis-specific SMC1β (light blue). The tripartite ring closes via the association of an α-kleisin subunit, of which three variants exist: the ubiquitous RAD21 (dark turquoise), and two meiosis-specific forms, REC8 and RAD21L (light turquoise). A third component, of which there are three variants, associates with the complex via binding to the α-kleisin: canonical SA1 or SA2 (dark orange) or the meiosis-specific STAG3 (SA3) (light orange). Loading of the cohesin complex onto chromosomes and its maintenance on chromosomes are controlled by loading factors, establishment and antiestablishment factors. Loading of the cohesin complex onto mitotic chromosomes is performed by a complex of SCC2–SCC4 (kollerin) and cohesin dissociation by PDS5-WAPL (releasin). Cohesin acetyltransferases (ESCO1 and ESCO2) are necessary to establish cohesion in mitotic S-phase through acetylation of SMC3, which recruits Sororin, a maintenance factor that counteracts releasin activity during mitotic S and G2 phases. (B) Scheme of the meiotic stages from (i) to (ix), showing the progression of one pair of homologous chromosomes (one red and the other blue, each drawn as two single lines representing sister chromatids without chromatin loops for illustration purposes) through the different stages. In reality, the progression is continuous. The presence of specific cohesin proteins as far as known is presented in the lower sections of the panels (neither the relative amounts nor the potential interactions between the subunits are taken into consideration). There are conflicting reports on the presence of cohesin proteins at some of the stages. In such cases, the cohesins are depicted in noncolored boxes. There is very little information about SA1. The data illustrated here are based on analyses of spermatocytes. (i) During premeiotic S-phase, newly formed sister chromatids (red or blue) are held together by cohesin complexes (not shown). The mitotic cohesin subunits are present. The meiosis-specific SMC1β is not yet present; however, REC8, RAD21L as well as perhaps STAG3 in a few cells are already beginning to be expressed; (ii) during leptotene, the chromosomes begin to condense and axial elements form, STAG3 and SMC1β are now present on the chromosomes; (iii) synapsis of homologous chromosomes begins during zygotene, facilitated by frequent DNA DSBs, of which one is represented in the inset, DSBs are initiated in leptotene; (iv) formation of the SC is complete in pachytene with all homologs fully synapsed, meiotic recombination proceeds as indicated in the inset; (v) crossovers (two examples are shown) which have been formed between homologs during pachytene, physically link the homologs together in diplotene. At this stage, the SC has largely dissolved; however, sister chromatid cohesion is maintained. Oocytes will arrest shortly after this stage, at a stage called dictyate arrest (not shown)—for many years in humans—and cohesion must be maintained during this time; (vi) at metaphase I, spindle attachments are formed at mono-oriented centromeres of homologs and chiasmata still resist microtubule pulling forces; (vii) cleavage of the α-kleisin subunit of cohesin by separase results in separation of homologs as chiasmata resolve in absence of arm cohesion. Centromeric cohesion is protected by Shugoshin/PP2A (not shown) and a pool of dephosphorylated cohesin subunits; (viii) sister chromatids align on the metaphase plate during metaphase II and spindle microtubules attach to bioriented kinetochores; (ix) cohesion is lost and sister chromatids are pulled apart in anaphase II, creating haploid gametes. (See Color Insert.)

assembly of DNA replication factories, DNA double-stranded break (DSB) repair, chromosome condensation and morphology, transcriptional control, T cell receptor rearrangement, and mitotic spindle assembly (for recent reviews, see Haering & Jessberger, 2012; Merkenschlager, 2010; Nasmyth, 2011; Nasmyth & Haering, 2009; Wood, Severson, & Meyer, 2010). Cohesins are essential for meiosis, where they play several roles, which are discussed in this review. The cohesin core complex (Fig. 1.1A) is based on a heterodimer of two SMC proteins (structural maintenance of chromosomes), SMC1 and SMC3, which associate with each other with high affinity through their central hinge domains. An α-kleisin protein (SCC1, also called RAD21/MCD1) closes the ring through interaction with the globular terminal domains of the SMC proteins. Cleavage of the α-kleisin at the metaphase-to-anaphase transition resolves cohesion and allows chromosome segregation. A fourth protein named SA (stromal antigen, also named SCC3) associates with the α-kleisin component of the tripartite ring. The exact functions of SA proteins remain unclear, but they are involved in a phosphorylation-dependent cohesin release pathway (see Section 4). In mammalian somatic cells, two different SA proteins, SA1 and SA2, are expressed from two distinct genes and were shown to account for some of the functional diversity of cohesin complexes. Loss of SA1 was very recently shown to cause embryonic lethality, chromosome segregation defects, aneuploidy, and specific changes in transcription patterns, whereas centromeric cohesion depends on SA2 (Remeseiro, Cuadrado, Carretero, et al., 2012; Remeseiro, Cuadrado, Gomez-Lopez, Pisano, & Losada, 2012). In addition to these two different SA subunits, meiotic cells express a third SA protein (SA3, also called STAG3), again from another gene, providing meiotic cells with an even greater number of different cohesin complexes to perform various functions. However, the diversity in meiocytes is even larger: one additional gene encoding an SMC1-type protein (SMC1β) and two other genes encoding α-kleisin proteins (RAD21L and REC8) are expressed exclusively in meiocytes, expanding the possible combination to at least 18 different cohesin core complexes during meiosis. Considering cohesin-associated and/or regulatory factors, about which very little is known in meiotic cells, this number is likely to increase even more; for example, two paralogs of the cohesin-associated factor PDS5 (PDS5A and PDS5B) coexist in somatic cells (Losada, Yokochi, & Hirano, 2005). Experimental data have confirmed the existence of at least six complexes (Jessberger, 2011; Uhlmann, 2011).

2. MEIOSIS I

In germ stem cells, which divide mitotically, the canonical somatic cohesin complex is expressed—at least there is no evidence for expression of meiotic versions of cohesin proteins in spermatogonia, the only premeiotic cells analyzed in this respect. Also, as mice deficient for REC8, RAD21L, or SMC1β all develop early prophase I meiocytes, there is no reason to assume that meiotic cohesins are required before entry into meiosis. However, the exact timing of the onset of expression of the meiosis-specific cohesin subunits in mammalian germ cells is still debated and more systematic analyses are needed (Fig. 1.1). REC8 is expressed and present on chromosomes during premeiotic S-phase in yeast (Watanabe & Nurse, 1999) and in mouse spermatocytes (Eijpe, Offenberg, Jessberger, Revenkova, & Heyting, 2003), and SMC1β and STAG3 appear in leptotene (Eijpe et al., 2003; Prieto et al., 2001), although in a small number of spermatogonia STAG3 may also be present (Prieto et al., 2002). It appears as if during premeiotic S-phase, sister chromatids are held together by SMC1α-type cohesin.

As complexes of SMC1α and REC8 were observed (Revenkova et al., 2004) and RAD21L is present in premeiotic S-phase (Gutierrez-Caballero et al., 2011; Ishiguro, Kim, Fujiyama-Nakamura, Kato, & Watanabe, 2011; Lee & Hirano, 2011), it is possible that REC8 and/or RAD21L associate initially with SMC1α-type cohesin and later with SMC1β-type cohesin. This hypothesis raises questions, such as for the existence and mechanism of postreplicative loading of newly formed cohesin complexes onto meiotic chromosomes. Replacement of a cohesin subunit such as an SA protein or even a kleisin by another kleisin or SA protein appears difficult without cohesin dissociation from the chromosome but cannot be entirely excluded. The existence of numerous distinct cohesin complexes in meiocytes and the distinct expression patterns for individual cohesin proteins renders postreplicative loading not unlikely, and postreplicative loading of RAD21-containing complexes during pachytene was recently suggested (Ishiguro et al., 2011; Lee & Hirano, 2011) (see Section 4).

Upon entry into meiosis, chromosomes start to condense and form specific axis–loop structures (Fig. 1.1B; for reviews on chromosomal aspects of meiosis see Costa & Cooke, 2007; Handel & Schimenti, 2010; Keeney, 2008; Kleckner, 2006; Lichten & de Massy, 2011; Yanowitz, 2010; Youds & Boulton, 2011). Axial elements (AEs) begin to assemble on sister chromatids in leptotene, and then AEs of homologous pairs of sister chromatids align and start to synapse in zygotene, and form the fully paired synaptonemal complex (SC) in pachytene. Formation of AEs allows the chromosomes to initiate the homology search required for synapsis of the homologs. After a period of "trial and error," the homologs find each other and align, greatly facilitated by the introduction of DNA DSBs by a meiosis-specific topoisomerase II-like enzyme called SPO11. The failure to generate these DSBs or to synapse triggers apoptosis (see Section 5). Gene expression generally declines as chromosomes condense and as long as the two pairs of sister chromatids (the "univalents" or "homologs") remain unsynapsed (Page et al., 2012). Upon synapsis, fully achieved in pachytene, gene expression dramatically increases.

Cohesins are essential for formation of a regular axis–loop structure of meiotic chromosomes. In absence of SMC1β, the AEs and SCs are greatly shortened to about half the length of those in wild-type mice (Revenkova et al., 2004). In parallel, the loops that emanate from the axes are extended by up to twofold. In contrast, mice which lack the AE component SYCP3 showed much longer axes than wild-type mice (Yuan et al., 2002). Together, this suggested that AE/SC components contribute significantly to axis compaction, which is restricted by cohesin. This hypothesis was supported by analyses of the $Smc1\beta^{-/-}Sycp3^{-/-}$ "double-knockout" strain, which features meiotic chromosome axes that are longer than in the $Smc1\beta^{-/-}$ strain, almost as long as in wild-type mice (Novak et al., 2008). Cohesin apparently forms an axis parallel to and interacting with the AE/SC and may hold the chromatin loops at their bases, thereby determining the number and location of the loops. This does not exclude, however, the presence of cohesins on the loops themselves. Shortened AEs/SCs were also observed in mice deficient in either REC8 (Bannister, Reinholdt, Munroe, & Schimenti, 2004; Xu, Beasley, Warren, van der Horst, & McKay, 2005) or RAD21L (Herran et al., 2011). Mice that lack both of these two meiosis-specific kleisins do not form any AEs and their spermatocytes arrest in leptotene (Llano et al., 2012), suggesting that different types of cohesin complexes collectively play an essential role in AE formation.

The role of cohesin in prophase I meiocytes is not limited to determining the axis–loop architecture of chromosomes. Synapsis itself also depends on cohesin. Mice lacking either SMC1β, REC8, or RAD21L display distinct synaptic failures.

Loss of SMC1β affects spermatocytes more dramatically than oocytes, for spermatocytes arrest in midpachytene with a failure to complete synapsis (Revenkova et al., 2004). Oocyte chromosome synapsis is also impaired, but the cells progress to metaphase II, although they suffer from loss of centromeric cohesion and eventually die. As oocytes do not possess the X/Y pair of sex chromosomes, which in spermatocytes forms a characteristic chromatin domain called the sex body with largely unsynapsed X and Y chromosomes, quality surveillance mechanisms that rely on the X/Y pair cannot act in oocytes, which explains in part their permissivity (see discussion of checkpoints in Section 5).

In the absence of REC8, the initial stages of leptotene and early zygotene appear normal, however, homologs largely fail to synapse (Bannister et al., 2004; Xu et al., 2005). In *Rec8*$^{-/-}$ mice, AEs form and sister chromatids appear to be held together, indicating that REC8 is not essential for this process during or after premeiotic S-phase. Aberrant localization of SC proteins, both SYCP1 and SYCP3, to the sister chromatids was reported, suggesting that REC8 cohesin directly or indirectly supports proper localization of these proteins (Xu et al., 2005). Deposition of SC proteins in absence of a proper chromosome architecture, that is, an SC, may be unspecific as these proteins tend to form aggregates and filaments even if expressed in somatic cells (Yuan, Brundell, & Hoog, 1996). Lack of the second meiosis-specific kleisin RAD21L in male mice also causes a defect in synapsis (Gutierrez-Caballero et al., 2011; Ishiguro et al., 2011; Lee & Hirano, 2011). Fragmented AEs and SCs are observed through double staining of the AE protein SYCP3 and the transverse element protein SYCP1 and reveal aberrant synapsis in comparison to wild-type mice. In oocytes, no synapsis deficiency was observed. What are the specific requirements of spermatocyte autosomes for synapsis versus oocyte chromosome synapsis? This question remains currently unanswered. Some redundancy may exist between meiosis-specific cohesin subunits and that redundancy may differ between the sexes—no systematic comparison of expression of all cohesin proteins in spermatocytes and oocytes has been presented—and double mutant mouse models may help to further elucidate the role of cohesins in synapsis. This role and the underlying mechanisms are still largely unclear. It is possible that cohesin rings on one pair of sister chromatids interact with rings on the other homologous pair of sister chromatids. Cohesins may also support synapsis more indirectly by generating an appropriate axis–loop architecture as discussed earlier.

How about the best-known role of cohesin, its function in *cohesion*? Analysis of the first mammalian meiosis-specific cohesin deficiency model, *Smc1β*$^{-/-}$ mice, showed complete loss of SCC in oocytes at metaphase II (Revenkova et al., 2004). At that stage, centromeric cohesion would have to be maintained to allow proper alignment of chromosomes on the metaphase plate, biorientation of kinetochores, and segregation of sister chromatids in a mitosis-like fashion. All sister chromatids fell apart in these mutants and the oocytes died. However, there was substantial cohesion in SMC1β-deficient meiosis I oocytes, at least in young mice, indicating that SMC1α-type cohesin complexes can provide sufficient cohesion to pass through the meiosis I spindle assembly checkpoint (see below). SMC1β-deficient spermatocytes die in midpachytene and thus metaphase I cohesion could only be analyzed by artificially driving prophase I spermatocytes into a metaphase I-like stage by okadaic acid treatment. This revealed loss of much, but not all cohesion, suggesting a significant contribution of SMC1β to metaphase I cohesion, at least under these artificial conditions. Further experiments are required to determine the exact contribution of SMC1α versus SMC1β cohesin to prophase I cohesion.

Loss of the meiosis-specific kleisin REC8 in mice does not impair establishment of SCC but impaired arm cohesion was observed even before the cells die in midpachytene, that is, before loss of centromeric cohesion would become apparent (Bannister et al., 2004; Xu et al., 2005). The authors argue that at least initially RAD21 could compensate for the lack of REC8. The recent discovery of the third kleisin, RAD21L, indicates that a second kleisin may play an additional or even a sole role in the compensation for the loss of REC8.

The absence of RAD21L prevents synapsis in spermatocytes but sister chromatid arm cohesion does not appear to be compromised (Herran et al., 2011). Driving these spermatocytes through prophase I and into metaphase I using okadaic acid treatment showed the persistence of centromeric cohesin. This indicates that RAD21L-containing cohesin is not required for cohesion, at least in prophase I.

How is *cohesion resolved*? Breakdown of the nuclear envelope signals the end of prophase and the homologous chromosomes align on the metaphase plate. At that stage, the SC is dissolved and only the chiasmata hold the homologs together. Cohesion along the chromosome arms is resolved, but centromeric SCC is maintained, and homologs are pulled to opposite poles by mono-oriented sister kinetochore spindle attachments (see below). To resolve arm cohesion, phosphorylation of REC8 targets it for proteolytic cleavage by separase (Buonomo et al., 2000; Kitajima, Miyazaki, Yamamoto, & Watanabe, 2003; Kudo et al., 2009, 2006). The kinases responsible for phosphorylation of REC8 have been identified, originally in yeast, as Casein kinase 1 and Dbf4/Cdc7 kinase (Ishiguro, Tanaka, Sakuno, & Watanabe,

2010; Katis et al., 2010; Rumpf et al., 2010), but their mammalian functional counterparts have not yet been described.

These kinases were identified in screens that were designed considering that dephosphorylation of REC8 was involved in protecting centromeric cohesion. Shugoshin (Japanese for guardian spirit) localizes to centromeres but not along chromosome arms. Shugoshin protects cohesin complexes bound at the centromeres of sister chromatids from separase cleavage via its positive interaction with the phosphatase PP2A (Kitajima et al., 2006; Riedel et al., 2006), recently reviewed in Gutierrez-Caballero, Cebollero, and Pendas (2012). Whether RAD21L is also targeted for phosphorylation (or dephosphorylation) by the same kinases (or phosphatase) has yet to be determined. RAD21L, whose exact spatiotemporal behavior is debated, appears to be removed from chromosomes arms at mid or late prophase I or metaphase I, and if removed in prophase I may not be a target of separase (Herran et al., 2011; Ishiguro et al., 2011; Lee & Hirano, 2011). However, the apparent replacement of RAD21L by RAD21 at the end of prophase I still requires further investigation, as does the alternate staining pattern of REC8 and RAD21L on pachytene chromosomes, suggested to represent a "barcode" for homolog recognition, since the results reported by the three groups (Hirano, Pendas, Watanabe) are in part conflicting.

One key feature of meiosis I is *recombination* between the two pairs of sister chromatids. Cohesins are involved here as well. There is ample evidence for an important contribution of cohesins to DNA repair and recombination in somatic cells, that is, primarily homologous recombination between individual sister chromatids, not—as in meiosis—between pairs of sister chromatids (reviewed in Cortes-Ledesma, de Piccoli, Haber, Aragon, & Aguilera, 2007; Strom & Sjogren, 2007; Watrin & Peters, 2006). Cohesin in somatic cells facilitates recombination by juxtaposing the two sister chromatids but may also recruit other proteins to sites of DNA repair and recombination. Upon DNA damage, the SMC1 and SMC3 proteins as well as the α-kleisin become phosphorylated, and in *Saccharomyces cerevisiae*, new cohesion is established both at the DSB site and in a genome-wide fashion (Heidinger-Pauli, Unal, Guacci, & Koshland, 2008), although in human cells cohesin loading is moderate and restricted to the vicinity of DSBs (Caron et al., 2012). The SMC proteins are phosphorylated by ATM or ATR kinases in mammalian cells (Kim, Xu, & Kastan, 2002; Yazdi et al., 2002), and in yeast, Rad21 (Scc1, Mcd1) becomes phosphorylated by Chk1 (Heidinger-Pauli et al., 2008). In meiocytes, such pathways have not been described, but phosphorylation of SMC3 was reported (Fukuda et al., 2012), although its relationship to ATM or ATR, both active in prophase I, and its function remain to be described.

Meiotic recombination is initiated by SPO11-dependent DSBs, with which RAD51 and DMC1 associate to allow strand invasion and subsequent repair of these breaks. Thereby, crossovers can be formed and chiasmata generated. Chiasmata are required to keep the homologs together until segregation in anaphase I. Thus, recombination between sisters should be avoided at this stage, and interhomolog recombination preferred, that is, the so-called interhomolog bias should be ensured. In yeast, the role of REC8 in the interhomolog bias or in the subsequent recombination reaction is controversial (Callender & Hollingsworth, 2010; Kim et al., 2010), and there are no published data for mammalian germ cells. The processing of RAD51 and DMC1 repair foci appears to be delayed in absence of SMC1β (Uddipta Biswas et al., submitted). Whether this is a consequence of a direct role of cohesin in the repair reaction, or whether this is an indirect effect of, for example, shortened axes or asynapsis remains to be determined. Some mouse cohesins such as SMC3 and SMC1β localize to chiasmata, but others like REC8 do not and rather seem to be specifically excluded from these structures (Eijpe et al., 2003). While the involvement of cohesin in the recombination/repair reaction itself is therefore uncertain—largely due to the substantial difficulty to separate individual activities of cohesins from each other in meiocytes—it is very likely that cohesin is important for the maintenance of chiasmata as discussed below.

Yet another function of cohesin in meiotic prophase concerns *telomeres*. SMC1β cohesin protects telomeres, for in the absence of this cohesin, the telomeres suffer from a range of abnormalities (Adelfalk et al., 2009). During bouquet formation, which occurs at the leptotene–zygotene transition, 15–20% of the telomeres fail to attach to the nuclear membrane. Foci of nuclear envelope protein SUN-1 are reduced, suggesting that SMC1β-protected telomeres are required for their establishment, however, no bias was observed in SUN-1 binding to normal or aberrant telomeres. The relationship between telomere clustering and synapsis is still debated. If it does increase the likelihood of successful synapsis, it is not a universal mechanism, for other organisms have developed other mechanisms to achieve synapsis without telomere clustering. For example, the fission yeast *Schizosaccharomyces pombe* oscillate their nuclear content back and forth apparently to enhance pairing and recombination (Vogel, Pavin, Maghelli, Julicher, & Tolic-Norrelykke, 2009). In *Rad21L*$^{-/-}$ spermatocytes, telomeres were examined by staining the telomere-specific protein RAP1 in cells that were fixed and squashed, which helps to preserve nuclear structure (Herran et al., 2011). In wild-type cells almost all telomere signals were found associated with the nuclear membrane. In the majority of mutant cells, between one and six telomeres were found internally, suggesting a similar effect of RAD21L and SMC1β in telomere attachment, but no evidence for aberrant telomere structure in absence of RAD21L was presented. *Rec8*$^{-/-}$ meiocytes were not yet specifically analyzed for telomere deficiencies. In *S. pombe*, REC8 staining was slightly weaker on the telomeres during the stage of "horseshoe tail movements," a characteristic prophase I movement of chromosomes where telomeres and centromeres are far separated (Watanabe & Nurse, 1999).

In somatic cells, it has been shown that SA1 interacts with the telomere-binding protein TRF2 (Canudas et al., 2007). Here, SA1, but not SA2, is involved in telomere cohesion (Canudas & Smith, 2009). Knockdown of SA1 in HeLa cells results in loss of establishment and maintenance of telomeric sister cohesion during S-phase. This was not observed in the SA2 knockdown, rather SA2 appears to function at centromeres. A role of STAG3 at telomeres during meiosis has not been described. No telomeric repeat binding domain is known for SMC1β; however, it may localize to telomeres via one or more intermediate protein(s). The precise mechanism by which SMC1β cohesin contributes to telomere maintenance remains to be determined. It may be speculated that the highly repetitive and recombination-prone telomeric repeats require a unique protein structure, facilitated or supported by cohesin, to prevent the hazardous event of telomere end-to-end joining or other aberrations.

3. MEIOSIS II

At the onset of meiosis II, centromeric cohesion is still maintained by REC8 cohesin complexes, protected by PP2A phosphatase. The centromeric protein Shugoshin is responsible for this protection as it recruits PP2A. Arm cohesion between sister chromatids has been lost via separase activity at the onset of anaphase of meiosis I. In meiosis II, the chromosomes are separated as in mitosis, that is, sister chromatids are pulled apart in opposite directions via bipolar centromeric interaction with kinetochores (Fig. 1.1). Therefore, cohesion between sister centromeres must be resolved at meiosis II. How cohesin complexes are targeted for proteolytic cleavage at this stage is still unknown. In fission yeast, both Shugoshin and PP2A are no longer present at the centromeres at metaphase of meiosis II and expression of *Sgo1* at this stage is inhibited by its 3′UTR (Kitajima, Kawashima, & Watanabe, 2004; Rabitsch et al., 2004). However, artificial expression of SGO1-GFP during meiosis II by removing its 3′UTR did not hinder sister chromatid separation, indicating that an as yet unknown mechanism may be involved (Gregan, Rumpf, Li, & Cipak, 2008; Gregan, Spirek, & Rumpf, 2008), consistent with the possible release of meiosis II cohesion in the presence of Shugoshin (Clarke, Tang, Ooi, & Orr-Weaver, 2005). In mammals, both Shugoshin (SGO2) and PP2A are present at the centromeres in meiosis II. It is not clear how this complex can protect centromeric cohesin cleavage in meiosis I but not in meiosis II, when cohesion can be released even in the presence of Shugoshin (Clift & Marston, 2011; Gregan, Spirek et al., 2008). Studies of SGO2 in mouse oocytes and spermatocytes have resulted in the proposal of a tension model, where bioriented centromeres cause a small shift in the localization of SGO2 from the inner centromere to the kinetochore, therefore eliminating cohesin protection (Gomez et al., 2007; Lee et al., 2008). Although there are conflicting reports—some of these studies date to before the identification of RAD21L, rendering the distinction between RAD21 and RAD21L questionable—it appears as if neither RAD21 nor RAD21L localize to chromosomes in meiosis II and are therefore not likely to be involved (Gomez et al., 2007; Gutierrez-Caballero et al., 2011; Ishiguro et al., 2011; Lee & Hirano, 2011).

Are there cohesins in *postmeiotic cells*? Spermatocytes and oocytes lose most cohesin from their centromeres at the metaphase II–anaphase II transition (Eijpe, Heyting, Gross, & Jessberger, 2000; Eijpe et al., 2003). In these studies, however, staining of fixed testis sections with anti-cohesin antibodies occasionally yielded signals for REC8 and SMC3 in spermatids, but the specificity of these signals remained uncertain as spermatids tend to generate high background in immunofluorescence experiments. Recently generated transgenic mouse strains expressing LAP (GFP)-tagged SMC3 or LAP-tagged SMC1β did not show signals in spermatids (Rolf Jessberger, unpublished observations). The situation is different in oocytes and zygotes. During oocyte maturation, the *Smc1α* gene becomes reexpressed, while the expression of *Smc1β* declines (Hodges, Revenkova, Jessberger, Hassold, & Hunt, 2005; Revenkova, Herrmann, Adelfalk, & Jessberger, 2010). This was interpreted as the preparation of the oocyte for postfertilization cell divisions, which are very rapid and require considerable amounts of cohesin. Upon fertilization, a switch in α-kleisin expression occurs; the meiotic REC8 disappears and the mitotic RAD21 reappears. If a specifically engineered RAD21/SCC1 is cleaved by TEV protease in zygotes, the initial embryonic cell division suffers from loss of SCC. Cleavage of REC8 by TEV in zygotes, however, left the embryonic cell division unaffected (Tachibana-Konwalski et al., 2010). This suggests a rapid switch from REC8-dependent to RAD21-dependent cohesin upon resumption of mitotic cell division after fertilization.

4. COHESIN LOADING AND ASSOCIATED FACTORS

To fulfill the functions discussed earlier, cohesins must not only be loaded onto chromosomes in a spatiotemporally controlled manner; their association, maintenance, and dissociation from chromosomes must also be tightly regulated. Thus, increasing attention should be given to the role of cohesin-regulatory factors in meiosis. Several factors interact with cohesins and regulate their various functions in somatic cells (Fig. 1.1A; reviewed in Haering & Jessberger, 2012; Merkenschlager, 2010; Nasmyth, 2011; Nasmyth & Haering, 2009; Wood et al., 2010). Here, we briefly discuss these factors and their roles in the mitotic cell cycle and discuss in more detail the current state of knowledge of their expression and functions in germ cells.

Cohesins are loaded onto chromatin throughout the cell cycle by a complex of at least two proteins, SCC2 (also known as NIPBL) and SCC4 (Bernard et al., 2006; Ciosk et al., 2000; Seitan et al., 2006; Watrin et al., 2006), recently termed kollerin (Nasmyth, 2011). In mitotic cells, the cohesin-associated factor PDS5 regulates cohesin function through the recruitment of additional factors such as WAPL and Sororin, which bind competitively to PDS5 and have anti- and pro-cohesion establishment activities, respectively (Nishiyama et al., 2010). During interphase and G1-phase, cohesin's residence time on chromatin is relatively short (Gerlich, Koch, Dupeux, Peters, & Ellenberg, 2006), presumably due to the activity of WAPL/PDS5, also called releasin (Nasmyth, 2011). In S-phase, cohesin complexes that are bound to chromatin somehow entrap the newly replicated sister chromatids and are converted into stably bound cohesins, a process referred to as cohesion establishment. This process depends on the cohesin acetyltransferases ESCO1 and ESCO2, or Eco1 in yeast, which acetylate SMC3 on two conserved lysine residues, K105/K106 in vertebrates and K112/K113 in yeast (Higashi et al., 2012; Rolef Ben-Shahar et al., 2008; Rowland et al., 2009; Unal et al., 2008; Zhang, Shi, Li et al., 2008). In vertebrates, acetylated cohesin recruits Sororin, which protects cohesin from releasin activity during S- and G2-phase until Sororin is phosphorylated in M-phase (Nishiyama et al., 2010; also recently reviewed by Zhang & Pati, 2012). In prophase, phosphorylation of the SA subunit by polo-like kinase 1 (PLK1) in concert with Aurora B enables releasin to remove most cohesin from chromosome arms without cleaving the α-kleisin subunit, possibly allowing the recycling of cohesin complexes for regulation of gene expression during the next interphase (Parelho et al., 2008; Peters, Tedeschi, & Schmitz, 2008; Wendt et al., 2008) and/or for SCC during the next cell division (Beckouet et al., 2010; Borges et al., 2010; Xiong, Lu, & Gerton, 2010), although the latter possibility would likely require deacetylation of SMC3, possibly by the mammalian HDAC8 (Deardorff et al., 2012). In yeast, although there is no clear prophase pathway, releasin was recently shown to remove cohesin from chromosomes by opening the cohesin ring at its Smc3-Scc1 interface (Chan, 2012). Once all chromosomes are aligned and bioriented on the metaphase plate and the spindle assembly checkpoint is satisfied, the anaphase-promoting complex/cyclosome (APC/C) is activated and targets securin and cyclin B for degradation, leaving the enzyme separase in an active form. Separase cleaves the α-kleisin subunit of cohesin, allowing the sister chromatids to be pulled apart and properly segregated during anaphase (Hauf, Waizenegger, & Peters, 2001; Uhlmann, Lottspeich, & Nasmyth, 1999; Uhlmann, Wernic, Poupart, Koonin, & Nasmyth, 2000).

An important and unanswered question in the field is whether meiotic cohesins are loaded onto chromosomes during meiotic prophase I and if so, by which mechanism. As mentioned earlier, one possibility is through DSBs: in mitotic cells, cohesins are loaded at DSB sites in G2/M in both yeast and human (Caron et al., 2012; Heidinger-Pauli et al., 2008; Strom, Lindroos, Shirahige, & Sjogren, 2004; Unal et al., 2004). However, loading of meiotic cohesins does not seem to depend on SPO11, and DSB formation is largely unperturbed in absence of SMC1β, REC8, or RAD21L (James, Schmiesing, Peters, Yokomori, & Disteche, 2002; RJ, unpublished data). The role of kollerin during gametogenesis remains poorly understood. In Drosophila, SCC2 localizes to the SC and contributes to the maintenance of meiotic chromosome cores (Gause et al., 2008). In *C. elegans*, loading of meiotic cohesin by SCC2 is required for DSB repair and for activation of the DNA damage checkpoint (Lightfoot, Testori, Barroso, & Martinez-Perez, 2011). In yeast, deletion of SCC2 right before meiosis induction showed that SCC2 was required for recruiting cohesin to meiotic chromosomes, and this loading was required not only for SCC but also for meiotic gene expression, thereby providing a positive feedback loop for REC8 expression (Lin, Jin, Liu, Hampton, & Yu, 2011).

Our understanding of the functions of the *cohesin-associated factors* in mammalian meiosis is very limited. In adult mice, *Pds5A* mRNA is expressed at low levels in all tissues, although expression seems to vary between tissues with testis showing relatively high levels, whereas *Pds5B* mRNA is highly expressed in adult brain and testis in comparison with other tissues. In the testis, high levels of *Pds5B* mRNA were detected by radioactive *in situ* hybridization (ISH) at the periphery of the seminiferous tubules in a pattern consistent with immature germ cells (Zhang et al., 2009, 2007). Both *Pds5A* and *Pds5B* are essential genes: mice in which either of these two genes was deleted exhibited several developmental anomalies similar to those found in Cornelia de Lange Syndrome patients and died shortly after birth (Zhang et al., 2009, 2007). While *Pds5A*$^{-/-}$ mice showed normal germ cell development, a severely reduced number of germ cells in testes and ovaries of newborn *Pds5B*$^{-/-}$ were found and this loss appeared to be more severe in males. The function of PDS5A and PDS5B in spermatogenesis was assessed by transplanting embryonic testes subcutaneously onto the back or flank of wild-type mice; by 6 weeks after transplantation, the explanted *Pds5A*$^{-/-}$ testes contained testicular cords with the full spectrum of spermatogenesis, including mature spermatids, but the explanted *Pds5B*$^{-/-}$ testis tubules contained only Sertoli cells (Zhang et al., 2009, 2007). Thus, it seems like PDS5B plays a much more important role than PDS5A in germ cell development and meiosis (at least in spermatocytes). In invertebrates and fungi, where only one Pds5 homolog per organism was found, Pds5 (Pds5p/Spo76/BIMD) was shown to play various roles from SC formation to DSB repair, SCC, and chromosome compaction during meiosis (Ding et al., 2006; Jin, Guacci, & Yu, 2009; Ren et al., 2005; Storlazzi et al., 2008; van Heemst, James, Poggeler, Berteaux-Lecellier, & Zickler, 1999; van Heemst et al., 2001; Wang, Read, & Norbury, 2002; Wang, Yoder, Antoshechkin, & Han, 2003).

Little is known about the meiotic functions of the factors that bind PDS5A/B. Neither the expression nor the function of Sororin has been described in germ cells. The WAPL protein is highly expressed in adult mouse testes and weakly expressed in every other mouse tissue tested (Kuroda et al., 2005). A similar pattern was found for *Wapl* mRNA expression in human tissues, although in this case expression at the protein level was not shown. Immunohistochemical analyses indicated that mouse WAPL was expressed abundantly in pachytene spermatocytes but was undetectable in condensing forms of sperm, suggesting that it is expressed predominantly in early stages of spermatogenesis (Kuroda et al., 2005). Immunocytological staining suggested that mouse WAPL localizes to the SC during the pachytene stage (Kuroda et al., 2005) and showed that it colocalizes with SYCP2 on the SC of pachytene oocytes (Zhang et al., 2008). Although these data suggest a role for WAPL in meiosis, experiments using conditional knockout mice will be necessary to determine what this role might be. Interestingly, the mouse *Wapl* gene was initially identified in a screening for genes whose expression in ES cells was affected by treatment with endocrine disruptors (2,3,7,8-tetrachlorodibenzo-*p*-dioxin or TCDD), and injection of TCDD into the peritoneal cavity of mice led to a decrease in *Wapl* expression in the testes, suggesting a potential role for mammalian WAPL in TCDD-induced reproductive toxicity (Kuroda et al., 2005). In Drosophila, WAPL was shown to play a role in meiotic segregation of achiasmate chromosomes in oocytes (Verni, Gandhi, Goldberg, & Gatti, 2000).

The expression, localization, and function of the cohesin acetyltransferase ESCO1 have not been described in germ cells. *Esco2* was recently suggested to be a potential regulator of the mammalian mitotic-to-meiotic transition based on its expression pattern in the embryonic gonad and postnatal testis, which is very similar to that of *Stra8* (Hogarth, Mitchell, Evanoff, Small, & Griswold, 2011), a target gene of retinoic acid (RA) whose upregulation at the onset of meiosis is essential for both oogenesis and spermatogenesis (for a recent review, see Griswold, Hogarth, Bowles, & Koopman, 2012). Similarly to *Stra8*, *Esco2* mRNA expression peaks at E14.5 in the embryonic ovary and day 10 dpp in the postnatal testis, which correlates with the beginning of meiosis in both sexes (Hogarth et al., 2011). However, except for the observation that *Esco2* expression is induced by RA (Evans et al., 2012; Leem et al., 2011), so far no evidence points to such a role. *Esco2* upregulation by RA might be indirect as it seems unclear whether its promoter contains RA response elements or not, although it does contain a potential site for transcriptional activation via SP1, but the induction of *Esco2* transcription by RA through SP1 remains to be demonstrated experimentally. Interestingly, ESCO2 is not expressed only at the time of meiosis entry but also during prophase I, a stage where SCC is assumed to be already established. *Esco2* mRNA and ESCO2 protein were detected by northern and western blotting in testes of 10- and 90-day-old mice (Hogarth et al., 2011). ISH experiments showed that *Esco2* was expressed in spermatocytes at 10, 20, and 90 dpp, and immunofluorescence performed on testis cross-sections suggested that ESCO2 was present diffusely throughout the nuclei of preleptotene and leptotene spermatocytes (10 dpp) and in discrete domains within the nuclei of pachytene spermatocytes (from 15 dpp on). Immunofluorescence on nuclear spreads from mouse spermatocytes showed that this distinct domain is in fact the XY body, where ESCO2 colocalizes with γH2AX on the sex chromatin loops

(Evans et al., 2012). This is intriguing because it suggests that SCC might be reinforced specifically on the loops emanating from the sex chromosome axes at that stage. Whether this is to counteract the action of releasin or other potential antiestablishment factors, for example, histone deacetylases, or to repair the DSBs on the sex chromosomes using the sister chromatid remains to be addressed. In oocytes, ESCO2 also colocalizes with γH2AX at DSB sites during prophase I. However, whether ESCO2 plays an important if any role in meiosis remains to be determined. In humans, mutations in *Esco2* cause a rare autosomal recessive disorder called Roberts syndrome/SC phocomelia (RBS/SC), which is characterized by a range of birth defects in both sexes (Vega et al., 2005). Very few pregnancies in RBS/SC patients have been documented and it is not clear whether they show an increased risk of pregnancy loss (Goh et al., 2010). There are no reports about fertility in male RBS/SC patients.

5. ROLE OF COHESIN IN CHECKPOINT MECHANISMS

During mitosis, checkpoint mechanisms ensure that one cell cycle event is completed before the next one starts. Given that meiosis consists of two sequential cell divisions, the first one being characterized by particularly complex events, it is not so surprising that it is also controlled by checkpoint mechanisms. So far, mechanisms that monitor DSB repair, chromosome synapsis, and proper spindle assembly have been described in meiocytes (reviewed by Burgoyne, Mahadevaiah, & Turner, 2009; Chiang, Schultz, & Lampson, 2012; MacQueen & Hochwagen, 2011; Sun & Kim, 2012). Recently, the formation of AEs very early in meiosis was suggested to be required for satisfying a checkpoint, perhaps of a new kind (Llano et al., 2012). Given the key roles that cohesins play in AE formation, synapsis, centromere organization, and their association with microtubules at the spindle pole, they are likely to be directly or indirectly involved in most of these checkpoint mechanisms.

Checkpoint or quality surveillance mechanisms are in place during the first meiotic prophase to ensure that only those cells in which all pairs of sister chromatids have found their homolog and are properly synapsed and in which most DSBs have been repaired can proceed further into gametogenesis (Bolcun-Filas et al., 2007, 2009; Burgoyne et al., 2009; Hamer et al., 2008; MacQueen & Hochwagen, 2011). It was recently shown that asynapsis and DSBs are monitored by distinct mechanisms involving HORMAD proteins (Wojtasz et al., 2012, 2009), although they share a common effector, the DNA damage response (DDR) sensor kinase ATR. In mouse meiocytes, ATR is recruited to unsynapsed chromatin and DSB sites where it phosphorylates a large set of substrates, including the checkpoint kinases CHK1 and CHK2 and the histone variant H2AX (MacQueen & Hochwagen, 2011). While CHK1 and CHK2 phosphorylation cause a cell cycle delay, providing additional time for the cell to complete synapsis and DSB repair (or undergo apoptosis), phosphorylated H2AX (γH2AX) causes transcriptional silencing, a phenomenon called meiotic silencing of unsynapsed chromosomes (MSUC) (Baarends et al., 2005; Burgoyne et al., 2009; Turner et al., 2005). In midpachytene, when most DSBs have been repaired and all autosomes are properly synapsed, MSUC is restricted to the sex chromatin in spermatocytes. This silencing, called meiotic sex chromosome inactivation (MSCI), is a quality control system unique to spermatocytes; in pathological conditions, ATR is attracted to unrepaired DSBs and/or unsynapsed regions, resulting in a failure of MSCI and expression of the "killer genes" *Zfy1/2* located on the Y chromosome (Royo et al., 2010). Such spermatocytes undergo apoptosis and are eliminated at stage IV of the testicular epithelial cycle (also referred to as the midpachytene checkpoint). In females, the meiotic prophase quality control mechanisms are different; defective oocytes are eliminated at later, more variable stages of prophase I, if at all. ATR activity is also recruited to unsynapsed chromosome regions and causes transcriptional silencing (Turner et al., 2005). It has been postulated that apoptosis in defective oocytes might be triggered by the inappropriate silencing of essential genes and/or sustained ATR activation (Burgoyne et al., 2009; Toth & Jessberger, 2010).

In several cohesin-deficient mouse models, spermatocytes are eliminated by apoptosis in early/midpachytene (Bannister et al., 2004; Herran et al., 2011; Revenkova et al., 2004; Xu et al., 2005). All of these deficiencies cause asynapsis of autosomes and impaired formation of the sex body, suggesting that autosomal silencing within the MSUC response occurs and silencing of sex chromosomes fails. This is supported by recent data from our lab showing *Zfy1/2* expression in $Smc1\beta^{-/-}$ spermatocytes (Biswas et al., submitted). The oocyte pool of $Smc1\beta^{-/-}$, $Rec8^{-/-}$, and $Rad21L^{-/-}$ mice is more rapidly depleted than that of wild-type mice, but many oocytes can be matured up to metaphase II. This reflects an observation made in many other mouse models with meiotic deficiencies: while the prophase I checkpoints in males are very stringent, those of females are not. In $Rec8^{-/-}$ $Rad21L^{-/-}$ mice, spermatocytes die in a leptotene-like stage, although the tubular stage correlate is difficult to define, and thus the existence of an additional early checkpoint remains uncertain (Llano et al., 2012).

Recent data suggest that cohesin is not only necessary for DSB repair and SC formation/synapsis as discussed earlier, but also for the activation of the meiotic DNA damage checkpoint. *C. elegans* carrying a null mutation for the cohesin loading factor SCC2 showed a complete absence of cohesin on their meiotic chromosomes, which resulted in the inability to form crossovers and the persistence of unrepaired recombination intermediates (Lightfoot et al., 2011). However, these unrepaired DSBs failed to activate the DNA damage checkpoint, suggesting that the meiotic cohesin complex is also required for this process. HUS-1, a member of the conserved 9-1-1 complex, was not recruited to meiotic chromosomes in these mutants, suggesting that cohesin is involved in the very early steps of DNA damage activation. Furthermore, partial knockdown of Scc-2 by RNA interference resulted in some cohesin loading onto meiotic chromosomes, which was not sufficient to efficiently repair DSBs but sufficient to activate the DDR. Cohesin was suggested to act as a platform for the recruitment and activation of checkpoint and DNA repair proteins (Jessberger, 2002, 2009). In HeLa cells, cohesin is required for the DNA damage checkpoint through recruitment of the mediator protein 53BP1 and activation of CHK2, a function that is independent of SCC (Watrin & Peters, 2009). How this relates to the activation of the DNA damage checkpoint in vertebrate meiocytes needs to be verified. Also, whether and how cohesin is involved in HORMAD-dependent asynapsis surveillance currently is an open question.

Accurate chromosome segregation during cell division requires sister kinetochores to attach to microtubules from opposite spindle poles. Kinetochore–microtubule attachment is a trial-and-error process that depends on the chromosome passenger complex (CPC), which is targeted to the inner centromere. The current model in mitotic cells proposes that in the absence of tension across centromeres, the catalytic subunit of the CPC, Aurora B kinase, activates a microtubule depolymerase (MCAK) which destabilizes the attachment, whereas proper tension causes kinetochores to stretch and spatially separates the attachment sites from Aurora B, thereby stabilizing the attachment (reviewed by Gutierrez-Caballero et al., 2012; Sakuno, Tanaka, Hauf, & Watanabe, 2011). Several indications suggest that both the CPC and MCAK regulate the SAC during the two meiotic cell divisions (reviewed by Sun & Kim, 2012). The lack of tension caused by improperly attached kinetochores is also sensed by SAC proteins (MAD1, MAD2, BUB1, BUB3, BUBR1, and MPS1), which, similarly to mitosis, localize to meiotic kinetochores, and cause a metaphase arrest by inhibiting the activity of the APC/C (through targeting its cofactor CDC20) until all chromosomes are properly aligned. Cohesin participates in generating this necessary tension during both meiotic divisions. In meiosis I, homologs are pulled apart, and sister chromatids must be mono-oriented to generate this tension, a process in which centromeric REC8-cohesin is essential (Watanabe & Nurse, 1999). SMC1β-cohesin also maintains chiasmata (Hodges et al., 2005), which until their resolution shortly before anaphase contribute to tension. In meiosis II, cohesin acts as in mitosis: it provides part of the necessary tension by keeping the two sister chromatids together and thereby resisting the pulling forces from the spindle poles. Whether interaction of cohesin with microtubule-associated proteins localized at the spindle poles (such as NuMA; Wong & Blobel, 2008) also contributes to this tension in meiotic cells remains to be determined. Topoisomerase II removes chromosome entanglements, catenanes, during mitosis and does so efficiently if spindle tension builds up (Baxter et al., 2011), but topoisomerase II may also generate catenation, which may help SCC and is supported by cohesin (Farcas, Uluocak, Helmhart, & Nasmyth, 2011). Whether a similar functional relationship between topoisomerase II and cohesin exists in meiosis is not known but could have important implications, for example, for the long-term stability of cohesion in oocytes.

In addition to this mechanical role in the SAC, cohesin was recently shown to play a biochemical one; phosphorylated histone H3-threonine 3 (H3-pT3) is one of two histone marks (together with H2A-pS121) necessary for targeting the CPC to the inner centromere in mitotic cells, and this histone modification is made by Haspin, a cohesin-associated kinase which interacts with PDS5 (Yamagishi, Honda, Tanno, & Watanabe, 2010). In mouse male meiosis, the CPC proteins are loaded sequentially at the centromeric heterochromatin during prophase I and appear to form a platform that can recruit other proteins involved in the SAC (Parra et al., 2009), but whether the same chromatin marks recruit the CPC during meiosis is unknown.

6. CENTROMERES, KINETOCHORES, AND CENTRIOLES

As discussed earlier, in metaphase I, centromeric cohesion must be maintained to allow sister chromatids to stay connected until the metaphase II/anaphase II transition, when centromeric cohesion is dissolved and the two sister chromatids segregate in a mitosis-like manner. Thus, in meiosis I, sister kinetochores must be mono-oriented.

Mono-orientation is intimately linked to cohesion and is achieved through activity of several proteins including monopolin, a casein kinase, condensin, and others. Monopolin was initially identified in *S. cerevisiae* as a protein essential for monopolar attachment of microtubules to the meiosis I kinetochores (Toth et al., 2000). Other proteins like Csm1 and Lrs4 associate with monopolin, and the complex requires a polo-like kinase (Cdc5) to be localized properly at the kinetochores (Clyne et al., 2003). Protein Spo13 is required to maintain the monopolin at kinetochores but also protects centromeric cohesion from dissolution (Katis et al., 2004; Lee, Kiburz, & Amon, 2004). The components of the monopolin complex, however, are not well conserved in higher eukaryotes, and corresponding protein complexes were not yet described for mammalian meiocytes. However, the casein kinase 1 Hrr25 is another subunit of the monopolin complex, is required for monopolar attachment, and is conserved from yeast to man (Petronczki et al., 2006). REC8 is a substrate for Hrr25, and possibly REC8 phosphorylation by this kinase may enhance centromeric protection and mono-orientation, in which kleisins are implicated. Other kinases are involved such as Aurora B, whose activity in mono-orientation is supported by monopolin (Monje-Casas, Prabhu, Lee, Boselli, & Amon, 2007), and the SMC2/SMC4-based complex condensin also supports monopolin-dependent mono-orientation in meiosis I (Brito, Yu, & Amon, 2010). Functionally, the monopolin complex seems to serve as a clamp that links the two sister kinetochores (Corbett et al., 2010) and a microtubule fusion activity was described in plants (Li & Dawe, 2009). Thus a complex network of activities and proteins, including cohesin, exists at least in yeast to ensure mono-orientation. Little, however, is yet known about these networks and proteins, if conserved, in mammalian meiocytes.

Recent evidence indicates an additional role of cohesin in *centriole cohesion* and duplication. Initially, a shorter splice variant of the cohesion protecting protein Shugoshin (SGO1) was observed at centrioles and depeletion of SGO1 caused separation of centrioles. This suggested that SGO1 provides the "glue" that keeps pairs of centrioles together (Wang et al., 2008), and cohesin components such as SMC1 were then also described at the centrioles (Guan, Ekwurtzel, Kvist, & Yuan, 2008). Separase was found to be required for timely centriole disengagement, and formation of new centrioles was delayed in absence of separase. However, centriole disengagement could only be completely blocked if separase and PLK1 were depleted (Tsou et al., 2009). Activation of separase triggered centriole disengagement, expression of a noncleavable SCC1 prevented it, and cleavage of an engineered SMC3 also caused centriole disengagement (Schockel, Mockel, Mayer, Boos, & Stemmann, 2011).

It is unknown whether cohesins play a similar role in meiocytes. Mammalian oocytes lack recognizable centrioles, and the proximal centriole brought into the human oocyte by the sperm upon fertilization duplicates and is thought to set up the first mitotic spindles. It is unclear how the mammalian sperm centriole establishes the new pairs of centrioles and whether cohesins are required. In rodents, the sperm centrioles degenerate and are not recognizable at the time of fertilization, however, and only at the blastocyst stage are new centrioles formed, suggesting some contribution of maternal centriole inheritance, at least in rodents. Overexpression of stable centriole components in Drosophila eggs leads to formation of many centrioles (Peel, Stevens, Basto, & Raff, 2007), and parthenogenetically activated rabbit oocytes also form centrioles (Szollosi & Ozil, 1991), indicating that it is rather due to the lack of stable centriole precursors than a principal inability of oocytes to form centrioles. Also, whether cohesin is involved in meiosis I centriole dynamics remains to be shown.

7. ERRORS AND AGE EFFECTS IN OOCYTES

Aneuploidy, caused by missegregation of chromosomes, is a predominantly oocyte-specific problem, with 90% of human trisomies having originated from a maternal error (Hassold, Hall, & Hunt, 2007; Jessberger, 2012). Chromosome segregation through meiosis I and II occurs in broadly the same manner in both males and females, where two rounds of cell division follow one round of DNA replication, resulting in haploid gametes. An important difference between the two sexes can be found in the temporal control of meiosis. In the male, meiosis in spermatocytes is a relentless process throughout the individual's lifetime, with mature sperm being continuously replenished from a pool of spermatogonial stem cells. In the female, all oocytes are produced at the fetal stage and arrest shortly after diplotene of prophase I, before birth. They will then remain quiescent until they mature, in humans up to several decades later. The striking effect of age can be seen in the statistics of the percentage of aneuploidies in humans, which begins to increase dramatically at the age of 35 years with one-third of pregnancies being affected (Hassold & Hunt, 2001). The challenge for oocytes may be in keeping the chromosomes, and all the necessary proteins, engaged and functional, over such a long period of time. Recent evidence has shed light on the role of cohesion in this process, resulting from the loss of cohesin complexes, or proteins involved in protecting cohesin complexes (Lister et al., 2010; Revenkova, Herrmann, et al., 2010). Initial data came from the analysis of *Smc1β*$^{-/-}$ oocytes (Hodges et al., 2005; Revenkova et al., 2004), which in the adult female show an increased incidence of single homologs, when they should normally be aligned and held together by chiasmata. Strikingly, this phenotype was found to correlate with age; the older the mouse, the higher the number of single homologs observed. The increase in single homologs is thought to occur via chiasma terminalization, where the chiasmata move along the chromosomes and eventually slip off their ends. The distance of the remaining chiasmata from the chromosome ends also decreased with increasing age, and the tendency for smaller chromosomes to be more likely to lose chiasmata completely was observed. This indicated that the meiotic cohesin SMC1β was required to maintain chiasmata over an extended period of time. The question then arose as to whether *Smc1β*, which is still being transcribed at low levels in aging oocytes, could be reloaded onto the chromosomes? To address this, one key experiment was to delete the *Smc1β* gene just after birth, by expression of Cre recombinase from the GDF9 promoter in a *Smc1β-floxed* strain (Revenkova, Adelfalk et al., 2010; Revenkova, Herrmann et al., 2010). No new *Smc1β* transcripts would be produced, and therefore no reloading of newly translated protein could occur. Remarkably, no difference to wild-type mice was found in oocyte survival and terminalization at high age, suggesting that reloading is not required. Loss of cohesin, and therefore cohesion, may explain the age-dependent increase in oocyte aneuploidy. However, these experiments did not prove that reloading cannot happen. To test this, an elaborate approach employing TEV protease-cleavable REC8 or RAD21 (SCC1) was used in oocytes (Tachibana-Konwalski et al., 2010). This allowed removal of REC8- or RAD21-type cohesin after microinjection of the protease, resulting in loss of cohesion. If a non-cleavable cohesin was expressed through Cre-mediated removal of a stop cassette upon Cre expression driven by the Zp3 promoter in growing oocytes, no rescue of cohesion was observed, indicating that under these conditions no or only insufficient reloading took place (Tachibana-Konwalski et al., 2010) .

Still, these experiments do not exclude reloading at follicular stages prior to Zp3 expression, and/or a low level of turnover, although these levels are not sufficient. Two studies indeed showed weakening of cohesion in aging mice, seen as an increase in interkinetochore distance and loss of Shugoshin from oocyte centromeres (Chiang, Duncan, Schindler, Schultz, & Lampson,

2010; Lister et al., 2010), as well as increased sensitivity of cohesion in aging oocytes to separase activation (Chiang, Schultz, & Lampson, 2011).

Another factor to be considered relevant for the aging problem is the maintenance of cohesion. Acetylation of SMC3 by ESCO1 and ESCO2, after cohesin has been loaded during S-phase, results in a more stable association with DNA (Rolef Ben-Shahar et al., 2008; Unal et al., 2008; Zhang, Shi, et al., 2008). How acetylation is controlled over the duration of oocyte arrest is still unknown. Deacetylation could play a role, and perhaps the balance between these two opposing forces could have an age-dependent effect. Sororin acts to antagonize releasin via its interaction with PDS5 (Rankin, Ayad, & Kirschner, 2005; Schmitz, Watrin, Lenart, Mechtler, & Peters, 2007) and therefore may affect this balance also in meiosis. The interplay between loading, establishment, and antiestablishment or release factors in mitotic cells has yet to be fully investigated, and for meiocytes remains largely elusive (Jessberger, 2012; Nasmyth, 2011).

8. PERSPECTIVE/QUESTIONS

The two meiotic cell divisions are highly complex and in many respects unique if compared with somatic cell divisions, posing great challenges for analysis. Mammalian meiosis research suffers from the lack of cell culture systems that faithfully and completely recapitulate meiosis I and II, and thus experimental approaches are mostly restricted to microscopic analyses—often highly sophisticated—and genetics, such as the use of transgenic animal models. Therefore, it does not seem surprising that despite many very important insights into the role of cohesins in mammalian gametogenesis, key questions remain to be answered.

Obviously, the surprising multitude of cohesin proteins and thus cohesin complexes in meiocytes poses the challenge of determining the composition of individual complexes, their localization, their appearance and disappearance during gametogenesis, and their functions. Despite significant progress in demonstrating the existence of certain cohesin complexes and in showing the functions of individual cohesin proteins, we are far from understanding this complexity. Why do meiocytes feature three kleisins, two SMC1-type proteins, three SA proteins, four of all of them meiosis-specific? In addition, it has become clear that many cohesin proteins are posttranslationally modified, however, the structural and functional meaning of these modifications in meiosis remain largely elusive. Another challenge is to describe the intricate relationships between meiotic processes such as synapsis and cohesion, synapsis and recombination, recombination and cohesion, and the difficulty to cleanly separate phenotypes, cause and consequence, direct and indirect effects, from each other.

To tease apart some of these intertwinings, genetic manipulations allowing stage-specific cohesin gene deletions, cohesin protein cleavage, cohesin gene expression, or prevention of cohesin ring opening, are invaluable tools and are increasingly being used, made possible by the development of new technologies such as the stage-specific Cre-expressing transgenic mice, the introduction of protease cleavage sites at precise positions within cohesins, the mutation of sites of posttranslational modification or cleavage, and many others.

Not only cohesins themselves, but also cohesin-associated factors need to be thoroughly understood. Loading and possibly slow reloading of cohesin, stabilization and weakening of cohesin, and the amazing long-term stability of cohesin are key points not only for deciphering the essential cell biology and biochemistry of meiocytes but also to comprehend the major medical problems associated with meiosis, aneuploidies, and sterility syndromes. Likewise, the interaction of cohesin with the kinetochores and even the centrioles, the role of cohesin in checkpoints such as in tension-generation for the SAC, the interplay between cohesin and condensin, cohesin and the SMC5/6 DNA repair complex, as well as cohesin and topoisomerase II, or the relationship of cohesin to the elusive mono-orientation proteins and mechanisms in mammalian meiocytes need to be determined.

REFERENCES

Adelfalk, C., Janschek, J., Revenkova, E., Blei, C., Liebe, B., Gob, E., et al. (2009). Cohesin SMC1beta protects telomeres in meiocytes. *The Journal of Cell Biology*, *187*, 185–199.

Baarends, W. M., Wassenaar, E., van der Laan, R., Hoogerbrugge, J., Sleddens-Linkels, E., Hoeijmakers, J. H. J., et al. (2005). Silencing of unpaired chromatin and histone H2A ubiquitination in mammalian meiosis. *Molecular and Cellular Biology*, *25*, 1041–1053.

Bannister, L. A., Reinholdt, L. G., Munroe, R. J., & Schimenti, J. C. (2004). Positional cloning and characterization of mouse mei8, a disrupted allelle of the meiotic cohesin Rec8. *Genesis*, *40*, 184–194.

Baxter, J., Sen, N., Martinez, V. L., De Carandini, M. E., Schvartzman, J. B., Diffley, J. F., et al. (2011). Positive supercoiling of mitotic DNA drives decatenation by topoisomerase II in eukaryotes. *Science*, *331*, 1328–1332.

Beckouet, F., Hu, B., Roig, M. B., Sutani, T., Komata, M., Uluocak, P., et al. (2010). An Smc3 acetylation cycle is essential for establishment of sister chromatid cohesion. *Molecular Cell*, *39*, 689–699.

Bernard, P., Drogat, J., Maure, J. F., Dheur, S., Vaur, S., Genier, S., et al. (2006). A screen for cohesion mutants uncovers Ssl3, the fission yeast counterpart of the cohesin loading factor Scc4. *Current Biology*, *16*, 875–881.

Bolcun-Filas, E., Costa, Y., Speed, R., Taggart, M., Benavente, R., De Rooij, D. G., et al. (2007). SYCE2 is required for synaptonemal complex assembly, double strand break repair, and homologous recombination. *The Journal of Cell Biology*, *176*, 741–747.

Bolcun-Filas, E., Hall, E., Speed, R., Taggart, M., Grey, C., de Massy, B., et al. (2009). Mutation of the mouse Syce1 gene disrupts synapsis and suggests a link between synaptonemal complex structural components and DNA repair. *PLoS Genetics*, *5*, e1000393.

Borges, V., Lehane, C., Lopez-Serra, L., Flynn, H., Skehel, M., Rolef Ben-Shahar, T., et al. (2010). Hos1 deacetylates Smc3 to close the cohesin acetylation cycle. *Molecular Cell*, *39*, 677–688.

Brito, I. L., Yu, H. G., & Amon, A. (2010). Condensins promote coorientation of sister chromatids during meiosis I in budding yeast. *Genetics*, *185*, 55–64.

Buonomo, S. B., Clyne, R. K., Fuchs, J., Loidl, J., Uhlmann, F., & Nasmyth, K. (2000). Disjunction of homologous chromosomes in meiosis I depends on proteolytic cleavage of the meiotic cohesin Rec8 by separin. *Cell*, *103*, 387–398.

Burgoyne, P. S., Mahadevaiah, S. K., & Turner, J. M. (2009). The consequences of asynapsis for mammalian meiosis. *Nature Reviews. Genetics*, *10*, 207–216.

Callender, T. L., & Hollingsworth, N. M. (2010). Mek1 suppression of meiotic double-strand break repair is specific to sister chromatids, chromosome autonomous and independent of Rec8 cohesin complexes. *Genetics*, *185*, 771–782.

Canudas, S., Houghtaling, B. R., Kim, J. Y., Dynek, J. N., Chang, W. G., & Smith, S. (2007). Protein requirements for sister telomere association in human cells. *The EMBO Journal*, *26*, 4867–4878.

Canudas, S., & Smith, S. (2009). Differential regulation of telomere and centromere cohesion by the Scc3 homologues SA1 and SA2, respectively, in human cells. *The Journal of Cell Biology*, *187*, 165–173.

Caron, P., Aymard, F., Iacovoni, J. S., Briois, S., Canitrot, Y., Bugler, B., et al. (2012). Cohesin protects genes against gammaH2AX Induced by DNA double-strand breaks. *PLoS Genetics*, *8*, e1002460.

Chan, K. L., Roig, M. B., Hu, B., Beckouët, F., Metson, J., & Nasmyth, K. (2012). Cohesin's DNA exit gate is distinct from its entrance gate and is regulated by acetylation. *Cell*, *150*(5), 961–974.

Chiang, T., Duncan, F. E., Schindler, K., Schultz, R. M., & Lampson, M. A. (2010). Evidence that weakened centromere cohesion is a leading cause of age-related aneuploidy in oocytes. *Current Biology*, *20*, 1522–1528.

Chiang, T., Schultz, R. M., & Lampson, M. A. (2011). Age-dependent susceptibility of chromosome cohesion to premature separase activation in mouse oocytes. *Biology of Reproduction*, *85*, 1279–1283.

Chiang, T., Schultz, R. M., & Lampson, M. A. (2012). Meiotic origins of maternal age-related aneuploidy. *Biology of Reproduction*, *86*, 1–7.

Ciosk, R., Shirayama, M., Shevchenko, A., Tanaka, T., Toth, A., & Nasmyth, K. (2000). Cohesin's binding to chromosomes depends on a separate complex consisting of Scc2 and Scc4 proteins. *Molecular Cell*, *5*, 243–254.

Clarke, A. S., Tang, T. T., Ooi, D. L., & Orr-Weaver, T. L. (2005). POLO kinase regulates the Drosophila centromere cohesion protein MEI-S332. *Developmental Cell*, *8*, 53–64.

Clift, D., & Marston, A. L. (2011). The role of shugoshin in meiotic chromosome segregation. *Cytogenetic and Genome Research*, *133*, 234–242.

Clyne, R. K., Katis, V. L., Jessop, L., Benjamin, K. R., Herskowitz, I., Lichten, M., et al. (2003). Polo-like kinase Cdc5 promotes chiasmata formation and cosegregation of sister centromeres at meiosis I. *Nature Cell Biology*, *5*, 480–485.

Corbett, K. D., Yip, C. K., Ee, L. S., Walz, T., Amon, A., & Harrison, S. C. (2010). The monopolin complex crosslinks kinetochore components to regulate chromosome-microtubule attachments. *Cell*, *142*, 556–567.

Cortes-Ledesma, F., de Piccoli, G., Haber, J. E., Aragon, L., & Aguilera, A. (2007). SMC proteins, new players in the maintenance of genomic stability. *Cell Cycle*, *6*, 914–918.

Costa, Y., & Cooke, H. (2007). Dissecting the mammalian synaptonemal complex using targeted mutations. *Chromosome Research*, *15*, 579.

Deardorff, M. A., Bando, M., Nakato, R., Watrin, E., Itoh, T., Minamino, M., Saitoh, K., Komata, M., Katou, Y., Clark, D., Cole, K. E., De Baere, E., Decroos, C., Di Donato, N., Ernst, S., Francey, L. J., Gyftodimou, Y., Hirashima, K., Hullings, M., Ishikawa, Y., Jaulin, C., Kaur, M., Kiyono, T., Lombardi, P. M., Magnaghi-Jaulin, L., Mortier, G. R., Nozaki, N., Petersen, M. B., Seimiya, H., Siu, V. M., Suzuki, Y., Takagaki, K., Wilde, J. J., Willems, P. J., Prigent, C., Gillessen-Kaesbach, G., Christianson, D. W., Kaiser, F. J., Jackson, L. G., Hirota, T., Krantz, I. D., & Shirahige, K. (2012). HDAC8 mutations in Cornelia de Lange syndrome affect the cohesin acetylation cycle. *Nature*, *489*(7415), 313–317.

Ding, D. Q., Sakurai, N., Katou, Y., Itoh, T., Shirahige, K., Haraguchi, T., et al. (2006). Meiotic cohesins modulate chromosome compaction during meiotic prophase in fission yeast. *The Journal of Cell Biology*, *174*, 499–508.

Eijpe, M., Heyting, C., Gross, B., & Jessberger, R. (2000). Association of mammalian SMC1 and SMC3 proteins with meiotic chromosomes and synaptonemal complexes. *Journal of Cell Science*, *113*(Pt 4), 673–682.

Eijpe, M., Offenberg, H., Jessberger, R., Revenkova, E., & Heyting, C. (2003). Meiotic cohesin REC8 marks the axial elements of rat synaptonemal complexes before cohesins SMC1beta and SMC3. *The Journal of Cell Biology*, *160*, 657–670.

Evans, E. B., Hogarth, C., Evanoff, R. M., Mitchell, D., Small, C., & Griswold, M. D. (2012). Localization and regulation of murine Esco2 during male and female meiosis. *Biology of Reproduction*, *87*, 1–8.

Farcas, A. M., Uluocak, P., Helmhart, W., & Nasmyth, K. (2011). Cohesin's concatenation of sister DNAs maintains their intertwining. *Molecular Cell*, *44*, 97–107.

Fukuda, T., Pratto, F., Schimenti, J. C., Turner, J. M., Camerini-Otero, R. D., & Hoog, C. (2012). Phosphorylation of chromosome core components may serve as axis Marks for the status of chromosomal events during mammalian meiosis. *PLoS Genetics*, *8*, e1002485.

Gause, M., Webber, H. A., Misulovin, Z., Haller, G., Rollins, R. A., Eissenberg, J. C., et al. (2008). Functional links between Drosophila Nipped-B and cohesin in somatic and meiotic cells. *Chromosoma*, *117*, 51–66.

Gerlich, D., Koch, B., Dupeux, F., Peters, J. M., & Ellenberg, J. (2006). Live-cell imaging reveals a stable cohesin-chromatin interaction after but not before DNA replication. *Current Biology*, *16*, 1571–1578.

Goh, E. S., Li, C., Horsburgh, S., Kasai, Y., Kolomietz, E., & Morel, C. F. (2010). The Roberts syndrome/SC phocomelia spectrum—A case report of an adult with review of the literature. *American Journal of Medical Genetics. Part A*, *152A*, 472–478.

Gomez, R., Valdeolmillos, A., Parra, M. T., Viera, A., Carreiro, C., Roncal, F., et al. (2007). Mammalian SGO2 appears at the inner centromere domain and redistributes depending on tension across centromeres during meiosis II and mitosis. *EMBO Reports*, *8*, 173–180.

Gregan, J., Rumpf, C., Li, Z., & Cipak, L. (2008). What makes centromeric cohesion resistant to separase cleavage during meiosis I but not during meiosis II? *Cell Cycle*, *7*, 151–153.

Gregan, J., Spirek, M., & Rumpf, C. (2008). Solving the shugoshin puzzle. *Trends in Genetics*, *24*, 205–207.

Griswold, M. D., Hogarth, C. A., Bowles, J., & Koopman, P. (2012). Initiating meiosis: The case for retinoic acid. *Biology of Reproduction*, *86*, 35.

Guan, J., Ekwurtzel, E., Kvist, U., & Yuan, L. (2008). Cohesin protein SMC1 is a centrosomal protein. *Biochemical and Biophysical Research Communications*, *372*, 761–764.

Gutierrez-Caballero, C., Cebollero, L. R., & Pendas, A. M. (2012). Shugoshins: From protectors of cohesion to versatile adaptors at the centromere. *Trends in Genetics*, *28*, 351–360.

Gutierrez-Caballero, C., Herran, Y., Sanchez-Martin, M., Suja, J. A., Barbero, J. L., Llano, E., et al. (2011). Identification and molecular characterization of the mammalian alpha-kleisin RAD21L. *Cell Cycle*, *10*, 1477–1487.

Haering, C. H., & Jessberger, R. (2012). Cohesin in determining chromosome architecture. *Experimental Cell Research*, *318*, 1386–1393.

Hamer, G., Wang, H., Bolcun-Filas, E., Cooke, H. J., Benavente, R., & Hoog, C. (2008). Progression of meiotic recombination requires structural maturation of the central element of the synaptonemal complex. *Journal of Cell Science*, *121*, 2445–2451.

Handel, M. A., & Schimenti, J. C. (2010). Genetics of mammalian meiosis: Regulation, dynamics and impact on fertility. *Nature Reviews. Genetics*, *11*, 124–136.

Hassold, T., Hall, H., & Hunt, P. (2007). The origin of human aneuploidy: Where we have been, where we are going. *Human Molecular Genetics*, *16* (Spec No. 2), R203–R208.

Hassold, T., & Hunt, P. (2001). To err (meiotically) is human: The genesis of human aneuploidy. *Nature Reviews. Genetics*, *2*, 280–291.

Hauf, S., Waizenegger, I. C., & Peters, J. M. (2001). Cohesin cleavage by separase required for anaphase and cytokinesis in human cells. *Science*, *293*, 1320–1323.

Heidinger-Pauli, J. M., Unal, E., Guacci, V., & Koshland, D. (2008). The kleisin subunit of cohesin dictates damage-induced cohesion. *Molecular Cell*, *31*, 47–56.

Herran, Y., Gutierrez-Caballero, C., Sanchez-Martin, M., Hernandez, T., Viera, A., Barbero, J. L., et al. (2011). The cohesin subunit RAD21L functions in meiotic synapsis and exhibits sexual dimorphism in fertility. *The EMBO Journal*, *30*, 3091–3105.

Higashi, T. L., Ikeda, M., Tanaka, H., Nakagawa, T., Bando, M., Shirahige, K., Kubota, Y., Takisawa, H., Masukata, H., & Takahashi, T. S. (2012). The prereplication complex recruits XEco2 to chromatin to promote cohesin acetylation in *Xenopus* egg extracts. *Curr Biol.*, *22*(11), 977–988.

Hodges, C. A., Revenkova, E., Jessberger, R., Hassold, T. J., & Hunt, P. A. (2005). SMC1beta-deficient female mice provide evidence that cohesins are a missing link in age-related nondisjunction. *Nature Genetics*, *37*, 1351–1355.

Hogarth, C. A., Mitchell, D., Evanoff, R., Small, C., & Griswold, M. (2011). Identification and expression of potential regulators of the mammalian mitotic-to-meiotic transition. *Biology of Reproduction*, *84*, 34–42.

Ishiguro, K., Kim, J., Fujiyama-Nakamura, S., Kato, S., & Watanabe, Y. (2011). A new meiosis-specific cohesin complex implicated in the cohesin code for homologous pairing. *EMBO Reports*, *12*, 267–275.

Ishiguro, T., Tanaka, K., Sakuno, T., & Watanabe, Y. (2010). Shugoshin-PP2A counteracts casein-kinase-1-dependent cleavage of Rec8 by separase. *Nature Cell Biology*, *12*, 500–506.

James, R. D., Schmiesing, J. A., Peters, A. H., Yokomori, K., & Disteche, C. M. (2002). Differential association of SMC1alpha and SMC3 proteins with meiotic chromosomes in wild-type and SPO11-deficient male mice. *Chromosome Research*, *10*, 549–560.

Jessberger, R. (2002). The many functions of SMC proteins in chromosome dynamics. *Nature Reviews. Molecular Cell Biology*, *3*, 767–778.

Jessberger, R. (2009). Cohesin's dual role in the DNA damage response: Repair and checkpoint activation. *The EMBO Journal*, *28*, 2491–2493.

Jessberger, R. (2011). Cohesin complexes get more complex: The novel kleisin RAD21L. *Cell Cycle*, *10*(13), 1–2.

Jessberger, R. (2012). Age-related aneuploidy through cohesion exhaustion. *EMBO Reports*, *13*, 539–546.

Jin, H., Guacci, V., & Yu, H. G. (2009). Pds5 is required for homologue pairing and inhibits synapsis of sister chromatids during yeast meiosis. *The Journal of Cell Biology*, *186*, 713–725.

Katis, V. L., Lipp, J. J., Imre, R., Bogdanova, A., Okaz, E., Habermann, B., et al. (2010). Rec8 phosphorylation by casein kinase 1 and Cdc7-Dbf4 kinase regulates cohesin cleavage by separase during meiosis. *Developmental Cell, 18*, 397–409.

Katis, V. L., Matos, J., Mori, S., Shirahige, K., Zachariae, W., & Nasmyth, K. (2004). Spo13 facilitates monopolin recruitment to kinetochores and regulates maintenance of centromeric cohesion during yeast meiosis. *Current Biology, 14*, 2183–2196.

Keeney, S. (2008). Spo11 and the formation of DNA double-strand breaks in meiosis. *Genome Dynamics and Stability, 2*, 81–123.

Kim, K. P., Weiner, B. M., Zhang, L., Jordan, A., Dekker, J., & Kleckner, N. (2010). Sister cohesion and structural axis components mediate homolog bias of meiotic recombination. *Cell, 143*, 924–937.

Kim, S. T., Xu, B., & Kastan, M. B. (2002). Involvement of the cohesin protein, Smc1, in Atm-dependent and independent responses to DNA damage. *Genes & Development, 16*, 560–570.

Kitajima, T. S., Kawashima, S. A., & Watanabe, Y. (2004). The conserved kinetochore protein shugoshin protects centromeric cohesion during meiosis. *Nature, 427*, 510–517.

Kitajima, T. S., Miyazaki, Y., Yamamoto, M., & Watanabe, Y. (2003). Rec8 cleavage by separase is required for meiotic nuclear divisions in fission yeast. *The EMBO Journal, 22*, 5643–5653.

Kitajima, T. S., Sakuno, T., Ishiguro, K., Iemura, S., Natsume, T., Kawashima, S. A., et al. (2006). Shugoshin collaborates with protein phosphatase 2A to protect cohesin. *Nature, 441*, 46–52.

Kleckner, N. (2006). Chiasma formation: Chromatin/axis interplay and the role(s) of the synaptonemal complex. *Chromosoma, 115*, 175–194.

Kudo, N. R., Anger, M., Peters, A. H., Stemmann, O., Theussl, H. C., Helmhart, W., et al. (2009). Role of cleavage by separase of the Rec8 kleisin subunit of cohesin during mammalian meiosis I. *Journal of Cell Science, 122*, 2686–2698.

Kudo, N. R., Wassmann, K., Anger, M., Schuh, M., Wirth, K. G., Xu, H., et al. (2006). Resolution of chiasmata in oocytes requires separase-mediated proteolysis. *Cell, 126*, 135–146.

Kuroda, M., Oikawa, K., Ohbayashi, T., Yoshida, K., Yamada, K., Mimura, J., et al. (2005). A dioxin sensitive gene, mammalian WAPL, is implicated in spermatogenesis. *FEBS Letters, 579*, 167–172.

Lee, J., & Hirano, T. (2011). RAD21L, a novel cohesin subunit implicated in linking homologous chromosomes in mammalian meiosis. *The Journal of Cell Biology, 192*, 263–276.

Lee, B. H., Kiburz, B. M., & Amon, A. (2004). Spo13 maintains centromeric cohesion and kinetochore coorientation during meiosis I. *Current Biology, 14*, 2168–2182.

Lee, J., Kitajima, T. S., Tanno, Y., Yoshida, K., Morita, T., Miyano, T., et al. (2008). Unified mode of centromeric protection by shugoshin in mammalian oocytes and somatic cells. *Nature Cell Biology, 10*, 42–52.

Leem, Y. E., Choi, H. K., Jung, S. Y., Kim, B. J., Lee, K. Y., Yoon, K., et al. (2011). Esco2 promotes neuronal differentiation by repressing Notch signaling. *Cellular Signalling, 23*, 1876–1884.

Li, X., & Dawe, R. K. (2009). Fused sister kinetochores initiate the reductional division in meiosis I. *Nature Cell Biology, 11*, 1103–1108.

Lichten, M., & de Massy, B. (2011). The impressionistic landscape of meiotic recombination. *Cell, 147*, 267–270.

Lightfoot, J., Testori, S., Barroso, C., & Martinez-Perez, E. (2011). Loading of meiotic cohesin by SCC-2 is required for early processing of DSBs and for the DNA damage checkpoint. *Current Biology, 21*, 1421–1430.

Lin, W., Jin, H., Liu, X., Hampton, K., & Yu, H. G. (2011). Scc2 regulates gene expression by recruiting cohesin to the chromosome as a transcriptional activator during yeast meiosis. *Molecular Biology of the Cell, 22*, 1985–1996.

Lister, L. M., Kouznetsova, A., Hyslop, L. A., Kalleas, D., Pace, S. L., Barel, J. C., et al. (2010). Age-related meiotic segregation errors in Mammalian oocytes are preceded by depletion of cohesin and sgo2. *Current Biology*, *20*, 1511–1521.

Llano, E., Herran, Y., Garcia-Tunon, I., Gutierrez-Caballero, C., de Alava, E., Barbero, J. L., et al. (2012). Meiotic cohesin complexes are essential for the formation of the axial element in mice. *The Journal of Cell Biology*, *197*, 877–885.

Losada, A., Yokochi, T., & Hirano, T. (2005). Functional contribution of Pds5 to cohesin-mediated cohesion in human cells and Xenopus egg extracts. *Journal of Cell Science*, *118*, 2133–2141.

MacQueen, A. J., & Hochwagen, A. (2011). Checkpoint mechanisms: The puppet masters of meiotic prophase. *Trends in Cell Biology*, *21*, 393–400.

Merkenschlager, M. (2010). Cohesin: A global player in chromosome biology with local ties to gene regulation. *Current Opinion in Genetics & Development*, *20*, 555–561.

Monje-Casas, F., Prabhu, V. R., Lee, B. H., Boselli, M., & Amon, A. (2007). Kinetochore orientation during meiosis is controlled by Aurora B and the monopolin complex. *Cell*, *128*, 477–490.

Nasmyth, K. (2011). Cohesin: A catenase with separate entry and exit gates? *Nature Cell Biology*, *13*, 1170–1177.

Nasmyth, K., & Haering, C. H. (2009). Cohesin: Its roles and mechanisms. *Annual Review of Genetics*, *43*, 525–558.

Nishiyama, T., Ladurner, R., Schmitz, J., Kreidl, E., Schleiffer, A., Bhaskara, V., et al. (2010). Sororin mediates sister chromatid cohesion by antagonizing Wapl. *Cell*, *143*, 737–749.

Novak, I., Wang, H., Revenkova, E., Jessberger, R., Scherthan, H., & Hoog, C. (2008). Cohesin Smc1beta determines meiotic chromatin axis loop organization. *The Journal of Cell Biology*, *180*, 83–90.

Page, J., de la Fuente, R., Manterola, M., Parra, M. T., Viera, A., Berrios, S., et al. (2012). Inactivation or non-reactivation: What accounts better for the silence of sex chromosomes during mammalian male meiosis? *Chromosoma*, *121*, 307–326.

Parelho, V., Hadjur, S., Spivakov, M., Leleu, M., Sauer, S., Gregson, H. C., et al. (2008). Cohesins functionally associate with CTCF on mammalian chromosome arms. *Cell*, *132*, 422–433.

Parra, M. T., Gomez, R., Viera, A., Llano, E., Pendas, A. M., Rufas, J. S., et al. (2009). Sequential assembly of centromeric proteins in male mouse meiosis. *PLoS Genetics*, *5*, e1000417.

Peel, N., Stevens, N. R., Basto, R., & Raff, J. W. (2007). Overexpressing centriole-replication proteins in vivo induces centriole overduplication and de novo formation. *Current Biology*, *17*, 834–843.

Peters, J. M., Tedeschi, A., & Schmitz, J. (2008). The cohesin complex and its roles in chromosome biology. *Genes & Development*, *22*, 3089–3114.

Petronczki, M., Matos, J., Mori, S., Gregan, J., Bogdanova, A., Schwickart, M., et al. (2006). Monopolar attachment of sister kinetochores at meiosis I requires casein kinase 1. *Cell*, *126*, 1049–1064.

Prieto, I., Suja, J. A., Pezzi, N., Kremer, L., Martinez, A. C., Rufas, J. S., & Barbero, J. L. (2001). Mammalian STAG3 is a cohesin specific to sister chromatid arms in meiosis I. *Nature Cell Biol.*, *3*, 761–766.

Prieto, I., Pezzi, N., Buesa, J. M., Kremer, L., Barthelemy, I., Carreiro, C., et al. (2002). STAG2 and Rad21 mammalian mitotic cohesins are implicated in meiosis. *EMBO Reports*, *3*, 543–550.

Rabitsch, K. P., Gregan, J., Schleiffer, A., Javerzat, J. P., Eisenhaber, F., & Nasmyth, K. (2004). Two fission yeast homologs of Drosophila Mei-S332 are required for chromosome segregation during meiosis I and II. *Current Biology*, *14*, 287–301.

Rankin, S., Ayad, N. G., & Kirschner, M. W. (2005). Sororin, a substrate of the anaphase-promoting complex, is required for sister chromatid cohesion in vertebrates. *Molecular Cell, 18*, 185–200.

Remeseiro, S., Cuadrado, A., Carretero, M., Martinez, P., Drosopoulos, W. C., Canamero, M., et al. (2012). Cohesin-SA1 deficiency drives aneuploidy and tumourigenesis in mice due to impaired replication of telomeres. *The EMBO Journal, 31*, 2076–2089.

Remeseiro, S., Cuadrado, A., Gomez-Lopez, G., Pisano, D. G., & Losada, A. (2012). A unique role of cohesin-SA1 in gene regulation and development. *The EMBO Journal, 31*, 2090–2102.

Ren, Q., Yang, H., Rosinski, M., Conrad, M. N., Dresser, M. E., Guacci, V., et al. (2005). Mutation of the cohesin related gene PDS5 causes cell death with predominant apoptotic features in Saccharomyces cerevisiae during early meiosis. *Mutation Research, 570*, 163–173.

Revenkova, E., Adelfalk, C., & Jessberger, R. (2010). Cohesin in oocytes—Tough enough for mammalian meiosis? *Genes, 1*, 495–504.

Revenkova, E., Eijpe, M., Heyting, C., Hodges, C. A., Hunt, P. A., Liebe, B., et al. (2004). Cohesin SMC1 beta is required for meiotic chromosome dynamics, sister chromatid cohesion and DNA recombination. *Nature Cell Biology, 6*, 555–562.

Revenkova, E., Herrmann, K., Adelfalk, C., & Jessberger, R. (2010). Oocyte cohesin expression restricted to predictyate stages provides full fertility and prevents aneuploidy. *Current Biology, 20*, 1529–1533.

Riedel, C. G., Katis, V. L., Katou, Y., Mori, S., Itoh, T., Helmhart, W., et al. (2006). Protein phosphatase 2A protects centromeric sister chromatid cohesion during meiosis I. *Nature, 441*, 53–61.

Rolef Ben-Shahar, T., Heeger, S., Lehane, C., East, P., Flynn, H., Skehel, M., et al. (2008). Eco1-dependent cohesin acetylation during establishment of sister chromatid cohesion. *Science, 321*, 563–566.

Rowland, B. D., Roig, M. B., Nishino, T., Kurze, A., Uluocak, P., Mishra, A., et al. (2009). Building sister chromatid cohesion: Smc3 acetylation counteracts an antiestablishment activity. *Molecular Cell, 33*, 763–774.

Royo, H., Polikiewicz, G., Mahadevaiah, S. K., Prosser, H., Mitchell, M., Bradley, A., et al. (2010). Evidence that meiotic sex chromosome inactivation is essential for male fertility. *Current Biology, 20*, 2117–2123.

Rumpf, C., Cipak, L., Dudas, A., Benko, Z., Pozgajova, M., Riedel, C. G., et al. (2010). Casein kinase 1 is required for efficient removal of Rec8 during meiosis I. *Cell Cycle, 9*, 2657–2662.

Sakuno, T., Tanaka, K., Hauf, S., & Watanabe, Y. (2011). Repositioning of aurora B promoted by chiasmata ensures sister chromatid mono-orientation in meiosis I. *Dev. Cell, 13*, 534–545.

Schmitz, J., Watrin, E., Lenart, P., Mechtler, K., & Peters, J. M. (2007). Sororin is required for stable binding of cohesin to chromatin and for sister chromatid cohesion in interphase. *Current Biology, 17*, 630–636.

Schockel, L., Mockel, M., Mayer, B., Boos, D., & Stemmann, O. (2011). Cleavage of cohesin rings coordinates the separation of centrioles and chromatids. *Nature Cell Biology, 13*, 966–972.

Seitan, V. C., Banks, P., Laval, S., Majid, N. A., Dorsett, D., Rana, A., et al. (2006). Metazoan Scc4 homologs link sister chromatid cohesion to cell and axon migration guidance. *PLoS Biology, 4*, e242.

Storlazzi, A., Tesse, S., Ruprich-Robert, G., Gargano, S., Poggeler, S., Kleckner, N., et al. (2008). Coupling meiotic chromosome axis integrity to recombination. *Genes & Development, 22*, 796–809.

Strom, L., Lindroos, H. B., Shirahige, K., & Sjogren, C. (2004). Postreplicative recruitment of cohesin to double-strand breaks is required for DNA repair. *Molecular Cell*, *16*, 1003–1015.

Strom, L., & Sjogren, C. (2007). Chromosome segregation and double-strand break repair—A complex connection. *Current Opinion in Cell Biology*, *19*, 344–349.

Sun, S. C., & Kim, N. H. (2012). Spindle assembly checkpoint and its regulators in meiosis. *Human Reproduction Update*, *18*, 60–72.

Szollosi, D., & Ozil, J. P. (1991). De novo formation of centrioles in parthenogenetically activated, diploidized rabbit embryos. *Biology of the Cell*, *72*, 61–66.

Tachibana-Konwalski, K., Godwin, J., van der Weyden, L., Champion, L., Kudo, N. R., Adams, D. J., et al. (2010). Rec8-containing cohesin maintains bivalents without turnover during the growing phase of mouse oocytes. *Genes & Development*, *24*, 2505–2516.

Toth, A., & Jessberger, R. (2010). Male meiosis: Y keep it silenced? *Current Biology*, *20*, R1022–R1024.

Toth, A., Rabitsch, K. P., Galova, M., Schleiffer, A., Buonomo, S. B., & Nasmyth, K. (2000). Functional genomics identifies monopolin: A kinetochore protein required for segregation of homologs during meiosis i. *Cell*, *103*, 1155–1168.

Tsou, M. F., Wang, W. J., George, K. A., Uryu, K., Stearns, T., & Jallepalli, P. V. (2009). Polo kinase and separase regulate the mitotic licensing of centriole duplication in human cells. *Developmental Cell*, *17*, 344–354.

Turner, J. M., Mahadevaiah, S. K., Fernandez-Capetillo, O., Nussenzweig, A., Xu, X., Deng, C. X., et al. (2005). Silencing of unsynapsed meiotic chromosomes in the mouse. *Nature Genetics*, *37*, 41–47.

Uhlmann, F. (2011). Cohesin subunit Rad21L, the new kid on the block has new ideas. *EMBO Reports*, *12*, 183–184.

Uhlmann, F., Lottspeich, F., & Nasmyth, K. (1999). Sister-chromatid separation at anaphase onset is promoted by cleavage of the cohesin subunit Scc1. *Nature*, *400*, 37–42.

Uhlmann, F., Wernic, D., Poupart, M. A., Koonin, E. V., & Nasmyth, K. (2000). Cleavage of cohesin by the CD clan protease separin triggers anaphase in yeast. *Cell*, *103*, 375–386.

Unal, E., Arbel-Eden, A., Sattler, U., Shroff, R., Lichten, M., Haber, J. E., et al. (2004). DNA damage response pathway uses histone modification to assemble a double-strand break-specific cohesin domain. *Molecular Cell*, *16*, 991–1002.

Unal, E., Heidinger-Pauli, J. M., Kim, W., Guacci, V., Onn, I., Gygi, S. P., et al. (2008). A molecular determinant for the establishment of sister chromatid cohesion. *Science*, *321*, 566–569.

van Heemst, D., James, F., Poggeler, S., Berteaux-Lecellier, V., & Zickler, D. (1999). Spo76p is a conserved chromosome morphogenesis protein that links the mitotic and meiotic programs. *Cell*, *98*, 261–271.

van Heemst, D., Kafer, E., John, T., Heyting, C., van Aalderen, M., & Zickler, D. (2001). BimD/SPO76 is at the interface of cell cycle progression, chromosome morphogenesis, and recombination. *Proceedings of the National Academy of Sciences of the United States of America*, *98*, 6267–6272.

Vega, H., Waisfisz, Q., Gordillo, M., Sakai, N., Yanagihara, I., Yamada, M., et al. (2005). Roberts syndrome is caused by mutations in ESCO2, a human homolog of yeast ECO1 that is essential for the establishment of sister chromatid cohesion. *Nature Genetics*, *37*, 468–470.

Verni, F., Gandhi, R., Goldberg, M. L., & Gatti, M. (2000). Genetic and molecular analysis of wings apart-like (wapl), a gene controlling heterochromatin organization in Drosophila melanogaster. *Genetics*, *154*, 1693–1710.

Vogel, S. K., Pavin, N., Maghelli, N., Julicher, F., & Tolic-Norrelykke, I. M. (2009). Self-organization of dynein motors generates meiotic nuclear oscillations. *PLoS Biology*, *7*, e1000087.

Wang, S. W., Read, R. L., & Norbury, C. J. (2002). Fission yeast Pds5 is required for accurate chromosome segregation and for survival after DNA damage or metaphase arrest. *Journal of Cell Science*, *115*, 587–598.

Wang, X., Yang, Y., Duan, Q., Jiang, N., Huang, Y., Darzynkiewicz, Z., et al. (2008). sSgo1, a major splice variant of Sgo1, functions in centriole cohesion where it is regulated by Plk1. *Developmental Cell, 14*, 331–341.
Wang, F., Yoder, J., Antoshechkin, I., & Han, M. (2003). Caenorhabditis elegans EVL-14/PDS-5 and SCC-3 are essential for sister chromatid cohesion in meiosis and mitosis. *Molecular and Cellular Biology, 23*, 7698–7707.
Watanabe, Y., & Nurse, P. (1999). Cohesin Rec8 is required for reductional chromosome segregation at meiosis. *Nature, 400*, 461–464.
Watrin, E., & Peters, J. M. (2006). Cohesin and DNA damage repair. *Experimental Cell Research, 312*, 2687–2693.
Watrin, E., & Peters, J. M. (2009). The cohesin complex is required for the DNA damage-induced G2/M checkpoint in mammalian cells. *The EMBO Journal, 28*, 2625–2635.
Watrin, E., Schleiffer, A., Tanaka, K., Eisenhaber, F., Nasmyth, K., & Peters, J. M. (2006). Human Scc4 is required for cohesin binding to chromatin, sister-chromatid cohesion, and mitotic progression. *Current Biology, 16*, 863–874.
Wendt, K. S., Yoshida, K., Itoh, T., Bando, M., Koch, B., Schirghuber, E., et al. (2008). Cohesin mediates transcriptional insulation by CCCTC-binding factor. *Nature, 451*, 796–801.
Wojtasz, L., Cloutier, J. M., Baumann, M., Daniel, K., Varga, J., Fu, J., et al. (2012). Meiotic DNA double-strand breaks and chromosome asynapsis in mice are monitored by distinct HORMAD2-independent and -dependent mechanisms. *Genes & Development, 26*, 958–973.
Wojtasz, L., Daniel, K., Roig, I., Bolcun-Filas, E., Xu, H., Boonsanay, V., et al. (2009). Mouse HORMAD1 and HORMAD2, two conserved meiotic chromosomal proteins, are depleted from synapsed chromosome axes with the help of TRIP13 AAA-ATPase. *PLoS Genetics, 5*, e1000702.
Wong, R. W., & Blobel, G. (2008). Cohesin subunit SMC1 associates with mitotic microtubules at the spindle pole. *Proceedings of the National Academy of Sciences of the United States of America, 105*, 15441–15445.
Wood, A. J., Severson, A. F., & Meyer, B. J. (2010). Condensin and cohesin complexity: The expanding repertoire of functions. *Nature Reviews. Genetics, 11*, 391–404.
Xiong, B., Lu, S., & Gerton, J. L. (2010). Hos1 is a lysine deacetylase for the Smc3 subunit of cohesin. *Current Biology, 20*, 1660–1665.
Xu, H., Beasley, M. D., Warren, W. D., van der Horst, G. T., & McKay, M. J. (2005). Absence of mouse REC8 cohesin promotes synapsis of sister chromatids in meiosis. *Developmental Cell, 8*, 949–961.
Yamagishi, Y., Honda, T., Tanno, Y., & Watanabe, Y. (2010). Two histone marks establish the inner centromere and chromosome bi-orientation. *Science, 330*, 239–243.
Yanowitz, J. (2010). Meiosis: Making a break for it. *Current Opinion in Cell Biology, 22*, 744–751.
Yazdi, P. T., Wang, Y., Zhao, S., Patel, N., Lee, E. Y., & Qin, J. (2002). SMC1 is a downstream effector in the ATM/NBS1 branch of the human S-phase checkpoint. *Genes & Development, 16*, 571–582.
Youds, J. L., & Boulton, S. J. (2011). The choice in meiosis—Defining the factors that influence crossover or non-crossover formation. *Journal of Cell Science, 124*, 501–513.
Yuan, L., Brundell, E., & Hoog, C. (1996). Expression of the meiosis-specific synaptonemal complex protein 1 in a heterologous system results in the formation of large protein structures. *Experimental Cell Research, 229*, 272–275.
Yuan, L., Liu, J.-G., Hoja, M.-R., Wilbertz, J., Nordqvist, K., & Hoog, C. (2002). Female germ cell aneuploidy and embryo death in mice lacking the meiosis-specific protein SCP3. *Science, 296*, 1115–1118.

Zhang, B., Chang, J., Fu, M., Huang, J., Kashyap, R., Salavaggione, E., et al. (2009). Dosage effects of cohesin regulatory factor PDS5 on mammalian development: Implications for cohesinopathies. *PloS One*, *4*, e5232.

Zhang, J., Hakansson, H., Kuroda, M., & Yuan, L. (2008). Wapl localization on the synaptonemal complex, a meiosis-specific proteinaceous structure that binds homologous chromosomes, in the female mouse. *Reproduction in Domestic Animals*, *43*, 124–126.

Zhang, B., Jain, S., Song, H., Fu, M., Heuckeroth, R. O., Erlich, J. M., et al. (2007). Mice lacking sister chromatid cohesion protein PDS5B exhibit developmental abnormalities reminiscent of Cornelia de Lange syndrome. *Development*, *134*, 3191–3201.

Zhang, N., & Pati, D. (2012). Sororin is a master regulator of sister chromatid cohesion and separation. *Cell Cycle*, *11*, 2073–2083.

Zhang, J., Shi, X., Li, Y., Kim, B. J., Jia, J., Huang, Z., et al. (2008). Acetylation of Smc3 by Eco1 is required for S phase sister chromatid cohesion in both human and yeast. *Molecular Cell*, *31*, 143–151.

CHAPTER TWO

Genomic Imprinting Is a Parental Effect Established in Mammalian Germ Cells

Xiajun Li[*,†,‡,§,1]
*Black Family Stem Cell Institute, Mount Sinai School of Medicine, New York, USA
†Department of Developmental and Regenerative Biology, Mount Sinai School of Medicine, New York, USA
‡Department of Oncological Sciences, Mount Sinai School of Medicine, New York, USA
§Graduate School of Biological Sciences, Mount Sinai School of Medicine, New York, USA
[1]Corresponding author: e-mail address: xiajun.li@mssm.edu

Contents

Abstract

Genomic imprinting is an epigenetic phenomenon in which either the paternal or the maternal allele of imprinted genes is expressed in somatic cells. It is unique to eutherian mammals, marsupials, and flowering plants. It is absolutely required for normal mammalian development. Dysregulation of genomic imprinting can cause a variety of

Current Topics in Developmental Biology, Volume 102
ISSN 0070-2153
http://dx.doi.org/10.1016/B978-0-12-416024-8.00002-7

human diseases. About 150 imprinted genes have been identified so far in mammals and many of them are clustered such that they are coregulated by a *cis*-acting imprinting control region, called the ICR. One hallmark of the ICR is that it contains a germ line-derived differentially methylated region that is methylated on the paternal chromosome or on the maternal chromosome. The DNA methylation imprint is reset in the germ line and differential methylation at an ICR is restored upon fertilization. The DNA methylation imprint is resistant to a genome-wide demethylation process in early embryos and is stably maintained in postimplantation embryos. Maintenance of the DNA methylation imprint is dependent on two distinct maternal effect genes (*Zfp57* and *PGC7/Stella*). In germ cells, around midgestation, the DNA methylation imprint is erased and undergoes another round of the DNA methylation imprint cycle that includes erasure, resetting, restoration, and maintenance of differential DNA methylation.

1. INTRODUCTION

Parental effect on the development of the progeny is very common in invertebrate model systems. Lately, it is also found to be quite prevalent in vertebrate animals and plants. Genomic imprinting is a kind of parental effect that has been intensively studied over past several decades. It is unique to eutherian mammals, marsupials, and flowering plants. Unlike normal biallelic expression of the majority of genes in the genome, expression of imprinted genes is preferentially expressed either from the paternal or from the maternal allele. This review will focus on mammalian genomic imprinting. For genomic imprinting in flowering plants, refer to these papers (Feil & Berger, 2007; Gehring, Reik, & Henikoff, 2009; Law & Jacobsen, 2010).

2. DISCOVERY OF GENOMIC IMPRINTING

Originally, imprinting was referred to as the selective elimination of one paternal chromosome in female embryos or both paternal chromosomes in male embryos in *Sciara* by Helen Crouse in 1960 (Li, 2010). In 1970, imprinting of a specific gene was first discovered in plants (Ferguson-Smith, 2011). Several years later, imprinted genes were also found to be present in mouse based on genetic studies involving deletion, duplication, or translocation of mouse chromosomal regions. These results indicate that some regions of the paternal or the maternal chromosomes have distinctive functions and they could not substitute for each other (Bartolomei & Ferguson-Smith, 2011; Ferguson-Smith, 2011; Li, 2010). It was also found that no live progeny were obtained from parthenogenetic or gynogenetic embryos carrying only one or two complete sets of maternal chromosomes. Neither did androgenetic embryos containing only the paternal chromosomes survive past implantation in mouse (Graham, 1974; Li, 2010; Surani & Barton, 1983). In 1984, two landmark studies involving nuclear transfer experiments demonstrated that maternal and paternal genomes are not identical for their contributions to normal mammalian embryonic development (McGrath & Solter, 1984; Surani, Barton, & Norris, 1984). Thus, the term imprinting was adopted to indicate that these marks are established in the male or female germ line and passed on to the progeny upon fertilization. In 1991, *Igf2r* at the *Tme* locus was the first mammalian imprinted gene identified by positional cloning that was found to be imprinted based on the original studies on the mouse deletion mutant (Bartolomei & Ferguson-Smith, 2011; Ferguson-Smith, 2011). In the same year, two other imprinted genes, *H19* and *Igf2*, were respectively determined by allelic expression based on strain-specific polymorphisms and inheritance of mutant phenotypes caused by gene targeting (Bartolomei & Ferguson-Smith, 2011; Ferguson-Smith, 2011). Roughly 150 imprinted genes have been identified in the mouse so far (http://www.mousebook.org/catalog.php?catalog=imprinting). It is estimated that there may be as many as several hundred imprinted genes in the mammalian genome and more imprinted genes are expected to be found (Koerner, Pauler, Huang, & Barlow, 2009). In addition, some imprinted genes were found to be imprinted in a tissue-specific manner (Bartolomei, 2009; Ferguson-Smith, 2011; Kelsey & Bartolomei, 2012; Li, 2010). Indeed, a few recent studies have reported some potential novel imprinted genes with different genome-wide approaches; although the numbers of potential novel imprinted genes varied greatly among different studies (Babak et al., 2008; Deveale, van der Kooy, & Babak, 2012; Gregg et al., 2010; Pollard et al., 2008; Wang et al., 2008; Xie et al., 2012). The discrepancy is probably caused by limitations in novel technologies at their infancy stages and the total number of imprinted genes is likely to be within the range of a few hundred at most (Kelsey & Bartolomei, 2012).

3. EXPRESSION OF IMPRINTED GENES

Imprinted genes are monoallelically expressed despite the fact that both copies of the imprinted genes are intact within a cell (Fig. 2.1). They are expressed either from the paternal allele (paternally expressed) or from the maternal allele (maternally expressed). However, it is found that monoallelic expression of imprinted genes is not absolute. In many cases,

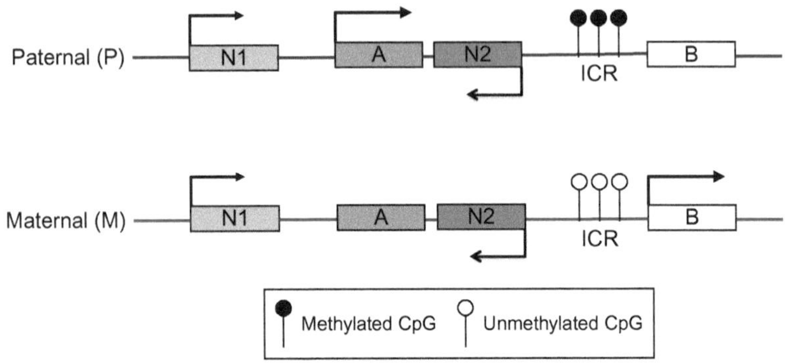

Figure 2.1 Imprinted genes are preferentially expressed either from the maternal or from the paternal allele. Unlike nonimprinted genes that display biallelic expression, imprinted genes are usually monoallelically expressed. As shown in this diagram, N1 and N2 are two nonimprinted genes that are biallelically expressed. The imprinted gene A is only expressed from its allele located on the paternal (P) chromosome. In contrast, the imprinted gene B in this imprinted region is preferentially expressed from the maternal (M) chromosome. This kind of parental origin-specific expression of the imprinted genes is controlled by a *cis*-acting imprinting control region (ICR) located in this imprinted domain. One hallmark of the ICR is that it has germ line-derived differential methylation at the CpG sites located within this ICR. For this ICR, it is only methylated on the paternal chromosome with methylated CpG sites indicated by the filled lollipops, whereas it is unmethylated on the maternal chromosome with unmethylated CpG sites indicated by unfilled lollipops. The line and the arrow above the genes indicate the transcriptional start site and direction of transcriptional elongation. (For color version of this figure, the reader is referred to the online version of this chapter.)

basal levels of expression from the other allele have been observed. So, it may be more appropriate to use the term "preferential expression of the imprinted genes" in some instances (Koerner et al., 2009). Interestingly, monoallelic expression of the imprinted genes could also be tissue-specific. Indeed, some imprinted genes only display an imprinted expression pattern exclusively in the placenta (Abramowitz & Bartolomei, 2012). These placenta-specific imprinted genes include *Ascl2*, *Plda2*, *Slc22a2*, and *Slc22a3*. The mouse insulin genes (*Ins1* and *Ins2*) which are adjacent to the *Igf2–H19* imprinted cluster region are specifically imprinted only in the yolk sac (Bartolomei, 2009; Giddings, King, Harman, Flood, & Carnaghi, 1994). While both alleles for these two insulin genes are expressed in mouse embryos, only the paternal alleles of both genes are expressed in the yolk sac (Bartolomei, 2009; Giddings et al., 1994). In addition, biallelic expression of the *Dlk1* imprinted gene has been observed in neuronal stem cells and niche astrocytes, whereas it is monoallelically expressed in most other somatic tissues including muscle cells (Ferron et al., 2011).

4. COREGULATION OF THE IMPRINTED GENES IN CLUSTERS

Most imprinted genes are clustered and coregulated, although there are also some singleton imprinted genes spread all over the entire genome (Abramowitz & Bartolomei, 2012; Ben-Porath & Cedar, 2000; Edwards & Ferguson-Smith, 2007; Verona, Mann, & Bartolomei, 2003). Each cluster could contain as few as two imprinted genes or as many as several dozen imprinted genes. It is common that some imprinted genes are interspersed with some nonimprinted genes within a cluster (Fig. 2.1). The imprinted genes within each cluster are coregulated by a *cis*-acting imprinting control region, termed ICR (Fig. 2.1). An ICR can exert its long-range effect on imprinted genes up to several megabases away. In each ICR, there is a germ line-derived differentially methylated region (DMR) which is either methylated at the CpG sites on the maternal allele or on the paternal allele (Fig. 2.1). Deletion of the DMR or loss of DNA methylation at the DMR will lead to abnormal expression of all imprinted genes in this cluster.

5. DNA METHYLATION

In mammalian cells, DNA methylation usually occurs at the CpG sites (Chen & Li, 2004; Klose & Bird, 2006; Ooi, O'Donnell, & Bestor, 2009). Lately, non-CpG methylation has been found at a small percentage of cytosine residues in embryonic stem (ES) cells as well as in oocytes (Chen, Feng, Joo, Jacobsen, & Pellegrini, 2011; Lister et al., 2009; Tomizawa et al., 2011). The transfer of the methyl group from *S*-adenosyl methionine to the 5′ position of cytosine is catalyzed by DNA methyltransferases (Chen & Riggs, 2011). The resultant product is 5-methylcytosine (5mC). There are two types of DNA methyltransferases: the maintenance DNA methyltransferase and the *de novo* DNA methyltransferase (Chen & Li, 2004; Chen & Riggs, 2011; Klose & Bird, 2006; Ooi et al., 2009). *De novo* methyltransferases are involved in the conversion of unmethylated cytosine to methylated cytosine. DNMT3A and DNMT3B are both *de novo* methyltransferases. DNMT3L does not have an active DNA methyltransferase domain but it can stimulate the activity of DNMT3A and DNMT3B. DNMT1 is the maintenance DNA methyltransferase and it can copy the methylation pattern from the existing strand to the CpG sites of a newly synthesized DNA strand after replication. DNMT1 also has both maternal and zygotic functions for maintaining DNA methylation in preimplantation embryos (Hirasawa et al., 2008).

6. DNA DEMETHYLATION

Despite its relative stability, DNA methylation is a dynamic process (Wu & Zhang, 2010). It is subject to both active and passive demethylation (Gehring et al., 2009). Several proteins involved in DNA repair pathway such as GADD45, MBD4, and TDG were found to play a role in the DNA demethylation process (Gehring et al., 2009; He et al., 2011; Ma, Guo, Ming, & Song, 2009; Wu & Zhang, 2010). AID, an enzyme that is required for class switch in immunoglobulin proteins, was also required for DNA demethylation during the reprogramming as well as in the germ line (Bhutani et al., 2010; Popp et al., 2010). ELP3, a protein in the elongation complex, was found to play a role in active demethylation of male pronuclei in early embryos (Okada, Yamagata, Hong, Wakayama, & Zhang, 2010). Lately, a lot of attention has been focused on three TET proteins that can convert 5mC to 5-hydroxymethylcytosine (5hmC) (Guo, Su, Zhong, Ming, & Song, 2011; Wu & Zhang, 2011). Indeed, recent studies prove that TET proteins are important players during the demethylation process (Gu et al., 2011; He et al., 2011; Ito et al., 2011). However, their functions in DNA methylation imprinting have not been defined until a recent paper provided some evidence suggesting that TET3-mediated conversion of 5mC to 5hmC may be involved in demethylation of the DNA methylation imprint in early embryos (Nakamura et al., 2012). However, more studies are needed to elucidate the demethylation process at the ICRs, especially during resetting of the DNA methylation imprint in germ cells.

7. CYCLE OF DMR METHYLATION

The cycle of DMR methylation begins during gametogenesis (Tilghman, 1999). In the male or female germ line, the existing DNA methylation imprint on the original paternal (blue line (or thin line) in Fig. 2.2) or maternal (red line (or thick line) in Fig. 2.2) chromosomes will be erased and a new DNA methylation imprint will be laid down on the new paternal or maternal chromosomes. During spermatogenesis, a new paternal DNA methylation imprint will replace the original paternal DNA methylation imprint present on the paternal chromosome or the original maternal DNA methylation imprint present on the maternal chromosome in a diploid cell

(Fig. 2.2). In oogenesis, a new maternal DNA methylation imprint will be established after erasure of the original paternal DNA methylation imprint present on the paternal chromosome or the original maternal DNA methylation imprint present on the maternal chromosome in a diploid cell (Fig. 2.2).

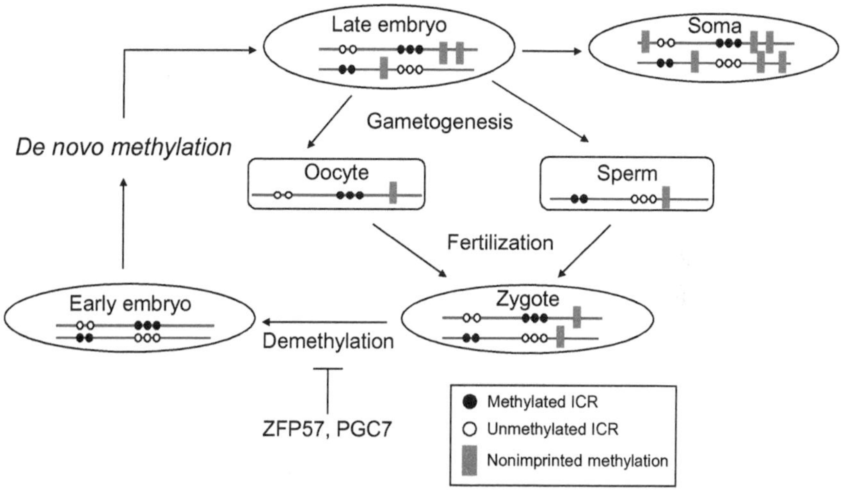

Figure 2.2 A diagram for the cycle of the DNA methylation imprint in mammalian embryos. Differential methylation at the ICRs of the imprinted regions is reset during gametogenesis. The original differential methylation mark is erased during this process in the germ line. New maternal-specific mark or paternal-specific mark at the ICRs is reestablished in the oocyte or the sperm, respectively. DNA methylation can also be established at the CpG sites at the nonimprinted regions during gametogenesis. Upon fertilization, the differential methylation pattern at the ICRs reemerges in the zygote after inheriting the maternal DNA methylation imprint from the oocyte and the paternal DNA methylation imprint from the sperm. DNA methylation at the nonimprinted regions is passed on from the oocyte or the sperm to the zygote as well. In early embryos, DNA methylation is subject to genome-wide demethylation in which DNA methylation in the paternal pronucleus is actively removed while the DNA methylation mark in the maternal pronucleus is gradually erased as well. Differential methylation at the imprinted regions is protected from this genome-wide demethylation process, whereas DNA methylation at nonimprinted regions is erased. Two distinct proteins (ZFP57 and PGC7/Stella) with maternal functions have been identified to maintain the DNA methylation imprint in early embryos. *De novo* methylation occurs at the nonimprinted regions in postimplantation embryos. Gametogenesis commences at midgestation embryos and the DNA methylation imprint is reestablished in the germ cells. *De novo* DNA methylation at nonimprinted regions can continue in somatic cells at late embryonic stages. Thick line (red in the web version), maternal chromosome; Thin line (blue in the web version), paternal chromosome; Filled circle, methylated CpG in an ICR; Unfilled circle, unmethylated CpG in an ICR; Shaded bar, methylated CpG at nonimprinted regions. (For interpretation of the references to color in this figure legend, the reader is referred to the online version of this chapter.)

The erasure of the original DNA methylation imprint occurs around the time (E10–12 in mouse) when primordial germ cells (PGCs) enter the gonad (Abramowitz & Bartolomei, 2012; Ferguson-Smith, 2011; Gehring et al., 2009). It is completed by E14 in both male and female germ cells in the mouse (Gehring et al., 2009; Lucifero, Mann, Bartolomei, & Trasler, 2004; Reik, Dean, & Walter, 2001). Establishment of new paternal DNA methylation imprint during spermatogenesis begins around E15–16 and is almost complete by E17–18 (Henckel, Chebli, Kota, Arnaud, & Feil, 2012; Lucifero et al., 2004; Reik et al., 2001). During oogenesis, the new maternal DNA methylation imprint begins to be established in growing oocytes after the new pups are born (Lucifero et al., 2004; Reik et al., 2001). Upon fertilization, the DNA methylation imprint is passed on from the sperm or oocyte to the zygote and asymmetric differential methylation at the ICRs is reconstituted (Tilghman, 1999; Fig. 2.2). In early preimplantation embryos, there is a genome-wide demethylation process (Fig. 2.2). While DNA methylation at nonimprinted regions is erased, the DNA methylation imprint at the imprinted regions is resistant to this process through the protective roles of *Zfp57* and *PGC7/Stella*, two distinct maternal effect genes in mammals (Li et al., 2008; Nakamura et al., 2007; Fig. 2.2). It is stably maintained during embryogenesis as well as in adult somatic cells (Fig. 2.2).

Although this model of DMR methylation cycle holds well for most imprinted regions, it was found that differential methylation at the *Snrpn* imprinted region can be reacquired in midgestational mouse embryos in an allele-specific manner despite the fact that it was absent in the oocytes or early preimplantation embryos (Li, 2010; Li et al., 2008). This suggests that there may be DNA methylation-independent memory that can reinitiate the acquisition of the DNA methylation imprint during embryogenesis (Li, 2010; Li et al., 2008). Alternatively, it means that the real germ line-derived differential methylation at the *Snrpn* imprinted region has yet to be precisely defined. The currently known *Snrpn* DMR region may not be solely responsible for the imprinting control of this domain even though it is also germ line-derived and differentially methylated. Interestingly, imprinting status can be maintained in a few genes that are specifically imprinted in the placenta in the absence of maintenance methylation by DNMT1 (Lewis et al., 2004; Umlauf et al., 2004). In addition, *de novo* DNA methylation is dispensable for establishing the maternal imprint in oocytes to mark the active X chromosome in preimplantation embryos as well as in the trophoblast of postimplantation embryos (Chiba et al., 2008). Despite the important role of the DNA methylation imprint, these findings suggest that there are some other regulatory mechanisms such as repressive histone marks in genomic imprinting (Chiba et al., 2008; Lewis et al., 2004; Umlauf et al., 2004). However, results from some of these studies may need to be reassessed because perdurance of maternal DNMT1 was not discovered until recently (Hirasawa et al., 2008). Putative roles for maternal DNMT1 in maintaining the DNA methylation imprint in *Dnmt1* zygotic mutant embryos derived from crosses with heterozygous female mice might not have been appropriately addressed in the previous studies (Barlow, 2011).

8. ESTABLISHMENT OF THE DNA METHYLATION IMPRINT

DNA methylation imprint is erased and reestablished in the male or female germ line (Tilghman, 1999). It has been shown that DNMT3A, DNMT3B, and DNMT3L are involved in the establishment of new DNA methylation imprinting in the germ line (Bourc'his, Xu, Lin, Bollman, & Bestor, 2001; Kaneda et al., 2004). KDM1B/AOF1, a histone lysine demethylase related to KDM1/LSD1, is also required for this process (Ciccone et al., 2009). ZFP57 is a KRAB zinc finger protein that maintains the DNA methylation imprint in embryos. It appears that the maternal function of ZFP57 is also necessary for the establishment of the maternal DNA methylation imprint at the *Snrpn* DMR region in the oocytes (Li, 2010; Li et al., 2008). Recently, it was found that PIWI-interacting RNA (piRNA) machinery is involved in establishing paternal DNA methylation imprinting at the *Rasgrf1* ICR in male germ cells (Tomizawa & Sasaki, 2012; Watanabe et al., 2011).

9. MAINTENANCE OF THE DNA METHYLATION IMPRINT

Most DNA methylation is erased in early preimplantation embryos (Tilghman, 1999). DNA methylation in male pronuclei is immediately subjected to a globally active demethylation process that is likely mediated by TET3 after fertilization (Edwards & Ferguson-Smith, 2007; Gu et al., 2011; Reik, 2007; Wu & Zhang, 2010). DNA methylation on chromosomes of female pronuclei is thought to be gradually lost in early preimplantation embryos (Edwards & Ferguson-Smith, 2007; Wu & Zhang, 2010). The DNA methylation imprint is protected from this genome-wide demethylation process (Wu & Zhang, 2010; Fig. 2.2). *PGC7/Stella*, a maternal effect gene, has been shown to play partially protective role during this process

(Nakamura et al., 2007; Fig. 2.2). *Zfp57*, the first identified mammalian maternal–zygotic effect gene, is a major player in maintaining the DNA methylation imprint in mouse embryos (Li et al., 2008; Fig. 2.2). Consistent with this, human ZFP57 plays a similar role in maintaining the DNA methylation imprint at multiple imprinted regions (Mackay et al., 2008).

Mouse ZFP57 also maintains the DNA methylation imprint in ES cells (Quenneville et al., 2011; Zuo et al., 2012). As a KRAB zinc finger protein, ZFP57 interacts with its obligate cofactor KAP1/Trim28/Tif1β (Li et al., 2008; Zuo et al., 2012). Indeed, ZFP57 mutant protein lacking its interaction domain with KAP1/Trim28/Tif1β is defective in the maintenance of the DNA methylation imprint in ES cells (Zuo et al., 2012). Further, loss of maternal KAP1/Trim28/Tif1β causes partial loss of DNA methylation at *H19* DMR as well as variable but only a slight loss of the DNA methylation imprint in other imprinted regions (Messerschmidt et al., 2012). ZFP57 and KAP1/Trim28/Tif1β have been shown to bind to similar imprinted domains (Quenneville et al., 2011). Further, ZFP57 can recruit DNA methyltransferases via KAP1/Trim28/Tif1β (Quenneville et al., 2011; Zuo et al., 2012). Thus, ZFP57 and its cofactor KAP1/Trim28/Tif1β play a major role in maintaining the DNA methylation imprint in both mouse embryos and ES cells (Li et al., 2008; Messerschmidt et al., 2012; Quenneville et al., 2011; Zuo et al., 2012).

10. IMPRINTING CONTROL MECHANISMS

Many of the known imprinted genes are clustered and coregulated by a *cis*-acting imprinting control region termed ICR (Bartolomei, 2009; Bartolomei & Ferguson-Smith, 2011; Ben-Porath & Cedar, 2000; Edwards & Ferguson-Smith, 2007). An ICR is also referred to as an ICE for imprinting control element (Santoro & Barlow, 2011). Differential methylation at the CpG sites within an ICR is required for the proper expression of imprinted genes (Bartolomei & Ferguson-Smith, 2011; Stoger et al., 1993). DNA methylation imprint at an ICR is established in the germ line and can be derived from the paternal or maternal chromosome. In mouse, at least 21 ICRs were discovered to contain a germ line-derived DMR (Tomizawa et al., 2011). In a recent genome-wide study, 55 DMRs were found to be germ line derived and indeed some of these novel DMRs may be located in previously unknown imprinted regions (Xie et al., 2012). There are four well-defined ICRs inherited on the paternal chromosomes. These reside in the *Igf2–H19*, *Dlk1–Gtl2*, *Rasgrf1*, and *Gpr1–Zdbf2* imprinted regions (Hiura et al., 2010; Watanabe et al., 2011). At least 17

DMRs are established at the ICRs of the maternal chromosomes during oogenesis (Tomizawa et al., 2011). These maternally inherited ICRs include the *Snrpn*, *Igf2r*, *Peg1*, *Peg3*, *Peg5*, *Peg10*, *Peg13*, *Gnas1A*, *Nespas*, *Grb10*, *Zac1*, *Mest*, *Mcts2*, *Impact*, *U2af1-rs1*, *Inpp5f*, and *Kcnq1* imprinted regions (Barlow, 2011; Tomizawa et al., 2011). Some of the imprinted regions have been more intensively studied than others and their molecular mechanisms in imprinting control are somewhat more understood, as described below.

11. *Igf2–H19* IMPRINTED REGION

The insulator transcription factor CTCF that is usually involved in boundary elements is a key regulator in the *Igf2–H19* imprinted region (Bartolomei, 2009; Fig. 2.3A). When the ICR located between *Igf2* and *H19* imprinted genes is methylated on the paternal allele, CTCF cannot bind to the ICR and allow the enhancer located downstream of the *H19* gene to interact with the promoter of the *Igf2* imprinted gene (Hark et al., 2000; Fig. 2.3A). Accordingly, the *Igf2* gene will be transcribed from the paternal allele (Schoenherr, Levorse, & Tilghman, 2003). In contrast, CTCF can bind to the ICR region of the maternal allele because it is not methylated. Binding of CTCF prevents the interaction between the enhancer and the *Igf2* promoter (Engel, West, Felsenfeld, & Bartolomei, 2004; Fig. 2.3A). The enhancer interacts with the promoter of the *H19* imprinted gene instead and thus drives the transcription of *H19* from the maternal allele (Bartolomei, 2009).

12. *Igf2r* IMPRINTED REGION

Another well-studied region is the *Igf2r* imprinted region (Santoro & Barlow, 2011; Fig. 2.3B). A long noncoding RNA, *Airn*, is transcribed from the chromosome carrying the unmethylated ICE or ICR located in the second intron of the *Igf2r* gene (Koerner & Barlow, 2010; Koerner et al., 2012; Wutz et al., 1997). *Airn* overlaps with *Igf2r* transcript in an antisense orientation (Sleutels, Zwart, & Barlow, 2002; Fig. 2.3B). Transcription of *Airn* or *Airn* noncoding RNA (ncRNA) itself is required for *cis*-acting imprinting control of this *Igf2r* imprinted region including transcription of the overlapping *Igf2r* gene and nonoverlapping *Slc22a3* and *Slc22a2* imprinted genes in the same cluster (Bartolomei, 2009; Bartolomei & Ferguson-Smith, 2011; Koerner & Barlow, 2010; Pauler, Koerner, & Barlow, 2007; Fig. 2.3B).

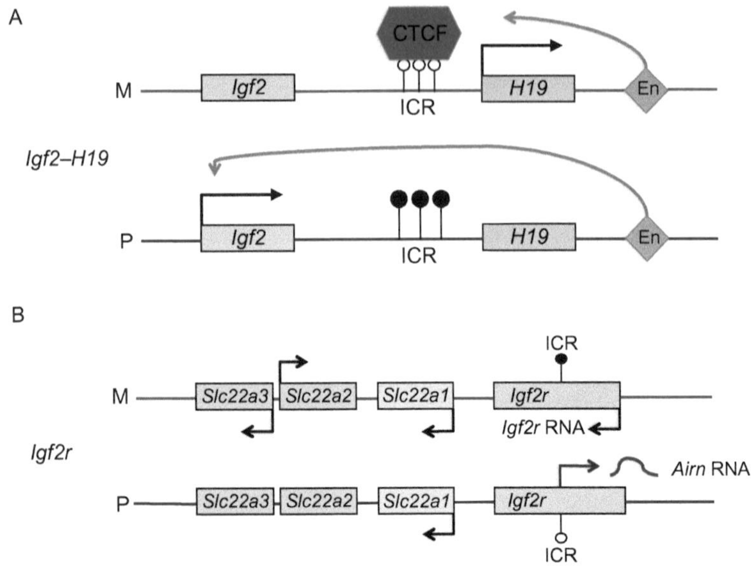

Figure 2.3 Diagrams are shown for the *Igf2–H19* and *Igf2r* imprinted regions. The relative distances for these genes are not drawn to scale. (A) The *Igf2–H19* imprinted region. There is an ICR located between the imprinted *Igf2* gene and imprinted *H19* gene. This ICR is methylated (filled lollipops) on the paternal (P) chromosome but not methylated (unfilled lollipops) on the maternal (M) chromosome. There is an enhancer located downstream of the imprinted *H19* gene which can activate the transcription of the *H19* gene or the *Igf2* gene depending on whether the ICR is methylated or not. As the ICR is not methylated on the maternal chromosome, the transcription factor CTCF will bind to this ICR. Association of CTCF with this ICR will block the interaction between the enhancer downstream of the *H19* gene in this region and the promoter sequence for the *Igf2* gene. Instead, this enhancer will interact with the promoter sequence of the *H19* gene and drive its transcription on the maternal chromosome. In contrast, the ICR is methylated on the paternal chromosome and, consequently, CTCF cannot bind to this ICR. The enhancer element located downstream of the *H19* gene can interact with the promoter sequence of the imprinted *Igf2* gene. Therefore, *Igf2* on the paternal chromosome will be transcribed and the imprinted *H19* gene will be repressed on the paternal chromosome. Accordingly, the imprinted *Igf2* gene is expressed from the paternal chromosome, whereas the imprinted *H19* gene is specifically expressed from the maternal chromosome in normally imprinted wild-type somatic cells. The line and the arrow above the genes indicate the transcriptional start site and direction of the transcriptional elongation. (B) The imprinted *Igf2r* imprinted region. There is an ICR located in the second intron of the imprinted *Igf2r* gene. This ICR is methylated (filled lollipop) on the maternal (M) chromosome but not methylated (unfilled lollipop) on the paternal (P) chromosome. When the ICR is not methylated on the paternal chromosome, a noncoding RNA, *Airn*, is transcribed in an antisense orientation to the *Igf2r* gene. Airn noncoding RNA or transcription of this noncoding RNA is required for the *cis*-acting imprinting control of this *Igf2r* imprinted region. When this ICR is methylated on the maternal chromosome, *Igf2r* is transcribed. Besides maternal-specific expression of the *Igf2r* imprinted genes, two other imprinted genes (*Slc22a3* and *Slc22a2*) are shown here that are both expressed from the maternal chromosome as well. In contrast, *Slc22a1* is not imprinted and is biallelically expressed despite that it is located in between the imprinted *Igf2r* gene and imprinted *Slc22a2* gene. The line and the arrow above the genes indicate the transcriptional start site and direction of the transcriptional elongation. (For color version of this figure, the reader is referred to the online version of this chapter.)

13. *Dlk1–Gtl2* IMPRINTED REGION

The *Dlk1–Gtl2* imprinted domain, also called *Dlk1–Dio3* imprinted domain, is located on mouse chromosome 12 (Edwards & Ferguson-Smith, 2007; Fig. 2.4A). *Dlk1* is expressed from the paternal chromosome whereas *Gtl2* is expressed from the maternal chromosome (Fig. 2.4A). They are coregulated by a *cis*-acting ICR which is located between *Dlk1* and *Gtl2* (Schmidt, Matteson, Jones, Guan, & Tilghman, 2000; Takada et al., 2000; Fig. 2.4A). This ICR is methylated on the paternal chromosome but not on the maternal chromosome (Fig. 2.4A). When this ICR is deleted and maternally transmitted, expression of all imprinted genes are affected in this domain (Lin et al., 2003). There appears to be a maternal to paternal epigenetic phenotype switch for this imprinted domain based on expression levels of the imprinted genes. Specifically, a twofold increase is observed for paternally expressed protein-coding genes *Dlk1* and *Dio3*, whereas diminished expression is observed for maternally expressed ncRNA genes such as *Gtl2* and some *snoRNA* genes (Lin et al., 2003).

14. *Snrpn* IMPRINTED REGION

Snrpn is another well-studied imprinted region that is associated with Prader–Willi syndrome (PWS) and Angelman syndrome (AS), two clinically distinctive parental origin-dependent disorders (Jiang, Tsai, Bressler, & Beaudet, 1998; Nicholls, Saitoh, & Horsthemke, 1998; Fig. 2.4B). PWS is caused by loss of expression of the imprinted genes on the paternal chromosome, whereas AS results from loss of the maternally expressed imprinted genes at the same imprinted domain. The *Snrpn* imprinted domain covers a large region of approximately 2 megabases in human chromosome 15 and mouse chromosome 7 (Barlow, 2011; Edwards & Ferguson-Smith, 2007; Horsthemke & Wagstaff, 2008; Kantor, Shemer, & Razin, 2006). The ICR for this imprinted region in humans consists of two separate elements, PWS-IC and AS-IC (Edwards & Ferguson-Smith, 2007; Kantor et al., 2006; Rabinovitz, Kaufman, Ludwig, Razin, & Shemer, 2012). PWS-IC is conserved in mouse. However, the corresponding AS-IC region has not been found in mouse so far (Edwards & Ferguson-Smith, 2007). When the PWS-IC region is unmethylated on the paternal chromosome, the paternally expressed genes, including protein-coding genes *Snrpn*, *Magel2*, and *Ndn*, as well as some ncRNA genes, are expressed specifically or preferentially from alleles on the paternal chromosome (Fig. 2.4B).

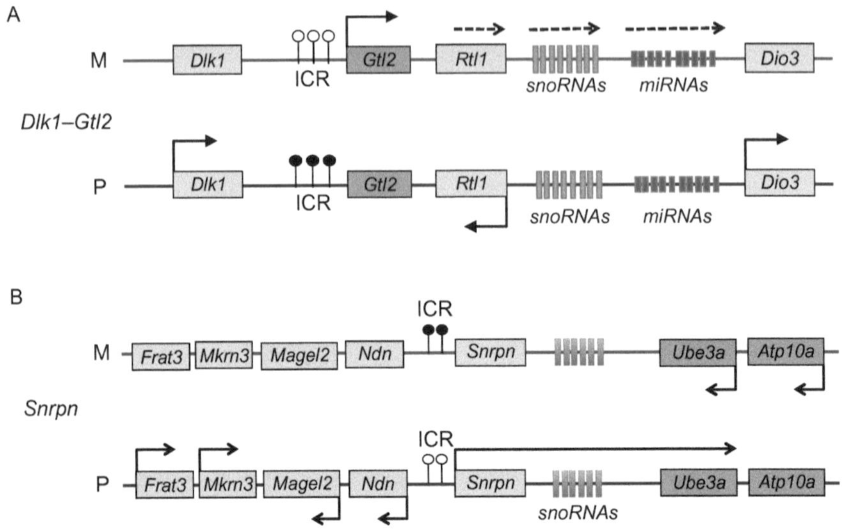

Figure 2.4 Diagrams are for the *Dlk1–Gtl2* and *Snrpn* imprinted regions. The relative distances for these genes are not drawn to scale. (A) The imprinted *Dlk1–Gtl2* region. There is an ICR located between the imprinted *Dlk1* gene and the imprinted *Gtl2* gene. This ICR is methylated (filled lollipops) on the paternal (P) chromosome but not methylated (unfilled lollipops) on the maternal (M) chromosome. When this ICR is methylated, the imprinted *Dlk1*, *Rtl1*, and *Dio3* genes are expressed from the paternal chromosome. These are all protein-coding genes. In contrast, the imprinted noncoding RNA *Gtl2* gene, many clustered *snoRNA* genes and *miRNA* genes, as well as the antisense transcript for *Rtl1* are expressed from the maternal chromosome that contains an unmethylated ICR. The line and the arrow above the genes indicate the transcriptional start site and direction of the transcriptional elongation. The dashed arrow indicates the transcriptional direction for the noncoding RNA transcript with uncertain transcriptional start sites. (B) The imprinted *Snrpn* region. There is an ICR near the *Snrpn* imprinted gene. When this ICR is methylated (filled lollipops) on the maternal (M) chromosome, *Ube3a* and *Atp10* are only expressed from the maternal chromosome. In contrast, the ICR on the paternal (P) chromosome is not methylated (unfilled lollipops). A long transcript is initiated at the beginning of the imprinted *Snrpn* gene and elongated on this imprinted region to cover a cluster of *snoRNA* genes. This transcript is extended to the imprinted *Ube3a* region on the paternal chromosome that is in the antisense orientation to the *Ube3a* transcript that is normally expressed from the maternal chromosome in brain. In addition, the imprinted genes including *Frat3*, *Mkrn3*, *Magel2*, and *Ndn* will also be expressed from the paternal chromosome but not from the maternal chromosome. The line above the genes indicates the transcriptional start site. The arrow points to the direction of the transcriptional elongation. (See Color Insert.)

In contrast, *Ube3a* and *Atp10a* are expressed from alleles on the maternal chromosome in the brain when PWS-IC is methylated. *Ube3a* is biallelically expressed in other tissues except brain (Horsthemke & Wagstaff, 2008). In humans, AS-IC is required for the methylation of PWS-IC in the female germ line, although the mouse counterpart has either not been found or does not exist (Edwards & Ferguson-Smith, 2007). If the latter is the case, it means that PWS-IC in the mouse could be properly methylated in the female germ line even in the absence of AS-IC.

15. IMPRINTING-RELATED HUMAN DISEASES

Dysregulation of imprinting is associated with many human diseases such as diabetes, cancer, and neurological disorders (Bartolomei & Ferguson-Smith, 2011; Li, 2010). Abnormally low expression of the imprinted *Igf2* genes as a result of hypomethylation at the ICR of the *Igf2–H19* domain can lead to Silver–Russel syndrome, a growth retardation disease (Bartolomei & Ferguson-Smith, 2011). In contrast, increased expression of the imprinted *Igf2* gene due to hypermethylation at the ICR of the *Igf2–H19* domain can cause Beckwith–Weidmann syndrome and such patients are susceptible to cancer (Bartolomei & Ferguson-Smith, 2011). The *Snrpn* imprinted region is associated with PWS and AS. Loss of paternally expressed imprinted genes is associated with PWS, whereas loss of maternally expressed imprinted genes leads to AS (Butler, 2011; Horsthemke & Wagstaff, 2008; Mabb, Judson, Zylka, & Philpot, 2011).

Loss of differential DNA methylation at imprinted regions such as the *PLAGL1* imprinted domain is commonly observed in patients with transient neonatal diabetes (Mackay, Temple, Shield, & Robinson, 2005). Mutations in human ZFP57 can cause transient neonatal diabetes and are associated with hypomethylation at multiple imprinted domains as well (Mackay et al., 2008). These mutations have also been linked with defects in the cardiovascular system, nervous system, and sensory organs (Mackay et al., 2008).

16. MACRO ncRNA IN IMPRINTING CONTROL

ncRNAs have been proved to be important players in transcriptional and posttranscriptional regulation in mammals. They can be divided into short ncRNAs and long ncRNAs (also called macro ncRNAs) (Barlow, 2011; Koerner et al., 2009). They can act both in *cis* and in *trans* and are usually associated with long ncRNAs and short ncRNAs, respectively. Transacting short ncRNAs include siRNAs in the RNAi pathway, miRNAs, and piRNAs that are around 20–30 nt in length. Short nucleolar (sno) RNAs are another class of short ncRNAs that usually range from 60 to 300 nt. In contrast, the *cis*-acting ncRNAs are large in size and can be as long as several hundred thousand nucleotides. The best known macro ncRNAs in mammals are *Xist* and its antisense transcript *Tsix* in X-chromosome inactivation in females (Gendrel & Heard, 2011; Lee, 2011). The other classes of well-studied mammalian macro ncRNAs are those transcribed in the clusters of imprinted genes (Barlow, 2011; Bartolomei & Ferguson-Smith, 2011). At least two macro ncRNAs have been experimentally proven to be required for controlling expression of imprinted genes in the whole cluster (Koerner et al., 2009; Santoro & Barlow, 2011). The *Airn* ncRNA is transcribed from the unmethylated ICE of the *Igf2r* imprinted region on the paternal chromosome (Mancini-Dinardo, Steele, Levorse, Ingram, & Tilghman, 2006; Sleutels et al., 2002). Similarly, *Kcnq1ot1* macro ncRNA is monoallelically expressed from the unmethylated ICE of the *Kcnq1* cluster on the paternal chromosome (Smilinich et al., 1999). These macro ncRNAs are required for transcriptional silencing of neighboring imprinted genes in *cis* so that these neighboring imprinted genes will only be transcribed from the other alleles on the maternal chromosomes (Mancini-Dinardo et al., 2006; Santoro & Barlow, 2011).

17. IMPRINTING IN PLURIPOTENT STEM CELLS

ES cells are usually derived from blastocysts and therefore they retain many characteristics of early embryos including genomic imprinting status (Mann, 2001; Rugg-Gunn, Ferguson-Smith, & Pedersen, 2005; Weng, Magnuson, & Storb, 1995). ES cells can also form embryoid bodies (EBs) which can be induced to differentiate into different cell lineages (Keller, 2005). As EBs can give rise to all three germ layers upon differentiation, they can be used as a model system to mimic developmental programs *in vivo* (Murry & Keller, 2008). *In vitro* differentiation of female ES cells has been widely used to elucidate molecular mechanisms in X-chromosome inactivation in mammals (Wutz, 2007). Given some similarities between X-inactivation and imprinted expression, it has been proposed that ES cells might offer a useful model for genomic imprinting (Reik & Lewis, 2005). Several studies have already shown that *in vitro* differentiation of ES cells into EBs indeed can be used to study some key steps during the establishment and maintenance of asymmetric expression of imprinted genes (Latos et al., 2009; Sun et al., 2006). Even expression of some tissue-specific imprinted genes can be recapitulated with the ES cell differentiation system *in vitro* (Kohama et al., 2012). Consistent with these hypotheses, the key regulator in genomic imprinting ZFP57 and its cofactor KAP1/Trim28/Tif1β have been shown to maintain the DNA methylation genomic imprint in ES cells (Quenneville et al., 2011; Zuo et al., 2012), similar to their roles in mouse embryos (Li et al., 2008; Messerschmidt et al., 2012).

Genomic imprinting has also been found to be relatively unstable in pluripotent stem cells, although some imprinted regions are more stable than others (Humpherys et al., 2001; Schumacher & Doerfler, 2004). Mouse induced pluripotent stem cells (iPS cells) also have a tendency to lose the DNA methylation imprint and allelic expression of imprinted genes (Sun et al., 2012). It was reported that genomic imprinting at the *Dlk1–Gtl2* imprinted region was not properly reprogrammed in a majority of mouse iPS clones (Stadtfeld et al., 2010; Sun et al., 2012). Interestingly, a different study suggested that the level and stoichiometry of reprogramming factors during the reprogramming process could influence the outcome of the imprinting status at the *Dlk1–Gtl2* imprinted region in the resultant iPS clones (Carey et al., 2011). Abnormal imprinting at this imprinted region in mouse iPS cells could even be attenuated by the addition of ascorbic acid during the reprogramming process (Stadtfeld et al., 2012). Human ES and iPS cells also exhibit variable loss of genomic imprinting at different imprinted regions in culture (Pick et al., 2009; Rugg-Gunn, Ferguson-Smith, & Pedersen, 2007).

18. EVOLUTION OF IMPRINTING

Imprinting is only present in eutherian mammals, marsupials, and flowering plants (Barlow, 2011; Bartolomei & Ferguson-Smith, 2011). Many imprinted genes have homologs in the lower organisms. However, their homologs are not imprinted like their mammalian counterparts. It is still unknown how imprinting has evolved to be quite prevalent in mammals. Several different theories have been proposed to explain evolution of genomic imprinting as indicated below.

The host defense theory hypothesizes that genomic imprinting is part of the host defense system against foreign viruses or retrotransposons (Barlow, 1993; Renfree, Hore, Shaw, Graves, & Pask, 2009). DNA methylation and histone modification machinery are recruited to silence the foreign DNA elements integrated into the genome. Genomic imprinting has evolved as a consequence of this kind of silencing of a particular gene. Supporting evidence for this theory is that DNMT3L is involved in both the establishment of the maternal DNA methylation imprint in oocytes and silencing of retrotransposons in the male germ cells (Bourc'his & Bestor, 2004; Bourc'his et al., 2001).

Another popular theory regarding the evolution of imprinting is the parental conflict or kinship hypothesis (Haig, 2004; Renfree et al., 2009; Wilkins & Haig, 2003). Paternally expressed genes are evolutionally selected to promote the transfer of nutrient from the mother to the fetus. In contrast, the mother may limit the growth of each individual progeny during pregnancy by silencing the expression of growth-promoting genes on the maternal chromosome that are normally expressed on the paternal chromosome. This parental conflict hypothesis is consistent with many observations of imprinted genes. Indeed, many paternally expressed imprinted genes such as *Igf2* and *Dlk1* promote the growth of the fetus and its placenta that supplies the nutrients to the fetus (DeChiara, Robertson, & Efstratiadis, 1991; Moon et al., 2002), whereas some maternally expressed imprinted genes such as *Igf2r* have an inhibitory role on growth of the progeny (Ludwig et al., 1996). Although the parental conflict theory could explain the functions of many imprinted genes, it cannot account for the functions of some other imprinted genes (Renfree et al., 2009). This hypothesis is also inconsistent with the recent findings that two maternal effect genes, *Zfp57* and *PGC7/Stella*, maintain both paternally as well as maternally inherited DNA methylation imprints in early mouse embryos (Li, 2010; Li et al., 2008; Nakamura et al., 2007).

Recently, a third hypothesis was proposed (Renfree et al., 2009). The coadaptation theory argues that genomic imprinting has evolved as coadaptation between mammalian embryonic development and reproductive behavior (Keverne & Curley, 2008). This theory is based on the observations that a subset of paternally expressed imprinted genes are expressed in the placenta as well as in the maternal hypothalamus. Thus, it is hypothesized that the placenta and the maternal hypothalamus may be coregulated and function as one unit in controlling both development of mammalian embryos and reproductive behavior (Renfree et al., 2009).

All three popular theories regarding genomic imprinting can explain some aspects of the imprinting phenomena, but not all of them. It is possible that genomic imprinting has evolved through multiple different mechanisms, with each one accounting for some imprinted genes and/or a portion of the imprinting phenomena. It also remains to be determined why genomic imprinting is only present in eutherian mammals, marsupials, and flowering plants.

19. CONCLUSIONS

Genomic imprinting is a unique epigenetic regulation in which one copy of the gene is active and the other copy of the gene is silenced in a diploid cell in a parental origin-dependent manner. The epigenetic mark is established in the germ line and stably maintained in somatic cells. For an imprinted domain consisting of a cluster of imprinted genes, this germ line-derived genomic imprint serves as the control element for both paternally expressed and maternally expressed genes within this imprinted domain. Thus, genomic imprinting is a special model system for studying both active and repressive epigenetic marks for gene regulation within the same diploid cell. Some findings based on studies in genomic imprinting have been applied to other systems in epigenetic regulation. As many imprinted genes are involved in fetal growth and embryonic development, proper regulation of genomic imprinting is essential for mammalian embryos to develop normally. Abnormal expression of imprinted genes will lead to a variety of human diseases including cancer, diabetes, cardiovascular defects, and neurological disorders. Therefore, it is important to study how genomic imprinting is properly regulated in mammals. Future efforts aimed at understanding both normal and abnormal patterns of genomic imprinting will not only advance our knowledge in this interesting epigenetic phenomenon but also may lead to potential therapeutic applications in the future.

ACKNOWLEDGMENTS

The work in the author's laboratory is supported by grants from NIH (GM093335), New York State (NYSTEM Contract #C026434), and American Heart Association (09SDG2400151). The author apologizes to others in the genomic imprinting field whose papers may not have been cited here due to space limitations.

REFERENCES

Abramowitz, L. K., & Bartolomei, M. S. (2012). Genomic imprinting: Recognition and marking of imprinted loci. *Current Opinion in Genetics & Development*, *22*, 72–78.

Babak, T., Deveale, B., Armour, C., Raymond, C., Cleary, M. A., van der Kooy, D., et al. (2008). Global survey of genomic imprinting by transcriptome sequencing. *Current Biology*, *18*, 1735–1741.

Barlow, D. P. (1993). Methylation and imprinting: From host defense to gene regulation? *Science*, *260*, 309–310.

Barlow, D. P. (2011). Genomic imprinting: A mammalian epigenetic discovery model. *Annual Review of Genetics*, *45*, 379–403.

Bartolomei, M. S. (2009). Genomic imprinting: Employing and avoiding epigenetic processes. *Genes & Development*, *23*, 2124–2133.

Bartolomei, M. S., & Ferguson-Smith, A. C. (2011). Mammalian genomic imprinting. *Cold Spring Harbor Perspectives in Biology, 3*.

Ben-Porath, I., & Cedar, H. (2000). Imprinting: Focusing on the center. *Current Opinion in Genetics & Development, 10*, 550–554.

Bhutani, N., Brady, J. J., Damian, M., Sacco, A., Corbel, S. Y., & Blau, H. M. (2010). Reprogramming towards pluripotency requires AID-dependent DNA demethylation. *Nature, 463*, 1042–1047.

Bourc'his, D., & Bestor, T. H. (2004). Meiotic catastrophe and retrotransposon reactivation in male germ cells lacking Dnmt3L. *Nature, 431*, 96–99.

Bourc'his, D., Xu, G. L., Lin, C. S., Bollman, B., & Bestor, T. H. (2001). Dnmt3L and the establishment of maternal genomic imprints. *Science, 294*, 2536–2539.

Butler, M. G. (2011). Prader-Willi Syndrome: Obesity due to genomic imprinting. *Current Genomics, 12*, 204–215.

Carey, B. W., Markoulaki, S., Hanna, J. H., Faddah, D. A., Buganim, Y., Kim, J., et al. (2011). Reprogramming factor stoichiometry influences the epigenetic state and biological properties of induced pluripotent stem cells. *Cell Stem Cell, 9*, 588–598.

Chen, P. Y., Feng, S., Joo, J. W., Jacobsen, S. E., & Pellegrini, M. (2011). A comparative analysis of DNA methylation across human embryonic stem cell lines. *Genome Biology, 12*, R62.

Chen, T., & Li, E. (2004). Structure and function of eukaryotic DNA methyltransferases. *Current Topics in Developmental Biology, 60*, 55–89.

Chen, Z. X., & Riggs, A. D. (2011). DNA methylation and demethylation in mammals. *The Journal of Biological Chemistry, 286*, 18347–18353.

Chiba, H., Hirasawa, R., Kaneda, M., Amakawa, Y., Li, E., Sado, T., et al. (2008). De novo DNA methylation independent establishment of maternal imprint on X chromosome in mouse oocytes. *Genesis, 46*, 768–774.

Ciccone, D. N., Su, H., Hevi, S., Gay, F., Lei, H., Bajko, J., et al. (2009). KDM1B is a histone H3K4 demethylase required to establish maternal genomic imprints. *Nature, 461*, 415–418.

DeChiara, T. M., Robertson, E. J., & Efstratiadis, A. (1991). Parental imprinting of the mouse insulin-like growth factor II gene. *Cell, 64*, 849–859.

Deveale, B., van der Kooy, D., & Babak, T. (2012). Critical evaluation of imprinted gene expression by RNA-Seq: A new perspective. *PLoS Genetics, 8*, e1002600.

Edwards, C. A., & Ferguson-Smith, A. C. (2007). Mechanisms regulating imprinted genes in clusters. *Current Opinion in Cell Biology, 19*, 281–289.

Engel, N., West, A. G., Felsenfeld, G., & Bartolomei, M. S. (2004). Antagonism between DNA hypermethylation and enhancer-blocking activity at the H19 DMD is uncovered by CpG mutations. *Nature Genetics, 36*, 883–888.

Feil, R., & Berger, F. (2007). Convergent evolution of genomic imprinting in plants and mammals. *Trends in Genetics, 23*, 192–199.

Ferguson-Smith, A. C. (2011). Genomic imprinting: The emergence of an epigenetic paradigm. *Nature Reviews. Genetics, 12*, 565–575.

Ferron, S. R., Charalambous, M., Radford, E., McEwen, K., Wildner, H., Hind, E., et al. (2011). Postnatal loss of Dlk1 imprinting in stem cells and niche astrocytes regulates neurogenesis. *Nature, 475*, 381–385.

Gehring, M., Reik, W., & Henikoff, S. (2009). DNA demethylation by DNA repair. *Trends in Genetics, 25*, 82–90.

Gendrel, A. V., & Heard, E. (2011). Fifty years of X-inactivation research. *Development, 138*, 5049–5055.

Giddings, S. J., King, C. D., Harman, K. W., Flood, J. F., & Carnaghi, L. R. (1994). Allele specific inactivation of insulin 1 and 2, in the mouse yolk sac, indicates imprinting. *Nature Genetics, 6*, 310–313.

Graham, C. F. (1974). The production of parthenogenetic mammalian embryos and their use in biological research. *Biological Reviews of the Cambridge Philosophical Society*, *49*, 399–424.
Gregg, C., Zhang, J., Weissbourd, B., Luo, S., Schroth, G. P., Haig, D., et al. (2010). High-resolution analysis of parent-of-origin allelic expression in the mouse brain. *Science*, *329*, 643–648.
Gu, T. P., Guo, F., Yang, H., Wu, H. P., Xu, G. F., Liu, W., et al. (2011). The role of Tet3 DNA dioxygenase in epigenetic reprogramming by oocytes. *Nature*, *477*, 606–610.
Guo, J. U., Su, Y., Zhong, C., Ming, G. L., & Song, H. (2011). Emerging roles of TET proteins and 5-hydroxymethylcytosines in active DNA demethylation and beyond. *Cell Cycle*, *10*, 2662–2668.
Haig, D. (2004). Genomic imprinting and kinship: How good is the evidence? *Annual Review of Genetics*, *38*, 553–585.
Hark, A. T., Schoenherr, C. J., Katz, D. J., Ingram, R. S., Levorse, J. M., & Tilghman, S. M. (2000). CTCF mediates methylation-sensitive enhancer-blocking activity at the H19/Igf2 locus. *Nature*, *405*, 486–489.
He, Y. F., Li, B. Z., Li, Z., Liu, P., Wang, Y., Tang, Q., et al. (2011). Tet-mediated formation of 5-carboxylcytosine and its excision by TDG in mammalian DNA. *Science*, *333*, 1303–1307.
Henckel, A., Chebli, K., Kota, S. K., Arnaud, P., & Feil, R. (2012). Transcription and histone methylation changes correlate with imprint acquisition in male germ cells. *The EMBO Journal*, *31*, 606–615.
Hirasawa, R., Chiba, H., Kaneda, M., Tajima, S., Li, E., Jaenisch, R., et al. (2008). Maternal and zygotic Dnmt1 are necessary and sufficient for the maintenance of DNA methylation imprints during preimplantation development. *Genes & Development*, *22*, 1607–1616.
Hiura, H., Sugawara, A., Ogawa, H., John, R. M., Miyauchi, N., Miyanari, Y., et al. (2010). A tripartite paternally methylated region within the Gpr1-Zdbf2 imprinted domain on mouse chromosome 1 identified by meDIP-on-chip. *Nucleic Acids Research*, *38*, 4929–4945.
Horsthemke, B., & Wagstaff, J. (2008). Mechanisms of imprinting of the Prader-Willi/Angelman region. *American Journal of Medical Genetics. Part A*, *146A*, 2041–2052.
Humpherys, D., Eggan, K., Akutsu, H., Hochedlinger, K., Rideout, W. M., 3rd, Biniszkiewicz, D., et al. (2001). Epigenetic instability in ES cells and cloned mice. *Science*, *293*, 95–97.
Ito, S., Shen, L., Dai, Q., Wu, S. C., Collins, L. B., Swenberg, J. A., et al. (2011). Tet proteins can convert 5-methylcytosine to 5-formylcytosine and 5-carboxylcytosine. *Science*, *333*, 1300–1303.
Jiang, Y., Tsai, T. F., Bressler, J., & Beaudet, A. L. (1998). Imprinting in Angelman and Prader-Willi syndromes. *Current Opinion in Genetics & Development*, *8*, 334–342.
Kaneda, M., Okano, M., Hata, K., Sado, T., Tsujimoto, N., Li, E., et al. (2004). Essential role for de novo DNA methyltransferase Dnmt3a in paternal and maternal imprinting. *Nature*, *429*, 900–903.
Kantor, B., Shemer, R., & Razin, A. (2006). The Prader-Willi/Angelman imprinted domain and its control center. *Cytogenetic and Genome Research*, *113*, 300–305.
Keller, G. (2005). Embryonic stem cell differentiation: Emergence of a new era in biology and medicine. *Genes & Development*, *19*, 1129–1155.
Kelsey, G., & Bartolomei, M. S. (2012). Imprinted genes ... and the number is? *PLoS Genetics*, *8*, e1002601.
Keverne, E. B., & Curley, J. P. (2008). Epigenetics, brain evolution and behaviour. *Frontiers in Neuroendocrinology*, *29*, 398–412.
Klose, R. J., & Bird, A. P. (2006). Genomic DNA methylation: The mark and its mediators. *Trends in Biochemical Sciences*, *31*, 89–97.
Koerner, M. V., & Barlow, D. P. (2010). Genomic imprinting-an epigenetic gene-regulatory model. *Current Opinion in Genetics & Development*, *20*, 164–170.

Koerner, M. V., Pauler, F. M., Huang, R., & Barlow, D. P. (2009). The function of non-coding RNAs in genomic imprinting. *Development, 136*, 1771–1783.
Koerner, M. V., Pauler, F. M., Hudson, Q. J., Santoro, F., Sawicka, A., Guenzl, P. M., et al. (2012). A downstream CpG island controls transcript initiation and elongation and the methylation state of the imprinted Airn macro ncRNA promoter. *PLoS Genetics, 8*, e1002540.
Kohama, C., Kato, H., Numata, K., Hirose, M., Takemasa, T., Ogura, A., et al. (2012). ES cell differentiation system recapitulates the establishment of imprinted gene expression in a cell-type-specific manner. *Human Molecular Genetics, 21*, 1391–1401.
Latos, P. A., Stricker, S. H., Steenpass, L., Pauler, F. M., Huang, R., Senergin, B. H., et al. (2009). An in vitro ES cell imprinting model shows that imprinted expression of the Igf2r gene arises from an allele-specific expression bias. *Development, 136*, 437–448.
Law, J. A., & Jacobsen, S. E. (2010). Establishing, maintaining and modifying DNA methylation patterns in plants and animals. *Nature Reviews. Genetics, 11*, 204–220.
Lee, J. T. (2011). Gracefully ageing at 50, X-chromosome inactivation becomes a paradigm for RNA and chromatin control. *Nature Reviews. Molecular Cell Biology, 12*, 815–826.
Lewis, A., Mitsuya, K., Umlauf, D., Smith, P., Dean, W., Walter, J., et al. (2004). Imprinting on distal chromosome 7 in the placenta involves repressive histone methylation independent of DNA methylation. *Nature Genetics, 36*, 1291–1295.
Li, X. (2010). Extending the maternal-zygotic effect with genomic imprinting. *Molecular Human Reproduction, 16*, 695–703.
Li, X., Ito, M., Zhou, F., Youngson, N., Zuo, X., Leder, P., et al. (2008). A maternal-zygotic effect gene, Zfp57, maintains both maternal and paternal imprints. *Developmental Cell, 15*, 547–557.
Lin, S. P., Youngson, N., Takada, S., Seitz, H., Reik, W., Paulsen, M., et al. (2003). Asymmetric regulation of imprinting on the maternal and paternal chromosomes at the Dlk1-Gtl2 imprinted cluster on mouse chromosome 12. *Nature Genetics, 35*, 97–102.
Lister, R., Pelizzola, M., Dowen, R. H., Hawkins, R. D., Hon, G., Tonti-Filippini, J., et al. (2009). Human DNA methylomes at base resolution show widespread epigenomic differences. *Nature, 462*, 315–322.
Lucifero, D., Mann, M. R., Bartolomei, M. S., & Trasler, J. M. (2004). Gene-specific timing and epigenetic memory in oocyte imprinting. *Human Molecular Genetics, 13*, 839–849.
Ludwig, T., Eggenschwiler, J., Fisher, P., D'Ercole, A. J., Davenport, M. L., & Efstratiadis, A. (1996). Mouse mutants lacking the type 2 IGF receptor (IGF2R) are rescued from perinatal lethality in Igf2 and Igf1r null backgrounds. *Developmental Biology, 177*, 517–535.
Ma, D. K., Guo, J. U., Ming, G. L., & Song, H. (2009). DNA excision repair proteins and Gadd45 as molecular players for active DNA demethylation. *Cell Cycle, 8*, 1526–1531.
Mabb, A. M., Judson, M. C., Zylka, M. J., & Philpot, B. D. (2011). Angelman syndrome: Insights into genomic imprinting and neurodevelopmental phenotypes. *Trends in Neurosciences, 34*, 293–303.
Mackay, D. J., Callaway, J. L., Marks, S. M., White, H. E., Acerini, C. L., Boonen, S. E., et al. (2008). Hypomethylation of multiple imprinted loci in individuals with transient neonatal diabetes is associated with mutations in ZFP57. *Nature Genetics, 40*, 949–951.
Mackay, D. J., Temple, I. K., Shield, J. P., & Robinson, D. O. (2005). Bisulphite sequencing of the transient neonatal diabetes mellitus DMR facilitates a novel diagnostic test but reveals no methylation anomalies in patients of unknown aetiology. *Human Genetics, 116*, 255–261.
Mancini-Dinardo, D., Steele, S. J., Levorse, J. M., Ingram, R. S., & Tilghman, S. M. (2006). Elongation of the Kcnq1ot1 transcript is required for genomic imprinting of neighboring genes. *Genes & Development, 20*, 1268–1282.

Mann, J. R. (2001). Deriving and propagating mouse embryonic stem cell lines for studying genomic imprinting. *Methods in Molecular Biology*, *181*, 21–39.

McGrath, J., & Solter, D. (1984). Completion of mouse embryogenesis requires both the maternal and paternal genomes. *Cell*, *37*, 179–183.

Messerschmidt, D. M., de Vries, W., Ito, M., Solter, D., Ferguson-Smith, A., & Knowles, B. B. (2012). Trim28 is required for epigenetic stability during mouse oocyte to embryo transition. *Science*, *335*, 1499–1502.

Moon, Y. S., Smas, C. M., Lee, K., Villena, J. A., Kim, K. H., Yun, E. J., et al. (2002). Mice lacking paternally expressed Pref-1/Dlk1 display growth retardation and accelerated adiposity. *Molecular and Cellular Biology*, *22*, 5585–5592.

Murry, C. E., & Keller, G. (2008). Differentiation of embryonic stem cells to clinically relevant populations: Lessons from embryonic development. *Cell*, *132*, 661–680.

Nakamura, T., Arai, Y., Umehara, H., Masuhara, M., Kimura, T., Taniguchi, H., et al. (2007). PGC7/Stella protects against DNA demethylation in early embryogenesis. *Nature Cell Biology*, *9*, 64–71.

Nakamura, T., Liu, Y. J., Nakashima, H., Umehara, H., Inoue, K., Matoba, S., et al. (2012). PGC7 binds histone H3K9me2 to protect against conversion of 5mC to 5hmC in early embryos. *Nature*, *486*, 415–419.

Nicholls, R. D., Saitoh, S., & Horsthemke, B. (1998). Imprinting in Prader-Willi and Angelman syndromes. *Trends in Genetics*, *14*, 194–200.

Okada, Y., Yamagata, K., Hong, K., Wakayama, T., & Zhang, Y. (2010). A role for the elongator complex in zygotic paternal genome demethylation. *Nature*, *463*, 554–558.

Ooi, S. K., O'Donnell, A. H., & Bestor, T. H. (2009). Mammalian cytosine methylation at a glance. *Journal of Cell Science*, *122*, 2787–2791.

Pauler, F. M., Koerner, M. V., & Barlow, D. P. (2007). Silencing by imprinted noncoding RNAs: Is transcription the answer?. *Trends in Genetics*, *23*, 284–292.

Pick, M., Stelzer, Y., Bar-Nur, O., Mayshar, Y., Eden, A., & Benvenisty, N. (2009). Clone- and gene-specific aberrations of parental imprinting in human induced pluripotent stem cells. *Stem Cells*, *27*, 2686–2690.

Pollard, K. S., Serre, D., Wang, X., Tao, H., Grundberg, E., Hudson, T. J., et al. (2008). A genome-wide approach to identifying novel-imprinted genes. *Human Genetics*, *122*, 625–634.

Popp, C., Dean, W., Feng, S., Cokus, S. J., Andrews, S., Pellegrini, M., et al. (2010). Genome-wide erasure of DNA methylation in mouse primordial germ cells is affected by AID deficiency. *Nature*, *463*, 1101–1105.

Quenneville, S., Verde, G., Corsinotti, A., Kapopoulou, A., Jakobsson, J., Offner, S., et al. (2011). In embryonic stem cells, ZFP57/KAP1 recognize a methylated hexanucleotide to affect chromatin and DNA methylation of imprinting control regions. *Molecular Cell*, *44*, 361–372.

Rabinovitz, S., Kaufman, Y., Ludwig, G., Razin, A., & Shemer, R. (2012). Mechanisms of activation of the paternally expressed genes by the Prader-Willi imprinting center in the Prader-Willi/Angelman syndromes domains. *Proceedings of the National Academy of Sciences of the United States of America*, *109*, 7403–7408.

Reik, W. (2007). Stability and flexibility of epigenetic gene regulation in mammalian development. *Nature*, *447*, 425–432.

Reik, W., Dean, W., & Walter, J. (2001). Epigenetic reprogramming in mammalian development. *Science*, *293*, 1089–1093.

Reik, W., & Lewis, A. (2005). Co-evolution of X-chromosome inactivation and imprinting in mammals. *Nature Reviews. Genetics*, *6*, 403–410.

Renfree, M. B., Hore, T. A., Shaw, G., Graves, J. A., & Pask, A. J. (2009). Evolution of genomic imprinting: Insights from marsupials and monotremes. *Annual Review of Genomics and Human Genetics*, *10*, 241–262.

Rugg-Gunn, P. J., Ferguson-Smith, A. C., & Pedersen, R. A. (2005). Epigenetic status of human embryonic stem cells. *Nature Genetics*, *37*, 585–587.

Rugg-Gunn, P. J., Ferguson-Smith, A. C., & Pedersen, R. A. (2007). Status of genomic imprinting in human embryonic stem cells as revealed by a large cohort of independently derived and maintained lines. *Human Molecular Genetics*, *16*(Spec No. 2), R243–R251.

Santoro, F., & Barlow, D. P. (2011). Developmental control of imprinted expression by macro non-coding RNAs. *Seminars in Cell & Developmental Biology*, *22*, 328–335.

Schmidt, J. V., Matteson, P. G., Jones, B. K., Guan, X. J., & Tilghman, S. M. (2000). The Dlk1 and Gtl2 genes are linked and reciprocally imprinted. *Genes & Development*, *14*, 1997–2002.

Schoenherr, C. J., Levorse, J. M., & Tilghman, S. M. (2003). CTCF maintains differential methylation at the Igf2/H19 locus. *Nature Genetics*, *33*, 66–69.

Schumacher, A., & Doerfler, W. (2004). Influence of in vitro manipulation on the stability of methylation patterns in the Snurf/Snrpn-imprinting region in mouse embryonic stem cells. *Nucleic Acids Research*, *32*, 1566–1576.

Sleutels, F., Zwart, R., & Barlow, D. P. (2002). The non-coding Air RNA is required for silencing autosomal imprinted genes. *Nature*, *415*, 810–813.

Smilinich, N. J., Day, C. D., Fitzpatrick, G. V., Caldwell, G. M., Lossie, A. C., Cooper, P. R., et al. (1999). A maternally methylated CpG island in KvLQT1 is associated with an antisense paternal transcript and loss of imprinting in Beckwith-Wiedemann syndrome. *Proceedings of the National Academy of Sciences of the United States of America*, *96*, 8064–8069.

Stadtfeld, M., Apostolou, E., Akutsu, H., Fukuda, A., Follett, P., Natesan, S., et al. (2010). Aberrant silencing of imprinted genes on chromosome 12qF1 in mouse induced pluripotent stem cells. *Nature*, *465*, 175–181.

Stadtfeld, M., Apostolou, E., Ferrari, F., Choi, J., Walsh, R. M., Chen, T., et al. (2012). Ascorbic acid prevents loss of Dlk1-Dio3 imprinting and facilitates generation of all-iPS cell mice from terminally differentiated B cells. *Nature Genetics*, *44*(398–405), S391–S392.

Stoger, R., Kubicka, P., Liu, C. G., Kafri, T., Razin, A., Cedar, H., et al. (1993). Maternal-specific methylation of the imprinted mouse Igf2r locus identifies the expressed locus as carrying the imprinting signal. *Cell*, *73*, 61–71.

Sun, B., Ito, M., Mendjan, S., Ito, Y., Brons, I. G., Murrell, A., et al. (2012). Status of genomic imprinting in epigenetically distinct pluripotent stem cells. *Stem Cells*, *30*, 161–168.

Sun, B. W., Yang, A. C., Feng, Y., Sun, Y. J., Zhu, Y., Zhang, Y., et al. (2006). Temporal and parental-specific expression of imprinted genes in a newly derived Chinese human embryonic stem cell line and embryoid bodies. *Human Molecular Genetics*, *15*, 65–75.

Surani, M. A., & Barton, S. C. (1983). Development of gynogenetic eggs in the mouse: Implications for parthenogenetic embryos. *Science*, *222*, 1034–1036.

Surani, M. A., Barton, S. C., & Norris, M. L. (1984). Development of reconstituted mouse eggs suggests imprinting of the genome during gametogenesis. *Nature*, *308*, 548–550.

Takada, S., Tevendale, M., Baker, J., Georgiades, P., Campbell, E., Freeman, T., et al. (2000). Delta-like and gtl2 are reciprocally expressed, differentially methylated linked imprinted genes on mouse chromosome 12. *Current Biology*, *10*, 1135–1138.

Tilghman, S. M. (1999). The sins of the fathers and mothers: Genomic imprinting in mammalian development. *Cell*, *96*, 185–193.

Tomizawa, S., Kobayashi, H., Watanabe, T., Andrews, S., Hata, K., Kelsey, G., et al. (2011). Dynamic stage-specific changes in imprinted differentially methylated regions during early mammalian development and prevalence of non-CpG methylation in oocytes. *Development*, *138*, 811–820.

Tomizawa, S., & Sasaki, H. (2012). Genomic imprinting and its relevance to congenital disease, infertility, molar pregnancy and induced pluripotent stem cell. *Journal of Human Genetics*, *57*, 84–91.

Umlauf, D., Goto, Y., Cao, R., Cerqueira, F., Wagschal, A., Zhang, Y., et al. (2004). Imprinting along the Kcnq1 domain on mouse chromosome 7 involves repressive histone methylation and recruitment of Polycomb group complexes. *Nature Genetics*, *36*, 1296–1300.

Verona, R. I., Mann, M. R., & Bartolomei, M. S. (2003). Genomic imprinting: Intricacies of epigenetic regulation in clusters. *Annual Review of Cell and Developmental Biology*, *19*, 237–259.

Wang, X., Sun, Q., McGrath, S. D., Mardis, E. R., Soloway, P. D., & Clark, A. G. (2008). Transcriptome-wide identification of novel imprinted genes in neonatal mouse brain. *PloS One*, *3*, e3839.

Watanabe, T., Tomizawa, S., Mitsuya, K., Totoki, Y., Yamamoto, Y., Kuramochi-Miyagawa, S., et al. (2011). Role for piRNAs and noncoding RNA in de novo DNA methylation of the imprinted mouse Rasgrf1 locus. *Science*, *332*, 848–852.

Weng, A., Magnuson, T., & Storb, U. (1995). Strain-specific transgene methylation occurs early in mouse development and can be recapitulated in embryonic stem cells. *Development*, *121*, 2853–2859.

Wilkins, J. F., & Haig, D. (2003). What good is genomic imprinting: The function of parent-specific gene expression. *Nature Reviews. Genetics*, *4*, 359–368.

Wu, S. C., & Zhang, Y. (2010). Active DNA demethylation: Many roads lead to Rome. *Nature Reviews. Molecular Cell Biology*, *11*, 607–620.

Wu, H., & Zhang, Y. (2011). Mechanisms and functions of Tet protein-mediated 5-methylcytosine oxidation. *Genes & Development*, *25*, 2436–2452.

Wutz, A. (2007). Xist function: Bridging chromatin and stem cells. *Trends in Genetics*, *23*, 457–464.

Wutz, A., Smrzka, O. W., Schweifer, N., Schellander, K., Wagner, E. F., & Barlow, D. P. (1997). Imprinted expression of the Igf2r gene depends on an intronic CpG island. *Nature*, *389*, 745–749.

Xie, W., Barr, C. L., Kim, A., Yue, F., Lee, A. Y., Eubanks, J., et al. (2012). Base-resolution analyses of sequence and parent-of-origin dependent DNA methylation in the mouse genome. *Cell*, *148*, 816–831.

Zuo, X., Sheng, J., Lau, H. T., McDonald, C. M., Andrade, M., Cullen, D. E., et al. (2012). Zinc finger protein ZFP57 requires its co-factor to recruit DNA methyltransferases and maintains DNA methylation imprint in embryonic stem cells via its transcriptional repression domain. *The Journal of Biological Chemistry*, *287*, 2107–2118.

CHAPTER THREE

The Identity and Fate Decision Control of Spermatogonial Stem Cells: Where Is the Point of No Return?

Makoto C. Nagano[1], Jonathan R. Yeh
Department of Obstetrics and Gynecology, Division of Experimental Medicine, McGill University, Montreal, Quebec, Canada
[1]Corresponding author: e-mail address: makoto.nagano@muhc.mcgill.ca

Contents

Current Topics in Developmental Biology, Volume 102
ISSN 0070-2153
http://dx.doi.org/10.1016/B978-0-12-416024-8.00003-9

Abstract

Spermatogonial stem cells (SSCs) are stem cells of the male germ line and support spermatogenesis for a lifetime after puberty by continuously self-renewing and generating committed progenitors. Accordingly, SSCs are defined functionally by their ability to regenerate and maintain spermatogenesis and are detected unequivocally based on their regenerative capacity. Here, we summarize past achievements of morphological and functional studies of SSCs and discuss issues to be addressed in future investigations. Using the mouse as a model organism, our particular foci are the heterogeneity of primitive spermatogonia and the maintenance of and exit from the stem cell state. By comparing to the biology of other stem cell types and organisms, we also propose possibilities and hypotheses for potential mechanisms of SSC fate decision control, involving stochastic entry into the commitment process and the interplay between SSCs and their descendants that coordinates SSC self-renewal and differentiation.

1. INTRODUCTION

Stem cells are defined universally by their function to maintain and regenerate the anatomy and function of an adult tissue (Weissman, 2000). Spermatogonial stem cells (SSCs) are stem cells of the male germ line and support the production of numerous sperm on a daily basis and throughout the adult life of a male. Their ability to maintain steady-state spermatogenesis and to regenerate spermatogenesis after damage is the only unequivocal parameter that defines SSCs. Accordingly, it has been impossible to prospectively identify or purify SSCs, thereby limiting SSC research.

Using the mouse as a model species, we here attempt to discuss critical issues that need to be addressed in future research of SSCs. We intend to place particular foci on the heterogeneity of early stage male germ cells and regulation of maintenance of and exit from the stem cell state. Readers are encouraged to refer to excellent review articles published previously to gain broader insight into SSC biology (de Rooij & Russell, 2001, de Rooij, van Dissel-Emiliani, & van Pelt, 1989; Lee & Shinohara, 2011; Oatley & Brinster, 2008, 2012; Phillips, Gassei, & Orwig, 2010; Yoshida, 2010).

2. HETEROGENEITY OF ISOLATED TYPE A SPERMATOGONIA AND THE SSC POPULATION SIZE IN THE MOUSE TESTIS

2.1. Morphological and phenotypic heterogeneity in undifferentiated spermatogonia

SSCs represent a subpopulation of diploid male germ cells (spermatogonia), which reside on the basement membrane in the seminiferous epithelium. Spermatogonia undergo a series of morphological differentiation from Type A to Intermediate to Type B spermatogonia, before entering meiosis (de Rooij & Russell, 2001). These subtypes are defined by nuclear morphology, or more specifically by the amount of heterochromatin. A developmental scheme of Type A spermatogonia that is widely accepted at present was originally proposed by Huckins (1971) and Oakberg (1971). In this "A_s" model, the most primitive spermatogonia are present as isolated, single cells (Type A-isolated or -single spermatogonia: A_s). One division of an A_s cell generates either two A_s cells or a pair of cells connected by a cytoplasmic bridge (A-paired: A_{pr}), which results from incomplete cytokinesis. The formation of the cytoplasmic bridge is the first morphological characteristic of spermatogonial differentiation. Incomplete cytokinesis occurs in descendants of A_{pr} cells, forming generally a "chain" of 2^n cells. This clonal development as a cell chain continues throughout spermatogenesis. A clone of 4–16 or 32 cells is called A-aligned (A_{al}) spermatogonia. Type A spermatogonia that belong to the A_s, A_{pr}, and A_{al} stages have been traditionally called "undifferentiated" and are indistinguishable by morphology, except for the length of the cell chain. A_{al} spermatogonia in chains of eight cells or longer then transform into A_1 spermatogonia without cell division and, following three subsequent mitotic divisions, become A_4 spermatogonia before developing into Intermediate spermatogonia. A_1–A_4 spermatogonia are called "differentiating" spermatogonia (de Rooij & Russell, 2001).

Among undifferentiated A spermatogonia, A_s cells have been considered to be stem cells (de Rooij & Russell, 2001). This concept is logical as a clear morphological difference is observed upon the formation of the cytoplasmic bridge. In theory, however, not all A_s cells can be stem cells, as some A_s cells that form A_{pr} cells eventually become spermatozoa (de Rooij & Russell, 2001; Fig. 3.1A). In other words, some A_s cells are destined to exit the A_s state or have already exited the stem cell state. Thus, a "retrospective" interpretation of a cellular process is required to determine the stem cell state

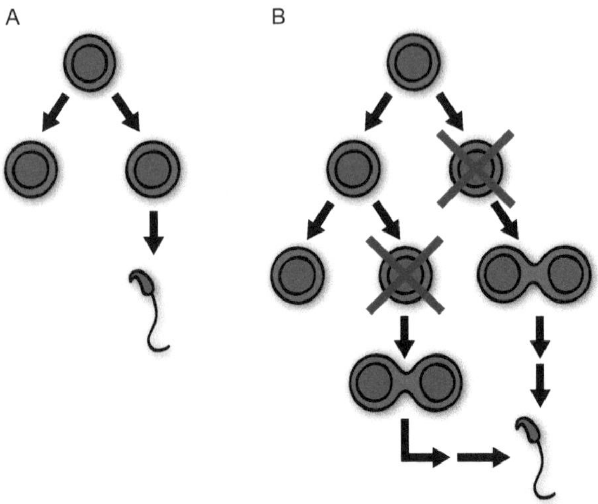

Figure 3.1 Schemes representing consequences of A_s spermatogonial divisions. (A) One A_s cell divides into two A_s cells. Both are morphologically indistinguishable but one is destined to become spermatozoa and is recognized "retrospectively" as a committed cell. (B) Four single cells are descendants of an A_s cell. Despite morphological similarities, two (marked by X) were destined to become A_{pr} cells; these cells had exited the A_s state. In this way, spermatogonial transplantation detects only the original SSCs functionally but retrospectively by ignoring all descendants. (For color version of this figure, the reader is referred to the online version of this chapter.)

and the exit from it (Fig. 3.1B), which is the inherently confusing nature of stem cell biology. On this basis, A_s cells cannot be homogenous cells and should be a mixture of stem cells and progenitors.

Accumulating data support this theoretical basis of heterogeneous "undifferentiated" spermatogonia. For example, using whole-mount immunostaining of mouse seminiferous tubules, Nakagawa, Sharma, Nabeshima, Braun, and Yoshida (2010) examined expression patterns of multiple molecules in spermatogonia at the A_s to A_{al} stages. E-cadherin (Ecad) is known to be expressed in undifferentiated spermatogonia, but not in more advanced germ cells (Tokuda, Kadokawa, Kurahashi, & Marunouchi, 2007). Immunostaining analysis of Ecad(+) A_s spermatogonia showed that the majority (~80%) expressed glial-cell-line-derived neurotrophic factor (GDNF) family receptor α1 (GFRα1) but not neurogenin 3 (Ngn3): that is, Ecad(+)GFRα1(+)Ngn3(−). Remaining Ecad(+) A_s cells expressed both molecules (Ecad(+)GFRα1(+)Ngn3(+)) or Ngn3 alone (Ecad(+)GFRα1(−)Ngn3(+)). In addition, the expression of c-Kit was observed in a small number of Ngn3(+) A_s spermatogonia (Nakagawa et al., 2010). These results demonstrate that

A_s spermatogonia exhibit different gene expression patterns and are thus recognized as a heterogeneous cell population.

2.2. Functional detection of SSCs

Two approaches are currently available to detect the regenerative capacity of SSCs. One classical approach is to damage spermatogenesis, for example, by using an alkylating reagent, followed by analyzing the regeneration of endogenous spermatogenesis. Although this approach was used in the past (Bucci & Meistrich, 1987; van den Aardweg, de Ruiter-Bootsma, Kramer, & Davids, 1983), it does not ensure the clonal origin of regenerated spermatogenesis or allow for further analysis of the cells of origin.

The second approach is spermatogonial transplantation (Brinster & Avarbock, 1994; Brinster & Zimmermann, 1994), which unequivocally detects functional SSCs. The transplantation technique selectively allows only those cells that continuously self-renew and produce progenitor cells to regenerate complete spermatogenesis in recipient testes. Spermatogenic colonies arising from donor SSCs can be readily detected in whole-mount testes if donor cells express a transgene, such as LacZ or GFP (Nagano, Avarbock, & Brinster, 1999; Ohta, Yomogida, Yamada, Okabe, & Nishimune, 2000). An advantage of this approach is that it determines absolute numbers of functional SSCs because a colony of donor-derived spermatogenesis arises from a single SSC (Dobrinski, Ogawa, Avarbock, & Brinster, 1999; Kanatsu-Shinohara et al., 2006; Zhang, Ebata, & Nagano, 2003).

A weakness of the transplantation approach is that it forces us to detect SSCs retrospectively. It is unique, however, in that it does not rely on morphology, such as the length of chains, or on gene expression patterns as readout measures (Fig. 3.1B). On the other hand, analyses based on cell morphology and marker expression represent attempts to detect SSCs prospectively. However, stem cells are defined by function. Hence, prospective identification of SSCs is technically demanding, as it is difficult to trace a cell lineage up to spermatozoa from the cell of origin and to confirm that the process continues over time (long-term self-renewal and differentiation). To overcome this, genetic labeling approaches have been employed (genetic lineage-tracing), which provide an effective tool to retrospectively analyze regeneration of labeled, endogenous spermatogenesis. A strength of this approach is that SSCs can be studied in the steady state (Klein, Nakagawa, Ichikawa, Yoshida, & Simons, 2010; Nakagawa, Nabeshima, & Yoshida, 2007; Sada, Suzuki, Suzuki, & Saga, 2009) (see Section 3). However, it is crucial to ensure the faithfulness of a marker and labeling technique, because the data interpretation relies on the specificity and effectiveness of the promoter driving marker gene expression in the intended cell of origin as well as on our knowledge about the biological properties of the targeted cells.

2.3. Estimating the proportion of functional SSCs in A_s spermatogonia

Does phenotypic heterogeneity in A_s spermatogonia imply functional heterogeneity? To answer this question, it is useful to assess the proportion of functional SSCs in A_s spermatogonia. Tegelenbosch and de Rooij (1993) has reported that an adult mouse testis contains 35,000 A_s spermatogonia and 330,000 undifferentiated spermatogonia (A_s, A_{pr}, and A_{al} cells combined) in a hybird mouse strain (C3H/101 F_1).

The absolute number of functional SSCs in mouse testis was measured using spermatogonial transplantation. To achieve this, the efficiency of SSC homing after transplantation needed to be first determined, as not all SSCs in donor cells colonize a recipient testis. This efficiency was determined using a serial transplantation technique, where donor cells were transplanted into the testes of two successive generations of recipient mice (Nagano, 2003), and was calculated by dividing colony numbers in the secondary recipient testes by those in the primary recipient testes; note that both numbers were functionally determined. It was found that $\sim$12% of total SSCs successfully colonized recipient testes. This value, together with results of several transplantation experiments (Nagano, 2003; Nagano et al., 1999; Shinohara, Orwig, Avarbock, & Brinster, 2000), gives 3000–6000 SSCs in an adult mouse testis, representing 0.01–0.02% of total cells in the seminiferous tubules.

SSCs are therefore estimated to represent 8.5–17% of A_s spermatogonia (3000–6000/35,000), suggesting that 80% or more of A_s spermatogonia are not SSCs, assuming that all SSCs are A_s spermatogonia. Likewise, functional SSCs are calculated to be 0.9–1.8% of A_s, A_{pr}, and A_{al} spermatogonia combined. These results support the notion that A_s spermatogonia are functionally heterogeneous and that SSCs are a small subpopulation of undifferentiated spermatogonia.

2.4. Issues involved with spermatogonial heterogeneity

Considering the heterogeneity of undifferentiated spermatogonia, one crucial issue in SSC research is how to detect exit from SSC status—when and where does irreversible commitment of SSCs to differentiation, "the point of no return," occur and how can we distinguish SSCs from progenitors? We intend to discuss this issue throughout this article and particularly in Section 5. Our essential argument is that we need a better understanding of SSC cell-surface markers to address this issue.

Another issue in SSC research is the terminology. The term, "undifferentiated" spermatogonia, is admittedly confusing (de Rooij & Russell, 2001). As described above, this term was proposed based on the nuclear morphology of spermatogonia, whereas SSCs are defined functionally, regardless of cell morphology. We thus need an alternative term to reconcile this confusion. By following the precedent established in studies of somatic stem cells, we propose a term with a functional implication, such as "progenitors" or "progenitor spermatogonia." Hereafter, we will use these terms to indicate primitive spermatogonia that lost SSC activity; otherwise, we will indicate specific morphological stages (i.e., A_s, A_{pr}, or A_{al}), as proposed previously (de Rooij & Russell, 2001).

3. CELL-SURFACE AND INTRACELLULAR SSC MARKERS

Due to the clonal development of male germ cells in cell chains, lineage-specific differentiation can be traced visually by cell morphology. In this context, spermatogenesis is perhaps the best studied self-renewing process in mammals. It is therefore ideal to link such knowledge, generated since 1950s (Leblond & Clermont, 1952), to the regenerative function of SSCs. Such a linkage should help answer the question, "where is the point of no return in SSC commitment?" Morphology can be associated with function via SSC markers that facilitate prospective cell isolation/identification followed by functional analyses of isolated/identified cells.

Various markers have been reported for SSCs and progenitor spermatogonia; a few of them are listed in Fig. 3.2. As SSC markers have been discussed extensively (Oatley & Brinster, 2008, 2012; Phillips et al., 2010; Yoshida, 2010), we focus on whether currently available SSC markers can associate SSC function with the morphology of spermatogonia.

3.1. Cell-surface SSC markers

SSC function is readily linked to marker expression patterns using a cell-surface molecule. For instance, integrin β1 and integrin α6 are cell-surface receptors of the extracellular matrix (Shinohara, Avarbock, & Brinster, 1999, Shinohara et al., 2000). Following immunological cell sorting and transplantation of testis cells expressing these molecules, more colonies of

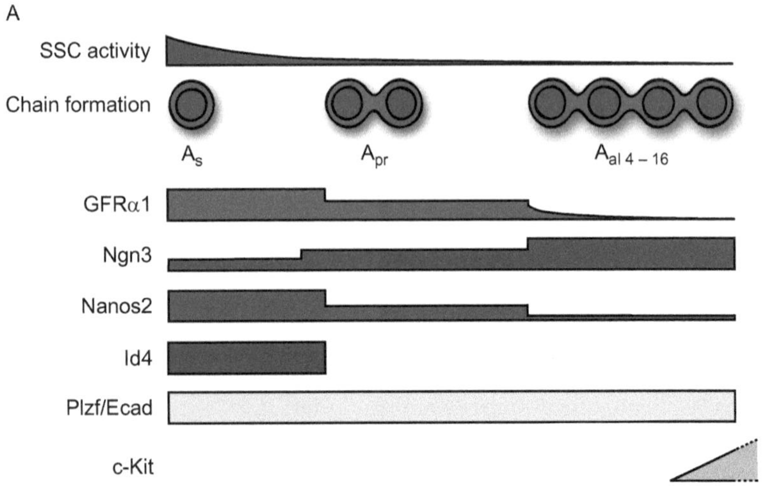

Figure 3.2 A schematic representation of a few SSC markers. SSC activity is thought to be lost gradually as spermatogonial commitment proceeds. It is currently unknown where in spermatogonial development SSCs reach irreversible commitment. Plzf and Ecad reportedly show near identical staining patterns on spermatogonia (Nakagawa et al., 2010). (For color version of this figure, the reader is referred to the online version of this chapter.)

spermatogenesis were generated in recipient testes compared to unsorted cells. This result indicates that these two molecules allow for enrichment for functional SSCs, and thus, are SSC markers.

This approach combining immunological cell sorting and spermatogonial transplantation has determined the most cell-surface markers reported thus far. However, these markers have not greatly aided us to morphologically distinguish functional SSCs from progenitors. For instance, Thy1 (CD90) is the marker that allows for the highest efficiency of SSC enrichment to date (Kubota, Avarbock, & Brinster, 2003; Oatley & Brinster, 2008) but Thy1(+) cells have not been visualized in mouse testis sections or in whole-mount seminiferous tubules. SSCs can be enriched using GFRα1 as a cell-surface marker, but enrichment is successful using only pup testis cells but not adult cells (Ebata, Zhang, & Nagano, 2005). Additionally, although GFRα1 expression is mostly found in A_s and A_{pr} spermatogonia, it is also seen in A_{al} cells (Grasso et al., 2012; Nakagawa et al., 2010; Schlesser et al., 2008; Tokuda et al., 2007; Fig. 3.2). As described above, A_s spermatogonia can be either GFRα1-positive or negative (Grasso et al., 2012; Nakagawa et al., 2010; Suzuki, Sada, Yoshida, & Saga, 2009). Likewise, although CD9 is an effective marker to enrich testis cells for SSCs, it is expressed in A_s through Type B spermatogonia (Hamra et al., 2004; Kanatsu-Shinohara, Toyokuni, & Shinohara, 2004; Phillips et al., 2010).

Among cell-surface markers used in SSC research, c-Kit, a negative SSC marker, has been most useful to associate SSC activity with cell morphology. c-Kit is expressed mostly by A_{al} to early spermatocytes (van Pelt & de Rooij, 1990, 1991), and mice defective for c-Kit signaling show spermatogenic arrest at the transition from A_{al} to A_1 (de Rooij, Okabe, & Nishimune, 1999). Functionally, excluding c-Kit(+) cells improves mouse SSC enrichment (Kubota et al., 2003). These observations suggest that A_{al} spermatogonia and farther likely do not possess SSC activity. It remains unknown, however, whether c-Kit expression is a cause or a consequence of differentiation.

Therefore, cell-surface SSC markers are expressed in a wide range of spermatogonia and have not allowed us to distinguish SSCs from progenitors.

3.2. Intracellular molecules as SSC markers

Various intracellular molecules have been reported to be SSC markers (Phillips, Gassei & Orwig, 2010). Among them, we focus on four molecules: Plzf, Ngn3, Nanos2, and Id4. Plzf is expressed in A_s, A_{pr}, and A_{al} spermatogonia (Buaas et al., 2004; Costoya et al., 2004). Plzf is generally considered a transcriptional repressor and inhibits c-Kit expression (David et al., 1998; Filipponi et al., 2007), but it can also activate gene expression (Doulatov et al., 2009; Labbaye et al., 2002). The role of Plzf in spermatogenesis was initially studied using a spontaneous mutant mouse strain (*luxoid*) or targeted deletion mutants (Buaas et al., 2004; Costoya et al., 2004). Plzf-null mice show normal spermatogenesis in the first cycle after birth but become infertile with age, presenting complete loss of germ cells in adults. Regenerative activity is not found in Plzf−/− testis cells upon spermatogonial transplantation. These results suggest that Plzf is required for SSC maintenance.

Ngn3 belongs to a family of basic helix–loop–helix transcription factors and its expression by SSCs was confirmed using spermatogonial transplantation in transgenic mice. However, Ngn3(+) SSCs are estimated to represent ~12% of total SSCs, while the majority of Ngn3(+) cells are committed progenitors (Nakagawa et al., 2007). Notably, this result exemplifies heterogeneity within the SSC population. In general, the expression of Ngn3 is associated with the initial commitment of SSCs and correlates with the progress of spermatogonial differentiation. For example, it is expressed in A_s, A_{pr}, and A_{al} spermatogonia but the proportion of Ngn3(+) cells increases as differentiation advances, an expression pattern that inversely correlates with GFRα1 expression (Nakagawa et al., 2010). The withdrawal of GDNF, a crucial factor to maintain SSC self-renewal, has been shown to upregulate Ngn3 expression in stem/progenitor spermatogonia *in vitro* (Oatley, Avarbock, Telaranta, Fearon, & Brinster, 2006). Thus, Ngn3 expression has been proposed to indicate SSCs that may be predisposed to differentiation commitment (Yoshida, 2010).

Nanos2 encodes a RNA-binding protein that represses translation through RNA-destabilization (Suzuki, Igarashi, Aisaki, Kanno, & Saga, 2010). It promotes SSC maintenance and blocks spermatogonial differentiation (Sada et al., 2009; Suzuki et al., 2009). Nanos2 expression in SSCs was examined in mice carrying tamoxifen-inducible Cre driven by the Nanos2 promoter (Sada et al., 2009). Genetic lineage-tracing experiments using a pulse-chase approach showed that labeled spermatogenic colonies were observed 3 months after tamoxifen treatment, indicating that the labeled cells of origin possessed long-term regeneration ability, the functional hallmark of SSCs. Conditional deletion of Nanos2 led to a gradual loss of Plzf(+) spermatogonia while its constitutive expression blocked spermatogonial differentiation at a pre-c-Kit stage. Nanos2 also blocks differentiation of fetal germ cells and suppresses the expression of Stra8, thereby blocking meiotic entry (Tsuda et al., 2009). Although Nanos2 may control SSC self-renewal, it may also influence the activity of progenitors, as the majority of Nanos2(+) cells are A_s and A_{pr} spermatogonia with the remainder belonging to A_{al} spermatogonia (Suzuki et al., 2009). Thus, although Nanos2 does not identify SSCs, morphology of spermatogonia can be linked to regenerative capacity through Nanos2.

Id4 shows a unique expression pattern in spermatogonia and promotes SSC self-renewal (Oatley, Kaucher, Racicot, & Oatley, 2011). It belongs to a class of helix–loop–helix molecules and acts as a transcriptional repressor. Interestingly, immunostaining of mouse testes shows that Id4-expressing cells are solely A_s spermatogonia; thus, Id4 is the only molecule reported that labels exclusively A_s spermatogonia. Id4 transcript expression is upregulated *in vitro* by GDNF while mutant mice lacking Id4 show gradual loss of Plzf(+) spermatogonia in adulthood. Furthermore, suppression of Id4 using small interfering RNA abolishes SSC expansion *in vitro* without affecting the cell survival. Thus, Id4 is apparently an exclusive marker for A_s spermatogonia and plays an important role in SSC maintenance.

3.3. Marker–function linkage

A "stem cell" marker must be linked to the cell's function, as stem cells are functionally defined. Both cell-surface and intracellular markers have advantages and disadvantages. Cell-surface markers are advantageous as they allow for recovery of live cells with multiparameter labeling; thus, recovered cells can be readily applied to functional SSC analyses. Nonetheless, they are limited in their ability to associate function with cell morphology at present. Intracellular markers have provided effective tools to link cell function to morphology. They can also be associated more readily with intracellular signaling and transcriptional/translational gene regulation, an advantage for dissecting mechanisms of SSC fate decision control. However, they are limited in analyses of live cells, particularly with multiple parameters, and require laborious and careful preparation for linage-tracing to ensure faithful labeling of cells of origin.

Although marker–function associations have not been developed well at the SSC level and we do not yet have definitive SSC markers, studies based on spermatogonial transplantation, mutant mice, genetic lineage-tracing, and immunological staining have generated the foundation to establish this association. Combining the data acquired, it is currently believed that stem cell capacity is lost gradually as spermatogonia advance from the A_s to A_{al} stages and this progression is reflected in part by marker expression patterns (Fig. 3.2; Phillips et al., 2010; Yoshida, 2010).

4. BALANCE BETWEEN SELF-RENEWAL AND COMMITMENT

4.1. Cell-extrinsic factors

The supportive effects of cell-extrinsic factors on SSC self-renewal are most clearly observed in culture. Mouse SSCs can be maintained *in vitro* virtually indefinitely and expanded exponentially over time (Kanatsu-Shinohara et al., 2003; Kubota, Avarbock, & Brinster, 2004). To accomplish this long-term survival and self-renewal, SSCs need to be exposed continuously to GDNF and fibroblast growth factor 2 (FGF2). Supported by these two growth factors, SSCs divide and form three-dimensional aggregates of germ cells on mitotically inactivated embryonic fibroblast feeder cells. These aggregating germ cells are spermatogonia at primitive stages, expressing various SSC markers, such as Plzf, c-Ret (the signal-transducing subunit of the

GDNF receptor), and Oct4 (Ebata, Yeh, Zhang, & Nagano, 2011; Kubota et al., 2004).

GDNF is a Sertoli-cell-derived growth factor and is the most crucial cell-extrinsic factor known to promote SSC self-renewal. The role of GDNF in spermatogenesis was first identified in mutant mice; GDNF overexpression resulted in the accumulation of spermatogonia that did not express c-Kit, while its haploinsufficiency caused the loss of the stem cell reserve (Meng et al., 2000). *In vitro*, spermatogonial aggregates do not develop or SSCs perish without GDNF. GDNF signaling is mediated by the PI3K/Akt pathway and Src family kinases (Braydich-Stolle, Kostereva, Dym, & Hofmann, 2007; Lee et al., 2007; Oatley, Avarbock, & Brinster, 2007). Activation of PI3K signaling promotes survival and self-renewal of stem/progenitor spermatogonia and its constitutive activation induces SSC expansion in the absence of GDNF but requiring FGF2 (Lee et al., 2007). While the PI3K/Akt pathway is the best studied pathway activated by GDNF, a recent report shows that the MAPK pathway also mediates GDNF signaling (Ishii, Kanatsu-Shinohara, Toyokuni, & Shinohara, 2012).

At least five genes have been identified functionally as the downstream targets of GDNF signaling: Bcl6b, Etv5, Lhx1, POU3F1, and Nanos2. The biological significance of these factors in supporting SSC self-renewal has been demonstrated in RNA interference experiments (Oatley et al., 2006, 2007; Wu et al., 2010) or using gain- and loss-of-function mutant mice (Sada, Hasegawa, Pin, & Saga, 2012). In addition, constitutive activation of Ras allows for *in vitro* expansion of spermatogonia without GDNF as long as FGF2 is supplied or even without GDNF and FGF2, depending on the Ras isoform that is mutated (Lee et al., 2009). Upon transplantation, cultured cells with activated Ras exhibit proper spermatogenic differentiation but can also cause seminomatous tumors, suggesting that germ cell tumors can result from excessive promotion of self-renewal activity. Similarly, cultured germ cells expressing Cyclin D2 and Cyclin E1 proliferate *in vitro* without growth factors and show tumorigenic activity upon transplantation (Lee et al., 2009).

FGF2 is expressed by various cell types in the testis, such as Sertoli cells, Leydig cells, and differentiating germ cells (Han et al., 1993; Mullaney & Skinner, 1992). *In vitro*, FGF2 supports GDNF to expand SSCs, but FGF2 alone is insufficient. A recent study shows that FGF2 phosphorylates MAPK1/3 and constitutive activation of downstream signaling supports spermatogonial proliferation *in vitro* without FGF2 and with GDNF (Ishii et al., 2012). This study also reports that Etv5 is a target gene of FGF2 signaling and in turn, Etv5 stimulates c-Ret and Bcl6b expression. Similar to constitutive activation of MAPK signaling, constitutive expression of Etv5 or Bcl6b allows for long-term maintenance of spermatogonial aggregates *in vitro*. Interestingly, however, cells with activated MAPK signaling and Etv5-overexpression do not regenerate spermatogenesis upon transplantation, while those overexpressing Bcl6b form germ cell tumors. These results indicate that SSC self-renewal and differentiation need to be balanced properly to support spermatogenesis and to avoid tumorigenesis.

Recent studies identified two additional cell-extrinsic factors that contribute to the maintenance of SSC self-renewal: colony-stimulating factor 1 (CSF1) and Wnt5a. When added to SSC cultures, CSF1 augments SSC proliferation induced by GDNF and FGF2, as assessed by spermatogonial transplantation (Oatley, Oatley, Avarbock, Tobias, & Brinster, 2009). CSF1 expression is observed in Leydig cells and a subfraction of myoid cells by immunostaining. Yeh, Zhang, and Nagano (2011) show that Wnt5a is expressed in Sertoli cells and supports SSC maintenance *in vitro* partly by stimulating SSC survival. Furthermore, the effects of Wnt5a result from activation of a JNK cascade, rather than canonical, Wnt/β-catenin signaling.

Although our knowledge regarding cell-extrinsic factors and signaling pathways influencing SSCs is increasing, it is unlikely that these factors control SSCs exclusively. For instance, cell types that express GDNF receptors are not confined to SSCs (Section 3). Potential Wnt5a receptors are widely expressed in spermatogonial aggregates *in vitro*, while only a minor fraction of these cells exhibit regenerative activity. Thus, it is logical to consider that committed progenitors are also controlled by these factors. This issue becomes important when discussing the role of the somatic niche in controlling SSC fate (Section 4.4).

4.2. Struggle between self-renewal and commitment

SSC self-renewal and commitment need to be finely balanced to sustain normal spermatogenesis and avoid tumorigenesis. The mechanism underlying this balance is largely unknown, because we cannot distinguish SSCs from progenitors—we do not know where the point of no return is—due to the lack of appropriate markers. This causes difficulty in building a clear-cut experimental paradigm to dissect the mechanism. Thus, it may be useful to look at other stem cell types where exit from the stem cell state is defined more clearly. Here, we discuss the commitment process of mouse embryonic stem cells (mESCs), the stem cells derived from the inner cell mass (ICM) of blastocysts that are pluripotent but tumorigenic.

The transcriptional network that sustains mESC self-renewal involves Oct4, Nanog, and Sox2 (herein called collectively "ESC transcription factors"). Each factor stimulates the expression of itself and the others, constituting a positive feedback loop (Boyer et al., 2005). Interestingly, many target genes of the ESC transcription factors encode proteins that induce lineage-specific differentiation, and their expression is suppressed by the ESC transcription factors. As mESCs exit from the stem cell state, ESC transcription factor expression declines and is eventually lost. From this perspective, the ESC commitment process can be regarded as "the loss of the stem cell state," rather than "the induction of differentiation." Conversely, maintenance of the stem cell state in ESCs can be regarded as "the blockade of differentiation." In terms of SSC commitment, the "loss of the stem cell state as release from differentiation blockade" is an interesting concept. It has been reported that reduction of Id4 leads to SSC loss without affecting overall cell survival (Oatley et al., 2011), suggestive of differentiation through the loss of the SSC state.

What triggers the differentiation of mESCs? A definitive answer has not been derived but it is known that FGF4, which is secreted by mESCs themselves, "predisposes" mESCs to differentiation (Fig. 3.3A). *In vivo*, FGF4 is

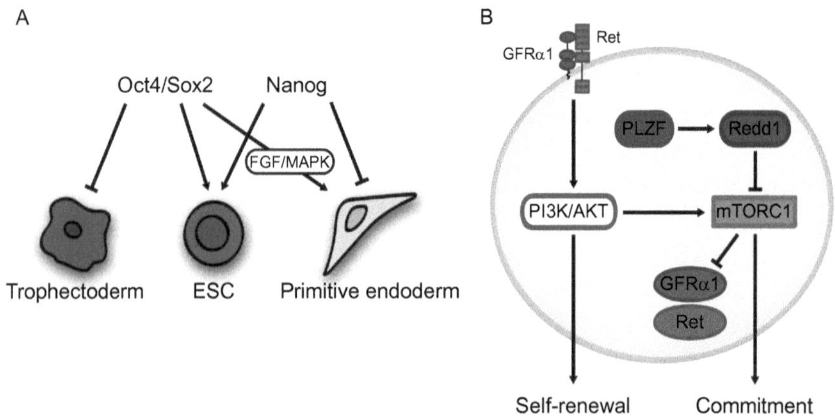

Figure 3.3 (A) Mouse ESC fate decision control. Oct4 and Sox2 upregulate FGF4, which initiates lineage commitment. This action is counteracted by Nanog to sustain "stemness." (B) GDNF–Plzf cross-talk controls the balance between self-renewal and differentiation of SPCs. GDNF signaling not only promotes self-renewal but also initiates commitment via mTORC1. Plzf upregulates Redd1, which suppresses mTORC1, thereby sustaining the SPC state. (For color version of this figure, the reader is referred to the online version of this chapter.)

expressed from one-cell stage embryos and in the ICM it promotes differentiation to primitive endoderm (Rappolee, Basilico, Patel, & Werb, 1994). Within the mESC population, Oct4 and Sox 2 stimulate FGF4 expression, which signals through MAPK and directs cells toward lineage commitment (Kunath et al., 2007; Silva & Smith, 2008). Indeed, upregulation of Oct4 induces differentiation of ICM and ESCs to primitive endoderm (Niwa, Miyazaki, & Smith, 2000). Under the mESC ground state, the preconditioning effects of Oct4 and Sox2 are counteracted by Nanog, thereby sustaining the stem cell state (Kunath et al., 2007; Silva & Smith, 2008). Hence, mESCs appear to be struggling continuously within their own cell community under a fragile balance between conflicting self-renewal and commitment forces. Accordingly, ESCs can be established and maintained by blocking the MAPK pathway without additional growth factors, serum, and feeder cells (Ying et al., 2008). As mESCs are not homogeneous cells (e.g., see Zhou et al., 2001), it would be interesting to know if mESCs primed for commitment are at a distinctive commitment stage or rather at a short-lived transient phase.

Conflicting forces seem to be at work, mediated by GDNF signaling, mammalian target of rapamycin complex 1 (mTORC1), and PLZF, in the maintenance of SSCs or more precisely, spermatogonial stem/progenitor cells (SPCs; Hobbs, Seandel, Falciatori, Rafii, & Pandolfi, 2010; Fig. 3.3B). GDNF activates the PI3K/Akt pathway, a mediator of which is mTORC1; the activity of mTORC1 is suppressed by rapamycin (Sabatini, 2006). Using Plzf knockout mice and the SSC culture system, Hobbs et al. (2010) report that the balance between self-renewal and commitment of SPCs is controlled by cross-talk between the PI3K/Akt pathway and Plzf–Redd1 signaling, with mTORC1 sitting at the junction of the cross-talk. Under the ground state of SPCs, GDNF signaling promotes SPC self-renewal through PI3K/Akt, which activates mTORC1. mTORC1 activation leads to suppression of GFRα1 and c-Ret, thereby limiting the response of SPCs to GDNF. mTORC1 activation is also known to inhibit Akt activity (Sabatini, 2006). Therefore, there is a negative feedback loop among GDNF, Akt, and mTORC1. On the other hand, Plzf induces Redd1 expression through direct binding to the Redd1 promoter; in turn Redd1 suppresses mTORC1 and sustains GDNF-stimulated self-renewal. Accordingly, Plzf deletion increases mTORC1 activity while reducing GDNF signaling, leading to a decline in SPC numbers and an increase in the proportion of c-Kit(+) cells. The administration of rapamycin restores at least partially the expression of GFRα1 and c-Ret and the number of SPCs.

Another mode of Plzf action that controls spermatogonial differentiation was reported recently (Hobbs et al., 2012). Plzf physically associates with Sall4, a zing-finger transcription factor, and displaces it from chromatin, thereby antagonizing Sall4 actions. Conversely, Sall4 confines Plzf in heterochromatin, inhibiting Plzf functions. *In vitro*, upon induction of SPC differentiation by retinoic acid (RA), Sall4 expression is upregulated in wild-type and Plzf-null SPCs, while Plzf expression is downregulated. Intriguingly, Sall4 upregulation was observed only transiently. *In vivo*, Sall4 expression increases from Plzf(+)Kit(−) to Plzf(+)Kit(+) stages of spermatogonia, suggesting that Sall4 can promote spermatogonial differentiation in mice. Upstream regulators of Sall4 are unknown for spermatogonia. As c-Kit was used as a read-out of commitment in this study, it is possible that transient Sall4 upregulation supports the transition of SPCs to A1 spermatogonia, rather than SSC commitment. Nonetheless, patterns of Plzf expression overlap those of Sall4 expression (Phillips et al., 2010), suggesting that the Plzf–Sall4 interaction may also participate in the fate decision control of SSCs.

Given the importance of Plzf in SPC fate control, an intriguing question is what regulates Plzf expression? Available evidence indicates that GDNF does not control Plzf expression. Identification of such regulators should improve our understanding of regulatory mechanisms of SSC/SPC fate. Additionally, it appears that GDNF is not a one-dimensional stimulator of SSC self-renewal. Just like Oct4 can induce both self-renewal and differentiation of ESCs, GDNF has two faces. *In vitro* studies suggest that GDNF supports both SSC self-renewal and commitment. GDNF promotes formation of spermatogonial aggregates and SSC self-renewal; yet, functional SSCs represent only 1–3% of the aggregate cells (Kanatsu-Shinohara et al., 2005; Oatley & Brinster, 2008; Yeh, Zhang, & Nagano, 2007). Hence, the majority of these cells are committed progenitors, which emerge only in the presence of GDNF. A recent morphological study *in vivo* supports this notion (Grasso et al., 2012). Additionally, in our unpublished observation, cells with activated Wnt/β-catenin signaling *in vitro*, which do not have regenerative activity (Yeh et al., 2011), emerge only when GDNF is present in culture; GDNF depletion rapidly eliminates these cells. Thus, cautions are apparently necessary when we interpret data involving manipulation of GDNF signaling *in vivo* and *in vitro*.

The above discussion dealt with growth factors, signaling cascades, and transcription factors, which contribute to SSC fate regulation, based on the concept of SSC commitment as a loss of the stem cell state. On the other hand, only a few regulators are reported to induce commitment of SSCs/SPCs. These regulators include cell-extrinsic factors, RA (Dann et al., 2008; Tong, Mitchell, Evanoff, & Griswold, 2011) and Neuregulin-1 (Hamra, Chapman, Nguyen, & Garbers, 2007), and cell-intrinsic factors, Stra8 (Giuili et al., 2002; Zhou, Li, et al., 2008; Zhou, Nie, et al., 2008), STAT3 via Ngn3 (Kaucher, Oatley, & Oatley, 2012; Oatley, Kaucher, Avarbock, & Brinster, 2010), Sohlh1, and Sohlh2 (Ballow, Meistrich, Matzuk, & Rajkovic, 2006; Barrios et al., 2012; Hao et al., 2008; Suzuki et al., 2012; Toyoda et al., 2009). Further efforts are expected to identify other factors that induce SSC commitment.

4.3. Posttranscriptional mechanisms that potentially support SSC commitment

Epigenetic control of the stem cell state has been reported with ESCs and is anticipated to contribute to SSC regulation. DNA methylation is a well studied epigenetic mechanism, but we do not describe it here, as it has been discussed recently (Lee & Shinohara, 2011). Rather, we briefly look at posttranscriptional control by microRNAs (miRNAs) and proapoptotic molecules.

Recent studies indicate that miRNAs participate in ESC fate control (reviewed in Subramanyam & Blelloch, 2011; Suh & Blelloch, 2011). Most miRNAs are generated through a two-step process. Primary miRNA is processed to precursor miRNA by the GCGR8/DROSHA complex, and then to mature miRNA by DICER. In ESCs, deletion of Dgcr8 or Dicer leads to a proliferation defect and differentiation block (Kanellopoulou et al., 2005; Murchison, Partridge, Tam, Cheloufi, & Hannon, 2005; Wang, Medvid, Melton, Jaenisch, & Blelloch, 2007). In Dgcr8-null ESCs, cell cycle activity is defective and ESC transcription factors fail to decline under differentiation-inducing conditions. The let-7 family of miRNAs, whose potential targets include nMyc, Sall4, and Lin28, can rescue these defects (Melton, Judson, & Blelloch, 2010).

Dicer-defective mice exhibit spermatogenic defects with poor proliferation activity in spermatogonia (Hayashi et al., 2008). The let-7 family of miRNAs is expressed in spermatogonia and upregulated during RA-induced differentiation, leading to LIN28 suppression (Tong et al., 2011). Transient inhibition of miR-21 significantly reduces the number of functional SSCs, and miR-21 expression is regulated by Etv5 in SSC-enriched spermatogonia (Niu et al., 2011). The studies of miRNA contribution to SSC fate control are still at an infant stage and more information is expected in the near future.

A recent study reports that Caspase3 induces differentiation of ESCs. Although Caspase3 is an inducer of programmed cell death, ESCs lacking this proapoptotic molecule fail to differentiate (Fujita et al., 2008). During RA-induced differentiation of ESCs, Caspase3 expression is upregulated without an increase in cell death. Further, induced expression of Caspase3 leads to ESC differentiation. While Caspase3 can cleave Nanog, cleavage-resistant Nanog enhances ESC self-renewal. Similarly, Caspase3 also promotes differentiation of hematopoietic stem cells (HSCs) and neural stem/progenitor cells (NSCs/NPCs) without increasing cell death (Fernando, Brunette, & Megeney, 2005; Janzen et al., 2008). These results suggest that molecules involved in programmed cell death can promote commitment of stem cells, rather than cell death. This is an interesting potential mechanism of SSC fate control that remains to be examined.

4.4. Does the niche dictate SSC fate?: Neighboring germ cells as a niche component

The stem cell niche is a surrounding microenvironment that influences stem cell actions. The concept of a stem cell niche was originally developed in fly gonads. In fly testis, somatic nursing cells (hub cells) exist at the distal tip in the cylindrically shaped testis and contribute to the maintenance of germline stem cells (GSCs) (Fuller & Spradling, 2007). Differentiation proceeds when GSCs migrate away from contact with hub cells, while those that sustain contact retain "stemness."

In contrast, mammalian testes do not have a distinct structure that suggests the location of SSC niches. The SSC niche is assumed to be composed of Sertoli cells, basal membrane, and peritubular myoid cells, which are all present in the vicinity of spermatogonia. Yoshida, Sukeno, and Nabeshima (2007) reports that SPCs preferentially locate to areas in the seminiferous epithelium that face the interstitial space and correspond to the vascular network. However, the location of SSC niches has not been identified, simply because we cannot visualize SSCs. Nonetheless, SPC localization along the interstitial space suggests that SSC niches may form a two-dimensional plane, rather than a pocket as in fly testes.

An intriguing question is if the somatic niche instructs the fate of SSCs. Somatic cells provide various growth factors that promote SSC self-renewal. Sertoli cells express GDNF, FGF2, and Wnt5a, while myoid cells and Leydig cells secrete CSF1. Upon spermatogonial death induced by busulfan, Sertoli cells upregulate the transcription of GDNF and Wnt5a (O'Shaughnessy, Hu, & Baker, 2008; Zohni, Zhang, Tan, Chan, & Nagano, 2012). They also secrete chemokines that attract GFRα1-expressing spermatogonia but not those expressing c-Kit (Simon et al., 2010). The decline of Sertoli cell functions is thought to contribute to the reduction of the SSC pool during aging (Ryu, Orwig, Oatley, Avarbock, & Brinster, 2006; Zhang, Ebata, Robaire, & Nagano, 2006). Furthermore, conditional and specific suppression of GDNF signaling through c-Ret leads to a rapid loss of SPCs (Savitt et al., 2012). In general, Sertoli cells play a number of important roles in influencing spermatogenesis; hence, they should play a critical role in SSC fate control.

However, it is noteworthy that these cell-extrinsic factors may well affect the activity of progenitors as well, since their receptors are expressed by both SSCs and progenitors (Sections 2 and 3), they should act on a "group" of spermatogonia but may not provide an efficient system to "fine-tune" the fragile balance between self-renewal and commitment of SSCs themselves. Further, no evidence currently exists that Sertoli cells indeed "instruct/dictate" SSC fate. Rather, evidence, albeit circumstantial, suggests that their role is a rather permissive one. When rat germ cells are transplanted into mouse testes, rat spermatogenesis regenerates completely (Clouthier, Avarbock, Maika, Hammer, & Brinster, 1996). However, even though rat SSCs and spermatogenesis are being supported by mouse Sertoli cells and somatic environment, the cycle of spermatogenesis follows the pattern characteristic to rats, not mice (França, Ogawa, Avarbock, Brinster, & Russell, 1998). This observation indicates that overall control of spermatogenesis is dominated by germ cells but not by somatic cells.

On the basis of these past studies, we hypothesize that the niche provides only a "supportive" environment in which SSCs determine their own fate as necessary (yet, not solely by themselves as discussed later); that is, "Somatic niche cells do not dictate the trajectory of SSC fate decision but allow SSCs to do what they need to do." This concept is supported in part by the studies of SPC fate control through two sets of competing forces described above: GDNF versus Plzf and Plzf versus Sall4. In these cases, particularly in GDNF–Plzf competition, the extrinsic stimulus emitted by Sertoli cells (i.e., GDNF) is interpreted and modulated by germ cells. GDNF promotes survival and self-renewal of SSCs but the outcome is decided among the receiving cells (Fig. 3.3B). As we do not yet know what stimulates Plzf expression, it is possible that this mechanism of seemingly germ-cell-intrinsic decision-making is in fact controlled by messages from the somatic niche. Nonetheless, the possibility that the niche is supportive, rather than instructive, has not been discussed or examined extensively and warrants our consideration toward better understanding SSC fate control mechanisms.

If germ cells "interpret and process" signals derived from the somatic niche, how are SSC self-renewal and commitment balanced? It is evident that the output of self-renewal (size of SSC pool) and the output of commitment (spermatozoa) need to be coordinated for proper spermatogenesis. Therefore, SSCs should need cues to accomplish this coordination. For instance, SSC fate control under steady-state spermatogenesis appears to be different from that during regeneration (Ebata, Zhang, & Nagano, 2007; Nakagawa et al., 2007; Zohni et al., 2012). If so, how do SSCs realize the difference? Here we consider the possibility that descendants of SSCs act as a niche component and contribute to SSC fate decision control.

In a recent study, we examined the effect of Wnt3a on SSC activity *in vitro* (Yeh, Zhang, & Nagano, 2012). Wnt3a increased SSC numbers in culture, as assessed by spermatogonial transplantation, and also increased the number of SPCs (cells in a spermatogonial aggregate) with activated Wnt/β-catenin signaling. Yet, Wnt/β-catenin signaling cells did not possess regenerative activity and the level of signaling activity inversely correlated to the number of functional SSCs. Importantly, Wnt3a stimulated cell divisions and encouraged the formation of spermatogonial aggregates under feeder-free conditions; without Wnt3a, spermatogonial aggregates rarely emerged even in the presence of GDNF and FGF2. Consequently, Wnt3a increased cell–cell interactions, or the number of neighboring cells physically associating in culture. Although circumstantial in nature, these results collectively suggest that Wnt3a increased SSC numbers indirectly by stimulating the division of non-SSCs, increasing their numbers, and thereby promoting the formation of spermatogonia communities (i.e., spermatogonial aggregates).

This study raises the possibility that interaction with descendants of SSCs contributes to the fate control of SSCs. Notably, the formation of a spermatogonial aggregate or "community" is not only an *in vitro* phenomenon but is observed *in vivo* shortly after transplantation (Fig. 3.4). It is also interesting to realize that there has been no precedent in stem cell research in general that normal or cancer stem cells are cultured solely by themselves with no accompanying progenitors—is it an inherent nature of stem cells to generate a community?

On the basis of our previous data of SSC culture experiments combined with spermatogonial transplantation, we find that SSC frequency *in vitro* is remarkably consistent: 0.7–1.7% of cultured SPCs/spermatogonial aggregates (Yeh et al., 2012), all of which express Plzf (Ebata et al., 2011). Interestingly, across studies that have employed SSC culture, the SSC frequency *in vitro* is 1–3% (Kanatsu-Shinohara et al., 2005; Oatley & Brinster, 2008;

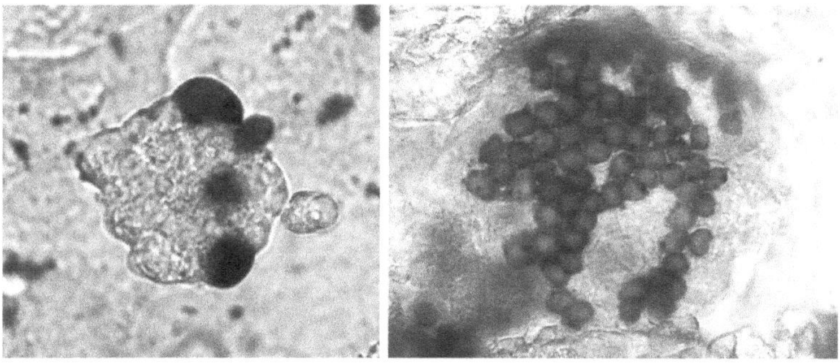

Figure 3.4 The germ cell community *in vitro* (A) and *in vivo* (B). (A) A spermatogonial aggregate that emerges in the presence of GDNF and FGF2. SSCs represent a small subpopulation of these cells. Blue staining (LacZ) indicates activation of Wnt/β-catenin signaling. (B) Aggregating germ cells observed in recipient testes 2 weeks after transplantation. Cells are stained for β-galactosidase activity. (See Color Insert.)

Yeh et al., 2007). It should be noted that this consistency is seen even when SSCs proliferate without GDNF and FGF2 due to a Ras mutation (0.1–2.7%) (Lee et al., 2009). Coincidentally, the frequency of functional SSCs in A_s, A_{pr}, and A_{al} spermatogonia combined, which express Plzf, is estimated to be 0.9–1.8% in a mouse testis (Section 2.3). Thus, this stable frequency of SSCs might not be a mere *in vitro* phenomenon, where the testicular somatic niche is absent, but might also apply to the situation *in vivo*. Regardless, it is intriguing that the proportion of SSCs relative to SPCs seems to be well coordinated, implicating the contribution of a SSC–progenitor interaction to SSC fate control.

Comparatively, evidence is available showing that interactions between stem cells and their descendants may regulate the size of the stem cell pool in a cell lineage. For instance, NSCs represent ~2% of aggregates of cultured neural cells arising from NSCs (neurospheres) (Reynolds & Rietze, 2005). *In vivo*, NSCs and NPCs communicate with each other through Notch and EGF signaling, thereby keeping the size of each cell population in check (Aguirre, Rubio, & Gallo, 2010). In the case of intestinal stem cells (ISCs), a single ISC produces crypt-like organoids *in vitro* composed of interacting ISCs and Paneth cells (Sato et al., 2011). Paneth cells arise from ISCs, yet act as the niche to sustain ISCs *in vivo*. In the hematopoietic system, "signals arising from differentiating cells are required for maintaining progenitor cell quiescence and function with the niche-derived signal in maintaining the progenitor state" (as quoted from Mondal et al., 2011).

Taken together, there appear to be sufficient observations to support the possibility that SSC fate is controlled and "fine-tuned" within a community of germ cells, while somatic cells supply a supportive environment where the germ cell community can survive and function. Importantly, this SSC–progenitor/descendant interplay should provide SSCs with cues to monitor in real time the development and condition of surrounding germ cells.

It is noted that such a "community effect" is at work within ESC colonies, in which differentiation stimulus mediated by FGF4 derived from some ESCs is interpreted by others, leading to the coexistence of ESCs and cells primed to commit. A question arises whether a change in nearby germ cells can be sensed by SSCs. A recent study reports that SPCs are continuously replaced by descendants of neighboring SSCs in steady-state spermatogenesis (Klein et al., 2010). Such a cell turnover likely requires communication among germ cells to sense neighboring cells transitioning towards commitment. Similarly, if spermatogonia or SSCs are killed by a cytotoxic agent, the mode of cell–cell communication among germ cells should change, which might accordingly adjust SSC behavior to regeneration from steady-state spermatogenesis (Ebata et al., 2007; Nakagawa et al., 2007; Zohni et al., 2012). The concept of SSC fate control by a community effect is yet speculative but incorporates the contribution of somatic cells in germ cell-dominated regulation of spermatogenesis and warrants further investigation.

5. THE POINT OF NO RETURN: WHEN AND HOW DOES SSC COMMITMENT START AND END?

We identify the lack of clear markers distinguishing the point of no return to be the greatest stumbling block in studying SSC fate control. It may therefore be useful to discuss biological systems where commitment is better understood. Here we consider meiotic commitment in budding yeast, vulva development in nematodes, and commitment of mouse HSCs.

5.1. Point of no return in the meiotic entry of yeast

In budding yeast (*Saccharomyces cerevisiae*), starvation induces meiosis. Hence, yeast enters meiosis when cultured in nutrient-poor medium lacking a nitrogen source or glucose. Irreversible commitment to meiosis is detected by transferring yeast cultured in a nutrient-rich medium (mitosis) to nutrient-poor medium (meiotic induction), followed by reexposure of the cells to rich medium at different phases of the meiotic cell cycle (Simchen, 2009; Simchen, Pinon, & Salts, 1972). "Commitment" is defined as the ability to continue with the developmental process (meiosis in this case) even in the presence of contradictory stimuli (rich medium). This experimental paradigm reveals three stages during the meiotic commitment process: readiness, partial commitment, and full commitment. The "readiness" stage implies that cells are undergoing meiosis but can return to mitosis when reexposed to rich medium. At partial commitment, cells are arrested in meiosis and fail to switch to mitosis in rich medium. Upon reaching the full commitment stage, cells complete meiotic division even in rich medium. Thus, meiotic commitment in budding yeast is accomplished within one cell cycle. The readiness stage starts just before the meiotic S phase, and the full commitment stage is reached at the end of Prophase I and before the first meiotic division (MI) (Fig. 3.5A); thus, the point of no return occurs just prior to MI.

It is important to note that the levels of readiness and commitment in the cell population increase with time in a gradual fashion but not in a discontinuous manner (simple on–off) until the cells reach the point of no return. Hence, meiotic commitment in yeast is an unstable and reversible process. This may be relevant to SSC studies, as the immunophenotype of A_s, A_{pr}, and A_{al} spermatogonia gradually changes during differentiation (Section 2). If this gradual change also applies to SSCs in terms of function, then, some A_s or even A_{pr} cells may be in a "readiness" stage and may retain the ability to reinitiate the SSC state (Nakagawa et al., 2007). One may also ask whether SSC commitment is related to the cell cycle, as in yeast, and if the process of cytoplasmic bridge formation, which occurs during the cell cycle (Greenbaum, Iwamori, Buchold, & Matzuk, 2011), could be involved in the SSC fate control mechanism.

An implication of this yeast example extends our thoughts to the possibility of reversion and dedifferentiation. Reversion can occur to a prior step as long as the point of no return has not been reached. On the other hand, dedifferentiation implies reprogramming of cells that have passed the point of no return, which rarely occurs in normal biological systems or which might lead to tumorigenesis. Therefore, reversion may take place during the SSC commitment process, but it is unlikely that committed cells dedifferentiate during normal spermatogenesis.

5.2. Development of nematode vulva

Development of *Caenorhabditis elegans* vulva starts from two "equivalent cells" that have the same potential to form the vulva (also called "equipotent cells" or "equivalence cell populations"; Henrique, 2001; Sommer, 2001).

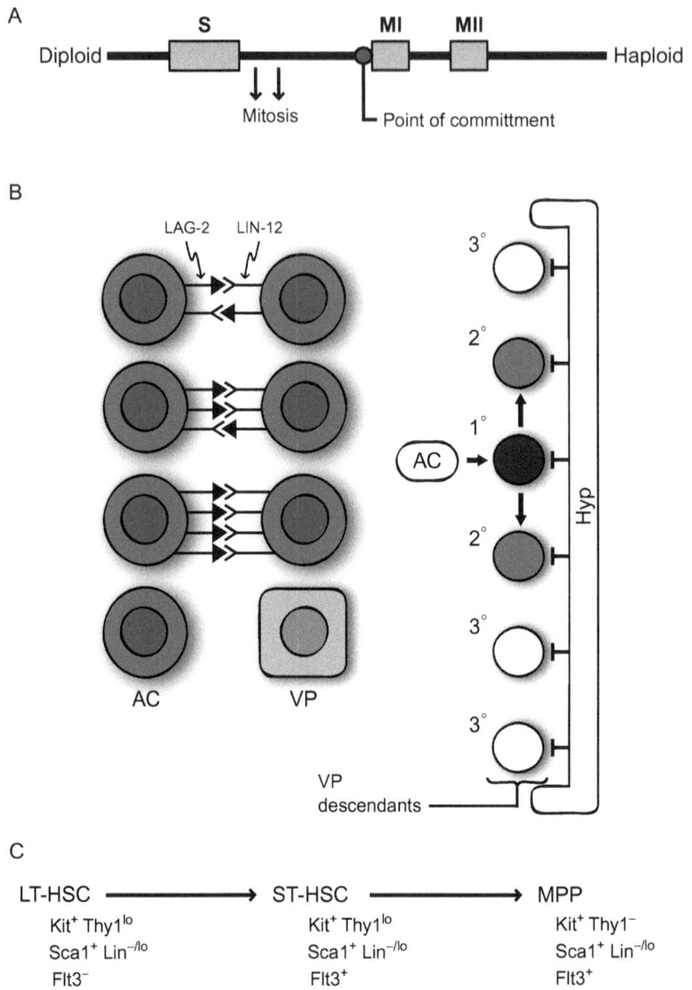

Figure 3.5 Schematic representations of commitment in yeast, nematodes, and mouse HSCs. (A) One meiotic division in budding yeast from one diploid cell to four haploid cells. Irreversible commitment to meiosis is reached just before the first division of meiosis (MI). Before this point, cells can either revert to mitosis or arrest at meiosis. (B) *Caenorhabditis elegans* vulval development. (Left) Two equivalent cells take different fates due to stochastic differences in expression of ligand (LAG-2) and receptor (LIN-12). AC, anchor cells; VP, ventral uterine precursor. (Right) VP cells take further differentiation paths by communicating among VP descendants and with niche cells (AC and hypoderm: Hyp). Note that the hypoderm equally provides all VP descendants with an inhibitory signal. (C) Mouse HSC commitment. Stage-specific cell-surface markers are shown. ST-HSCs can be recognized as committed progenitors. LT, long-term; ST, short-term; MPP, multipotent progenitors. (For color version of this figure, the reader is referred to the online version of this chapter.)

Each cell expresses LIN-12, a receptor molecule homologous to Notch, and LAG-2, a ligand homologous to a Notch ligand, Delta. Both cells communicate with one another via this signaling. As the expression of these molecules fluctuates within each cell, a small difference in expression levels occurs stochastically between the two cells. Importantly, the signaling cascade downstream of LIN-12 (receptor) stimulates the expression of itself while suppressing LAG-2 expression (ligand). Thus, if one cell receives slightly more LAG-2, it experiences slightly more activated LIN-12 signaling, which then leads to further ligand downregulation and receptor upregulation (Fig. 3.5B, left). In this way, the initial difference generated by chance between the two equivalent cells is amplified. The cell that becomes the receiver differentiates to the ventral uterine precursor (VP) while the other develops into the anchor cell (AC). It is important to note that the formation of VP cells is a "distinctive stage" of development, where the cells have established distinctive developmental characteristics, rather than a "transient phase."

Subsequently, VP cells further differentiate into three cell types, composed of a total of six cells (Fig. 3.5B, right). These three cell types are generated through communication among themselves under the influence from niche cells (AC and hypodermal cells). AC determines the 1° cell, which in turn controls the fate of the other cells, while hypodermal cells emit inhibitory signals to all six cells (Fig. 3.5B, right). The lack of signals from both AC and hypoderm results in alternative alignment of 1° and 2° cells, with no 3° cells (Henrique, 2001). Hence, *C. elegans* vulva develops through chance events within a cell–cell communication network, where descendants fine-tune the destiny of their own and neighboring cells, while receiving cues from their niche.

For SSCs, a recent study suggests that commitment occurs in a stochastic manner (Klein et al., 2010). In this context, an intriguing possibility is that a single SSC divides into two single cells that possess equivalent regenerative potential. One loses self-renewal capacity and commits to differentiation by stochastic communication between the two cells. Such communication might be influenced or modulated by signals from neighboring cells, including SSC-descendants and somatic cells, just like how VP cells differentiate in the nematode.

5.3. Commitment process in the HSC system

Gradual commitment, yet with distinctive differentiation stages, is seen in the mouse HSC system. Mouse bone marrow cells can be enriched for HSCs using cell-surface markers: c-Kit(+)Thy1(low)Sca1(+)Lineage(−/low) (KTSL cells).

This population is divided into long-term (LT) and short-term (ST) HSCs by the expression of Flt3; KTSL Flt3(−) are LT-HSCs while KTSL Flt3(+) are ST-HSCs (Adolfsson et al., 2001; Christensen & Weissman, 2001). Both cell populations are equivalent in their multipotency and produce the entire spectrum of terminally differentiated cell types but upon transplantation, ST-HSCs can sustain complete hematopoiesis for only 8–12 weeks; thus, ST-HSCs have limited self-renewal ability than LT-HSCs. In this sense, ST-HSCs can be called progenitors, which have passed the point of no return. Accordingly, the functional hierarchy of the HSC commitment process can be identified using cell-surface markers (Fig. 3.5C). It is interesting to note that the point of no return becomes detectable first in self-renewal potential; the point of no return for multipotency is seen when multipotent progenitors commit to lymphoid and myeloid lineages.

An important lesson for investigations into SSC commitment is that cell-surface markers indicate the point of no return by "prospectively" distinguishing the "function" of stem cells from progenitors. This is only possible when the marker–function linkage is established extensively. In the HSC commitment process, therefore, Flt3 expression can be used as a phenotypic readout of commitment, which is confirmed by transplantation experiments.

5.4. Toward identification of the point of no return in SSC commitment

As emphasized above, we need markers to distinguish the cells before and after the point of no return. Studies of HSCs exemplify the importance of identifying multiple cell-surface markers that allow for functional dissection of stem cell commitment. More efforts are necessary to determine function–marker linkages to eventually associate SSC function with cell morphology and strictly establish the identity of SSCs.

In our study of Wnt5a effects on SSC self-renewal (Yeh et al., 2011), we attempted to fractionate cells expressing Thy1, the most stringent SSC marker to date, by using Wnt5a receptors as a secondary marker. However, nearly all Thy1(+) cells in adult testes expressed a Wnt5a receptor. Conversely, nearly all Thy1(+) cells were c-Kit(−) (Kubota et al., 2003), indicating that the Thy1(+) population cannot be fractionated by c-Kit. On the other hand, a recent study reports that CD9-expressing cells can be separated into EPCAM(+) and EPCAM(low/−) cells and that SSCs are more concentrated in the CD9(+)EPCAM(low/−) population (Kanatsu-Shinohara, Takashima, Ishii, & Shinohara, 2011). As CD9 is a less stringent marker than

Thy1 (Oatley & Brinster, 2008), the success of the CD9-based approach suggests that the choice of a primary cell population may be important for fractionating SSC-enriched cells. Nonetheless, the CD9 study established a commitment level in SSC-enriched cells and provided a tool to search for the point of no return.

Lastly, we look at an achievement of the extensive marker–function linkage. Built on profound knowledge about HSC markers, Kiel et al. (2005) identified cell-surface receptors of the SLAM family (CD150, CD244, and CD48) that allow for an astonishing level of HSC enrichment. The study shows that HSC commitment advances from CD150(+)CD244(−)CD48(−) to CD150(−)CD244(+)CD48(−) to CD150(−)CD244(+)CD48(+). Using this small number of markers, HSCs can be visualized in the bone marrow. Furthermore, the CD150(+)CD48(−)Sca1(+)Lineage(−)c-Kit(+) population represents 0.0058% of nucleated mouse bone marrow cells. Although such strict cell fractionation inevitably leads to low cell recovery, one in two sorted cells successfully repopulates the bone marrow after transplantation and sustains long-term complete hematopoiesis. This is remarkable because the transplantation result implies that either HSCs are "purified" with 50% homing efficiency or are enriched to 50% purity with 100% homing efficiency (or the combination of values in between). Such a high homing efficiency has not been reported in HSC studies. Therefore, once stem cells are purified or near-purified, they may expose their remarkable and yet-to-be-discovered abilities.

6. CONCLUDING REMARKS

It is now clear that heterogeneous cell populations constitute both SSCs and progenitor spermatogonia, from a morphological and functional perspective. It is likely that the commitment of SSCs and progenitors proceeds gradually, perhaps initiated by stochastic events of SSC preconditioning, followed by "readiness" and "full commitment" stages. Heterogeneous cell populations are an inevitable consequence of such a gradual process. Before reaching full commitment, cells may retain self-renewal ability, regardless of their morphology, and revert from one stage to another. However, once the point of no return is reached, cells should be fixed in a lineage-specific commitment program and proceed toward spermatozoa. It is speculated that before full commitment, SSCs are in an unstable state, influenced by conflicting forces for self-renewal and commitment.

These forces should be coordinated in cell–cell communication among germ cells and somatic cells in the testis. We propose that the somatic niche provides a supportive environment and emits stimuli that influence a community of germ cells as a whole, rather than SSCs specifically. Within this heterogeneous cell community, a SSC may communicate with its own descendants and possibly with other SSCs in the vicinity. This SSC-descendant interplay may thus be the important mechanism that fine-tunes SSC fate decision control, under regional cues exerted by somatic cells onto the germ cell community. Alteration of the environment (germ and somatic cells), such as after testicular injuries, during aging, or in early postnatal development, may force SSCs to adjust themselves to an emerging communication network leading to different SSC behaviors from steady-state spermatogenesis. The root of germ-cell-dominated control of spermatogenesis may therefore be found in this SSC-descendant interplay within the germ cell community.

Although the potential mechanisms discussed above do not go beyond speculation at present, the most important task that we now face to better understand SSC fate control mechanisms is the identification of markers for SSCs and progenitors that define the point of no return. Such markers should also help us determine the characteristics of each cell population.

REFERENCES

Adolfsson, J., Borge, O. J., Bryder, D., Theilgaard-Mönch, K., Astrand-Grundström, I., Sitnicka, E., et al. (2001). Upregulation of Flt3 expression within the bone marrow Lin(-)Sca1(+)c-kit(+) stem cell compartment is accompanied by loss of self-renewal capacity. *Immunity*, *15*, 659–669.

Aguirre, A., Rubio, M. E., & Gallo, V. (2010). Notch and EGFR pathway interaction regulates neural stem cell number and self-renewal. *Nature*, *467*, 323–327.

Ballow, D., Meistrich, M. L., Matzuk, M., & Rajkovic, A. (2006). Sohlh1 is essential for spermatogonial differentiation. *Developmental Biology*, *294*, 161–167.

Barrios, F., Filipponi, D., Campolo, F., Gori, M., Bramucci, F., Pellegrini, M., et al. (2012). SOHLH1 and SOHLH2 control Kit expression during postnatal male germ cell development. *Journal of Cell Science*, *125*, 1455–1464.

Boyer, L. A., Lee, T. I., Cole, M. F., Johnstone, S. E., Levine, S. S., Zucker, J. P., et al. (2005). Core transcriptional regulatory circuitry in human embryonic stem cells. *Cell*, *122*, 947–956.

Braydich-Stolle, L., Kostereva, N., Dym, M., & Hofmann, M. C. (2007). Role of Src family kinases and N-Myc in spermatogonial stem cell proliferation. *Developmental Biology*, *304*, 34–45.

Brinster, R. L., & Avarbock, M. R. (1994). Germline transmission of donor haplotype following spermatogonial transplantation. *Proceedings of the National Academy of Sciences of the United States of America*, *91*, 11303–11307.

Brinster, R. L., & Zimmermann, J. W. (1994). Spermatogenesis following male germ-cell transplantation. *Proceedings of the National Academy of Sciences of the United States of America*, *91*, 11298–11302.

Buaas, F. W., Kirsh, A. L., Sharma, M., McLean, D. J., Morris, J. L., Griswold, M. D., et al. (2004). Plzf is required in adult male germ cells for stem cell self-renewal. *Nature Genetics*, *36*, 647–652.
Bucci, L. R., & Meistrich, M. L. (1987). Effects of busulfan on murine spermatogenesis: Cytotoxicity, sterility, sperm abnormalities, and dominant lethal mutations. *Mutation Research*, *176*, 259–268.
Christensen, J. L., & Weissman, I. L. (2001). Flk-2 is a marker in hematopoietic stem cell differentiation: A simple method to isolate long-term stem cells. *Proceedings of the National Academy of Sciences of the United States of America*, *98*, 14541–14546.
Clouthier, D. E., Avarbock, M. R., Maika, S. D., Hammer, R. E., & Brinster, R. L. (1996). Rat spermatogenesis in mouse testis. *Nature*, *381*, 418–421.
Costoya, J. A., Hobbs, R. M., Barna, M., Cattoretti, G., Manova, K., Sukhwani, M., et al. (2004). Essential role of Plzf in maintenance of spermatogonial stem cells. *Nature Genetics*, *36*, 653–659.
Dann, C. T., Alvarado, A. L., Molyneux, L. A., Denard, B. S., Garbers, D. L., & Porteus, M. H. (2008). Spermatogonial stem cell self-renewal requires OCT4, a factor downregulated during retinoic acid-induced differentiation. *Stem Cells*, *26*, 2928–2937.
David, G., Alland, L., Hong, S. H., Wong, C. W., DePinho, R. A., & Dejean, A. (1998). Histone deacetylase associated with mSin3A mediates repression by the acute promyelocytic leukemia-associated PLZF protein. *Oncogene*, *16*, 2549–2556.
de Rooij, D. G., Okabe, M., & Nishimune, Y. (1999). Arrest of spermatogonial differentiation in jsd/jsd, Sl17H/Sl17H, and cryptorchid mice. *Biology of Reproduction*, *61*, 842–847.
de Rooij, D. G., & Russell, L. D. (2001). All you wanted to know about spermatogonia but were afraid to ask. *Journal of Andrology*, *21*, 776–798.
de Rooij, D. G., van Dissel-Emiliani, F. M., & van Pelt, A. M. (1989). Regulation of spermatogonial proliferation. *Annals of the New York Academy of Sciences*, *564*, 140–153.
Dobrinski, I., Ogawa, T., Avarbock, M. R., & Brinster, R. L. (1999). Computer assisted image analysis to assess colonization of recipient seminiferous tubules by spermatogonial stem cells from transgenic donor mice. *Molecular Reproduction and Development*, *53*, 142–148.
Doulatov, S., Notta, F., Rice, K. L., Howell, L., Zelent, A., Licht, J. D., et al. (2009). PLZF is a regulator of homeostatic and cytokine-induced myeloid development. *Genes & Development*, *23*, 2076–2087.
Ebata, K. T., Yeh, J. R., Zhang, X., & Nagano, M. C. (2011). Soluble growth factors stimulate spermatogonial stem cell divisions that maintain a stem cell pool and produce progenitors *in vitro*. *Experimental Cell Research*, *317*, 1319–1329.
Ebata, K. T., Zhang, X., & Nagano, M. C. (2005). Expression patterns of cell-surface molecules on male germ line stem cells during postnatal mouse development. *Molecular Reproduction and Development*, *72*, 171–181.
Ebata, K. T., Zhang, X., & Nagano, M. C. (2007). Male germ line stem cells have an altered potential to proliferate and differentiate during postnatal development in mice. *Biology of Reproduction*, *76*, 841–847.
Fernando, P., Brunette, S., & Megeney, L. A. (2005). Neural stem cell differentiation is dependent upon endogenous caspase 3 activity. *The FASEB Journal*, *19*, 1671–1673.
Filipponi, D., Hobbs, R. M., Ottolenghi, S., Rossi, P., Jannini, E. A., Pandolfi, P. P., et al. (2007). Repression of kit expression by Plzf in germ cells. *Molecular and Cellular Biology*, *27*, 6770–6781.
França, L. R., Ogawa, T., Avarbock, M. R., Brinster, R. L., & Russell, L. D. (1998). Germ cell genotype controls cell cycle during spermatogenesis in the rat. *Biology of Reproduction*, *59*, 1371–1377.

Fujita, J., Crane, A. M., Souza, M. K., Dejosez, M., Kyba, M., Flavell, R. A., et al. (2008). Caspase activity mediates the differentiation of embryonic stem cells. *Cell Stem Cell, 2*, 595–601.

Fuller, M. T., & Spradling, A. C. (2007). Male and female Drosophila germline stem cells: Two versions of immortality. *Science, 316*, 402–404.

Giuili, G., Tomljenovic, A., Labrecque, N., Oulad-Abdelghani, M., Rassoulzadegan, M., & Cuzin, F. (2002). Murine spermatogonial stem cells: Targeted transgene expression and purification in an active state. *EMBO Reports, 3*, 753–759.

Grasso, M., Fuso, A., Dovere, L., de Rooij, D. G., Stefanini, M., Boitani, C., et al. (2012). Distribution of GFRA1-expressing spermatogonia in adult mouse testis. *Reproduction, 143*, 325–332.

Greenbaum, M. P., Iwamori, T., Buchold, G. M., & Matzuk, M. M. (2011). Germ cell intercellular bridges. *Cold Spring Harbor Perspectives in Biology, 3*, a005850.

Hamra, F. K., Chapman, K. M., Nguyen, D., & Garbers, D. L. (2007). Identification of neuregulin as a factor required for formation of aligned spermatogonia. *The Journal of Biological Chemistry, 282*, 721–730.

Hamra, F. K., Schultz, N., Chapman, K. M., Grellhesl, D. M., Cronkhite, J. T., Hammer, R. E., et al. (2004). Defining the spermatogonial stem cell. *Developmental Biology, 269*, 393–410.

Han, I. S., Sylvester, S. R., Kim, K. H., Schelling, M. E., Venkateswaran, S., Blanckaert, V. D., et al. (1993). Basic fibroblast growth factor is a testicular germ cell product which may regulate Sertoli cell function. *Molecular Endocrinology, 7*, 889–897.

Hao, J., Yamamoto, M., Richardson, T. E., Chapman, K. M., Denard, B. S., Hammer, R. E., et al. (2008). Sohlh2 knockout mice are male-sterile because of degeneration of differentiating type A spermatogonia. *Stem Cells, 26*, 1587–1597.

Hayashi, K., de Sousa, Chuva, Lopes, S. M., Kaneda, M., Tang, F., Hajkova, P., et al. (2008). MicroRNA biogenesis is required for mouse primordial germ cell development and spermatogenesis. *PloS One, 3*, e1738.

Henrique, D. (2001). On equivalence groups and the Notch/LN-12 communication system. In D. R. Marshak, R. L. Gardner & D. Gottlieb (Eds.), *Stem cell biology* (pp. 37–60). New York: Cold Spring Harbor Laboratory Press.

Hobbs, R. M., Fagoonee, S., Papa, A., Webster, K., Altruda, F., Nishinakamura, R., et al. (2012). Functional antagonism between Sall4 and Plzf defines germline progenitors. *Cell Stem Cell, 10*, 284–298.

Hobbs, R. M., Seandel, M., Falciatori, I., Rafii, S., & Pandolfi, P. P. (2010). Plzf regulates germline progenitor self-renewal by opposing mTORC1. *Cell, 142*, 468–479.

Huckins, C. (1971). The spermatogonial stem cell population in adult rats. I. Their morphology, proliferation and maturation. *The Anatomical Record, 169*, 533–557.

Ishii, K., Kanatsu-Shinohara, M., Toyokuni, S., & Shinohara, T. (2012). FGF2 mediates mouse spermatogonial stem cell self-renewal via upregulation of Etv5 and Bcl6b through MAP2K1 activation. *Development, 139*, 1734–1743.

Janzen, V., Fleming, H. E., Riedt, T., Karlsson, G., Riese, M. J., Lo Celso, C., et al. (2008). Hematopoietic stem cell responsiveness to exogenous signals is limited by caspase-3. *Cell Stem Cell, 2*, 584–594.

Kanatsu-Shinohara, M., Inoue, K., Miki, H., Ogonuki, N., Takehashi, M., Morimoto, T., et al. (2006). Clonal origin of germ cell colonies after spermatogonial transplantation in mice. *Biology of Reproduction, 75*, 68–74.

Kanatsu-Shinohara, M., Miki, H., Inoue, K., Ogonuki, N., Toyokuni, S., Ogura, A., et al. (2005). Long-term culture of mouse male germline stem cells under serum-or feeder-free conditions. *Biology of Reproduction, 72*, 985–991.

Kanatsu-Shinohara, M., Ogonuki, N., Inoue, K., Miki, H., Ogura, A., Toyokuni, S., et al. (2003). Long-term proliferation in culture and germline transmission of mouse male germline stem cells. *Biology of Reproduction, 69*, 612–616.

Kanatsu-Shinohara, M., Takashima, S., Ishii, K., & Shinohara, T. (2011). Dynamic changes in EPCAM expression during spermatogonial stem cell differentiation in the mouse testis. *PLoS One, 6*, e23663.
Kanatsu-Shinohara, M., Toyokuni, S., & Shinohara, T. (2004). CD9 is a surface marker on mouse and rat male germline stem cells. *Biology of Reproduction, 70*, 70–75.
Kanellopoulou, C., Muljo, S. A., Kung, A. L., Ganesan, S., Drapkin, R., Jenuwein, T., et al. (2005). Dicer-deficient mouse embryonic stem cells are defective in differentiation and centromeric silencing. *Genes & Development, 19*, 489–501.
Kaucher, A. V., Oatley, M. J., & Oatley, J. M. (2012). NEUROG3 is a critical downstream effector for STAT3-regulated differentiation of mammalian stem and progenitor spermatogonia. *Biology of Reproduction, 86*, 164–174.
Kiel, M. J., Yilmaz, O. H., Iwashita, T., Yilmaz, O. H., Terhorst, C., & Morrison, S. J. (2005). SLAM family receptors distinguish hematopoietic stem and progenitor cells and reveal endothelial niches for stem cells. *Cell, 121*, 1109–1121.
Klein, A. M., Nakagawa, T., Ichikawa, R., Yoshida, S., & Simons, B. D. (2010). Mouse germ line stem cells undergo rapid and stochastic turnover. *Cell Stem Cell, 7*, 214–224.
Kubota, H., Avarbock, M. R., & Brinster, R. L. (2003). Spermatogonial stem cells share some, but not all, phenotypic and functional characteristics with other stem cells. *Proceedings of the National Academy of Sciences of the United States of America, 100*, 6487–6492.
Kubota, H., Avarbock, M. R., & Brinster, R. L. (2004). Growth factors essential for self-renewal and expansion of mouse spermatogonial stem cells. *Proceedings of the National Academy of Sciences of the United States of America, 101*, 16489–16494.
Kunath, T., Saba-El-Leil, M. K., Almousailleakh, M., Wray, J., Meloche, S., & Smith, A. (2007). FGF stimulation of the Erk1/2 signalling cascade triggers transition of pluripotent embryonic stem cells from self-renewal to lineage commitment. *Development, 134*, 2895–2902.
Labbaye, C., Quaranta, M. T., Pagliuca, A., Militi, S., Licht, J. D., Testa, U., et al. (2002). PLZF induces megakaryocytic development, activatesTpo receptor expression and interacts with GATA1 protein. *Oncogene, 21*, 6669–6679.
Leblond, C. P., & Clermont, Y. (1952). Spermiogenesis of rat, mouse, hamster and guinea pig as revealed by the periodic acid-fuchsin sulfurous acid technique. *The American Journal of Anatomy, 90*, 167–215.
Lee, J., Kanatsu-Shinohara, M., Inoue, K., Ogonuki, N., Miki, H., Toyokuni, S., et al. (2007). Akt mediates self-renewal division of mouse spermatogonial stem cells. *Development, 134*, 1853–1859.
Lee, J., Kanatsu-Shinohara, M., Morimoto, H., Kazuki, Y., Takashima, S., Oshimura, M., et al. (2009). Genetic reconstruction of mouse spermatogonial stem cell self-renewal *in vitro* by Ras-cyclin D2 activation. *Cell Stem Cell, 5*, 76–86.
Lee, J., & Shinohara, T. (2011). Epigenetic modifications and self-renewal regulation of mouse germline stem cells. *Cell Research, 21*, 1164–1171.
Melton, C., Judson, R. L., & Blelloch, R. (2010). Opposing microRNA familiesregulate self-renewal in mouse embryonic stem cells. *Nature, 463*, 621–626.
Meng, X., Lindahl, M., Hyvönen, M. E., Parvinen, M., de Rooij, D. G., Hess, M. W., et al. (2000). Regulation of cell fate decision of undifferentiated spermatogonia by GDNF. *Science, 287*, 1489–1493.
Mondal, B. C., Mukherjee, T., Mandal, L., Evans, C. J., Sinenko, S. A., Martinez-Agosto, J. A., et al. (2011). Interaction between differentiating cell- and niche-derived signals in hematopoietic progenitor maintenance. *Cell, 147*, 1589–1600.
Mullaney, B. P., & Skinner, M. K. (1992). Basic fibroblast growth factor (bFGF) gene expression and protein production during pubertal development of the seminiferous tubule: Follicule-stimulating hormone-induced Sertoli cell bFGF expression. *Endocrinology, 131*, 2928–2934.

Murchison, E. P., Partridge, J. F., Tam, O. H., Cheloufi, S., & Hannon, G. J. (2005). Characterization of Dicer-deficient murine embryonic stem cells. *Proceedings of the National Academy of Sciences of the United States of America, 102*, 12135–12140.

Nagano, M. C. (2003). Homing efficiency and proliferation kinetics of male germ line stem cells following transplantation in mice. *Biology of Reproduction, 69*, 701–707.

Nagano, M., Avarbock, M. R., & Brinster, R. L. (1999). Pattern and kinetics of mouse donor spermatogonial stem cell colonization in recipient testes. *Biology of Reproduction, 60*, 1429–1436.

Nakagawa, T., Nabeshima, Y., & Yoshida, S. (2007). Functional identification of the actual and potential stem cell compartments in mouse spermatogenesis. *Developmental Cell, 12*, 195–206.

Nakagawa, T., Sharma, M., Nabeshima, Y., Braun, R. E., & Yoshida, S. (2010). Functional hierarchy and reversibility within the murine spermatogenic stem cell compartment. *Science, 328*, 62–67.

Niu, Z., Goodyear, S. M., Rao, S., Wu, X., Tobias, J. W., Avarbock, M. R., et al. (2011). MicroRNA-21 regulates the self-renewal of mouse spermatogonial stem cells. *Proceedings of the National Academy of Sciences of the United States of America, 108*, 12740–12745.

Niwa, H., Miyazaki, J., & Smith, A. G. (2000). Quantitative expression of Oct-3/4 defines differentiation, dedifferentiation or self-renewal of ES cells. *Nature Genetics, 24*, 372–376.

Oakberg, E. F. (1971). Spermatogonial stem-cell renewal in the mouse. *The Anatomical Record, 169*, 515–531.

Oatley, J. M., Avarbock, M. R., & Brinster, R. L. (2007). Glial cell line-derived neurotrophic factor regulation of genes essential for self-renewal of mouse spermatogonial stem cells is dependent on Src family kinase signaling. *The Journal of Biological Chemistry, 282*, 25842–25851.

Oatley, J. M., Avarbock, M. R., Telaranta, A. I., Fearon, D. T., & Brinster, R. L. (2006). Identifying genes important for spermatogonial stem cell self-renewal and survival. *Proceedings of the National Academy of Sciences of the United States of America, 103*, 9524–9529.

Oatley, J. M., & Brinster, R. L. (2008). Regulation of spermatogonial stem cell self-renewal in mammals. *Annual Review of Cell and Developmental Biology, 24*, 263–286.

Oatley, J. M., & Brinster, R. L. (2012). The germline stem cell niche unit in mammalian testes. *Physiological Reviews, 92*, 577–595.

Oatley, J. M., Kaucher, A. V., Avarbock, M. R., & Brinster, R. L. (2010). Regulation of mouse spermatogonial stem cell differentiation by STAT3 signaling. *Biology of Reproduction, 83*, 427–433.

Oatley, M. J., Kaucher, A. V., Racicot, K. E., & Oatley, J. M. (2011). Inhibitor of DNA binding 4 is expressed selectively by single spermatogonia in the male germline and regulates the self-renewal of spermatogonial stem cells in mice. *Biology of Reproduction, 85*, 347–456.

Oatley, J. M., Oatley, M. J., Avarbock, M. R., Tobias, J. W., & Brinster, R. L. (2009). Colony stimulating factor 1 is an extrinsic stimulator of mouse spermatogonial stem cell self-renewal. *Development, 136*, 1191–1199.

Ohta, H., Yomogida, K., Yamada, S., Okabe, M., & Nishimune, Y. (2000). Real-time observation of transplanted 'green germ cells': Proliferation and differentiation of stem cells. *Development, Growth & Differentiation, 42*, 105–112.

O'Shaughnessy, P. J., Hu, L., & Baker, P. J. (2008). Effect of germ cell depletion on levels of specific mRNA transcripts in mouse Sertoli cells and Leydig cells. *Reproduction, 135*, 839–850.

Phillips, B. T., Gassei, K., & Orwig, K. E. (2010). Spermatogonial stem cell regulation and spermatogenesis. *Philosophical Transactions of the Royal Society of London. Series B, Biological Sciences, 365*, 1663–1678.

Rappolee, D. A., Basilico, C., Patel, Y., & Werb, Z. (1994). Expression and function of FGF-4 in peri-implantation development in mouse embryos. *Development, 120*, 2259–2269.
Reynolds, B. A., & Rietze, R. L. (2005). Neural stem cells and neurospheres—Re-evaluating the relationship. *Nature Methods, 2*, 333–336.
Ryu, B. Y., Orwig, K. E., Oatley, J. M., Avarbock, M. R., & Brinster, R. L. (2006). Effects of aging and niche microenvironment on spermatogonial stem cell self-renewal. *Stem Cells, 24*, 1505–1511.
Sabatini, D. M. (2006). mTOR and cancer: Insights into a complex relationship. *Nature Reviews. Cancer, 6*, 729–734.
Sada, A., Hasegawa, K., Pin, P. H., & Saga, Y. (2012). NANOS2 acts downstream of glial cell line-derived neurotrophic factor signaling to suppress differentiation of spermatogonial stem cells. *Stem Cells, 30*, 280–291.
Sada, A., Suzuki, A., Suzuki, H., & Saga, Y. (2009). The RNA-binding protein NANOS2 is required to maintain murine spermatogonial stem cells. *Science, 325*, 1394–1398.
Sato, T., van Es, J. H., Snippert, H. J., Stange, D. E., Vries, R. G., Es, J. H., et al. (2011). Paneth cells constitute the niche for Lgr5 stem cells in intestinal crypts. *Nature, 469*, 415–418.
Savitt, J., Singh, D., Zhang, C., Chen, L. C., Folmer, J., Shokat, K. M., et al. (2012). The *in vivo* response of stem and other undifferentiated spermatogonia to the reversible inhibition of glial cell line-derived neurotrophic factor signaling in the adult. *Stem Cells, 30*, 732–740.
Schlesser, H. N., Simon, L., Hofmann, M. C., Murphy, K. M., Murphy, T., Hess, R. A., et al. (2008). Effects of ETV5 (ets variant gene 5) on testis and body growth, time course of spermatogonial stem cell loss, and fertility in mice. *Biology of Reproduction, 78*, 483–489.
Shinohara, T., Avarbock, M. R., & Brinster, R. L. (1999). beta1- and alpha6-integrin are surface markers on mouse spermatogonial stem cells. *Proceedings of the National Academy of Sciences of the United States of America, 96*, 5504–5509.
Shinohara, T., Orwig, K. E., Avarbock, M. R., & Brinster, R. L. (2000). Spermatogonial stem cell enrichment by multiparameter selection of mouse testis cells. *Proceedings of the National Academy of Sciences of the United States of America, 97*, 8346–8351.
Silva, J., & Smith, A. (2008). Capturing pluripotency. *Cell, 132*, 532–536.
Simchen, G. (2009). Commitment to meiosis: What determines the mode of division in budding yeast? *BioEssays, 31*, 169–177.
Simchen, G., Pinon, R., & Salts, Y. (1972). Sporulation in *Saccharomyces cerevisiae*: Premeiotic DNA synthesis, readiness and commitment. *Experimental Cell Research, 75*, 207–218.
Simon, L., Ekman, G. C., Garcia, T., Carnes, K., Zhang, Z., Murphy, T., et al. (2010). ETV5 regulates sertoli cell chemokines involved in mouse stem/progenitor spermatogonia maintenance. *Stem Cells, 28*, 1882–1892.
Sommer, R. J. (2001). As good as they get: Cells in nematode vulva development and evolution. *Current Opinion in Cell Biology, 13*, 715–720.
Subramanyam, D., & Blelloch, R. (2011). From microRNAs to targets: Pathway discovery in cell fate transitions. *Current Opinion in Genetics & Development, 21*, 498–503.
Suh, N., & Blelloch, R. (2011). Small RNAs in early mammalian development: From gametes to gastrulation. *Development, 138*, 1653–1661.
Suzuki, H., Ahn, H. W., Chu, T., Bowden, W., Gassei, K., Orwig, K., et al. (2012). SOHLH1 and SOHLH2 coordinate spermatogonial differentiation. *Developmental Biology, 361*, 301–312.
Suzuki, A., Igarashi, K., Aisaki, K., Kanno, J., & Saga, Y. (2010). NANOS2 interacts with the CCR4-NOT deadenylation complex and leads to suppression of specific RNAs. *Proceedings of the National Academy of Sciences of the United States of America, 107*, 3594–3599.

Suzuki, H., Sada, A., Yoshida, S., & Saga, Y. (2009). The heterogeneity of spermatogonia is revealed by their topology and expression of marker proteins including the germ cell-specific proteins Nanos2 and Nanos3. *Developmental Biology*, *336*, 222–231.

Tegelenbosch, R. A., & de Rooij, D. G. (1993). A quantitative study of spermatogonial multiplication and stem cell renewal in the C3H/101 F1 hybrid mouse. *Mutation Research*, *290*, 193–200.

Tokuda, M., Kadokawa, Y., Kurahashi, H., & Marunouchi, T. (2007). CDH1 is a specific marker for undifferentiated spermatogonia in mouse testes. *Biology of Reproduction*, *76*, 130–141.

Tong, M. H., Mitchell, D., Evanoff, R., & Griswold, M. D. (2011). Expression of Mirlet7 family microRNAs in response to retinoic acid-induced spermatogonial differentiation in mice. *Biology of Reproduction*, *85*, 189–197.

Toyoda, S., Miyazaki, T., Miyazaki, S., Yoshimura, T., Yamamoto, M., Tashiro, F., et al. (2009). Sohlh2 affects differentiation of KIT positive oocytes and spermatogonia. *Developmental Biology*, *325*, 238–248.

Tsuda, M., Sasaoka, Y., Kiso, M., Abe, K., Haraguchi, S., Kobayashi, S., et al. (2009). Conserved role of nanos proteins in germ cell development. *Science*, *301*, 1239–1241.

van den Aardweg, G. J., de Ruiter-Bootsma, A. L., Kramer, M. F., & Davids, J. A. (1983). Growth and differentiation of spermatogenetic colonies in the mouse testis after irradiation with fission neutrons. *Radiation Research*, *94*, 447–463.

van Pelt, A. M., & de Rooij, D. G. (1990). Synchronization of the seminiferous epithelium after vitamin A replacement in vitamin A-deficient mice. *Biology of Reproduction*, *43*, 363–367.

van Pelt, A. M., & de Rooij, D. G. (1991). Retinoic acid is able to reinitiate spermatogenesis in vitamin A-deficient rats and high replicate doses support the full development of spermatogenic cells. *Endocrinology*, *128*, 697–704.

Wang, Y., Medvid, R., Melton, C., Jaenisch, R., & Blelloch, R. (2007). DGCR8 is essential for microRNA biogenesis and silencing of embryonic stem cell self-renewal. *Nature Genetics*, *39*, 380–385.

Weissman, I. L. (2000). Translating stem and progenitor cell biology to the clinic: Barriers and opportunities. *Science*, *287*, 1442–1446.

Wu, X., Oatley, J. M., Oatley, M. J., Kaucher, A. V., Avarbock, M. R., & Brinster, R. L. (2010). The POU domain transcription factor POU3F1 is an important intrinsic regulator of GDNF-induced survival and self-renewal of mouse spermatogonial stem cells. *Biology of Reproduction*, *82*, 1103–1111.

Yeh, J. R., Zhang, X., & Nagano, M. C. (2007). Establishment of a short-term *in vitro* assay for mouse spermatogonial stem cells. *Biology of Reproduction*, *77*, 897–904.

Yeh, J. R., Zhang, X., & Nagano, M. C. (2011). Wnt5a is a cell-extrinsic factor that supports self-renewal of mouse spermatogonial stem cells. *Journal of Cell Science*, *124*, 2357–2366.

Yeh, J. R., Zhang, X., & Nagano, M. C. (2012). Indirect Effects of Wnt3a/β-Catenin Signalling Support Mouse Spermatogonial Stem Cells In Vitro. *PLoS One*, 7, e40002.

Ying, Q. L., Wray, J., Nichols, J., Batlle-Morera, L., Doble, B., Woodgett, J., et al. (2008). The ground state of embryonic stem cell self-renewal. *Nature*, *453*, 519–523.

Yoshida, S. (2010). Stem cells in mammalian spermatogenesis. *Development, Growth & Differentiation*, *52*, 311–317.

Yoshida, S., Sukeno, M., & Nabeshima, Y. (2007). A vasculature-associated niche for undifferentiated spermatogonia in the mouse testis. *Science*, *317*, 1722–1726.

Zhang, X., Ebata, K. T., & Nagano, M. C. (2003). Genetic analysis of the clonal origin of regenerating mouse spermatogenesis following transplantation. *Biology of Reproduction*, *69*, 1872–1878.

Zhang, X., Ebata, K. T., Robaire, B., & Nagano, M. C. (2006). Aging of male germ line stem cells in mice. *Biology of Reproduction*, *74*, 119–124.

Zhou, Q., Li, Y., Nie, R., Friel, P., Mitchell, D., Evanoff, R. M., et al. (2008). Expression of stimulated by retinoic acid gene 8 (Stra8) and maturation of murine gonocytes and spermatogonia induced by retinoic acid in vitro. *Biology of Reproduction*, *78*, 537–545.

Zhou, Q., Nie, R., Li, Y., Friel, P., Mitchell, D., Hess, R. A., et al. (2008). Expression of stimulated by retinoic acid gene 8 (Stra8) in spermatogenic cells induced by retinoic acid: An *in vivo* study in vitamin A-sufficient postnatal murine testes. *Biology of Reproduction*, *79*, 35–42.

Zhou, S., Schuetz, J. D., Bunting, K. D., Colapietro, A. M., Sampath, J., Morris, J. J., et al. (2001). The ABC transporter Bcrp1/ABCG2 is expressed in a wide variety of stem cells and is a molecular determinant of the side-population phenotype. *Nature Medicine*, *7*, 1028–1034.

Zohni, K., Zhang, X., Tan, S. L., Chan, P., & Nagano, M. C. (2012). The efficiency of male fertility restoration is dependent on the recovery kinetics of spermatogonial stem cells after cytotoxic treatment with busulfan in mice. *Human Reproduction*, *27*, 44–53.

CHAPTER FOUR

Germline Stem Cells

Ji Wu*,†,1, Huacheng Luo*,†, Hu Wang*,†

*Key laboratory for the Genetics of Developmental and Neuropsychiatric Disorders (Ministry of Education), Bio-X Institutes, Shanghai Jiao Tong University, Shanghai, China

†School of Life Sciences and Biotechnology, Shanghai Jiao University, Shanghai, China

[1]Corresponding author: e-mail address: jiwu@sjtu.edu.cn

Contents

Abstract

Stem cells support tissue maintenance by balancing self-renewal and differentiation. Germline stem cells (GSCs) belong to a kind of stem cells that are crucial to development and tissue homeostasis with their ability to both differentiate into gametes and maintain an undifferentiated state by their selfrenewal property. The GSCs are unique in that they transmit genetic information from parent to offspring. In this review, we concentrate on human and mammalian models including rat and mouse, which are used to study GSCs, and summarize current progress about the GSC origin, location, characteristics, regulatory mechanisms, application, and role of GSCs in gametogenesis.

Stem cells support tissue maintenance by balancing self-renewal and differentiation. Germline stem cells (GSCs) belong to a kind of stem cells which are crucial to development and tissue homeostasis with their ability to both differentiate into mature cell type (gamete) and maintain an undifferentiated state by their selfrenewal property. This balance can maintain the stem cell

Current Topics in Developmental Biology, Volume 102
ISSN 0070-2153
http://dx.doi.org/10.1016/B978-0-12-416024-8.00004-0

pool and need interact with its surrounding microenvironmental cells named "niche." Additionally, these unique features can be used in tissue regeneration and organ transplantation in clinical medicine. A significant property of GSCs is that they are unique in that they transmit genetic information from parent to offspring.

1. SPERMATOGONIAL STEM CELLS

1.1. Introduction

Spermatogonial stem cells (SSCs) can self-renew and produce daughter cells differentiated into spermatozoa throughout life (Nagano et al., 2001). According to their shape and property, the SSCs can be subdivided into Asingle (As) spermatogonia, Apaired (Apr) spermatogonia, and Aaligned (Aal) spermatogonia (Wong, Jin, & Xie, 2005). Besides that SSCs could be differentiated into sperm *in vitro* to treat infertility patients via the *in vitro* fertilization and embryo transfer technique (He, Kokkinaki, Jiang, Dobrinski, & Dym, 2010), more and more researches have recently reported that the SSCs can be converted into a kind of pluripotent cells like embryonic stem (ES) cells under appropriate culture conditions (Conrad et al., 2008; Guan et al., 2006; Kanatsu-Shinohara, Inoue, et al., 2004; Kossack et al., 2009). Some reports had showed that the application of human SSCs had provided new and promising therapeutic prospects for regenerative medicine without moral and ethical issues because of human ES cell limitation (Geijsen & Hochedlinger, 2009; Payne & Braun, 2008). Moreover, SSCs have been considered as a target for a male contraceptive. It is possible to preferentially target the SSCs with some small molecule RNAs (including siRNA, microRNA inhibitors, or piRNA) to selectively regulate SSC differentiation and development, thus obtaining a novel male contraceptive method (Cheng & Mruk, 2012). In addition, SSCs can be used to study the regulation mechanisms of stem cell self-renewal and differentiation both *in vitro* and *in vivo* besides generation of transgenic animals using SSCs (He et al., 2008).

1.2. The origin of SSCs

SSCs come from gonocytes in the postnatal testis, which are derived from primordial germ cells (PGCs) during fetal development. PGC is a specific cell population that arises around the gastrulation in the pluripotent proximal epiblast stage embryo at about 7–7.25 days postcoitum (dpc) (Ginsburg, Snow, & McLaren, 1990). Then the PGCs will reach at the indifferent gonad between 8.5 and 12.5 dpc in mice via the hindgut. In the male gonad, PGCs will migrate in the gonad and start to give rise to gonocytes surrounded by sertoli precursor cells and peritubular myoid cells at about 13.5 dpc (Phillips, Gassei, & Orwig, 2010). Generally, gonocyte has been subdivided into mitotic (M)-prospermatogonia, T1-prospermatogonia, and T2-propsermatogonia (McLaren, 2003). M-prospermatogonia is far away from the basal membrane and located in the center of the testicular cords. Afterward, they will become T1-prospermatogonia and arrest at G0 mitotic stage about 16.5 dpc of mouse development (McLaren, 2003). During a few days after birth, these gonocytes will transform to T2-prospermatogonia which will migrate into the seminiferous tubule basement membrane. Then T2-prospermatogonia divides to form undifferentiated type A spermatogonia (Yoshida, Sukeno, & Nabeshima, 2007). Therefore, SSCs derive from the undifferentiated precursor cells named gonocytes, which arise from PGCs immediately migrating to the genital ridge and taking a critical part in the formation of the embryonic gonad. The formation of SSCs from the gonocytes will take place from 0 to 6 days postpartum (dpp) in the male rodent testes (Hamra, Chapman, Wu, & Garbers, 2008).

1.3. The location, isolation, self-renewal, and pluripotency of SSCs

SSCs are generally located in a defined anatomical niche. In mice, SSCs reside at the basement membrane of seminiferous tubules in the testis (Brinster, 2002). SSCs are rare and occupy only about 0.02–0.03% of the total testis cell population (Tegelenbosch & de Rooij, 1993). However, two enzymatic digestion steps and magnetic-activated cell sorting (MACS) can efficiently isolate and purify SSCs from neonatal and adult testicular tissues and culture for a long time maintaining their spermatogenic properties (Kanatsu-Shinohara, Inoue, et al., 2004; Kubota, Avarbock, & Brinster, 2004; Xiong et al., 2011; Yuan, Hou, & Wu, 2009). Guan et al. had demonstrated that SSCs could be isolated from 6- to 8-day-old male mice (Guan et al., 2006). Seandel et al. have isolated and purified SSCs from testes of 4- to 6-week-old mice using SSC-specific expression gene marker sorting (Seandel et al., 2007). Mizrak and de Rooij had obtained the SSCs by isolating SSCs from testes of 3- to 35-week-old mice (de Rooij & Mizrak, 2008). Additionally, some researchers had suggested that SSCs can be isolated from adult rat testes (Hamra et al., 2008; van Pelt et al., 1996). Also, a series of recent papers have reported that SSCs could be raised from adult human testes (Conrad et al., 2008; He et al., 2010; Mirzapour et al., 2011).

An important property of stem cells including SSCs is their ability to self-renew. *In vitro* cultures of SSCs can contribute to self-renewing division of SSCs, which will grow logarithmically up with the capacity to maintain the spermatogenic activities in the mouse testes (Kanatsu-Shinohara et al., 2003; Ogawa et al., 2004). There are some reports which suggested that some important biological signaling factors, glial cell line-derived neurotrophic factor (GDNF)/ETV5 and short-type PB cadherin, could promote self-renewal of mouse and rat SSCs *in vitro* and *in vivo* (Wu, Goodyear, Tobias, Avarbock, & Brinster, 2011; Wu, Jester, & Orth, 2005, Wu et al., 2008). Oatley and Brinster had indicated that SSCs can self-renew *in vivo* in the specific niche or *in vitro* culture in the appropriate conditions in mammals. Also, Oatley and Brinster had found that mammal SSCs can self-renew undergoing the regulation (Oatley & Brinster, 2008). Dann et al. and Ryu et al. suggested that SSCs could self-renew with the requirement of OCT4 and depended on the effects of aging and niche microenvironment (Dann et al., 2008; Ryu, Orwig, Oatley, Avarbock, & Brinster, 2006). In addition, numerous researches have reported that there exists some molecular mechanism applied to explain the self-renewal of SSCs (Oatley, Avarbock, Telaranta, Fearon, & Brinster, 2006).

Under defined culture conditions, SSCs can be reprogrammed into embryonic stem cell (ESC)-like pluripotent stem cells termed ESC-like cells (ESC-LCs) that can differentiate into cells of all three germ layers. Shinohara et al. reported the establishment of ESC-LCs from cultured neonatal mouse SSCs (Kanatsu-Shinohara, Inoue, et al., 2004). Subsequently, Guan et al. demonstrated that adult mouse SSCs can dedifferentiate into ESC-LCs (Guan et al., 2006). Using growth factor-mediated neural lineage induction, ESC-LCs can differentiate into neural precursor cells. Further, Guan et al. reported the generation of functional cardiomyocytes from ESC-LCs derived from adult mouse SSCs (Guan et al., 2006). Recently, Conrad et al. and Kossack et al. showed that adult human SSCs can generate ESC-LCs that are pluripotent and comparable with that of mouse ESC-LCs derived from SSCs based on developmental potential (Conrad et al., 2008; Kossack et al., 2009) (Fig. 4.1).

1.4. Role of SSCs in spermatogenesis

The initiation process of spermatogenesis is critical to maintain the long-term source of sperm. However, the formation of the SSC population and associated microenvironment named "niche" in the testis is critical to

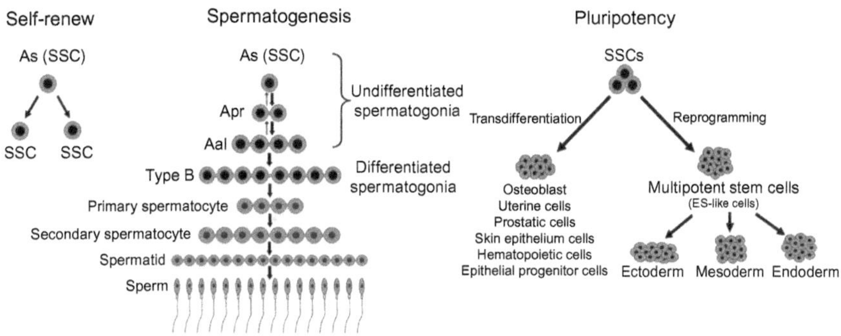

Figure 4.1 The process of self-renewal, spermatogenesis, and pluripotency of SSCs. (For color version of this figure, the reader is referred to the online version of this chapter.)

the initiation of spermatogenesis. Some reports had described that a significant increase of germ cells in the testis due to the mitotic activity of SSCs is because an SSC needs to undergo 8–9 divisions on average before differentiating into a spermatocyte (Lok & de Rooij, 1983; Tegelenbosch & de Rooij, 1993). Subsequently, a spermatocyte will proceed with meiotic divisions for further differentiation. Some experiments had confirmed that ultimately a spermatogonium is able to produce 2048 or sometimes 4096 spermatozoa, but it should depend on the number of mitosis and meiosis (Russell, Ettlin, Hikim, & Clegg, 1993). In addition, the numbers of germ cells can be regulated and accomplished in the spermatogonial population of cells, and the appropriate ratio of germ cells and Sertoli cells should be provided for maintaining the process of spermatogenesis (De Rooij & Lok, 1987). Several studies have showed that the spermatogenesis from As to A1 spermatogonia has referred to several regulatory signaling molecules, including stem cell factor (SCF)/c-kit receptor signaling pathway (de Rooij, Okabe, & Nishimune, 1999), some specific transcription factors (Bettegowda & Wilkinson, 2010), testosterone levels (Shuttlesworth et al., 2000), and retinoic acid (RA) (Van Pelt & De Rooij, 1990) (Fig. 4.1).

1.5. Expression of the specific genes in SSCs

It is rather necessary to study the expression of some specific genes in SSCs in order to maintain SSC self-renewal, spermatogenesis, or reprogramming into pluripotency. In rodents, a6-integrin (CD49F), b1-integrin (CD29), and THY-1 (CD90) are surface markers which are mainly expressed at mammal spermatogonial stem/progenitor cells (Kubota, Avarbock,

& Brinster, 2003; Shinohara, Avarbock, & Brinster, 1999). Also, Kanatsu-Shinohara et al. confirm that CD9 is a specific marker expressed at the mouse and rat spermatogonial stem/progenitor cells (Kanatsu-Shinohara, Toyokuni, & Shinohara, 2004). Some studies indicated that GFRA1 and RET are coreceptors for GDNF and expression markers for spermatogonial stem/progenitor cells (Hofmann, Braydich-Stolle, & Dym, 2005; Naughton, Jain, Strickland, Gupta, & Milbrandt, 2006). In humans, a6-integrin is expressed in spermatogonia and was applied to isolate and purify SSCs by MACS (Conrad et al., 2008). Based on several researches, some specific cell surface markers can also be expressed in the rodent, such as GFRa1, PLZF (Zbtb16), and CDH1 (Buaas et al., 2004; Costoya et al., 2004; Fujita et al., 2005; Kanatsu-Shinohara, Toyokuni, et al., 2004; Shinohara et al., 1999). Transgenic and conditional approaches can well identify that neurogenin 3 (Ngn3) is expressed at the early stage of spermatogonia (Yoshida et al., 2007). Sada et al. show that Nanos2 is expressed at the early stage of SSCs (Sada, Suzuki, Suzuki, & Saga, 2009). In addition, transgenic models demonstrated that Stra-8 (stimulated by RA-8) was expressed at undifferentiated spermatogonia (Giuili et al., 2002; Guan et al., 2006). Through the knockout and overexpression rodent models, researchers suggested that GDNF and its receptor GFRa1 play an important role in SSC self-renewal (Meng et al., 2000). Up to now, expression of CD49f (a6-integrin), CD133, GFR-a1, GPR-125, MAGE-4, PLZF (Zbtb16), SSEA-4, and CD90 (THY1) in human SSCs has been reported (Conrad et al., 2008; Costoya et al., 2004; Dym, Kokkinaki, & He, 2009; Izadyar et al., 2011) (Table 4.1).

1.6. The regulatory mechanisms of SSC self-renewal, pluripotency, and spermatogenesis in mammals

1.6.1 The key regulatory genes in SSC development

Recently, several genetic studies indicate that both transcriptional and translational regulation play an important role in controlling the mouse SSC development. Without expression of the gene encoding PLZF transcription factor would lead to an age-dependent depletion of SSCs (Buaas et al., 2004; Costoya et al., 2004). Some other transcription factors including ETV5, LHX1, TAF4b, BCL6b, KIT, and KITL have also been implicated in the regulation of SSC self-renewal (Falender et al., 2005; Matzuk & Lamb, 2008; Oatley et al., 2006). Several studies also suggested that the integrin alpha 6, THY1 (also CD90), and ubiquitin carboxyl-terminal hydrolase L1 molecules had been considered as specific surface markers for mouse SSCs and progenitor stem cells to maintain the proliferation of these cells

Table 4.1 Expression of specific SSC markers in the rodent/human testis

Specific gene markers	Expression at rodent (R) or human (H)	Undifferentiated spermatogonia	Differentiated spermatogonia	References
GFR-α1	R and H	+		Meng et al. (2000) and Conrad et al. (2008)
GPR125	R and H	+		Seandel et al. (2007) and Dym et al. (2009)
KIT	R		+	Yoshinaga et al. (1991) and Rajpert-De Meyts et al. (2003)
MAGE-4	H	+		Rajpert-De Meyts et al. (2003)
CDH1 (CD324)	R	+		Tokuda et al. (2007)
b1-integrin (CD29)	R and H	+	+	Schaller et al. (1993) and Shinohara et al. (1999)
a6-integrin (CD49f)	R and H	+	+	Shinohara et al. (1999) and Conrad et al. (2008)
CD9	R	+	+	Kanatsu-Shinohara, Toyokuni, et al. (2004)
Thy-1 (CD90)	R and H	+		Kubota et al. (2003) and Conrad et al. (2008)
RET	R	+		Naughton et al. (2006)
Blimp-1	R and H	+		Skutella (2009)
Dazl	R and H	+	+	Cooke et al. (1996)
Mvh (Ddx4)	R and H	+	+	Fujiwara et al. (1994)
Fragilis (Ifitm3)	R and H	+		Huang et al. (2009)
Pou5 fl (Oct4)	R and H	+		Dann et al. (2008) and Pesce et al. (1998)
Esg-1 (Dppa5a)	R and H	+		Guan et al. (2006)

Continued

Table 4.1 Expression of specific SSC markers in the rodent/human testis—cont'd

Specific gene markers	Expression at rodent (R) or human (H)	Undifferentiated spermatogonia	Differentiated spermatogonia	References
Plzf (Zbtb16)	R and H	+		Costoya et al. (2004)
Rex 1 (Zfg42)	R and H	+	+	Guan et al. (2006), Kanatsu-Shinohara, Inoue, et al. (2004), and Kanatsu-Shinohara, Toyokuni, et al. (2004)
Utfl	R and H	+		van Bragt et al. (2008)
STRA8	R and H	+	+	Giuili et al. (2002)
Nanos2	R and H	+		Sada et al. (2009)
Utp14b (jsd)	R	+		Boettger-Tong et al. (2000)
Bc16b	R and H	+		Oatley et al. (2006)
Lhx1	R and H			Oatley et al. (2007)
GCNA1	R and H	+	+	Enders and May (1994)
Epcam	R and H	+	+	Tokuda et al. (2007)
CD24	R and H	+		Kubota et al. (2003)
Nanos3	R and H	+		Suzuki et al. (2009) and Tsuda et al. (2003)
EGR3	R and H	+		Hamra et al. (2004)
Ngn3	R and H	+	+	Raverot et al. (2005) and Yoshida et al. (2007)
Sox3	R and H	+		Raverot et al. (2005)
TAF4B	R and H	+	+	Falender et al. (2005)
Numb	R and H	+	+	Corallini et al. (2006)
Lrp4	R and H	+	+	Yamaguchi et al. (2006)
Sall4	R and H	+		Wang et al. (2001)
Lin28 (Tex 17)	R and H	+		Zheng et al. (2009)
Sohlh1	R and H	+	+	Ballow et al. (2006a)
Sohlh2	R and H	+		Ballow et al. (2006b)

(He et al., 2010; Kubota et al., 2003; Shinohara et al., 1999). Recently, G protein-coupled receptor 125 (GPR125) is a potential stem and progenitor cell surface marker (GPR125) expressed on the adult testis, and its expression is restricted in undifferentiated spermatogonia cells to promote its proliferation (Seandel et al., 2007). Also, Kristensen et al. indicated that the transcription factor UTF1 may be involved in selfrenewal of SSCs in human testis (Kristensen et al., 2008). Some recent advances have put forward that mTORC1 can resist the GFRa1/c-Ret signaling pathway, which will inhibit the PI3K/AKT signaling pathway, leading to a change in balance from the process of SSC self-renewal to differentiation (Hobbs, Seandel, Falciatori, Rafii, & Pandolfi, 2010). However, mTORC1 levels had been considered to be negatively regulated by REDD1 expression, which is positive regulation by PLZF. In this case, mutation of PLZF will result in an increase of mTORC1 activity, which will enhance the antagonization of GFRa1/c-Ret signaling pathway, and decrease of SSC self-renewal (Costoya et al., 2004). Zhang et al. have reported that retinol (vitamin A) has a property to maintain self-renewal of SSCs in adult mouse testis (Zhang, Sun, et al., 2011). Moreover, BMP signaling factors, including multiple BMPs, BMP4, BMP7, and BMP8, play an important role in SSC maintenance and are also required for SSC self-renewal (Puglisi, Montanari, Chiarella, Stefanini, & Boitani, 2004; Zhao, Chen, Liu, Xu, & Qi, 2001).

Seminiferous tubules are composed of sertoli cells, leydig cells, and germ cells. SSCs including all kinds of spermatogonia colonize at the basal compartment, reside at the basal lamina of the epithelium, and have a close interaction with sertoli cells. Therefore, there are some essential signal pathways or signaling molecules from sertoli cells which can conduct the proliferation and differentiation of SSCs. GDNF, which is secreted from Sertoli cells, has referred to SSC self-renewal and proliferation (Meng et al., 2000). *In vivo*, lack of GDNF signaling molecules will lead to the age-dependent depletion of the SSC compartment (Meng et al., 2000). *In vitro* experiment had indicated that GDNF could contribute to the proliferation of SSCs by promoting their survival or self-renewal when adding GDNF (Tadokoro, Yomogida, Ohta, Tohda, & Nishimune, 2002). Also, the proliferation of SSCs depends on the interaction with GDNF by directly binding to its coreceptor c-Ret and its receptor GFRa1, which is a surface marker for SSCs and progenitor stem cells in the mouse testis (Meng et al., 2000; Naughton et al., 2006). Gfra1 silencing in mouse SSCs will lead to their differentiation via repressing the expression of RET tyrosine kinase (He, Jiang, Hofmann, & Dym, 2007; Oatley et al., 2006). Oatley and

Brinster had found that the KIT (CD117) ligand–receptor system and the niche factor, GDNF, may be involved in the process of SSC self-renewal (Oatley & Brinster, 2008). Numerous recent advances had indicated that some extrinsic signals also took a significant part in the division or differentiation of SSCs in mice. The SCF secreted by Sertoli cells can activate c-Kit (CD117) expression, which is a tyrosine kinase receptor for SCF, and can regulate the differentiation of SSCs (Ohta, Yomogida, Dohmae, & Nishimune, 2000; Yoshinaga et al., 1991). Caires et al. had pointed out that systemic endocrine and local cues will promote somatic niche cells to secrete some essential factors contributing to the homeostasis of SSCs in mammals (Caires, Broady, & McLean, 2010). Importantly, Sertoli cells can clearly affect the development of all SSCs in the epithelium and obviously play both instructive and permissive parts in the process of spermatogenesis. Therefore, promotion of the SSC population should contain the establishment and maintenance of a niche microenvironment in the seminiferous tubules of the testis (Greenbaum, Iwamori, Buchold, & Matzuk, 2011).

1.6.2 Epigenetic control in SSC development

More and more researches had paid attention to the epigenetic regulation mechanisms involved in the maintenance and reprogramming of the SSCs in various species, especially in health and disease fields. The epigenetic analysis includes DNA methylation/demethylation and modification of histones such as methylation/demethylation, acetylation/deacetylation, and phosphorylation/dephosphorylation (Jenuwein & Allis, 2001). During mammalian development, two periods (gametogenesis and early embryogenesis) are characterized by epigenetic reprogramming. In spermatogenesis, global sex-specific reprogramming of the epigenome occurrence will appear followed by DNA methylation and chromatin modifications (Holliday, 1989). Spermatogenesis is characterized by a particularly spectacular chromatin remodeling central to the epigenetic reprogramming process, in which somatic linker histones are sequentially replaced by testis-specific variants, followed by the replacement of most histones with protamines (Kimmins & Sassone-Corsi, 2005). The formation of spermatozoa from SSCs in a male's lifetime is due to the differentiation process controlled by particular transcriptional programs and widely chromatin remodeling modification (Kimmins & Sassone-Corsi, 2005). Genomic imprinting, including paternal and maternal genomes, is an epigenetic mechanism which plays a critical role in development, growth, and behavior of mammalian animals (Surani,

Barton, & Norris, 1984). Several recent studies have suggested that genomic imprinting memory will be eliminated in PGCs and reestablished during gametogenesis (Reik & Walter, 2001). As an important epigenetic mechanism, imprinted genes play a central role in controlling the transcription of many kinds of genes. Up to now, about 150 imprinted genes have been identified in mice (see http://www.mousebook.org) and 212 imprinted genes have been detected in humans (http://www.geneimprint.com/site/genes-by-species) (Reik & Walter, 2001). Several researches have reported that three important paternally imprinted domains, named H19, *Meg3IG,* and *Rasgrf1,* will be methylated during spermatogenesis in mammals (Li, Lees-Murdock, Xu, & Walsh, 2004). SSCs show an androgenetic DNA methylation pattern in animals, while it will perform a hypermethylation in DMRs of paternally imprinted regions including *H19* and *Meg3IG,* and show a hypomethylation in DMRs of maternally imprinted regions such as *Igf2r, Snrpn,* and *Peg10* (Kanatsu-Shinohara, Inoue, et al., 2004). Interestingly, ESC-LCs derived from SSCs showed distinct methylation patterns from those of SSCs, and the demethylation level of ESC-LCs in the DMRs of paternally imprinted genes such as *H19* and *Meg3* had been identified (Kanatsu-Shinohara, Inoue, et al., 2004). This result has indicated that the transformation of DNA methylation patterns may play a significant role in the reprogramming mechanisms to multipluripotency if conversion from SSCs to ESC-LCs takes place (Lee & Shinohara, 2011).

The process of spermatogenesis is precisely conducted. There is no doubt that subtle changes involved in epigenetic mechanisms in the development of SSCs would lead to disorder or infertility. Epigenetic reprogramming, which refers to the erasing and resetting of DNA methylation at imprinted regions, TEs, and repeat sequence domains of the genome, plays a significant role in the process of spermatogenesis and fertilization (Morgan, Santos, Green, Dean, & Reik, 2005). DNMT3L contributed to the methylation of some imprinted genes is referred to the silencing of TEs in the testis, as lack of DNMT3L will result in a distinct demethylation level at LINE-1 and the intracisternal A-particle transposons (Bourc'his & Bestor, 2004). Deficient expression of Dnmt3a and Dnmt3b in SSCs would induce an obvious variation in methylation patterns at paternally imprinted genes (Kato et al., 2007). In addition, lack of genomic methylation would occur in heterozygous or wild-type embryos during Dnmt1 deletion. Further, abnormal methylation patterns could be found at maternally and paternally imprinted regions in Dnmt1-deficient embryos (Li, Bestor, & Jaenisch, 1992).

Recently, Godmann et al. have shown that KDM1 (also named LSD1/AOF2/BHC110) is extensively expressed in the spermatogonia and performs some critical function in a repressor complex including histone deacetylase 1 (Godmann et al., 2007). Through inactivating mono- and dimethylation labels from H3K4, KDM1 would lead to the inhibition of androgen receptor target genes (Shi et al., 2004). Moreover, the results from Hofmann et al. affirmed that, by repression expression of chromatin modifiers including KDM1 and HDAC, the epigenetic profile of an SSC line would change obviously (Hofmann, Narisawa, Hess, & Millan, 1992). In addition, the histone H3 methylation and acetylation significantly mediated by HDAC-containing protein complex play a critical role in the regulation of gene expression of SSCs (Hayashita et al., 2005). For instance, expression of *Pou5f1* and *Gfra1* in SSCs was obviously regulated by the chromatin modification (Godmann, May, & Kimmins, 2010).

As described above, in the process of spermatogenesis, TP (transition protein) and PRM (protamine) function can be regulated because of epigenetic modifications. TP1 (transition protein 1)- or TP2-deficient mice show some slight abnormalities in chromatin condensation during the process of spermatogenesis (Zhao, Shirley, et al., 2001). Loss of histone demethylase Jhdm2a affecting the Tnp1 (transition nuclear protein 1) and Prm1 (protamine 1) transcription induced the chromatin condensation defects including the process of spermatogenesis with abnormal nuclear structure and elongation of spermatids (Okada, Scott, Ray, Mishina, & Zhang, 2007). Although most of the epigenetic modifications involved in regulating some important development properties of SSCs have been observed, there are still lots of questions to be clarified in the future.

1.6.3 Small RNAs in regulation of SSC development

MicroRNAs (miRNAs) is a family of small and short noncoding RNAs (about 19–23 nt), which play a critical part in posttranscriptionally regulating the target gene expression with mainly binding to the 3′-untranslated region (3′UTR) of mRNAs, including regulatory genes of self-renewal, cellular differentiation, proliferation, and pluripotency (Bartel, 2009). Moreover, miRNAs have been considered as important regulators of the process of spermatogenesis. A study has showed that the Mirlet7 family miRNAs are highly expressed in SSCs, which can be significantly induced by RA through repression of Lin28 expression. Both *in vitro* and *in vivo* studies confirmed that Mirlet7 could downregulate its target gene expression, including *Mycn*, *Ccnd1,* and *Col1a2* during the process of spermatogenesis.

Therefore, Mirlet7 family miRNAs have a critical function in RA-induced SSC differentiation (Tong, Mitchell, Evanoff, & Griswold, 2011). Recently, Lian et al. have suggested that miR-383 may affect at the early stage of spermatogenesis and thus may perform a significant regulation in the germ cell proliferation (Lian et al., 2010). Also, miR-21, along with miR-34c, -182, -183, -146a, -465a-3p, -465b-3p, -465c-3p, and -465c-5p, is preferentially expressed in the Thy1$^+$ SSC-enriched population. Importantly, transient inhibition of miR-21 in the SSC cultures increased the number of SSCs undergoing apoptosis and significantly reduced the number of donor-derived colonies of spermatogenesis, indicating that miR-21 is important in maintaining the SSC population (Kotaja et al., 2006). Further, miR-21 is functionally significant in maintaining the SSC population *in vitro*, interacting with the transcription factor ETV5, known to be critical for SSC self-renewal, which is a direct regulator of miR-21 expression (Niu et al., 2011). Wu et al. had provided some evidences that miR-184 could negatively control the expression of nuclear receptor corepressor 2 (Ncor2) by targeting its 3′UTR and repressing the NCOR2 protein expression. The result apparently indicated that miR-184 could engage in the post-transcription regulation of mRNAs, especially in the process of mammalian spermatogenesis (Wu, Bao, Wang, Hu, & Xu, 2011). Hayashita et al. reported that miR-17–92 and miR-290–295 clusters were highly expressed in developing germ cells. miR-17-92 cluster has been considered to contribute to the cell cycling, and miR-290 to -295 (also miR-290–295 cluster) had been extensively found in PGCs and SSCs (Hayashita et al., 2005). let7, miR-125a, and miR-9 have all been detected to regulate the expression of LIN28, a critical regulator of ESC-LCs (Zhong et al., 2010). In addition, miR-290 and the miR-302 cluster have been found to be highly expressed and involved in the maintenance of ESC-LC population (Zovoilis et al., 2008). Yan et al. demonstrated that miR-34b, miR-34c, and miR-449 were highly expressed in testis, which could target the NOTCH1 and BCL2 signaling regulating the germ cell differentiation and survival, respectively (Yan et al., 2009). Additionally, miR-18a could directly regulate the HSF2 (heat shock factor 2) transcription factor, which mainly controls the expression of many genes for the successful process of spermatogenesis (Bjork, Sandqvist, Elsing, Kotaja, & Sistonen, 2010). *In vitro* or *in vivo* researches have suggested that the members of Mir-17-92 and Mir-106b-25 family are significantly negatively regulated during RA-induced spermatogonial differentiation (Tong, Mitchell, McGowan, Evanoff, & Griswold, 2012) (Table 4.2).

Table 4.2 Regulation of miRNA molecules in the development of SSCs

miRNAs	Function	Predicted or confirmed targeting genes	References
Mirlet7	Spermatogenesis	*Mycn*, *Ccnd1*, and *Colla2*	Tong et al. (2011)
miR-383	Spermatogenesis	*Irf1*	Lian et al. (2010)
miR-21	Self-renewal	*Etv5*, *P12*	Niu et al. (2011)
miR-34c	Spermatogenesis	*Klf4*, *Notch1*, *Ccng1*, *Ccnb1*, *Ccnc*, *Ccne1*, *Cdk4*, and *Cdk6*	Bouhallier et al. (2010) and Yan et al. (2009)
miR-184	Spermatogenesis	*Akt3*, *Ncor2*	Niu et al. (2011) and Wu, Bao, et al. (2011)
miR-17-92	Self-renewal	*Stat3*, *E2f1*	Hayashita et al. (2005)
miR-290-295	Pluripotency	*Fbx15*	Hayashita et al. (2005)
miR-125a	Pluripotency	*Lin28*	Zhong et al. (2010)
miR-9	Pluripotency	*Lin28*	Hayashi et al. (2005) and Zhong et al. (2010)
miR-302	Pluripotency	*Nr2f2*	Zovoilis et al. (2008)
miR-34b	Differentiation	*Notch1*, *Lgr4*	Vogt et al. (2011) and Yan et al. (2009)
miR-449	Differentiation	*Bcl2*, *Notch1*	Yan et al. (2009)
miR-214	Spermatogenesis	*Tgif2*, *Notch2*	Luo et al. (2010)
miR-18a	Spermatogenesis	*Hsf2*	Bjork et al. (2010)
miR-10a	Differentiation	*Ncor2*	Niu et al. (2011)
miR-124a	Pluripotency	*Slug*, *Iqgap1*, and *Itgb1*	Yan et al. (2009)

1.6.4 piRNA in regulation of SSC development

piRNA (about 30 nt small RNA) can be derived from mammals' reproductive cells, which combined with Piwi subfamily proteins to form the piRNA complex protein (named piRC). Recently, study of the piRNA function has made significant progresses, especially in the regulation process of the spermatogenesis in the testis. Moreover, piRC can regulate the gene transcription and inhibit abnormal gene expression when spermatogenesis is proceeding. Alternatively, mammal Piwi subfamily proteins MIWI, MIWI2, and MILI are mainly present in the middle and the later period of spermatogonia, which play a significant role in spermatogenesis (Carmell et al., 2007; Deng & Lin, 2002). Grivna et al. reported that MIWI protein was expressed in spermatocyte and around spermatozoa, which had an important role on specific gene translation and maintaining the process of spermatogenesis (Grivna, Pyhtila, & Lin, 2006). Several recent advances had showed that piRNAs are present in the spermatogonia and play a critical part in the process of spermatogenesis. piRNA main expression in pachytene spermatocyte and round spermatozoa stage. Therefore, piRNAs mainly involve in the regulation in and after meiosis process, which carry out the function to inhibit the retrotransposons in the process of spermatogenesis (Aravin, Sachidanandam, Girard, Fejes-Toth, & Hannon, 2007; Carmell et al., 2007).

1.7. The applications of SSCs

Human ES cells could differentiate into tissues derived from all three germ layers, and the developmental potential had demonstrated that these cells could have clinical applications in the treatment of diabetes, Parkinson's disease, and other serious illnesses reported by Evans and Martin (Evans & Kaufman, 1981; Martin, 1981). However, recently, more and more research has suggested that the potential application of ES cells in regenerative medicine has caused lots of serious controversies including moral, legal, and ethical issues. Although numerous studies have demonstrated that the induced pluripotent cells have some potential in regenerative medicine, acquisition of these cells involves genetic modification and there are some critical obstacles before applied to the regenerative medicine and clinical research. Therefore, using appropriate approaches for isolation and transformation of adult human SSCs into other cell types, SSCs could be considered as an important source of pluripotent cells applied to human tissue repair and regeneration medicine (Simon, Hess, & Cooke, 2010).

The potential of SSCs which can perform spermatogenesis in some certain environment, but it does not mean that they have a potential ability to be limited to germ cell development. Izadyar et al. have revealed that there are two distinct populations of SSCs: one is multipotent cells with $OCT4^+$ and $c\text{-}KIT^+$, and another carries out the process of spermatogenesis with $OCT4^+$ and $c\text{-}KIT^-$ (Izadyar et al., 2008). A study had showed that SSCs could directly transdifferentiate into prostatic, uterine, and skin epithelium when recombined with the appropriate mesenchyme and transplanted *in vivo* (Simon et al., 2009). Moreover, Boulanger et al. had transplanted a mixture of mouse testis and mammary cells into the mammary fat pad and found the SSCs directly converting into the cells with the mammary epithelial progenitor cell characteristics (Boulanger, Mack, Booth, & Smith, 2007). Also, Ning et al. showed that SSCs could represent morphologic and functional properties of hematopoietic cells after transplantation into bone marrow in mice (Ning et al., 2010). Therefore, SSCs can transdifferentiate into some nonspermatogenic cell types when placed directly into an environment that will contribute to a particular differentiation pathway (de Rooij & Mizrak, 2008).

In addition, several studies have reported that SSCs can be isolated from the testis and induced in culture and converted to ESC-LCs with the capability to differentiate into multiple differentiated cell lineages (Guan et al., 2006; Kanatsu-Shinohara, Inoue, et al., 2004; Seandel et al., 2007).

The essential function of SSCs can be self-renewal and providing the developing gametes that are continuously produced from the testes. For humans, the testicular biopsy would be contributed to the SSCs for cryopreservation, storage, and transplantation. So the SSCs from affected individuals could offer the donor cells which may be transplanted back into the patient and also successfully recover fertility (He et al., 2010). Obviously, maintenance of fertility is an important aspect for SSC application (Kanatsu-Shinohara et al., 2005). Moreover, there is a potential application of SSCs to recover fertility in some reproduction-deficient patients (Brinster, 2007; Kanatsu-Shinohara et al., 2005; Ryu et al., 2006).

In addition, SSCs have been considered as a powerful tool for the production of transgenic and knockout animals, including animal model with targeting genomic mutations and genetic modification (Hamra et al., 2002; Kanatsu-Shinohara et al., 2003, 2005; Takehashi et al., 2007).

Therefore, abundance of recent advances has showed that the potential of SSCs has been used for biomedical applications in regenerative medicine and provides a model to study gametogenesis. One major advantage of SSCs is that SSCs can convert into ESC-LCs under appropriate culture condition. Thus, the human SSCs or their progeny have a huge potential for cell and tissue engineering-based medical regeneration treated for various human diseases. Moreover, it has been emphasized that the SSCs reprogrammed to pluripotency spontaneously involve in neither the application of exogenous genes nor the interference of retroviruses. Thus, SSCs have great potential to be considered as the most effective and safety approach to give rise to these specific cells which eventually can be applied to the clinical treatment of human diseases (Dym et al., 2009).

2. FEMALE GERMLINE STEM CELLS

2.1. Introduction

GSCs could continuously produce gametes throughout the life of various organisms, but in most mammalian species, long-persisting belief of reproductive biology is that GSCs only exist in postnatal or adult testis of male and that female ovary produced a defined stock of oocytes (Zuckerman, 1951; Zuckerman & Baker, 1977). Although this idea has been persistent for more than half a century, some researchers challenge the view of limited oocyte pool and debate with different voice. Mitotically active germ cells in the early stages of meiotic prophase have been reported in ovaries of adult prosimian primates (Ioannou, 1967). Four years later, similar studies also reported that using cytotoxic agents suggested the postnatal production of new follicles and the possibility of the existence of a GSC pool in the postnatal mouse ovary (Generoso, Huff, & Stout, 1971; Generoso, Stout, & Huff, 1971). Recently, Johnson et al. have shown that GSCs from mouse ovary can regenerate the follicle reserve in postnatal mice (Johnson, Canning, Kaneko, Pru, & Tilly, 2004). Then, Bukovsky et al. claimed that surface epithelium cells may be an origin of adult human germ cells and follicular renewal (Bukovsky, Caudle, Svetlikova, & Upadhyaya, 2004). One year after their report, Johnson et al. claimed that bone marrow is a potential source of germ cells that could sustain oocyte production in adult mice (Johnson et al., 2005). However, this conclusion brings controversies later on (Eggan, Jurga, Gosden, Min, & Wagers, 2006; Gosden, 2004). Data from Kerr et al. support postnatal follicle renewal in postnatal and adult ovaries of mice through the method of healthy follicles' quantification but show no evidence for ovarian GSCs (Kerr et al., 2006). We have succeeded in isolating and purifying the female germline stem cells (FGSCs) from neonatal and adult mice using mouse vasa homolog (MVH)-magnetic bead sorting and located these FGSCs in cortical surface of ovaries. Further, an FGSC line from neonatal mice was established and cultured for more than 15 months with normal karyotype and high telomerase activity. FGSCs from adult mice were also cultured for more than 6 months. These FGSCs of long-term culture maintained their capacity to produce normal oocytes and fertile offspring after transplantation into ovaries (Zou et al., 2009). In order to improve purification efficiency, several germ cell-specific markers were screened. Our results showed that the efficiency of FGSC purification was remarkably enhanced using the germline-specific protein Fragilis, compared with that using MVH (Zou, Hou, Sun, Xie, & Wu, 2011). In addition, we have demonstrated the successful generation of transgenic or gene knockdown mice using FGSCs. The efficiency of gene transfer or gene knockdown was 29–37%, and it took 2 months to produce transgenic offspring (Zhang, Yang, et al., 2011). Using MVH fluorescence-activated cell sorting, White et al. have succeeded in isolating and purifying FGSCs from adult mice and reproductive-age women (White et al., 2012).

2.2. Origin of FGSCs

Mitosis is the process through which a eukaryotic cell separates the chromosomes in its cell nucleus into two identical sets, in two separate nuclei, and results in two daughter cells with a full complement of chromosomes (diploid). Meiosis only exists in germ cell development which yields haploid daughter cell(s). Both spermatogonia and oogonia are capable of expanding by self-renewal through mitosis and differentiation through meiosis.

Development of female reproductive system starts with the migration of PGCs along the hindgut of the fetus to the gonadal ridge; the cells along the gonadal ridge are mostly mesenchymal cells. After arriving at the gonadal ridge, the PGCs are considered oogonia (Durcova-Hills, Wianny, Merriman, Zernicka-Goetz, & McLaren, 2003). The oogonia divide mitotically and differentiate into oocyte that undergoes the meiosis I and are arrested in this phase when complete. Traditional thinking believed that all of oogonia go to meiosis, but now, more and more evidence demonstrated that oogonia or FGSCs still exist throughout reproductive life of various species (Bukovsky et al., 2008) or whole life. Successive stages of differentiation for the female germ cells were shown in Fig. 4.2.

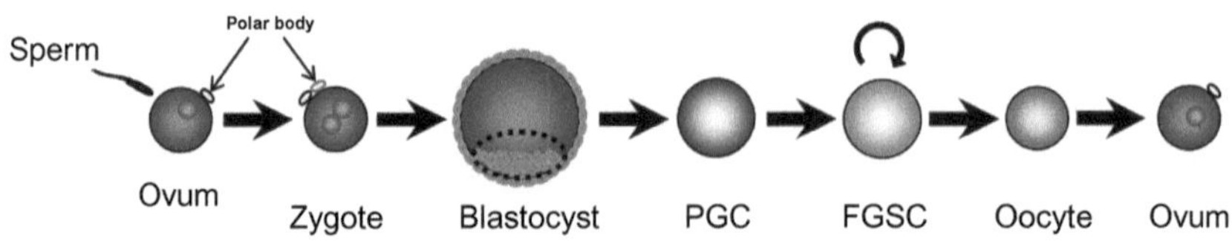

Figure 4.2 Origin and differentiation of FGSCs. PGC, primordial germ cell. (See Color Insert.)

2.3. Isolation, location, and characteristics of FGSCs

The FGSCs can be isolated by two enzymatic digestion steps as the same strategy as that of SSC isolation. Further, FGSCs can be mostly purified with MVH or Fragilis-magnetic bead sorting. For FGSC location, female mice were injected with BrdU and then double stained with BrdU–MVH antibody; the result indicated that FGSCs were present in cortical surface of ovaries (Zou et al., 2011, 2009). Interestingly, FGSCs were mostly similar to SSCs. First, the isolated FGSCs were morphologically similar to freshly isolated SSCs, showing large cell bodies with little cytoplasm, helical nuclei with slight staining, a large ratio of nuclear plasma, and nuclear diameter of 12–20 μm. Second, FGSCs expanded in the same culture conditions as that of SSCs, suggesting that they have similar requirements for their self-renewal (Wu et al., 2008; Yuan et al., 2009). Growth factors are important for expansion of SSCs and FGSCs *in vitro*, such as GDNF, EGF, bFGF, and also LIF; among these growth factors in culture, GDNF is a key factor for GSC proliferation. Moreover, growth pattern of FGSCs and SSCs is similar. For instance, most of them grow with a typical grape-like cluster (shown in Fig. 4.3).

For FGSC characteristic, we observed that FGSCs express germ cell-specific markers such as Dazl, MVH, Blimp-1, Fragilis, and Stella, and do not express pluripotent markers such as Nanog and Sox2. These long-term cultured FGSCs show high telomerase activity and have a normal karyotype (40, XX). In addition, the FGSC is positive for alkaline phosphatase and show that the maternally imprinted region is partially methylated and the paternally imprinted regions are demethylated, indicating a female imprinting pattern (Zou et al., 2011, 2009).

2.4. Role of FGSCs in oogenesis and menopause

Germ cells are specialized reproductive cells which are responsible for the transmission of genetic information from one generation to the next, and GSC have great potential in the future of science and medicine. In the past decades, it was reported that ESCs are able to differentiate into gametes, and it is worth mentioning that cultured mouse ESCs can develop into oocytes as the oogenesis in the ovary (Hu et al., 2012; Hübner et al., 2003; Kerkis et al., 2007; Qing et al., 2007; Yamauchi, Hasegawa, Chuma, Nakatsuji, & Suemori, 2009), which provides a better understanding of gametogenesis in mammals. However, reproductive therapy with ESCs has many limitations, including ethical issues, low efficiency of directed differentiation, etc. FGSC

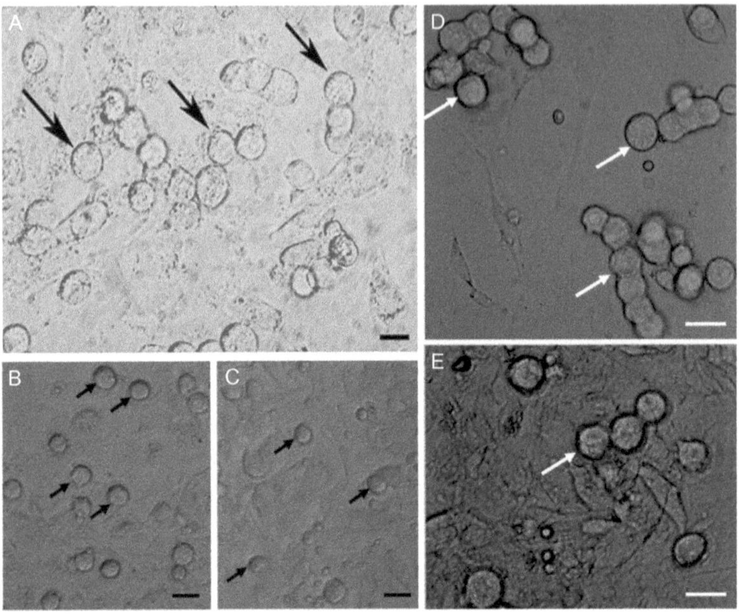

Figure 4.3 Characteristic of FGSC is similar to that of SSCs. (A) Morphology of adult mouse SSCs (arrows) cultured for 24 h after isolation (Yuan et al., 2009). (B, C) Morphology of FGSCs (arrows) from neonatal (B) and adult (C) ovaries cultured for 24 h after isolation. (D, E) The FGSCs isolated from neonatal (E) and adult (F) mice expanded and were undergoing mitosis (arrows) (Zhang, Yang, et al., 2011). Scale bars, 10 μm (A), 20 μm (B–E). (See Color Insert.)

offers a great potential for new approaches of infertility therapeutic and animal model generation, novel contraceptive design, and the acceleration of animal breeding.

The appealing outcomes reviewed above lead us to face the important questions of why are FGSCs apparently so rare in female ovary and why menopause is an inevitable outcome of advancing age when FGSCs still exist. Although the female process of oogenesis (Bukovsky, Svetlikova, & Caudle, 2005) is similar to the male process of spermatogenesis (shown in Fig. 4.4), the rarity of adult FGSCs and oogenesis begins before a female is born, maybe resulting from some differences in gametes production between male and female. FGSC is the member of stem cells, described as immortal symbol by functional definition. However, all of current aging models indicated that stem cell reserve and function depend on the homeostatic control and tissue regenerative potential. Stem cell population size and

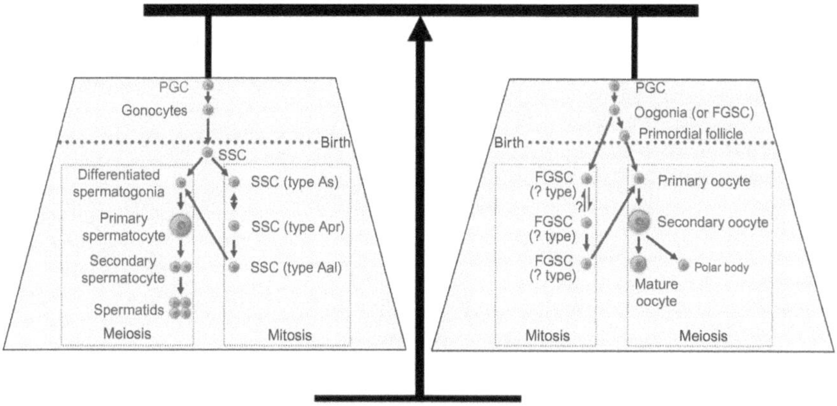

Figure 4.4 Gametogenesis process. Oogenesis and spermatogenesis have similar process. However, whether different types of FGSCs exist during their self-renewal processes is unknown. (For color version of this figure, the reader is referred to the online version of this chapter.)

function will be changed gradually as time goes by, probably in a cell-autonomous fashion and also in response to internal cues of aging surrounding environment (Bukovsky, 2011; Hosni & Bastu, 2012; Niikura, Niikura, & Tilly, 2009; Perheentupa & Huhtaniemi, 2009; Virant-Klun & Skutella, 2010).

Although the role of FGSC in oogenesis is still unclear, SSC in spermatogenesis is well researched; similar process of them will provide us new cues for further exploration, also the experience of ovary GSC research based on *Caenorhabditis elegans* and *Drosophila*.

REFERENCES

Aravin, A. A., Sachidanandam, R., Girard, A., Fejes-Toth, K., & Hannon, G. J. (2007). Developmentally regulated piRNA clusters implicate MILI in transposon control. *Science, 316*, 744–747.

Ballow, D., Meistrich, M. L., Matzuk, M., & Rajkovic, A. (2006a). Sohlh1 is essential for spermatogonial differentiation. *Developmental Biology, 294*, 161–167.

Ballow, D. J., Xin, Y., Choi, Y., Pangas, S. A., & Rajkovic, A. (2006b). Sohlh2 is a germ cell-specific bHLH transcription factor. *Gene expression patterns, 6*, 1014–1018.

Bartel, D. P. (2009). MicroRNAs: Target recognition and regulatory functions. *Cell, 136*, 215–233.

Bettegowda, A., & Wilkinson, M. F. (2010). Transcription and post-transcriptional regulation of spermatogenesis. *Philosophical Transactions of the Royal Society of London. Series B, Biological Sciences, 365*, 1637–1651.

Boettger-Tong, H. L., Johnston, D. S., Russell, L. D., Griswold, M. D., & Bishop, C. E. (2000). Juvenile spermatogonial depletion (jsd) mutant seminiferous tubules are capable of supporting transplanted spermatogenesis. *Biology of Reproduction, 63*, 1185–1191.

Bjork, J. K., Sandqvist, A., Elsing, A. N., Kotaja, N., & Sistonen, L. (2010). miR-18, a member of Oncomir-1, targets heat shock transcription factor 2 in spermatogenesis. *Development, 137*, 3177–3184.

Bouhallier, F., Allioli, N., Lavial, F., Chalmel, F., Perrard, M. H., Durand, P., et al. (2010). Role of miR-34c microRNA in the late steps of spermatogenesis. *RNA, 16*, 720–731.

Boulanger, C. A., Mack, D. L., Booth, B. W., & Smith, G. H. (2007). Interaction with the mammary microenvironment redirects spermatogenic cell fate in vivo. *Proceedings of the National Academy of Sciences of the United States of America, 104*, 3871–3876.

Bourc'his, D., & Bestor, T. H. (2004). Meiotic catastrophe and retrotransposon reactivation in male germ cells lacking Dnmt3L. *Nature, 431*, 96–99.

Brinster, R. L. (2002). Germline stem cell transplantation and transgenesis. *Science, 296*, 2174–2176.

Brinster, R. L. (2007). Male germline stem cells: From mice to men. *Science, 316*, 404–405.

Buaas, F. W., Kirsh, A. L., Sharma, M., McLean, D. J., Morris, J. L., Griswold, M. D., et al. (2004). Plzf is required in adult male germ cells for stem cell self-renewal. *Nature Genetics, 36*, 647–652.

Bukovsky, A. (2011). Ovarian stem cell niche and follicular renewal in mammals. *Anatomical Record (Hoboken), 294*, 1284–1306.

Bukovsky, A., Caudle, M. R., Svetlikova, M., & Upadhyaya, N. B. (2004). Origin of germ cells and formation of new primary follicles in adult human ovaries. *Reproductive Biology and Endocrinology, 2*, 20.

Bukovsky, A., Gupta, S. K., Virant-Klun, I., Upadhyaya, N. B., Copas, P., Van Meter, S. E., et al. (2008). Study origin of germ cells and formation of new primary follicles in adult human and rat ovaries. *Methods in Molecular Biology, 450*, 233–265.

Bukovsky, A., Svetlikova, M., & Caudle, M. R. (2005). Oogenesis in cultures derived from adult human ovaries. *Reproductive Biology and Endocrinology, 3*, 17.

Caires, K., Broady, J., & McLean, D. (2010). Maintaining the male germline: Regulation of spermatogonial stem cells. *The Journal of Endocrinology, 205*, 133–145.

Carmell, M. A., Girard, A., van de Kant, H. J., Bourc'his, D., Bestor, T. H., de Rooij, D. G., et al. (2007). MIWI2 is essential for spermatogenesis and repression of transposons in the mouse male germline. *Developmental Cell, 12*, 503–514.

Cheng, C. Y., & Mruk, D. D. (2012). The blood-testis barrier and its implications for male contraception. *Pharmacological Reviews, 64*, 16–64.

Conrad, S., Renninger, M., Hennenlotter, J., Wiesner, T., Just, L., Bonin, M., et al. (2008). Generation of pluripotent stem cells from adult human testis. *Nature, 456*, 344–349.

Cooke, H. J., Lee, M., Kerr, S., & Ruggiu, M. (1996). A murine homologue of the human DAZ gene is autosomal and expressed only in male and female gonads. *Human Molecular Genetics, 5*, 513–516.

Corallini, S., Fera, S., Grisanti, L., Falciatori, I., Muciaccia, B., Stefanini, M., et al. (2006). Expression of the adaptor protein m-Numb in mouse male germ cells. *Reproduction, 132*, 887–897.

Costoya, J. A., Hobbs, R. M., Barna, M., Cattoretti, G., Manova, K., Sukhwani, M., et al. (2004). Essential role of Plzf in maintenance of spermatogonial stem cells. *Nature Genetics, 36*, 653–659.

Dann, C. T., Alvarado, A. L., Molyneux, L. A., Denard, B. S., Garbers, D. L., & Porteus, M. H. (2008). Spermatogonial stem cell self-renewal requires OCT4, a factor downregulated during retinoic acid-induced differentiation. *Stem Cells, 26*, 2928–2937.

De Rooij, D. G., & Lok, D. (1987). Regulation of the density of spermatogonia in the seminiferous epithelium of the Chinese hamster: II. Differentiating spermatogonia. *The Anatomical Record, 217*, 131–136.

de Rooij, D. G., & Mizrak, S. C. (2008). Deriving multipotent stem cells from mouse spermatogonial stem cells: A new tool for developmental and clinical research. *Development, 135*, 2207–2213.

de Rooij, D. G., Okabe, M., & Nishimune, Y. (1999). Arrest of spermatogonial differentiation in jsd/jsd, Sl17H/Sl17H, and cryptorchid mice. *Biology of Reproduction, 61*, 842–847.

Deng, W., & Lin, H. (2002). miwi, a murine homolog of piwi, encodes a cytoplasmic protein essential for spermatogenesis. *Developmental Cell, 2*, 819–830.

Durcova-Hills, G., Wianny, F., Merriman, J., Zernicka-Goetz, M., & McLaren, A. (2003). Developmental fate of embryonic germ cells (EGCs), in vivo and in vitro. *Differentiation, 71*, 135–141.

Dym, M., Kokkinaki, M., & He, Z. (2009). Spermatogonial stem cells: Mouse and human comparisons. *Birth Defects Research. Part C, Embryo Today, 87*, 27–34.

Eggan, K., Jurga, S., Gosden, R., Min, I. M., & Wagers, A. J. (2006). Ovulated oocytes in adult mice derive from non-circulating germ cells. *Nature, 441*, 1109–1114.

Enders, G. C., & May, J. J., 2nd (1994). Developmentally regulated expression of a mouse germ cell nuclear antigen examined from embryonic day 11 to adult in male and female mice. *Developmental Biology, 163*, 331–340.

Evans, M. J., & Kaufman, M. H. (1981). Establishment in culture of pluripotential cells from mouse embryos. *Nature, 292*, 154–156.

Falender, A. E., Freiman, R. N., Geles, K. G., Lo, K. C., Hwang, K., Lamb, D. J., et al. (2005). Maintenance of spermatogenesis requires TAF4b, a gonad-specific subunit of TFIID. *Genes & Development, 19*, 794–803.

Fujita, K., Ohta, H., Tsujimura, A., Takao, T., Miyagawa, Y., Takada, S., et al. (2005). Transplantation of spermatogonial stem cells isolated from leukemic mice restores fertility without inducing leukemia. *The Journal of Clinical Investigation, 115*, 1855–1861.

Fujiwara, Y., Komiya, T., Kawabata, H., Sato, M., Fujimoto, H., Furusawa, M., et al. (1994). Isolation of a DEAD-family protein gene that encodes a murine homolog of Drosophila vasa and its specific expression in germ cell lineage. *Proceedings of the National Academy of Sciences, 91*, 12258–12262.

Geijsen, N., & Hochedlinger, K. (2009). gPS navigates germ cells to pluripotency. *Cell Stem Cell, 5*, 3–4.

Generoso, W. M., Huff, S. W., & Stout, S. K. (1971). Chemically induced dominant-lethal mutations and cell killing mouse oocytes in the advanced stages of follicular development. *Mutation Research, 11*, 411–420.

Generoso, W. M., Stout, S. K., & Huff, S. W. (1971). Effects of alkylating chemicals on reproductive capacity of adult female mice. *Mutation Research, 13*, 172–184.

Ginsburg, M., Snow, M. H., & McLaren, A. (1990). Primordial germ cells in the mouse embryo during gastrulation. *Development, 110*, 521–528.

Giuili, G., Tomljenovic, A., Labrecque, N., Oulad-Abdelghani, M., Rassoulzadegan, M., & Cuzin, F. (2002). Murine spermatogonial stem cells: Targeted transgene expression and purification in an active state. *EMBO Reports, 3*, 753–759.

Godmann, M., Auger, V., Ferraroni-Aguiar, V., Di Sauro, A., Sette, C., Behr, R., et al. (2007). Dynamic regulation of histone H3 methylation at lysine 4 in mammalian spermatogenesis. *Biology of Reproduction, 77*, 754–764.

Godmann, M., May, E., & Kimmins, S. (2010). Epigenetic mechanisms regulate stem cell expressed genes Pou5f1 and Gfra1 in a male germ cell line. *PLoS One, 5*, e12727.

Gosden, R. G. (2004). Germline stem cells in the postnatal ovary: Is the ovary more like a testis? *Human Reproduction Update, 10*, 193–195.

Greenbaum, M. P., Iwamori, T., Buchold, G. M., & Matzuk, M. M. (2011). Germ cell intercellular bridges. *Cold Spring Harbor Perspectives in Biology, 3*, a005850.

Grivna, S. T., Pyhtila, B., & Lin, H. (2006). MIWI associates with translational machinery and PIWI-interacting RNAs (piRNAs) in regulating spermatogenesis. *Proceedings of the National Academy of Sciences of the United States of America, 103*, 13415–13420.

Guan, K., Nayernia, K., Maier, L. S., Wagner, S., Dressel, R., Lee, J. H., et al. (2006). Pluripotency of spermatogonial stem cells from adult mouse testis. *Nature, 440*, 1199–1203.
Hamra, F. K., Chapman, K. M., Wu, Z., & Garbers, D. L. (2008). Isolating highly pure rat spermatogonial stem cells in culture. *Methods in Molecular Biology, 450*, 163–179.
Hamra, F. K., Gatlin, J., Chapman, K. M., Grellhesl, D. M., Garcia, J. V., Hammer, R. E., et al. (2002). Production of transgenic rats by lentiviral transduction of male germ-line stem cells. *Proceedings of the National Academy of Sciences of the United States of America, 99*, 14931–14936.
Hamra, F. K., Schultz, N., Chapman, K. M., Grellhesl, D. M., Cronkhite, J. T., Hammer, R. E., et al. (2004). Defining the spermatogonial stem cell. *Developmental Biology, 269*, 393–410.
Hayashita, Y., Osada, H., Tatematsu, Y., Yamada, H., Yanagisawa, K., Tomida, S., et al. (2005). A polycistronic microRNA cluster, miR-17-92, is overexpressed in human lung cancers and enhances cell proliferation. *Cancer Research, 65*, 9628–9632.
He, Z., Jiang, J., Hofmann, M. C., & Dym, M. (2007). Gfra1 silencing in mouse spermatogonial stem cells results in their differentiation via the inactivation of RET tyrosine kinase. *Biology of Reproduction, 77*, 723–733.
He, Z., Jiang, J., Kokkinaki, M., Golestaneh, N., Hofmann, M. C., & Dym, M. (2008). Gdnf upregulates c-Fos transcription via the Ras/Erk1/2 pathway to promote mouse spermatogonial stem cell proliferation. *Stem Cells, 26*, 266–278.
He, Z., Kokkinaki, M., Jiang, J., Dobrinski, I., & Dym, M. (2010). Isolation, characterization, and culture of human spermatogonia. *Biology of Reproduction, 82*, 363–372.
Hobbs, R. M., Seandel, M., Falciatori, I., Rafii, S., & Pandolfi, P. P. (2010). Plzf regulates germline progenitor self-renewal by opposing mTORC1. *Cell, 142*, 468–479.
Hofmann, M. C., Braydich-Stolle, L., & Dym, M. (2005). Isolation of male germ-line stem cells; influence of GDNF. *Developmental Biology, 279*, 114–124.
Hofmann, M. C., Narisawa, S., Hess, R. A., & Millan, J. L. (1992). Immortalization of germ cells and somatic testicular cells using the SV40 large T antigen. *Experimental Cell Research, 201*, 417–435.
Holliday, R. (1989). DNA methylation and epigenetic mechanisms. *Cell Biophysics, 15*, 15–20.
Hosni, W., & Bastu, E. (2012). Ovarian stem cells and aging. *Climacteric, 15*, 125–132.
Hu, Y., Sun, J., Wang, J., Wang, L., Bai, Y., Yu, M., et al. (2012). Characterization of female germ cells derived from mouse embryonic stem cells through expression of GFP under the control of Figla promoter. *Journal of Cellular Biochemistry, 113*, 1111–1121.
Huang, Y. H., Chin, C. C., Ho, H. N., Chou, C. K., Shen, C. N., Kuo, H. C., et al. (2009). Pluripotency of mouse spermatogonial stem cells maintained by IGF-1-dependent pathway. *The FASEB Journal, 23*, 2076–2087.
Hübner, K., Fuhrmann, G., Christenson, L. K., Kehler, J., Reinbold, R., De La Fuente, R., et al. (2003). Derivation of oocytes from mouse embryonic stem cells. *Science, 300*, 1251–1256.
Ioannou, J. M. (1967). Oogenesis in adult prosimians. *Journal of Embryology and Experimental Morphology, 17*, 139–145.
Izadyar, F., Pau, F., Marh, J., Slepko, N., Wang, T., Gonzalez, R., et al. (2008). Generation of multipotent cell lines from a distinct population of male germ line stem cells. *Reproduction, 135*, 771–784.
Izadyar, F., Wong, J., Maki, C., Pacchiarotti, J., Ramos, T., Howerton, K., et al. (2011). Identification and characterization of repopulating spermatogonial stem cells from the adult human testis. *Human Reproduction, 26*, 1296–1306.
Jenuwein, T., & Allis, C. D. (2001). Translating the histone code. *Science, 293*, 1074–1080.

Johnson, J., Bagley, J., Skaznik-Wikiel, M., Lee, H. J., Adams, G. B., Niikura, Y., et al. (2005). Oocyte generation in adult mammalian ovaries by putative germ cells in bone marrow and peripheral blood. *Cell, 122*, 303–315.

Johnson, J., Canning, J., Kaneko, T., Pru, J. K., & Tilly, J. L. (2004). Germline stem cells and follicular renewal in the postnatal mammalian ovary. *Nature, 428*, 145–150.

Kanatsu-Shinohara, M., Inoue, K., Lee, J., Yoshimoto, M., Ogonuki, N., Miki, H., et al. (2004). Generation of pluripotent stem cells from neonatal mouse testis. *Cell, 119*, 1001–1012.

Kanatsu-Shinohara, M., Ogonuki, N., Inoue, K., Miki, H., Ogura, A., Toyokuni, S., et al. (2003). Long-term proliferation in culture and germline transmission of mouse male germline stem cells. *Biology of Reproduction, 69*, 612–616.

Kanatsu-Shinohara, M., Ogonuki, N., Iwano, T., Lee, J., Kazuki, Y., Inoue, K., et al. (2005). Genetic and epigenetic properties of mouse male germline stem cells during long-term culture. *Development, 132*, 4155–4163.

Kanatsu-Shinohara, M., Toyokuni, S., & Shinohara, T. (2004). CD9 is a surface marker on mouse and rat male germline stem cells. *Biology of Reproduction, 70*, 70–75.

Kato, Y., Kaneda, M., Hata, K., Kumaki, K., Hisano, M., Kohara, Y., et al. (2007). Role of the Dnmt3 family in de novo methylation of imprinted and repetitive sequences during male germ cell development in the mouse. *Human Molecular Genetics, 16*, 2272–2280.

Kerkis, A., Fonseca, S. A., Serafim, R. C., Lavagnolli, T. M., Abdelmassih, S., Abdelmassih, R., et al. (2007). In vitro differentiation of male mouse embryonic stem cells into both presumptive sperm cells and oocytes. *Cloning and Stem Cells, 9*, 535–548.

Kerr, J. B., Duckett, R., Myers, M., Britt, K. L., Mladenovska, T., & Findlay, J. K. (2006). Quantification of healthy follicles in the neonatal and adult mouse ovary: Evidence for maintenance of primordial follicle supply. *Reproduction, 132*, 95–109.

Kimmins, S., & Sassone-Corsi, P. (2005). Chromatin remodelling and epigenetic features of germ cells. *Nature, 434*, 583–589.

Kossack, N., Meneses, J., Shefi, S., Nguyen, H. N., Chavez, S., Nicholas, C., et al. (2009). Isolation and characterization of pluripotent human spermatogonial stem cell-derived cells. *Stem Cells, 27*, 138–149.

Kotaja, N., Bhattacharyya, S. N., Jaskiewicz, L., Kimmins, S., Parvinen, M., Filipowicz, W., et al. (2006). The chromatoid body of male germ cells: Similarity with processing bodies and presence of Dicer and microRNA pathway components. *Proceedings of the National Academy of Sciences of the United States of America, 103*, 2647–2652.

Kristensen, D. M., Nielsen, J. E., Skakkebaek, N. E., Graem, N., Jacobsen, G. K., Rajpert-De Meyts, E., et al. (2008). Presumed pluripotency markers UTF-1 and REX-1 are expressed in human adult testes and germ cell neoplasms. *Human Reproduction, 23*, 775–782.

Kubota, H., Avarbock, M. R., & Brinster, R. L. (2003). Spermatogonial stem cells share some, but not all, phenotypic and functional characteristics with other stem cells. *Proceedings of the National Academy of Sciences of the United States of America, 100*, 6487.

Kubota, H., Avarbock, M. R., & Brinster, R. L. (2004). Growth factors essential for self-renewal and expansion of mouse spermatogonial stem cells. *Proceedings of the National Academy of Sciences of the United States of America, 101*, 16489–16494.

Lee, J., & Shinohara, T. (2011). Epigenetic modifications and self-renewal regulation of mouse germline stem cells. *Cell Research, 21*, 1164–1171.

Li, E., Bestor, T. H., & Jaenisch, R. (1992). Targeted mutation of the DNA methyltransferase gene results in embryonic lethality. *Cell, 69*, 915–926.

Li, J. Y., Lees-Murdock, D. J., Xu, G. L., & Walsh, C. P. (2004). Timing of establishment of paternal methylation imprints in the mouse. *Genomics, 84*, 952–960.

Lian, J., Tian, H., Liu, L., Zhang, X. S., Li, W. Q., Deng, Y. M., et al. (2010). Down-regulation of microRNA-383 is associated with male infertility and promotes testicular embryonal carcinoma cell proliferation by targeting IRF1. *Cell Death and Disease, 1*, e94.
Lok, D., & de Rooij, D. G. (1983). Spermatogonial multiplication in the Chinese hamster. I. Cell cycle properties and synchronization of differentiating spermatogonia. *Cell and Tissue Kinetics, 16*, 7–18.
Luo, L., Ye, L., Liu, G., Shao, G., Zheng, R., Ren, Z., et al. (2010). Microarray-based approach identifies differentially expressed microRNAs in porcine sexually immature and mature testes. *PLoS One, 5*, e11744.
Martin, G. R. (1981). Isolation of a pluripotent cell line from early mouse embryos cultured in medium conditioned by teratocarcinoma stem cells. *Proceedings of the National Academy of Sciences of the United States of America, 78*, 7634–7638.
Matzuk, M. M., & Lamb, D. J. (2008). The biology of infertility: Research advances and clinical challenges. *Nature Medicine, 14*, 1197–1213.
McLaren, A. (2003). Primordial germ cells in the mouse. *Developmental Biology, 262*, 1–15.
Meng, X., Lindahl, M., Hyvonen, M. E., Parvinen, M., de Rooij, D. G., Hess, M. W., et al. (2000). Regulation of cell fate decision of undifferentiated spermatogonia by GDNF. *Science, 287*, 1489–1493.
Mirzapour, T., Movahedin, M., Tengku Ibrahim, T. A., Haron, A. W., Nowroozi, M. R., & Rafieian, S. H. (2011). Effects of basic fibroblast growth factor and leukaemia inhibitory factor on proliferation and short-term culture of human spermatogonial stem cells. *Andrologia, 44*, 41–55.
Morgan, H. D., Santos, F., Green, K., Dean, W., & Reik, W. (2005). Epigenetic reprogramming in mammals. *Human Molecular Genetics, 1*, R47–R58.
Nagano, M., Brinster, C. J., Orwig, K. E., Ryu, B. Y., Avarbock, M. R., & Brinster, R. L. (2001). Transgenic mice produced by retroviral transduction of male germ-line stem cells. *Proceedings of the National Academy of Sciences of the United States of America, 98*, 13090–13095.
Naughton, C. K., Jain, S., Strickland, A. M., Gupta, A., & Milbrandt, J. (2006). Glial cell-line derived neurotrophic factor-mediated RET signaling regulates spermatogonial stem cell fate. *Biology of Reproduction, 74*, 314–321.
Niikura, Y., Niikura, T., & Tilly, J. L. (2009). Aged mouse ovaries possess rare premeiotic germ cells that can generate oocytes following transplantation into a young host environment. *Aging (Albany NY), 1*, 971–978.
Ning, L., Goossens, E., Geens, M., Van Saen, D., Van Riet, I., He, D., et al. (2010). Mouse spermatogonial stem cells obtain morphologic and functional characteristics of hematopoietic cells in vivo. *Human Reproduction, 25*, 3101–3109.
Niu, Z., Goodyear, S. M., Rao, S., Wu, X., Tobias, J. W., Avarbock, M. R., et al. (2011). MicroRNA-21 regulates the self-renewal of mouse spermatogonial stem cells. *Proceedings of the National Academy of Sciences of the United States of America, 108*, 12740–12745.
Oatley, J. M., Avarbock, M. R., Telaranta, A. I., Fearon, D. T., & Brinster, R. L. (2006). Identifying genes important for spermatogonial stem cell self-renewal and survival. *Proceedings of the National Academy of Sciences of the United States of America, 103*, 9524–9529.
Oatley, J. M., Avarbock, M. R., & Brinster, R. L. (2007). Glial cell line-derived neurotrophic factor regulation of genes essential for self-renewal of mouse spermatogonial stem cells is dependent on Src family kinase signaling. *The Journal of Biological Chemistry, 282*, 25842–25851.
Oatley, J. M., & Brinster, R. L. (2008). Regulation of spermatogonial stem cell self-renewal in mammals. *Annual Review of Cell and Developmental Biology, 24*, 263–286.
Ogawa, T., Ohmura, M., Tamura, Y., Kita, K., Ohbo, K., Suda, T., et al. (2004). Derivation and morphological characterization of mouse spermatogonial stem cell lines. *Archives of Histology and Cytology, 67*, 297–306.

Ohta, H., Yomogida, K., Dohmae, K., & Nishimune, Y. (2000). Regulation of proliferation and differentiation in spermatogonial stem cells: The role of c-kit and its ligand SCF. *Development, 127*, 2125–2131.

Okada, Y., Scott, G., Ray, M. K., Mishina, Y., & Zhang, Y. (2007). Histone demethylase JHDM2A is critical for Tnp1 and Prm1 transcription and spermatogenesis. *Nature, 450*, 119–123.

Payne, C. J., & Braun, R. E. (2008). Human adult testis-derived pluripotent stem cells: Revealing plasticity from the germline. *Cell Stem Cell, 3*, 471–472.

Pesce, M., Wang, X., Wolgemuth, D. J., & Scholer, H. (1998). Differential expression of the Oct-4 transcription factor during mouse germ cell differentiation. *Mechanisms of Development, 71*, 89–98.

Perheentupa, A., & Huhtaniemi, I. (2009). Aging of the human ovary and testis. *Molecular and Cellular Endocrinology, 299*, 2–13.

Phillips, B. T., Gassei, K., & Orwig, K. E. (2010). Spermatogonial stem cell regulation and spermatogenesis. *Philosophical Transactions of the Royal Society of London. Series B, Biological Sciences, 365*, 1663–1678.

Puglisi, R., Montanari, M., Chiarella, P., Stefanini, M., & Boitani, C. (2004). Regulatory role of BMP2 and BMP7 in spermatogonia and Sertoli cell proliferation in the immature mouse. *European Journal of Endocrinology, 151*, 511–520.

Qing, T., Shi, Y., Qin, H., Ye, X., Wei, W., Liu, H., et al. (2007). Induction of oocyte-like cells from mouse embryonic stem cells by co-culture with ovarian granulosa cells. *Differentiation, 75*, 902–911.

Rajpert-De Meyts, E., Jacobsen, G. K., Bartkova, J., Aubry, F., Samson, M., Bartek, J., et al. (2003). The immunohistochemical expression pattern of Chk2, p53, p19INK4d, MAGE-A4 and other selected antigens provides new evidence for the premeiotic origin of spermatocytic seminoma. *Histopathology, 42*, 217–226.

Raverot, G., Weiss, J., Park, S. Y., Hurley, L., & Jameson, J. L. (2005). Sox3 expression in undifferentiated spermatogonia is required for the progression of spermatogenesis. *Developmental Biology, 283*, 215–225.

Reik, W., & Walter, J. (2001). Genomic imprinting: Parental influence on the genome. *Nature Reviews. Genetics, 2*, 21–32.

Russell, L. D., Ettlin, R. A., Hikim, A. P. S., & Clegg, E. D. (1993). Histological and histopathological evaluation of the testis. *International Journal of Andrology, 16*, 83.

Ryu, B. Y., Orwig, K. E., Oatley, J. M., Avarbock, M. R., & Brinster, R. L. (2006). Effects of aging and niche microenvironment on spermatogonial stem cell self-renewal. *Stem Cells, 24*, 1505–1511.

Sada, A., Suzuki, A., Suzuki, H., & Saga, Y. (2009). The RNA-binding protein NANOS2 is required to maintain murine spermatogonial stem cells. *Science, 325*, 1394–1398.

Schaller, J., Glander, H. J., & Dethloff, J. (1993). Evidence of beta 1 integrins and fibronectin on spermatogenic cells in human testis. *Human Reproduction, 8*, 1873–1878.

Seandel, M., James, D., Shmelkov, S. V., Falciatori, I., Kim, J., Chavala, S., et al. (2007). Generation of functional multipotent adult stem cells from GPR125+ germline progenitors. *Nature, 449*, 346–350.

Shi, Y., Lan, F., Matson, C., Mulligan, P., Whetstine, J. R., Cole, P. A., et al. (2004). Histone demethylation mediated by the nuclear amine oxidase homolog LSD1. *Cell, 119*, 941–953.

Shinohara, T., Avarbock, M. R., & Brinster, R. L. (1999). beta1- and alpha6-integrin are surface markers on mouse spermatogonial stem cells. *Proceedings of the National Academy of Sciences of the United States of America, 96*, 5504–5509.

Shuttlesworth, G. A., de Rooij, D. G., Huhtaniemi, I., Reissmann, T., Russell, L. D., Shetty, G., et al. (2000). Enhancement of A spermatogonial proliferation and differentiation in irradiated rats by gonadotropin-releasing hormone antagonist administration. *Endocrinology, 141*, 37–49.

Skutella, T. (2009). Induced pluripotent stem cells from adult testis: A new source of stem cells? *Regenerative Medicine*, *4*, 3–5.

Simon, L., Ekman, G. C., Kostereva, N., Zhang, Z., Hess, R. A., Hofmann, M. C., et al. (2009). Direct transdifferentiation of stem/progenitor spermatogonia into reproductive and nonreproductive tissues of all germ layers. *Stem Cells*, *27*, 1666–1675.

Simon, L., Hess, R. A., & Cooke, P. S. (2010). Spermatogonial stem cells, in vivo transdifferentiation and human regenerative medicine. *Expert Opinion on Biological Therapy*, *10*, 519–530.

Surani, M. A., Barton, S. C., & Norris, M. L. (1984). Development of reconstituted mouse eggs suggests imprinting of the genome during gametogenesis. *Nature*, *308*, 548–550.

Suzuki, H., Sada, A., Yoshida, S., & Saga, Y. (2009). The heterogeneity of spermatogonia is revealed by their topology and expression of marker proteins including the germ cell-specific proteins Nanos2 and Nanos3. *Developmental Biology*, *336*, 222–231.

Tadokoro, Y., Yomogida, K., Ohta, H., Tohda, A., & Nishimune, Y. (2002). Homeostatic regulation of germinal stem cell proliferation by the GDNF/FSH pathway. *Mechanisms of Development*, *113*, 29–39.

Takehashi, M., Kanatsu-Shinohara, M., Miki, H., Lee, J., Kazuki, Y., Inoue, K., et al. (2007). Production of knockout mice by gene targeting in multipotent germline stem cells. *Developmental Biology*, *312*, 344–352.

Tegelenbosch, R. A., & de Rooij, D. G. (1993). A quantitative study of spermatogonial multiplication and stem cell renewal in the C3H/101 F1 hybrid mouse. *Mutation Research*, *290*, 193–200.

Tokuda, M., Kadokawa, Y., Kurahashi, H., & Marunouchi, T. (2007). CDH1 is a specific marker for undifferentiated spermatogonia in mouse testes. *Biology of Reproduction*, *76*, 130–141.

Tong, M. H., Mitchell, D., Evanoff, R., & Griswold, M. D. (2011). Expression of Mirlet7 family microRNAs in response to retinoic acid-induced spermatogonial differentiation in mice. *Biology of Reproduction*, *85*, 189–197.

Tong, M. H., Mitchell, D. A., McGowan, S. D., Evanoff, R., & Griswold, M. D. (2012). Two miRNA Clusters, Mir-17-92 (Mirc1) and Mir-106b-25 (Mirc3), are involved in the regulation of spermatogonial differentiation in mice. *Biology of Reproduction*, *86*, 72.

Tsuda, M., Sasaoka, Y., Kiso, M., Abe, K., Haraguchi, S., Kobayashi, S., et al. (2003). Conserved role of nanos proteins in germ cell development. *Science*, *301*, 1239–1241.

van Bragt, M. P., Roepers-Gajadien, H. L., Korver, C. M., Bogerd, J., Okuda, A., Eggen, B. J., et al. (2008). Expression of the pluripotency marker UTF1 is restricted to a subpopulation of early A spermatogonia in rat testis. *Reproduction*, *136*, 33–40.

Van Pelt, A. M., & De Rooij, D. G. (1990). The origin of the synchronization of the seminiferous epithelium in vitamin A-deficient rats after vitamin A replacement. *Biology of Reproduction*, *42*, 677–682.

van Pelt, A. M., Morena, A. R., van Dissel-Emiliani, F. M., Boitani, C., Gaemers, I. C., de Rooij, D. G., et al. (1996). Isolation of the synchronized A spermatogonia from adult vitamin A-deficient rat testes. *Biology of Reproduction*, *55*, 439–444.

Virant-Klun, I., & Skutella, T. (2010). Stem cells in aged mammalian ovaries. *Aging (Albany NY)*, *2*, 3–6.

Vogt, M., Munding, J., Gruner, M., Liffers, S. T., Verdoodt, B., Hauk, J., et al. (2011). Frequent concomitant inactivation of miR-34a and miR-34b/c by CpG methylation in colorectal, pancreatic, mammary, ovarian, urothelial, and renal cell carcinomas and soft tissue sarcomas. *Virchows Archiv*, *458*, 313–322.

Wang, P. J., McCarrey, J. R., Yang, F., & Page, D. C. (2001). An abundance of X-linked genes expressed in spermatogonia. *Nature Genetics*, *27*, 422–426.

White, Y. A., Woods, D. C., Takai, Y., Ishihara, O., Seki, H., & Tilly, J. L. (2012). Oocyte formation by mitotically active germ cells purified from ovaries of reproductive-age women. *Nature Medicine*, *18*, 413–421.

Wong, M. D., Jin, Z., & Xie, T. (2005). Molecular mechanisms of germline stem cell regulation. *Annual Review of Genetics*, *39*, 173–195.

Wu, J., Bao, J., Wang, L., Hu, Y., & Xu, C. (2011). MicroRNA-184 downregulates nuclear receptor corepressor 2 in mouse spermatogenesis. *BMC Developmental Biology*, *11*, 64.

Wu, X., Goodyear, S. M., Tobias, J. W., Avarbock, M. R., & Brinster, R. L. (2011). Spermatogonial stem cell self-renewal requires ETV5-mediated downstream activation of Brachyury in mice. *Biology of Reproduction*, *85*, 1114–1123.

Wu, J., Jester, W. F., Jr., & Orth, J. M. (2005). Short-type PB-cadherin promotes survival of gonocytes and activates JAK-STAT signaling. *Developmental Biology*, *284*, 437–450.

Wu, J., Zhang, Y., Tian, G. G., Zou, K., Lee, C. M., Yu, Q., et al. (2008). Short-type PB-cadherin promotes self-renewal of spermatogonial stem cells via multiple signaling pathways. *Cellular Signalling*, *20*, 1052–1060.

Xiong, J., Wang, H., Guo, G., Wang, S., He, L., Chen, H., et al. (2011). Male germ cell apoptosis and epigenetic histone modification induced by Tripterygium wilfordii Hook F. *PLoS One*, *6*, e20751.

Yamaguchi, Y. L., Tanaka, S. S., Kasa, M., Yasuda, K., Tam, P. P., & Matsui, Y. (2006). Expression of low density lipoprotein receptor-related protein 4 (Lrp4) gene in the mouse germ cells. *Gene Expression Patterns*, *6*, 607–612.

Yamauchi, K., Hasegawa, K., Chuma, S., Nakatsuji, N., & Suemori, H. (2009). In vitro germ cell differentiation from cynomolgus monkey embryonic stem cells. *PLoS One*, *4*, e5338.

Yan, N., Lu, Y., Sun, H., Qiu, W., Tao, D., Liu, Y., et al. (2009). Microarray profiling of microRNAs expressed in testis tissues of developing primates. *Journal of Assisted Reproduction and Genetics*, *26*, 179–186.

Yoshida, S., Sukeno, M., & Nabeshima, Y. (2007). A vasculature-associated niche for undifferentiated spermatogonia in the mouse testis. *Science*, *317*, 1722–1726.

Yoshinaga, K., Nishikawa, S., Ogawa, M., Hayashi, S., Kunisada, T., & Fujimoto, T. (1991). Role of c-kit in mouse spermatogenesis: Identification of spermatogonia as a specific site of c-kit expression and function. *Development*, *113*, 689–699.

Yuan, Z., Hou, R., & Wu, J. (2009). Generation of mice by transplantation of an adult spermatogonial cell line after cryopreservation. *Cell Proliferation*, *42*, 123–131.

Zhang, S., Sun, J., Pan, S., Zhu, H., Wang, L., Hu, Y., et al. (2011). Retinol (vitamin A) maintains self-renewal of pluripotent male germline stem cells (mGSCs) from adult mouse testis. *Journal of Cellular Biochemistry*, *112*, 1009–1021.

Zhang, Y., Yang, Z., Yang, Y., Wang, S., Shi, L., Xie, W., et al. (2011). Production of transgenic mice by random recombination of targeted genes in female germline stem cells. *Journal of Molecular Cell Biology*, *3*, 132–141.

Zhao, G. Q., Chen, Y. X., Liu, X. M., Xu, Z., & Qi, X. (2001). Mutation in Bmp7 exacerbates the phenotype of Bmp8a mutants in spermatogenesis and epididymis. *Developmental Biology*, *240*, 212–222.

Zhao, M., Shirley, C. R., Yu, Y. E., Mohapatra, B., Zhang, Y., Unni, E., et al. (2001). Targeted disruption of the transition protein 2 gene affects sperm chromatin structure and reduces fertility in mice. *Molecular and Cellular Biology*, *21*, 7243–7255.

Zheng, K., Wu, X., Kaestner, K. H., & Wang, P. J. (2009). The pluripotency factor LIN28 marks undifferentiated spermatogonia in mouse. *BMC Developmental Biology*, *9*, 38.

Zhong, X., Li, N., Liang, S., Huang, Q., Coukos, G., & Zhang, L. (2010). Identification of microRNAs regulating reprogramming factor LIN28 in embryonic stem cells and cancer cells. *The Journal of Biological Chemistry*, *285*, 41961–41971.

Zou, K., Hou, L., Sun, K., Xie, W., & Wu, J. (2011). Improved efficiency of female germline stem cell purification using fragilis-based magnetic bead sorting. *Stem Cells and Development, 20*, 2197–2204.

Zou, K., Yuan, Z., Yang, Z., Luo, H., Sun, K., Zhou, L., et al. (2009). Production of offspring from a germline stem cell line derived from neonatal ovaries. *Nature Cell Biology, 11*, 631–636.

Zovoilis, A., Nolte, J., Drusenheimer, N., Zechner, U., Hada, H., Guan, K., et al. (2008). Multipotent adult germline stem cells and embryonic stem cells have similar microRNA profiles. *Molecular Human Reproduction, 14*, 521–529.

Zuckerman, S. (1951). The number of oocytes in the mature ovary. *Recent Progress in Hormone Research, 6*, 63–108.

Zuckerman, S., & Baker, T. G. (1977). The development of the ovary and the process of oogenesis. In S. Zuckerman & B. J. Weir (Eds.), *Ovary* (pp. 41–67). New York: Academic Press.

CHAPTER FIVE

LINCing the Nuclear Envelope to Gametogenesis

Martin P. Kracklauer*, **Jana Link**†, **Manfred Alsheimer**†,[1]
*Department of Physiology, Wayne State University Medical School, Detroit, Michigan, USA
†Department of Cell and Developmental Biology, Biocenter, University of Würzburg, Würzburg, Germany
[1]Corresponding author: e-mail address: alsheimer@biozentrum.uni-wuerzburg.de

Contents

Abstract

Gametogenesis combines two important features: reduction of the genome content from diploid to haploid by carefully partitioning chromosomes, and the subsequent differentiation into fertilization-competent gametes, which in males is characterized by profound nuclear restructuring. These are quite difficult tasks and require a tight coordination of different cellular mechanisms. Recent studies in the field established a key role for LINC complexes in both meiosis and sperm head formation. LINC complexes comprise SUN and KASH domain proteins that form nuclear envelope (NE) bridges, linking the nucleoskeleton to the cytoskeleton. They are well known for their crucial roles in diverse cellular and developmental processes, such as nuclear positioning and cell polarization. In this review, we highlight key roles ascribed to LINC complexes and to the nucleocytoskeletal connection in gametogenesis. First, we give a short overview about the general features of LINC components and the profound reorganization of the NE in germ cells. We then focus on specific roles of LINC complexes in meiotic chromosome dynamics and their impact on pairing, synapsis, and recombination. Finally, we provide an update of the mechanisms controlling sperm head formation and discuss the role of sperm-specific LINC complexes in nuclear shaping and their relation to specialized cytoskeletal structures that form concurrently with nuclear restructuring and sperm elongation.

Current Topics in Developmental Biology, Volume 102
ISSN 0070-2153
http://dx.doi.org/10.1016/B978-0-12-416024-8.00005-2

1. INTRODUCTION

Active movement, anchorage, and precise positioning of the nucleus play fundamental roles during diverse cellular and developmental processes, including cell shaping, establishment of cell polarity, cell migration, fertilization, and many more. In addition to these, correct nuclear anchorage and positioning are decisive factors for proper differentiation of various cell types (Burke & Roux, 2009; Fridkin, Penkner, Jantsch, & Gruenbaum, 2009; Starr, 2009). These processes depend on a tightly regulated interaction between the nuclear envelope (NE) and the cytoskeleton. In this context, the so-called LINC complexes (linker of nucleoskeleton and cytoskeleton) have gained increasing attention, as they actually define the molecular basis of a physical connection of the nucleus to the peripheral cytoskeleton. In the past several years, numerous studies aimed at determining the actual function of LINC components have convincingly demonstrated that LINC complexes are pivotal for nuclear migration, anchoring, and positioning in somatic cells and, strikingly, for directed meiotic chromosome movements as well. Furthermore, they are an essential determinant for NE integrity and for maintaining general nuclear morphology and shape (Razafsky & Hodzic, 2009; Starr & Fridolfsson, 2010). Due in no small part to these studies, LINC complexes emerged as highly critical NE-bridging complexes formed across the NE by the interaction of members of two transmembrane (TM) protein families: SUN domain proteins located in the inner nuclear membrane (INM), interconnecting with the outer nuclear membrane (ONM) residing KASH domain partners (Crisp et al., 2006; Méjat & Misteli, 2010; Stewart-Hutchinson, Hale, Wirtz, & Hodzic, 2008).

Gametogenesis is a tightly regulated, goal-oriented developmental process by which diploid germ stem cells produce haploid fertilization-competent gametes. Groundbreaking studies in the past decade have provided substantial evidence that the NE occupies exceedingly important and, to some extent, unexpected roles in different steps of germ cell development. For example, during meiosis, the NE serves as a platform for the characteristic, evolutionarily conserved chromosome movements required for efficient haploidization of the genome (Alsheimer, 2009). Furthermore, there is a substantial evidence that the nucleocytoplasmic intersection functions at the very center of nuclear reorganization and directed shaping during sperm head formation (Göb, Schmitt, Benavente, & Alsheimer, 2010; Kierszenbaum, Rivkin, & Tres, 2011; Kracklauer et al., 2010). The underlying mechanisms, however, are highly complex and not yet completely understood. In this review, we provide an overview of recent advances that decisively influence our current view of the germ cell NE, that is, it represents a key factor for germ cell viability and development. To mark the recently ascertained significance of the NE bridges formed by SUN and KASH proteins, we put a major focus on the peculiarities and specific functions of LINC complexes and their components and discuss their relevance for germ cell differentiation and fertility. Furthermore, we point toward possible further avenues of research into LINC complex formation in gametogenesis.

2. LINC COMPLEXES ARE VERSATILE CONNECTORS

SUN domain proteins constitute an evolutionarily conserved family of proteins that share a common C-terminal motif, the so-called SUN (Sad1p/Unc84 homology) domain (Hagan & Yanagida, 1995; Malone, Fixsen, Horvitz, & Han, 1999). As emphasized by the InterPro Database (InterPro Entry: IPR012919; http://www.ebi.ac.uk/interpro), representatives of this family can be found in nearly all (if not all) eukaryotic taxa, from protists up to higher vertebrates (for selected examples, see Table 5.1). SUN domain proteins are integral membrane proteins of the INM with a single TM domain separating an N-terminal nucleoplasmic region from the C-terminal part that extends into the perinuclear space (PNS) (Hodzic, Yeater, Bengtsson, Otto, & Stahl, 2004; Padmakumar et al., 2005). Though their N-terminal nucleoplasmic domains are not conserved, most family members nonetheless show remarkable conserved binding to nucleoskeletal elements and chromatin. SUN proteins directly interact with the KASH (Klarsicht/Anc1/Syne1 homology) domain of respective KASH protein partners within the PNS via their C-terminal SUN domain to form a complex that bridges both nuclear membranes (McGee, Rillo, Anderson, & Starr, 2006; Sosa, Rothballer, Kutay, & Schwartz, 2012; Starr & Fischer, 2005; Starr & Fridolfsson, 2010). The KASH partners in turn are typically large proteins that contain a short, conserved C-terminal domain, the so-called KASH domain, which marks the defining feature of this protein family. KASH domains consist of a hydrophobic region spanning the ONM and a short, 6–30-residue-long sequence at the very C-terminal end that extends into the PNS and is designed to bind SUN domains. The large N-terminal domains of KASH proteins, however, are highly divergent and perform a variety of distinct functions, including centrosome attachment, nuclear migration, and tethering nuclei to different cytoskeletal elements (Razafsky & Hodzic, 2009; Starr, 2009; Starr & Fischer, 2005).

According to current sequence information from the rapidly growing genome and transcriptome databases, SUN and KASH proteins have undergone a remarkable diversification over the course of evolution. In particular, concomitant with the increase in complexity of the organism, the number of SUN proteins has significantly increased as well. While lower organisms, such as yeast and cellular slime molds, appear to cope with only one SUN domain protein, nematodes and flies, for example, both contain two genes for SUN proteins, and the mammalian genome even encodes five distinct members of the SUN protein family: Sun1, Sun2, Sun3, Sun4/SPAG4, and Sun5/SPAG4L (Table 5.1; Fig. 5.1; Starr & Fischer, 2005; Tzur, Wilson, & Gruenbaum, 2006b). This increase in the intraspecies SUN protein repertoire is augmented by the existence of various splice isoforms, as recently demonstrated for mammalian Sun1 and Sun5/SPAG4L (Frohnert et al., 2011; Göb, Meyer-Natus, Benavente, & Alsheimer, 2011; Göb et al., 2010; Liu et al., 2007), thus providing manifold options for assembling miscellaneous LINC complexes (Table 5.1; Fig. 5.1). Several studies demonstrate that expression of the individual SUN proteins and/or isoforms depends on the cell type, suggesting cell type-specific adaption of LINC complexes to meet distinct cellular and physiological requirements. For example, in *C. elegans*, UNC-84 is expressed in most cells, whereas SUN-1/Matefin expression is restricted to germ cells. Notably, both *C. elegans* SUN proteins show no overlapping activity, indicating that respective LINC complexes occupy distinct functions (Fridkin et al., 2004). Similar to the situation in *C. elegans*, the *Drosophila* SUN protein Klaroid is present in almost every cell type, while Spag4 appears to be strictly confined to the male germ line (Kracklauer et al., 2007, 2010; Technau & Roth, 2008). Consistently, the two major mammalian SUN proteins, Sun1 and Sun2, are widely expressed in different cell types (Crisp et al., 2006, Hodzic et al., 2004; Padmakumar et al., 2005). Notably, RNAi and knockout studies revealed that in somatic cells, these two mammalian SUN paralogs are partially redundant, but to some degree also have divergent properties and functions (Chi et al., 2009; Crisp et al., 2006; Ding et al., 2007; Haque et al., 2006; Lei et al., 2009). However, besides these two almost ubiquitously expressed SUN proteins, mammals have three additional genes coding for Sun3, Sun4 (SPAG4), and Sun5 (SPAG4L), respectively, that all show a much more restricted, testis-specific expression, suggesting

Table 5.1 SUN and KASH proteins of selected model organisms, their interaction partners and presumed functions

Organism	Nuclear partner	SUN protein	KASH protein	Cytoskeletal partner	Function of the UNC complex	References
Mammals	?	Sun1	Nesprin1/2	Actin	Anchorage of nuclei, maintenance of NE integrity	Crisp et al. (2006)
	?	Sun1	KASH5	Microtubules	Meiotic telomere attachment to the NE	Ding et al. (2007), Morimoto et al. (2012)
	?	Sun1η	Nesprin3	Actin (?)	Nuclear remodeling in spermiogenesis	Göb et al. (2010)
	?	Sun2	Nesprin1/2	Actin	Anchorage of nuclei, maintenance of NE integrity	Crisp et al. (2006)
	?	Sun2	KASH5	Microtubules	Meiotic telomere attachment to the NE	Schmitt et al. (2007), Morimoto et al. (2012)
	?	Sun3	Nesprin1	Microtubules (?)	Nuclear remodeling in spermiogenesis	Göb et al. (2010)
	?	Sun4	?	?	Spermiogenesis-specific expression, function	Shao, Tarnasky, Lee, Oko, and van der Hoorn (1999)
	?	Sun5	?	?	Spermiogenesis-specific expression, function unknown	Frohnert, Schweizer, and Hoyer-Fender (2011)
C. elegans	?	UNC-84	ANC-1	Actin	Nuclear anchorage	Starr and Han (2002), Starr et al. (2001)
	?	UNC-84	UNC-83	Microtubules	Nuclear migration	Meyerzon, Fridolfsson, Ly, McNally, and Starr (2009)
	HIM-8, ZIM-1-3	SUN-1	ZYG-12	Microtubules	Nuclear anchorage in germline, attachments of meiotic chromosomes, and centrosome anchorage	Malone et al. (2003), Penkner et al. (2009)
	?	SUN-1	KDP-1	?	Embryonic development, cell cycle progression	McGee, Stagljar, and Starr (2009)
	CED-4	SUN-1	?	?	Apoptosis pathway	Tzur, et al. (2006b)

Continued

Table 5.1 SUN and KASH proteins or selected model organisms, their interaction partners and presumed functions—cont'd

Organism	Nuclear partner	SUN protein	KASH protein	Cytoskeletal partner	Function of the UNC complex	References
D. melanogaster	?	Klaroid	Klarsicht	Microtubules	Nuclear anchorage and migration	Kracklauer, Banks, Xie, Wu, and Fischer (2007), Patterson et al. (2004)
	?	Spag4	?	Dynein-dynactin, Yuri Gagarin (?)	Centriole and nuclear anchorage in spermatogenesis	Kracklauer et al. (2010)
	?	?	Msp-300	Actin	Nuclear anchorage	Yu et al. (2006), Xie and Fischer (2008)
S. cerevisiae	Mps2	Mps3	?	?	Spindle pole duplication and anchorage	Jaspersen, Giddings, and Winey (2002), Jaspersen et al. (2006)
	Ndj1	Mps3	Csm4 (?)	Actin (?)	Meiotic telomere attachment to the NE	Conrad et al. (2008), Wanat et al. (2008)
S. pombe	Bqt1–4	Sad1	Km s1	Microtubules	Meiotic telomere attachment to the SPB	Miki et al. (2004), Chikashige et al. (2006)
	?	Sad1	Km s2	Microtubules	Centromere tethering and attachment	King, Drivas, and Blobel (2008)

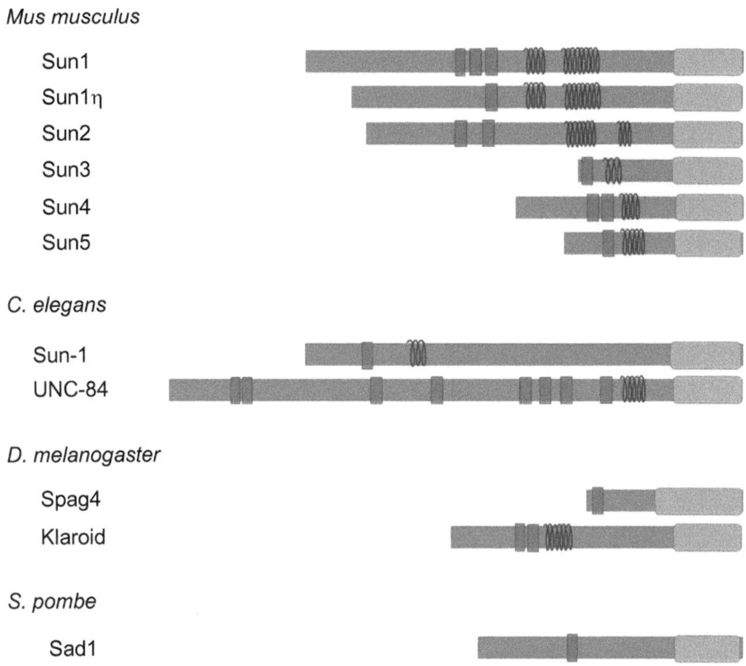

Figure 5.1 Modular composition of SUN domain proteins from selected model organisms. Green: predicted hydrophobic regions including presumptive transmembrane domains (TM), blue: coiled-coil domains, and yellow: SUN domains. (For interpretation of the references to color in this figure legend, the reader is referred to the online version of this chapter.)

that they play specific roles in germ cell development (Frohnert et al., 2011; Göb et al., 2010; Shao et al., 1999).

Although the primary function of KASH proteins in connecting SUN proteins to the cytoskeleton is well conserved, the KASH family comprises a large number of exceedingly diverse protein variants and isoforms. With the exception of the KASH domain itself, which is not only necessary but also sufficient to localize KASH proteins to the ONM, KASH proteins appear not to be homologous to one another (Starr & Fischer, 2005; Starr & Fridolfsson, 2010; Zhang et al., 2001). The highly variable cytoplasmic domains, however, are free to interact with a variety of proteins including actin, microtubule components, and plectin, which links to intermediate filaments (reviewed by Starr & Fridolfsson, 2010). To date, several KASH proteins have been identified in a variety of different model organisms (see Table 5.1 for selected examples). Prominent members are *C. elegans* ANC-1 and ZYG-12. ANC-1, like its functional homologs MSP-300 in

Drosophila and Syne-1/Nesprin-1 in mammals, is a large, actin-binding protein that functions in nuclear anchoring (Padmakumar et al., 2004; Starr & Han, 2002; Yu et al., 2006). ZYG-12 on the other hand, similar to fission yeast Kms1, binds dynein and mediates NE attachment of the centrosome. Beyond that, together with its partner SUN-1/Matefin, it forms a nucleocytoplasmic bridge to transfer cytoplasmic forces to move telomeres in meiosis (Malone et al., 2003; Miki et al., 2004; Penkner et al., 2007; see below). The mammalian genome encodes at least four KASH proteins (Padmakumar et al., 2004; Roux et al., 2009; Wilhelmsen et al., 2005; Zhang et al., 2001). Two of them, Nesprin-1 (Syne-1) and 2 (Syne-2), are large actin-binding proteins (Zhang et al., 2001; Zhen, Libotte, Munck, Noegel, & Korenbaum, 2002), whereas Nesprin-3, a smaller molecule, binds to plectin, which in turn links to the actin and/or the intermediate filament system (Ketema et al., 2007). Nesprin-4, however, is restricted to few cell types and binds to kinesin, which links to the microtubule system (Roux et al., 2009). Since their INM partners Sun1 and Sun2 both interact with A-type lamins and other components of the INM (i.e., emerin), the SUN–KASH interaction within the PNS forms a functional cross-linkage of the nucleoskeleton and the cytoskeleton (Crisp et al., 2006; Haque et al., 2010; Starr & Fridolfsson, 2010). The huge number of SUN and KASH protein isoforms and variants, multiple options to combine the INM and ONM partners, and the cell type-specific expression of individual LINC components offer a broad range of possibilities to form functional LINC complexes that can each be assigned to tissue- and developmental stage-specific duties.

Besides their primary function in connecting nucleoplasmic to cytoplasmic structures, LINC complexes are crucial not only for dynamic processes, such as anchorage and migration of entire nuclei, but also for positioning and movement of nuclear structures (Starr & Fridolfsson, 2010). Most striking is their role in meiosis, where SUN–KASH bridges serve as specific NE docking sites for meiotic telomeres, mediating the characteristic movements of chromosomes that are essential for meiotic chromosome pairing, synapsis, and recombination (Alsheimer, 2009; Chi et al., 2009; Ding et al., 2007; Penkner et al., 2007; Schmitt et al., 2007; Shimanuki et al., 1997). Beyond that, a number of recent studies have provided strong evidence that LINC complexes play a most central role in nuclear deformation and shaping (Oda & Fukuda, 2011; Olins et al., 2009; Xiong et al., 2008; Zhou, Graumann, Evans, & Meier, 2012). Previous reports suggested that during sperm differentiation, major forces guiding the characteristic elongation and shaping of the sperm head act from the cytoplasm (Kierszenbaum, Rivkin, & Tres, 2003a, 2003b; Mendoza-Lujambio et al., 2002). In light of this, LINC complexes are expected to be key players for well-directed nuclear shaping in that they function as primary connectors to transfer these forces to the nucleus (Göb et al., 2010; Kierszenbaum et al., 2011; Kracklauer et al., 2010).

3. GENERAL FEATURES OF THE NUCLEOCYTOPLASMIC INTERSECTION IN GERM CELLS

3.1. The meiotic NE

With first depiction of the meiotic bouquet configurations more than 100 years ago, it became evident that during meiosis, the NE is more than merely a cladding surrounding the chromosomes. Instead, it serves as a platform that vitally supports chromosomal processes in the interior (Eisen, 1900; Gelei, 1921; Platner, 1885; Wilcox, 1895). In meiotic prophase I, homologous chromosomes have to move to find, pair, synapse, and recombine with the right partner. Meiotic chromosomal movements follow a remarkable, evolutionarily highly conserved choreography, highlighting their importance in preparing the homologs for subsequent segregation (Alsheimer, 2009; Dresser, 2008; Parvinen & Söderström, 1976; Scherthan, 2001; Zickler & Kleckner, 1998). In premeiotic cells, telomeres usually are spread in the nucleus, but upon entry into meiosis, they dramatically change their spatial arrangement, migrating to the nuclear periphery where they structurally attach to the NE. Once attached, telomeres start to move on the surface of the NE and transiently cluster at the pole that faces the microtubule-organizing center, thus trailing the chromosomes into the typical "flower-like" bouquet configuration. As meiosis progresses, telomeres become dispersed over the surface of the NE, but remain tightly attached until late prophase (Fig. 5.2; for extensive review, see Harper, Golubovskaya, & Cande, 2004; Scherthan, 2001; Zickler & Kleckner, 1998). Telomere-led movements have been proposed to directly promote homolog search, pairing, synapsis, and recombination. Beyond this, they appear to be required for untangling faulty nonhomologous associations. (Koszul & Kleckner, 2009; Rasmussen, 1986; Rockmill & Roeder, 1998; Sato et al., 2009; Scherthan, Bahler, & Kohli, 1994, Scherthan et al., 1996). Consistent with this, failures in telomere attachment or movements lead to severe defects in homologous chromosome pairing and recombination (Cooper, Watanabe, & Nurse, 1998; Ding et al., 2007; Nimmo, Pidoux, Perry,

& Allshire, 1998; Penkner et al., 2007; Sonntag Brown, Zanders, & Alani, 2011; Trelles-Sticken, Dresser, & Scherthan, 2000; Wanat et al., 2008).

Chromosome movements are directed remotely from outside the nucleus and require coupling of the telomeres to the cytoskeleton (Fig. 5.2; Koszul, Kim, Prentiss, Kleckner, & Kameoka, 2008; Koszul & Kleckner, 2009; Sato et al., 2009; Scherthan et al., 2007; Trelles-Sticken, Adelfalk, Loidl, & Scherthan, 2005; Yamamoto, West, McIntosh, & Hiraoka, 1999; Zhou, Rolls, Hall, Malone, & Hanna-Rose, 2009). Importantly, with his detailed ultrastructural studies from 1956, Montrose Moses provided clear evidence that NE attachment of meiotic telomeres is much more than a loose association, instead being a tight structural connection formed early in meiotic prophase and preserved for nearly the entire meiotic prophase I (Moses, 1956a, 1956b). Meiotic telomere–NE association is mechanically exceedingly stable and actually withstands harsh spreading techniques (Moses, 1977; Moses & Solari, 1976). Further studies demonstrated that telomeres in effect attach to the NE through flat, disk-shaped attachment plates that are in intimate contact with the INM. Remarkably, at the INM contact sites, thin fibrils arise from these plates that traverse both nuclear membranes and extend into the cytoplasm, suggesting that they actually represent the points where cytoplasmic forces act to drive telomere movements (Esponda & Gimenez-Martin, 1972; Liebe, Alsheimer, Höög, Benavente, & Scherthan, 2004; Sato et al., 2009). While numerous studies in the past decade hypothesized that these filamentous structures comprise LINC complexes, the actual molecular nature of these fibrils remained obscure. First direct evidence came from immunogold localization performed on mammalian spermatocytes, which demonstrated that Sun2 is part of the membrane-spanning threads, suggesting that these fibrils indeed consist of SUN–KASH bridging complexes (Schmitt et al., 2007; for further discussion, see below).

Movement of telomeres within the NE apparently requires, at least in mammals, a more general reorganization of the NE. In particular, this reorganization occurs at the level of the nuclear lamina, a tight meshwork of nuclear intermediate filament proteins providing mechanical stability and structural integrity of the NE and of the entire nucleus and, moreover, is critically involved in many fundamental cellular and developmental processes (for recent overviews, see Cohen, Hernandez, & Stewart, 2008; Dechat, Adam, Taimen, Shimi, & Goldman, 2010; Gruenbaum, Margalit, Goldman, Shumaker, & Wilson, 2005). In mammalian meiotic cells, three of the four lamin isoforms typically expressed in differentiated somatic cells are absent, indicative of a general weakening of the nuclear lamina structure in the context of meiotic telomere dynamics (Vester, Smith, Krohne, & Benavente, 1993). In addition to the significant reduction of somatic lamin isoforms, meiotic prophase I cells express a novel lamin, lamin C2, which represents a short "N-terminally truncated" meiosis-specific isoform of the *Lmna* gene (Alsheimer & Benavente, 1996; Furukawa, Inagaki, & Hotta, 1994). While in somatic cells, lamins are distributed in a continuous ring-like pattern, lamin C2 forms discontinuous domains at the nuclear periphery, a phenomenon suggested to be a consequence of the missing N-terminal domain (Alsheimer, von Glasenapp, Hock, & Benavente, 1999; Alsheimer, von Glasenapp, Schnölzer, Heid, & Benavente, 2000; Jahn, Schramm, Benavente, & Alsheimer, 2010). Remarkably, lamin C2 is highly enriched at the sites of telomere attachment, and it was proposed that lamin C2 enhances local flexibility of the NE to enable movement of telomeres, thus supporting pairing, recombination, and synapsis (Alsheimer et al., 1999; Jahn et al., 2010). Accordingly, mice lacking lamin C2 show defective telomere dynamics that lead to massive failures in pairing, synapsis, and recombination of the homologs (Alsheimer et al., 2004; Link et al., 2012).

3.2. The nucleocytoplasmic intersection during sperm differentiation

Profound nuclear restructuring is a pre-eminent feature of the sperm differentiation process and, significantly, failures in formation of the sperm head are major causes of male infertility (Yan, 2009). A most prominent aspect is the characteristic, well-directed shaping of the sperm nucleus from spherical to elongated (Fig. 5.3; Hermo, Pelletier, Cyr, & Smith, 2010a, 2010b; Kierszenbaum & Tres, 2004; Kierszenbaum et al., 2003a, 2011; Russell, Russell, MacGregor, & Meistrich, 1991). The typical nuclear reshaping during sperm head formation requires elaborate cooperation of different cellular mechanisms and includes polarization of the cell and its components, assembly of sperm-specific cytoskeletal structures, and nuclear movement and chromatin compaction, resulting in the characteristic elongation of initially round cell nuclei.

A number of previous studies demonstrate that elongation and shaping of the sperm nucleus involve an extensive remodeling of the NE. In particular, this concerns both its general composition as well as the behavior of its individual components. For example, in mammals, similar to the situation in meiotic cells, lamins A, C, and B2 are essentially absent during postmeiotic sperm differentiation. Instead, lamin B3, a novel germ cell-specific lamin B2 isoform, is expressed. Together with lamin B1, it takes part in forming a specifically modified nuclear periphery (Schütz, Alsheimer, Öllinger, & Benavente, 2005a; Schütz, Benavente, & Alsheimer, 2005b). A further central characteristic of the NE during sperm head elongation is the extensive redistribution and polarization of a variety of NE components. This applies not only to both nuclear lamins expressed in spermatids, that is, lamins B1 and B3, but also to INM proteins LAP2 and lamin B receptor. All these proteins behave in a similar manner: they gradually relocate to the posterior pole of the nuclei (Alsheimer, Fecher, & Benavente, 1998; Mylonis et al., 2004; Schütz, et al., 2005a). Recent studies identified novel, spermiogenesis-specific LINC complexes that show different, quite striking polarization during sperm head formation. One of these LINC complexes, formed by Sun3 and Nesprin1, is exclusively found in posterior NE regions that are decorated by cytoplasmic microtubules, suggesting a role in nucleocytoskeletal linkage (Göb et al., 2010). Comparably, mammalian Spag4/Sun4 and its presumed *Drosophila* ortholog both localize to the posterior NE. In mammals, Sun4 localization completely overlaps with that of Sun3, indicative of potential cooperative function (Kracklauer et al., 2010; Shao et al., 1999; J.L & M.A. unpublished data). Behavior and spatial distribution significantly differ between posterior LINC complexes and other NE components. In round spermatids, lamins B1 and B3 and LAP2 initially localize throughout the entire NE, and as spermiogenesis progresses, gradually redistribute and congregate in a small region at the very posterior end. In contrast, as soon as cells are haploid, Sun3/Nesprin1 LINC complexes show complete posterior polarization but are excluded from the very posterior implantation fossa, a distribution retained throughout sperm differentiation (Fig. 5.3; Alsheimer et al., 1998; Göb et al., 2010). Of particular importance, however, was the finding that with Sun1η/Nesprin3 and Sun5 (Spag4L), distinct LINC components are present at the opposite, anterior spermatid pole as well, likely linking to the microtubule-opposing spermatid-specific actin cytoskeleton (Fig. 5.3; Frohnert et al., 2011; Göb et al., 2010; Kierszenbaum et al., 2011).

Polarization of NE components shows remarkable correlation with the formation of highly polarized cytoplasmic structures, that is, a posterior manchette consisting of bundles of microtubules that cover the posterior part of the nucleus, and the acroplaxome, a thin cytoplasmic sheet consisting mainly of F-actin at the anterior pole between the elongating nucleus and the acrosome (Kierszenbaum et al., 2003a, 2003b, 2011; Russell et al., 1991). The microtubule manchette and the acroplaxome both are in tight contact with the NE and form synchronously with elongation of the spermatid nucleus. Hence, they were suggested to be active players in directed spermatid nuclear shaping, possibly by generating orchestrated clutching forces that squeeze the nucleus into its designated form (Kierszenbaum & Tres, 2004; Kierszenbaum et al., 2003a, 2003b, 2011). Consistent with this, mutations impairing acroplaxome–manchette integrity provoke severe nuclear shape abnormalities (Meistrich, Trostle-Weige, & Russell, 1990; Mendoza-Lujambio et al., 2002). Until recently, the functional significance of the conspicuous general redistribution of NE proteins, their actual relation to the polarized cytoskeletal elements, and the mechanism of cytoskeletal–nuclear force transduction remained unclear. However, it has been repeatedly discussed that changes in spermatid nuclear shape require a particularly modified nucleocytoplasmic intersection that allows for—or even directs—nuclear shaping and sperm-specific chromatin reorganization (Alsheimer et al., 1998; Kierszenbaum et al., 2011; Mylonis et al., 2004; Schütz et al., 2005a; Schütz et al., 2005b). With the identification of LINC components in differentiating spermatids, this discussion has gained new impact, as the SUN–KASH bridging complexes essentially embody the missing link in current concepts of sperm head shaping (see below).

4. SUN–KASH BRIDGES DRIVE MEIOSIS

Sad1 was the first described SUN protein, identified by a genetic screen for proteins essential for spindle formation in *S. pombe* (Hagan & Yanagida, 1990, 1995). Sad1 is a component of the spindle pole body (SPB; the yeast microtubule-organizing center), is associated with the NE, has a crucial role in SPB function, and is essential for general cell viability (Hagan & Yanagida, 1995). In fission yeast meiosis, Sad1 interacts with KASH domain protein Kms1, forming a LINC complex required for telomere attachment and congregation at the SPB (Niwa, Shimanuki, & Miki, 2000; Shimanuki et al., 1997). Since bouquet formation is highly conserved, it was postulated that comparable mechanisms may exist in other eukaryotes as well (Tomita & Cooper, 2006). Indeed, inspired by this hypothesis and propelled by successive identification of a multitude of SUN proteins in many other eukaryotes, several subsequent studies demonstrated that the LINC complex function in meiotic telomere attachment is well conserved (Alsheimer, 2009; Fridkin et al., 2009; Fig. 5.3). In a first study, performed in mammals, Sun2 has been shown to selectively colocalize with NE-attached telomeres in meiotic cells, comprising part of the membrane-spanning fibrillar complex that connects to cytoplasmic structures (Schmitt et al., 2007). Shortly thereafter, Ding et al. provided direct functional evidence that mammalian Sun1 is essential for telomere attachment and bouquet formation. In *Sun1*$^{-/-}$ meiocytes, bouquet formation is severely disturbed, as most telomeres fail to attach to the NE and remain in the nuclear interior. As a consequence, efficient homolog pairing, synapsis, and recombination are hampered, leading to meiotic cell death and infertility (Ding et al., 2007). In the same year, Penkner et al. published their results on SUN1/Matefin function in *C. elegans* meiosis: in a SUN1/Matefin mutant carrying a missense mutation in the conserved SUN domain, binding to its KASH partner ZYG-12 is disrupted, leading to abrogation of chromosome movements. This in turn results in lack of presynaptic alignment of the homologs and gives rise to massive failures in synapsis and homologous recombination (Baudrimont et al., 2010; Penkner et al., 2007). Consistent results were further obtained in budding yeast: Conrad, Lee, Wilkerson, and Dresser (2007) demonstrated that telomere attachment and clustering in *S. cerevisiae* essentially depends on SUN protein Mps3. Several SUN proteins were identified in plants as well, some of which are expressed in meiotic stages (Murphy, Simmons, & Bass, 2010). Since telomere behavior in plants is quite similar to that of other organisms, and plant meiotic chromosome motility depends on cytoplasmic forces that involve both the actin and tubulin cytoskeletons, plants were suspected to have similar LINC complex-based mechanisms (Bass, Marshall, Sedat, Agard, & Cande, 1997; Murphy et al., 2010; Sheehan & Pawlowski, 2009). However, whether this is the case remains to be demonstrated.

Although, to date, the conserved role of LINC complexes in meiotic telomere and chromosome behavior is unquestioned, *Drosophila* presents an interesting deviation from other examples discussed so far. In some respects, *Drosophila* telomeres can be considered unusual and, in meiosis, behave in a manner not found in other species. Though they are often found close to the NE, they appear not to be structurally attached there. This is consistent with the observation that *Drosophila* does not have a bouquet stage, and it seems unlikely that telomere clustering, or any other telomere-related mechanism, plays a direct role in pairing, synapsis, and recombination in *Drosophila* (Lake & Hawley, 2012). Interestingly, the *D. melanogaster* genome encodes two KASH proteins, Klarsicht (Klar) and MSP-300, and two SUN domain proteins, Klaroid (Koi) and Spag4, that form functional LINC complexes (Kracklauer et al., 2007, 2010; Mosley-Bishop, Li, Patterson, & Fischer, 1999; Yu et al., 2006). Koi, Klar, and MSP-300 are all observed to localize to the NE of the oocyte nucleus. There, they are involved in nuclear anchoring but are not essential for meiosis (Kracklauer et al., 2007; Technau & Roth, 2008; Xie & Fischer, 2008; Yu et al., 2006). Likewise, LINC complexes involving Klar, MSP-300, and Koi are not essential in spermatogenesis, as male flies carrying homozygous mutations in either of these genes produce functional sperm and are fertile (Kracklauer et al., 2010; unpublished data). Spag4 is expressed in both male meiocytes and spermatids, but Spag4 appears not to be essential for meiosis, as primary spermatocytes in *spag4* null mutant male flies undergo meiosis to generate apparently normal round spermatids (Kracklauer et al., 2010). All these data suggest that in *Drosophila*, any role the telomeres normally play in supporting interhomolog interactions appears to have been discarded in evolution, and it has thus been postulated that a different, unique mode of homolog alignment and partner choice underlies homology search and chromosome pairing in meiosis in both sexes (Lake & Hawley, 2012; Tsai & McKee, 2011).

Upon entry into meiosis, the nucleoplasmic domains of SUN proteins provide specific NE docking sites for meiotic telomeres. Remarkably, for telomere attachment, all analyzed organisms thus far use SUN proteins, which are not restricted to first meiotic prophase, but are also expressed in other cell types, where they function in different processes independent of telomere binding (Starr & Fridolfsson, 2010). This, however, raises the following obvious questions: how in fact do meiotic telomeres recognize these docking sites, and how is SUN–telomere binding regulated? A further important aspect is how do these LINC–telomere complexes move to arrive at the typical clusters? While in higher eukaryotes, these issues remained rather speculative, recent studies in both fission and budding yeast shed some light onto this matter. For example, in *S. pombe*, coupling telomeres to Sad1 depends on meiosis-specific adaptor molecules. In vegetative cells of fission yeast, telomeres are dispersed along the NE and are tethered to the NE. This quite unusual telomere attachment in vegetative cells, however, is independent of SUN protein Sad1 and is conducted by ubiquitously expressed INM proteins, Bqt3 and 4 (Chikashige, Haraguchi, & Hiraoka, 2010; Chikashige et al., 2009). Upon entry into meiosis, mating pheromone induces expression of two meiosis-specific proteins, Bqt1 and Bqt2. Both proteins act in a complex, in which Bqt2 binds to telomeric Rap1, whereas Bqt1 directly interacts with Sad1, thus mediating telomere binding to the Sad1/Kms1 LINC complexes (Chikashige et al., 2006; Tang, Jin, & Cande, 2006). Once this switch has occurred, on the cytoplasmic side, Kms1 interacts with a dynein motor complex to tether telomeres to the SPB, a stable configuration that is

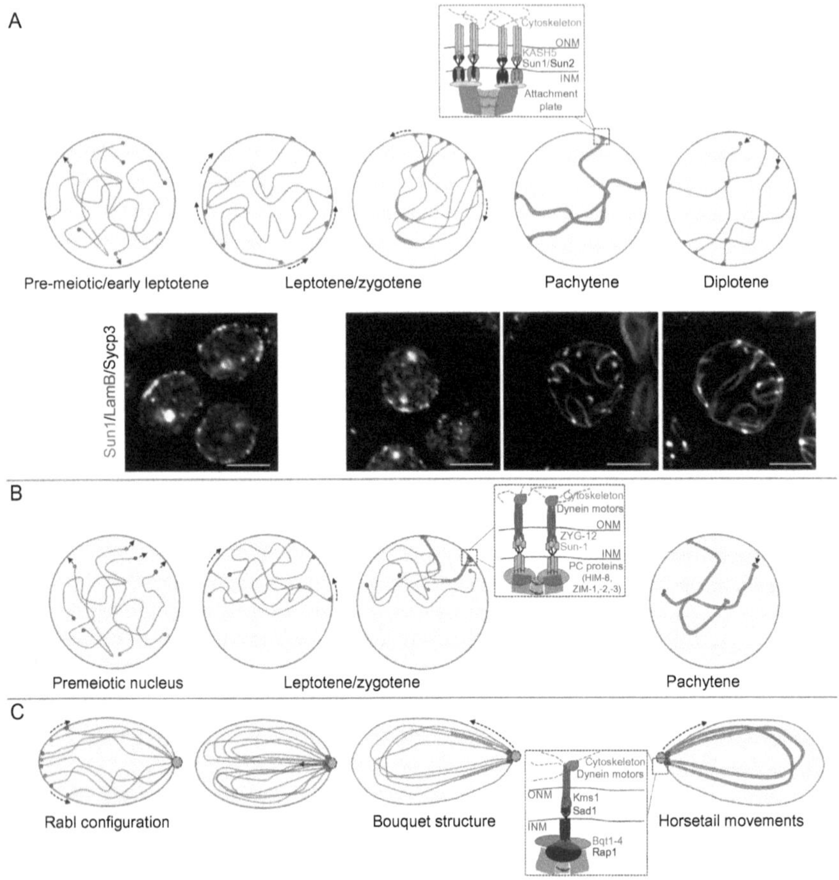

Figure 5.2 Schematic representation of meiotic telomere dynamics in different model organisms. (A) The meiotic LINC complex (see inset) responsible for telomere attachment in the mouse is composed of Sun1 and Sun2 interacting with KASH5 in the outer nuclear membrane (ONM). Corresponding immunofluorescence images of mouse spermatocytes stained with Sun1, Lamin B, and Sycp3 show meiotic telomere dynamics *in vivo*. Sun1 signals are located at telomere ends of synaptonemal complexes (labeled by Sycp3) and lie within the NE labeled by lamin B. Scale bars 5 μm. (B) Meiotic telomere attachment in *C. elegans* (see inset) is mediated by the INM protein SUN-1 interacting with the KASH domain protein ZYG-12. (C) Attachment and clustering of telomeres at the spindle pole body in *S. pombe* (see inset) is mediated by the SUN domain protein Sad1 binding to the KASH domain protein Kms1. (See Color Insert.)

preserved during the entire "horsetail" movement (Fig. 5.2; Chikashige, Haraguchi, & Hiraoka, 2007; Miki et al., 2002). In budding yeast, Ndj1 has been suggested to function comparably to the fission yeast Bqt1/2 complex. Ndj1 is a telomere binding protein that, like Bqt1 and 2, is expressed specifically during meiotic prophase. Ndj1 was shown to tether telomeres to LINC complexes via direct interaction with SUN protein Mps3 (Conrad, Dominguez, & Dresser, 1997; Conrad et al., 2007; Trelles-Sticken et al., 2000). Mps3, in turn, interacts with KASH protein Csm4 that links to the actin cytoskeleton, which in budding yeast appears to provide the driving force for bouquet formation (Conrad et al., 2008; Kosaka, Shinohara, & Shinohara, 2008; Koszul et al., 2008; Trelles-Sticken et al., 2005; Wanat et al., 2008).

As mentioned, the situation in higher organisms is less clear. Although it is known that LINC complexes in higher eukaryotes play a key role in anchoring telomeres and in linking these to the centrosome as well, the complete molecular connection remains elusive. In mammals, both Sun1 and Sun2 are almost ubiquitously expressed, but do not bind telomeres in somatic cells. Therefore, it must be postulated that the meiotic nuclear periphery has a cell type-specific signature allowing for establishing a stage-specific connection between SUN proteins and the telomeres. This, certainly, could rely on mechanisms comparable to those in fission and budding yeast. However, functional orthologs of fission yeast Bqt proteins or budding yeast Ndj1 have not been identified to date. In a different scenario, meiosis-specific modifications of either SUN or telomeric proteins may enable direct binding of telomeres to the nucleoplasmic domain of SUN proteins. Until recently, it was also not known how meiotic cells establish the link to the cytoskeleton. Previous attempts to detect any of the known mammalian KASH proteins in meiocytes had failed, suggesting that mammals use novel, not yet identified KASH or KASH-like proteins for linking telomeres to the cytoskeletal motor system (unpublished data). However, in a very recent publication, Morimoto and colleagues indeed report on a previously unknown, meiosis-specific KASH protein, which they termed KASH5. The KASH domain of KASH5 interacts with the SUN domains of both Sun1 and Sun2 and, via its cytoplasmic extension, binds the microtubule-associated dynein–dynactin complex, thus connecting LINC bound telomeres to the microtubule motor system (Morimoto et al., 2012).

5. LINCs IN SPERM HEAD FORMATION

Besides their unquestioned role in nuclear migration, positioning, anchoring, chromosome attachment, and directed movement (Razafsky & Hodzic, 2009; Starr & Fridolfsson, 2010), recent studies demonstrated that both SUN and KASH proteins are crucial for NE integrity and for maintaining nuclear morphology (Khatau et al., 2009; Oda & Fukuda, 2011; Olins et al., 2009; Xiong et al., 2008; Zhou et al., 2012). This makes it entirely conceivable that LINC complexes could also have central functions in nuclear shaping during sperm head formation, and thus could be major determinants for fertility and reproduction. Consistent with this hypothesis, it was found that spermatid nuclear elongation and reshaping closely correlate with the appearance of novel LINC components, which, in addition, show exceptional and meaningful polarization (Frohnert et al., 2011; Göb et al., 2010; Kracklauer et al., 2010). Besides the ubiquitously expressed Sun1, mammalian spermatids express a variety of distinct SUN proteins, Sun1η, Sun3, Sun4 (Spag4), as well as two isoforms of Sun5 (Spag4L), which are exclusively and specifically used in this particular differentiation process (Frohnert et al., 2011; Göb et al., 2010; Shao et al., 1999; Fig. 5.1). Interestingly, *Drosophila*, which is evolutionarily distant from vertebrates, contains a distinct gene encoding for male germ line-specific Spag4, a presumptive ortholog of mammalian Sun4. Spag4 was proven to be essential for spermiogenesis, suggesting an evolutionarily conserved LINC-based strategy for sperm differentiation (Kracklauer et al., 2010). Mammalian spermatids form at least two different LINC complexes consisting either of germ cell-specific Sun3 and Nesprin1 or of Sun1η, a sperm-specific isoform of Sun1, which forms a distinct LINC complex with Nesprin3 (Table 5.1). During sperm head formation, both sperm-specific LINC complexes polarize, but to opposite poles, likely connecting to different sperm-specific cytoskeletal structures. Sun3/Nesprin1 localizes to the posterior pole, but only to NE regions that are covered by the posterior microtubule manchette, suggesting a role in interconnecting the NE with the microtubule cytoskeleton (Göb et al., 2010; Kierszenbaum et al., 2011). In a previous study, Russell and colleagues reported that rod-like filaments link the innermost microtubules of the manchette to the outer, and intriguingly, across the PNS to the inner leaflet of the NE. Because of their general properties and prospective functions, Sun3/Nesprin1 could actually comprise these filaments, which in turn are thought to provide the structural basis for transmission of cytoplasmic forces that elongate nuclei (Dadoune, 2003; Göb et al., 2010; Russell et al., 1991; Toshimori & Ito, 2003; Fig. 5.3). On the other hand, LINC complexes consisting of Sun1η and Nesprin3 show a different localization. In the course of nuclear elongation, they accumulate at the opposite, anterior pole, where they are part of the anterior acrosomal membrane system (Göb et al., 2010; Fig. 5.3). Interestingly, Nesprin3 was shown to bind plectin, which for its part is able to interact with actin

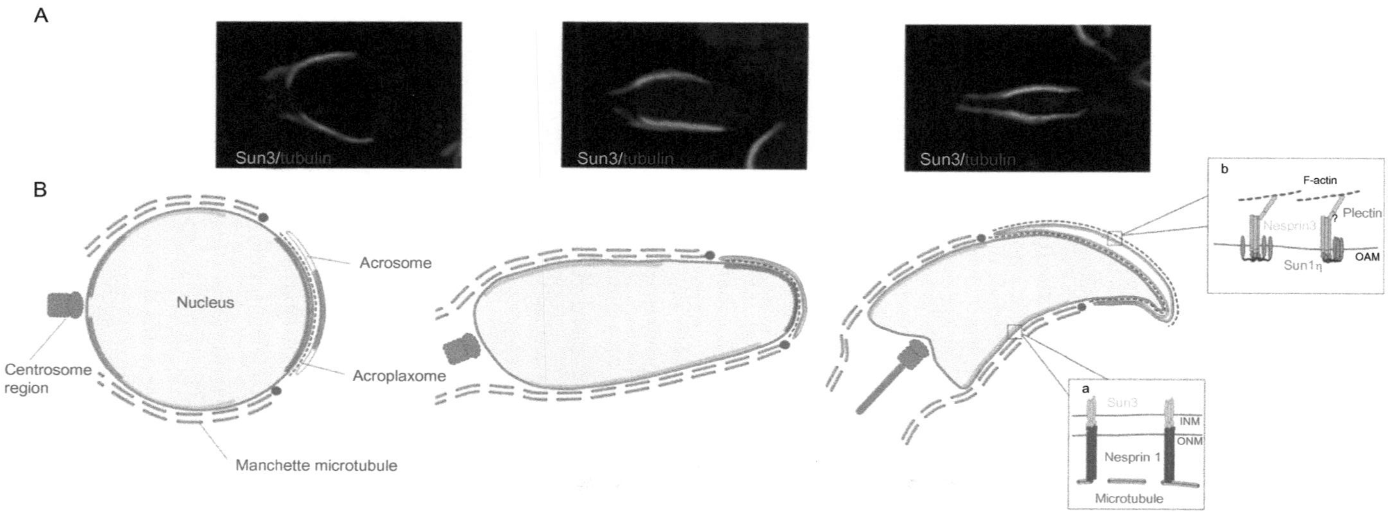

Figure 5.3 Distribution of LINC complex proteins during mouse spermiogenesis. (A) Sun3 (in green) is localized within the NE only at sites, where the cytoplasmic manchette (in red) is in close contact with the NE. (B) Schematic representation of the redistribution of spermiogenic SUN domain proteins during nuclear elongation. Yellow: Sun3, blue: Sun1 and Sun1η, and red: Spag4L. Posterior and anterior spermiogenesis-specific LINC complexes are shown in insets a and b. (See Color Insert.)

(Ketema et al., 2007; Seifert, Lawson, & Wiche, 1992). In spermatids, actin is enriched at the acroplaxome and at specialized membranous junctions at the anterior spermatid/Sertoli cell interface, which both have key functions in directed shaping of the sperm head (Gliki, Ebnet, Aurrand-Lions, Imhof, & Adams, 2004; Kierszenbaum et al., 2003a, 2003b; Kierszenbaum & Tres, 2004; Mueller, Rosenquist, Takai, Bronson, & Wimmer, 2003). The acrosome is tightly bound to the acroplaxome plate, which in turn is anchored to the spermatid NE, most likely via novel integral membrane proteins such as the recently identified Dpy19l2 (Kierszenbaum & Tres, 2004; Pierre et al., 2012). Thus, in connecting the acrosome to the anterior actin cytoskeleton, Sun1η/Nespin3 LINC complexes provide a continuous, mechanically stable actin-based network, which may be used to transfer actin-dependent, microtubule-opposing forces to the nucleus, even from the distant anterior spermatid/Sertoli cell interface (Göb et al., 2010; Fig. 5.3).

Besides Sun1η and Sun3, mammalian spermatids express two further germ cell-specific SUN proteins, Sun4 and Sun5. Like Sun3, Sun4 (Spag4) localizes to the posterior NE of round and elongating spermatids, which in principle resembles the distribution of its presumed ortholog in *Drosophila* (Kracklauer et al., 2010; Shao et al., 1999; J.L & M.A. unpublished data). Intriguingly, recent colocalization experiments reveal that Sun4 distribution within the NE entirely overlaps with that of Sun3, suggesting that Sun4 may be part of either the same or a different LINC complex linking to the microtubule manchette (unpublished data). As recently demonstrated by Frohnert et al., Sun5 exists in two closely related isoforms (Spag4L/4L-2) that both are co-expressed in spermiogenesis. In contrast to Sun3 and Sun4, Sun5 localizes to the apical NE of round spermatids facing the acrosome, and it was suggested that it may function in a protein complex that links the acrosomic vesicle to the spermatid nucleus (Frohnert et al., 2011; Pierre et al., 2012). However, whether Sun5 assumes this role remains to be demonstrated.

Recent analysis of *Drosophila* Spag4 provided convincing evidence for a direct role of LINC components in sperm head formation (Kracklauer et al., 2010). During initial stages of spermatid elongation, Spag4 is transiently associated with the centriolar hemisphere of the round spermatid nuclei and then forms a “stripe” in the groove of so-called canoe stage nuclei, in which microtubules and actin filaments are also found. As nuclear elongation and condensation progress, Spag4 forms a focus at a “dent” on the nuclei where the basal body embeds in the nucleus (Tokuyasu, 1974, 1975). This dent structure persists throughout spermatogenesis, and Spag4 localizes there even in mature sperm in the seminal vesicle (Kracklauer et al., 2010).

Homozygous *spag4* mutations result in sterility and are associated with two striking cellular phenotypes: the detachment of the basal body from the elongating, condensing sperm nuclei, and frequent deformations of the elongated condensed sperm nuclei. The process of plasma membrane investment of the elongated spermatids termed individualization commences, but is never completed, in the *spag4* mutant, and mature sperm are not produced. The essential function of Spag4 thus appears to be attachment of the basal body to the nucleus, and the deformed nuclei and the failures in individualization may be secondary to the failure of the basal bodies to attach to the nuclei (Kracklauer et al., 2010). While Spag4 interacts neither with Klar nor with MSP-300 to fulfill its function, it interacts genetically both with p50/dynamitin, a component of the dynein complex, and with the coiled-coil protein Yuri Gagarin (Yuri), mutations in which show defects in gravitaxis (Armstrong, Texada, Munjaal, Baker, & Beckingham, 2006), and which is also essential for male fertility (Kracklauer et al., 2010; Texada, Simonette, Johnson, Deery, & Beckingham, 2008). Subcellular localization of Yuri is similar to Spag4, and aspects of the *yuri* mutant phenotype in the testis are parallel to that of *spag4*: specifically, elongated spermatid nuclei are often deformed, the basal bodies are disarrayed with respect to the elongated nuclei, and individualization largely fails to occur (Texada et al., 2008). The behavior of Yuri and Spag4 thus evokes that of a typical KASH–SUN pair of proteins. However, none of the Yuri isoforms has an obvious KASH domain. Therefore, it remains open whether the attachment of the basal body to the spermatid nucleus actually involves a typical LINC complex in *Drosophila*.

Together, the recent findings suggest that spermiogenesis-specific LINC or LINC-like complexes connect the differentiating spermatid nucleus to the surrounding cytoskeletal structures to transfer forces for its well-directed shaping and elongation, which in turn is a most critical parameter for male fertility (Göb et al., 2010; Kracklauer et al., 2010). Previous studies have localized several cytoplasmic motor proteins to the sperm-specific cytoskeletal structures, where they function in vesicle transport and/or sperm head shaping (for overview, see Kierszenbaum et al., 2011). Interestingly, in *Drosophila*, the absence of Dlc90F, a core component of the sperm-specific dynein motor complex, which in turn is associated with Spag4, results in failure of centriole–nucleus association. In addition, Dlc90F-dependent dynein complexes appear to have a key role in nuclear shaping, as in the *dlc90F* mutant background, spermatid nuclei frequently fail to elongate (Li, Serr, Newman, & Hays, 2004). Consistent with a role for motor complexes, in rat and *Octopus*, KIF1C or related motor proteins have been found tightly associated with NE at the posterior manchette–nuclear interface of elongating spermatids, and it was suggested that they contribute to generation and transmission of forces needed for the shaping of nuclei (Wang, Zhu, Yu, Tan, & Yang, 2010; Yang, Jefferson, & Sperry, 2006; Yang & Sperry, 2003). Furthermore, mammalian Nesprin1 was shown to interact with the Kif3B subunit of kinesin-2. Kif3B appears to be highly expressed during spermiogenesis and thus may be involved in transferring forces to the NE that are generated by the kinesin-2 complex (Dang et al., 2012; Fan & Beck, 2004).

6. CONCLUDING REMARKS

Several mechanisms control different aspects of germ cell development and differentiation. In this respect, a number of recent studies in the field have established a central role of LINC complexes in the key processes of genome haploidization and sperm head formation. Conclusive evidence has been provided that in meiosis, the SUN/KASH-based NE bridges act at the front line of the tightly coordinated, highly dynamic chromosome movements, which are required for faithful homolog pairing, synapsis, and recombination. Furthermore, LINC complexes have been demonstrated to have essential functions in sperm formation in that they anchor the nucleus to polarized cytoskeletal elements and transfer forces to shape nuclei. Although much progress has been made in this matter, some important issues remain open, and some recent findings have even raised new questions. For example, as exemplified in Table 5.1, many of the key players are still unknown in various model organisms and remain to be identified to give a complete picture of the central, conserved mechanisms and their species-specific deviations. A further, quite challenging task involves the identification of the key factors that regulate telomere–SUN attachment in multicellular organisms. Interestingly, two recent studies reported that the nucleoplasmic domain of *C. elegans* SUN-1/Matefin is target for both CHK-2- and PLK-2-dependent meiosis-specific phosphorylation. Phosphorylation of particular serine residues in the nucleoplasmic domain of SUN-1/Matefin is dispensable for telomere attachment, but is essentially required for telomere movement and attachment to the centrosome (Labella, Woglar, Jantsch, & Zetka, 2011; Penkner et al., 2009). CHK-2-dependent SUN-1 phosphorylation is induced by DNA damage, which is generated by Spo11 (Penkner et al., 2009). On the other hand, PLK-2, which phosphorylates SUN-1 at a different site, also coordinates pairing and synapsis (Labella et al., 2011). Therefore, it is tempting to speculate that selected kinases operating in the nucleus assume responsibility for fine-tuning of telomere-led chromosome motions and coordination of processes at the NE with the events in the nuclear interior, an important aspect that definitely needs to be followed up in the future.

An astonishing outcome in past years is that sperm differentiation, in particular in mammals, relies on a number of unique, cell type-specific LINC components. This raises a number of yet unsolved questions: Why could sperm head formation require such a divergent set of novel SUN proteins, and does not cope with the common SUN proteins? What are the peculiarities and specific functions of the individual LINC complexes formed by these proteins? These are only a few of the remaining points and, thus, future work will need to address these and other interesting issues.

ACKNOWLEDGMENTS

We apologize to those colleagues whose important studies in this field were not discussed in detail because of space limitations. Our research received financial support from the German Research Foundation (DFG), grant Al 1090/1-1, SPP1384 "Mechanisms of Genome Haploidization" and the Graduate School 1048 "Molecular basis of organ development in vertebrates."

REFERENCES

Alsheimer, M. (2009). The dance floor of meiosis: Evolutionary conservation of nuclear envelope attachment and dynamics of meiotic telomeres. *Genome Dynamics*, *5*, 81–93.

Alsheimer, M., & Benavente, R. (1996). Change of karyoskeleton during mammalian spermatogenesis: Expression pattern of nuclear lamin C2 and its regulation. *Experimental Cell Research*, *228*, 181–188.

Alsheimer, M., Fecher, E., & Benavente, R. (1998). Nuclear envelope remodelling during rat spermiogenesis: Distribution and expression pattern of LAP2/thymopoietins. *Journal of Cell Science*, *111*, 2227–2234.

Alsheimer, M., Liebe, B., Sewell, L., Stewart, C. L., Scherthan, H., & Benavente, R. (2004). Disruption of spermatogenesis in mice lacking A-type lamins. *Journal of Cell Science*, *117*, 1173–1178.

Alsheimer, M., von Glasenapp, E., Hock, R., & Benavente, R. (1999). Architecture of the nuclear periphery of rat pachytene spermatocytes: Distribution of nuclear envelope proteins in relation to synaptonemal complex attachment sites. *Molecular Biology of the Cell*, *10*, 1235–1245.

Alsheimer, M., von Glasenapp, E., Schnölzer, M., Heid, H., & Benavente, R. (2000). Meiotic lamin C2: The unique aminoterminal hexapeptide GNAEGR is essential for nuclear envelope association. *Proceedings of the National Academy of Sciences of the United States of America*, *97*, 13120–13125.

Armstrong, J. D., Texada, M. J., Munjaal, R., Baker, D. A., & Beckingham, K. M. (2006). Gravitaxis in *Drosophila melanogaster*: A forward genetic screen. *Genes, Brain, and Behavior*, *5*, 222–239.

Bass, H. W., Marshall, W. F., Sedat, J. W., Agard, D. A., & Cande, W. Z. (1997). Telomeres cluster *de novo* before the initiation of synapsis: A three-dimensional spatial analysis of telomere positions before and during meiotic prophase. *The Journal of Cell Biology, 137*, 5–18.

Baudrimont, A., Penkner, A., Woglar, A., Machacek, T., Wegrostek, C., Gloggnitzer, J., et al. (2010). Leptotene/zygotene chromosome movement via the SUN/KASH protein bridge in *Caenorhabditis elegans*. *PLoS Genetics, 6*, e1001219.

Burke, B., & Roux, K. J. (2009). Nuclei take a position: Managing nuclear location. *Developmental Cell, 17*, 587–597.

Chi, Y. H., Cheng, L. I., Myers, T., Ward, J. M., Williams, E., Su, Q., et al. (2009). Requirement for Sun1 in the expression of meiotic reproductive genes and piRNA. *Development, 136*, 965–973.

Chikashige, Y., Haraguchi, T., & Hiraoka, Y. (2007). Another way to move chromosomes. *Chromosoma, 116*, 497–505.

Chikashige, Y., Haraguchi, T., & Hiraoka, Y. (2010). Nuclear envelope attachment is not necessary for telomere function in fission yeast. *Nucleus, 1*, 481–486.

Chikashige, Y., Tsutsumi, C., Yamane, M., Okamasa, K., Haraguchi, T., & Hiraoka, Y. (2006). Meiotic proteins bqt1 and bqt2 tether telomeres to form the bouquet arrangement of chromosomes. *Cell, 125*, 59–69.

Chikashige, Y., Yamane, M., Okamasa, K., Tsutsumi, C., Kojidani, T., Sato, M., et al. (2009). Membrane proteins Bqt3 and -4 anchor telomeres to the nuclear envelope to ensure chromosomal bouquet formation. *The Journal of Cell Biology, 187*, 413–427.

Cohen, T. V., Hernandez, L., & Stewart, C. L. (2008). Functions of the nuclear envelope and lamina in development and disease. *Biochemical Society Transactions, 36*, 1329–1334.

Conrad, M. N., Dominguez, A. M., & Dresser, M. E. (1997). Ndj1p, a meiotic telomere protein required for normal chromosome synapsis and segregation in yeast. *Science, 276*, 1252–1255.

Conrad, M. N., Lee, C. Y., Chao, G., Shinohara, M., Kosaka, H., Shinohara, A., et al. (2008). Rapid telomere movement in meiotic prophase is promoted by NDJ1, MPS3, and CSM4 and is modulated by recombination. *Cell, 133*, 1175–1187.

Conrad, M. N., Lee, C. Y., Wilkerson, J. L., & Dresser, M. E. (2007). MPS3 mediates meiotic bouquet formation in Saccharomyces cerevisiae. *Proceedings of the National Academy of Sciences of the United States of America, 104*, 8863–8868.

Cooper, J. P., Watanabe, Y., & Nurse, P. (1998). Fission yeast *Taz1* protein is required for meiotic telomere clustering and recombination. *Nature, 392*, 828–831.

Crisp, M., Liu, Q., Roux, K., Rattner, J. B., Shanahan, C., Burke, B., et al. (2006). Coupling of the nucleus and cytoplasm: Role of the LINC complex. *The Journal of Cell Biology, 172*, 41–53.

Dadoune, J.-P. (2003). Expression of mammalian spermatozoal nucleoproteins. *Microscopy Research and Technique, 61*, 56–75.

Dang, R., Zhu, J. Q., Tanm, F. Q., Wangm, W., Zhoum, H., & Yang, W. X. (2012). Molecular characterization of a KIF3B-like kinesin gene in the testis of *Octopus tankahkeei* (Cephalopoda, Octopus). *Molecular Biology Reports, 39*, 5589–5598.

Dechat, T., Adam, S. A., Taimen, P., Shimi, T., & Goldman, R. D. (2010). Nuclear lamins. *Cold Spring Harbor Perspectives in Biology, 2*, a000547.

Ding, X., Xu, R., Yu, J., Xu, T., Zhuang, Y., & Han, M. (2007). SUN1 is required for telomere attachment to nuclear envelope and gametogenesis in mice. *Developmental Cell, 12*, 863–872.

Dresser, M. E. (2008). Chromosome mechanics and meiotic engine maintenance. *PLoS Genetics, 4*, e1000210.

Eisen, G. (1900). The spermatogenesis of Batrachoseps. *Journal of Morphology, 17*, 1–117.

Esponda, P., & Gimenez-Martin, G. (1972). The attachment of the synaptonemal complex to the nuclear envelope. An ultrastructural and cytochemical analysis. *Chromosoma, 38*, 405–417.

Fan, J., & Beck, K. A. (2004). A role for the spectrin superfamily member Syne-1 and kinesin II in cytokinesis. *Journal of Cell Science, 117*, 619–629.

Fridkin, A., Mills, E., Margalit, A., Neufeld, E., Lee, K. K., Feinstein, N., et al. (2004). Matefin, a *Caenorhabditis elegans* germ line-specific SUN-domain nuclear membrane protein, is essential for early embryonic and germ cell development. *Proceedings of the National Academy of Sciences of the United States of America, 101*, 6987–6992.

Fridkin, A., Penkner, A., Jantsch, V., & Gruenbaum, Y. (2009). SUN-domain and KASH-domain proteins during development, meiosis and disease. *Cellular and Molecular Life Sciences, 66*, 1518–1533.

Frohnert, C., Schweizer, S., & Hoyer-Fender, S. (2011). SPAG4L/SPAG4L-2 are testis-specific SUN domain proteins restricted to the apical nuclear envelope of round spermatids facing the acrosome. *Molecular Human Reproduction, 17*, 207–218.

Furukawa, K., Inagaki, H., & Hotta, Y. (1994). Identification and cloning of an mRNA coding for a germ cell-specific A-type lamin in mice. *Experimental Cell Research, 212*, 426–430.

Gelei, J. (1921). Weitere Studien über die Oogenese des Dendrocoelum lacteum. II. Die Längskonjugation der Chromosomen. *Archiv Für Zellforschung, 16*, 88–169.

Gliki, G., Ebnet, K., Aurrand-Lions, M., Imhof, B. A., & Adams, R. H. (2004). Spermatid differentiation requires the assembly of a cell polarity complex downstream of junctional adhesion molecule-C. *Nature, 431*, 320–324.

Göb, E., Meyer-Natus, E., Benavente, R., & Alsheimer, M. (2011). Expression of individual mammalian Sun1 isoforms depends on the cell type. *Communicative and Integrative Biology, 4*, 440–442.

Göb, E., Schmitt, J., Benavente, R., & Alsheimer, M. (2010). Mammalian sperm head formation involves different polarization of two novel LINC-complexes. *PLoS One, 5*, e12072.

Gruenbaum, Y., Margalit, A., Goldman, R. D., Shumaker, D. K., & Wilson, K. L. (2005). The nuclear lamina comes of age. *Nature Reviews. Molecular Cell Biology, 6*, 21–31.

Hagan, I., & Yanagida, M. (1990). Novel potential mitotic motor protein encoded by the fission yeast *cut7*$^{+}$ gene. *Nature, 347*, 563–566.

Hagan, I., & Yanagida, M. (1995). The product of the spindle formation gene sad1 + associates with the fission yeast spindle pole body and is essential for viability. *The Journal of Cell Biology, 129*, 1033–1047.

Haque, F., Lloyd, D. J., Smallwood, D. T., Dent, C. L., Shanahan, C. M., Fry, A. M., et al. (2006). SUN1 interacts with nuclear lamin A and cytoplasmic nesprins to provide a physical connection between the nuclear lamina and the cytoskeleton. *Molecular and Cellular Biology, 26*, 3738–3751.

Haque, F., Mazzeo, D., Patel, J. T., Smallwood, D. T., Ellis, J. A., Shanahan, C. M., et al. (2010). Mammalian SUN protein interaction networks at the inner nuclear membrane and their role in laminopathy disease processes. *The Journal of Biological Chemistry, 285*, 3487–3498.

Harper, L., Golubovskaya, I., & Cande, W. Z. (2004). A bouquet of chromosomes. *Journal of Cell Science, 117*, 4025–4032.

Hermo, L., Pelletier, R. M., Cyr, D. G., & Smith, C. E. (2010a). Surfing the wave, cycle, life history, and genes/proteins expressed by testicular germ cells. Part 1: Background to spermatogenesis, spermatogonia and spermatocytes. *Microscopy Research and Technique, 73*, 241–278.

Hermo, L., Pelletier, R. M., Cyr, D. G., & Smith, C. E. (2010b). Surfing the wave, cycle, life history and genes/proteins expressed by testicular germ cells. Part 2: Changes in spermatid organelles associated with development of spermatozoa. *Microscopy Research and Technique*, *73*, 279–319.

Hodzic, D. M., Yeater, D. B., Bengtsson, L., Otto, H., & Stahl, P. D. (2004). Sun2 is a novel mammalian inner nuclear membrane protein. *The Journal of Biological Chemistry*, *279*, 25805–25812.

Jahn, D., Schramm, S., Benavente, R., & Alsheimer, M. (2010). Dynamic properties of meiosis-specific lamin C2 and its impact on nuclear envelope integrity. *Nucleus*, *1*, 273–283.

Jaspersen, S. L., Giddings, T. H., & Winey, M. (2002). Mps3p is a novel component of the yeast spindle pole body that interacts with the yeast centrin homologue Cdc31p. *The Journal of Cell Biology*, *159*, 945–956.

Jaspersen, S. L., Martin, A. E., Glazko, G., Giddings, T. H., Morgan, G., Mushegian, A., et al. (2006). The Sad1-UNC-84 homology domain in Mps3 interacts with Mps2 to connect the spindle pole body with the nuclear envelope. *The Journal of Cell Biology*, *174*, 665–675.

Ketema, M., Wilhelmsen, K., Kuikman, I., Janssen, H., Hodzic, D., & Sonnenberg, A. (2007). Requirements for the localization of nesprin-3 at the nuclear envelope and its interaction with plectin. *Journal of Cell Science*, *120*, 3384–3394.

Khatau, S. B., Hale, C. M., Stewart-Hutchinson, P. J., Patel, M. S., Stewart, C. L., Searson, P. C., et al. (2009). A perinuclear actin cap regulates nuclear shape. *Proceedings of the National Academy of Sciences of the United States of America*, *106*, 19017–19022.

Kierszenbaum, A. L., Rivkin, E., & Tres, L. L. (2003a). Acroplaxome, an F-actin-keratin-containing plate, anchors the acrosome to the nucleus during shaping of the spermatid head. *Molecular Biology of the Cell*, *14*, 4628–4640.

Kierszenbaum, A. L., Rivkin, E., & Tres, L. L. (2003b). The actin-based motor myosin Va is a component of the acroplaxome, an acrosome-nuclear envelope junctional plate, and of manchette-associated vesicles. *Cytogenetic and Genome Research*, *103*, 337–344.

Kierszenbaum, A. L., Rivkin, E., & Tres, L. L. (2011). Cytoskeletal track selection during cargo transport in spermatids is relevant to male fertility. *Spermatogenesis*, *1*, 221–230.

Kierszenbaum, A. L., & Tres, L. L. (2004). The acrosome-acroplaxome-manchette complex and the shaping of the spermatid head. *Archives of Histology and Cytology*, *67*, 271–284.

King, M. C., Drivas, T. G., & Blobel, G. (2008). A network of nuclear envelope membrane proteins linking centromeres to microtubules. *Cell*, *134*, 427–438.

Kosaka, H., Shinohara, M., & Shinohara, A. (2008). Csm4-dependent telomere movement on nuclear envelope promotes meiotic recombination. *PLoS Genetics*, *4*, e1000196.

Koszul, R., Kim, K. P., Prentiss, M., Kleckner, N., & Kameoka, S. (2008). Meiotic chromosomes move by linkage to dynamic actin cables with transduction of force through the nuclear envelope. *Cell*, *133*, 1188–1201.

Koszul, R., & Kleckner, N. (2009). Dynamic chromosome movements during meiosis: A way to eliminate unwanted connections? *Trends in Cell Biology*, *19*, 716–724.

Kracklauer, M. P., Banks, S. M., Xie, X., Wu, Y., & Fischer, J. A. (2007). Drosophila klaroid encodes a SUN domain protein required for Klarsicht localization to the nuclear envelope and nuclear migration in the eye. *Fly*, *1*, 75–85.

Kracklauer, M. P., Wiora, H. M., Deery, W. J., Chen, X., Bolival, B., Romanowicz, D., et al. (2010). The Drosophila SUN protein Spag4 cooperates with the coiled-coil protein Yuri Gagarin to maintain association of the basal body and spermatid nucleus. *Journal of Cell Science*, *123*, 2763–2772.

Labella, S., Woglar, A., Jantsch, V., & Zetka, M. (2011). Polo kinases establish links between meiotic chromosomes and cytoskeletal forces essential for homolog pairing. *Developmental Cell*, *21*, 948–958.

Lake, C. M., & Hawley, R. S. (2012). The molecular control of meiotic chromosomal behavior: Events in early meiotic prophase in Drosophila oocytes. *Annual Review of Physiology*, *74*, 425–451.

Lei, K., Zhang, X., Ding, X., Guo, X., Chen, M., Zhu, B., et al. (2009). SUN1 and SUN2 play critical but partially redundant roles in anchoring nuclei in skeletal muscle cells in mice. *Proceedings of the National Academy of Sciences of the United States of America, 104*, 7426–7431.

Li, M.-G., Serr, M., Newman, E. A., & Hays, T. S. (2004). The Drosophila tctex-1 light chain is dispensable for essential cytoplasmic dynein functions but is required during spermatid differentiation. *Molecular Biology of the Cell, 15*, 3005–3014.

Liebe, B., Alsheimer, M., Höög, C., Benavente, R., & Scherthan, H. (2004). Telomere attachment, meiotic chromosome condensation, pairing, and bouquet stage duration are modified in spermatocytes lacking axial elements. *Molecular Biology of the Cell, 15*, 827–837.

Link, J., Jahn, D., Schmitt, J., Göb, E., Baar, J., Ortega, S., et al. (2012). The nuclear lamina regulates chromosome dynamics and promotes efficient homologous recombination during mouse meiosis (submitted for publication).

Liu, Q., Pante, N., Misteli, T., Elsagga, M., Crisp, M., Hodzic, D., Burke, B., & Roux, K. J. (2007). Functional association of Sun1 with nuclear pore complexes. *J. Cell Biol., 178*, 785–798.

Malone, C. J., Fixsen, W. D., Horvitz, H. R., & Han, M. (1999). UNC-84 localizes to the nuclear envelope and is required for nuclear migration and anchoring during C. elegans development. *Development, 126*, 3171–3181.

Malone, C. J., Misner, L., Le Bot, N., Tsai, M. C., Campbell, J. M., Ahringer, J., et al. (2003). The C. elegans hook protein, ZYG-12, mediates the essential attachment between the centrosome and nucleus. *Cell, 115*, 825–836.

McGee, M. D., Rillo, R., Anderson, A. S., & Starr, D. A. (2006). UNC-83 is a KASH protein required for nuclear migration and is recruited to the outer nuclear membrane by a physical interaction with the SUN protein UNC-84. *Molecular Biology of the Cell, 17*, 1790–1801.

McGee, M. D., Stagljar, I., & Starr, D. A. (2009). KDP-1 is a nuclear envelope KASH protein required for cell-cycle progression. *Journal of Cell Science, 122*, 2895–2905.

Meistrich, M. L., Trostle-Weige, P. K., & Russell, L. D. (1990). Abnormal manchette development in spermatids of azh/azh mutant mice. *The American Journal of Anatomy, 188*, 74–86.

Méjat, A., & Misteli, T. (2010). LINC complexes in health and disease. *Nucleus, 1*, 40–52.

Mendoza-Lujambio, I., Burfeind, P., Dixkens, C., Meinhardt, A., Hoyer-Fender, S., Engel, W., et al. (2002). The Hook1 gene is non-functional in the abnormal spermatozoon head shape (azh) mutant mouse. *Human Molecular Genetics, 11*, 1647–1658.

Meyerzon, M., Fridolfsson, H. N., Ly, N., McNally, F. J., & Starr, D. A. (2009). UNC-83 is a nuclear-specific cargo adaptor for kinesin-1-mediated nuclear migration. *Development, 136*, 2725–2733.

Miki, F., Kurabayashi, A., Tange, Y., Okazaki, K., Shimanuki, M., & Niwa, O. (2004). Two-hybrid search for proteins that interact with Sad1 and Kms1, two membrane-bound components of the spindle pole body in fission yeast. *Molecular Genetics and Genomics, 270*, 449–461.

Miki, F., Okazaki, K., Shimanuki, M., Yamamoto, A., Hiraoka, Y., & Niwa, O. (2002). The 14 kDa dynein light chain-family protein Dlc1 is required for regular oscillatory nuclear movement and efficient recombination during meiotic prophase in fission yeast. *Molecular Biology of the Cell, 13*, 930–946.

Morimoto, A., Shibuya, H., Zhu, X., Kim, J., Ishiguro, K., Han, M., et al. (2012). A conserved KASH domain protein associates with telomeres, SUN1, and dynactin during mammalian meiosis. *The Journal of Cell Biology, 198*, 165–172.

Moses, M. J. (1956a). Chromosomal structures in crayfish spermatocytes. *The Journal of Biophysical and Biochemical Cytology, 2*, 215–218.

Moses, M. J. (1956b). Studies on nuclei using correlated cytochemical, light, and electron microscope techniques. *The Journal of Biophysical and Biochemical Cytology, 2* (Suppl.), 397–406.

Moses, M. J. (1977). Synaptonemal complex karyotyping in spermatocytes of the chinese hamster (*Cricetulus griseus*). *Chromosoma, 60*, 99–125.

Moses, M. J., & Solari, A. J. (1976). Positive contrast staining and protected drying of surface spreads: Electron microscopy of the synaptonemal complex by a new method. *Journal of Ultrastructure Research, 54*, 109–114.

Mosley-Bishop, K. L., Li, Q., Patterson, L., & Fischer, J. A. (1999). Molecular analysis of the *klarsicht* gene and its role in nuclear migration within differentiating cells of the *Drosophila* eye. *Current Biology, 9*, 1211–1220.

Mueller, S., Rosenquist, T. A., Takai, Y., Bronson, R. A., & Wimmer, E. (2003). Loss of nectin-2 at Sertoli-spermatid junctions leads to male infertility and correlates with severe spermatozoan head and midpiece malformation, impaired binding to the zona pellucida, and oocyte penetration. *Biology of Reproduction, 69*, 1330–1340.

Murphy, S. P., Simmons, C. R., & Bass, H. W. (2010). Structure and expression of the maize (Zea mays L.) SUN-domain protein gene family: Evidence for the existence of two divergent classes of SUN proteins in plants. *BMC Plant Biology, 10*, 269.

Mylonis, I., Drosou, V., Brancorsini, S., Nikolakaki, E., Sassone-Corsi, P., & Giannakouros, T. (2004). Temporal association of protamine 1 with the inner nuclear membrane protein lamin B receptor during spermiogenesis. *The Journal of Biological Chemistry, 279*, 11626–11631.

Nimmo, E. R., Pidoux, A. L., Perry, P. E., & Allshire, R. C. (1998). Defective meiosis in telomere-silencing mutants of *Schizosaccharomyces pombe*. *Nature, 392*, 825–828.

Niwa, O., Shimanuki, M., & Miki, F. (2000). Telomere-led bouquet formation facilitates homologous chromosome pairing and restricts ectopic interaction in fission yeast meiosis. *The EMBO Journal, 19*, 3831–3840.

Oda, Y., & Fukuda, H. (2011). Dynamics of Arabidopsis SUN proteins during mitosis and their involvement in nuclear shaping. *The Plant Journal, 66*, 629–641.

Olins, A. L., Hoang, T. V., Zwerger, M., Herrmann, H., Zentgraf, H., Noegel, A. A., et al. (2009). The LINC-less granulocyte nucleus. *European Journal of Cell Biology, 88*, 203–214.

Padmakumar, V. C., Abraham, S., Braune, S., Noegel, A. A., Tunggal, B., Karakesisoglou, I., et al. (2004). Enaptin, a giant actin-binding protein, is an element of the outer nuclear membrane and the actin cytoskeleton. *Experimental Cell Research, 295*, 330–339.

Padmakumar, V. C., Libotte, T., Lu, W., Zaim, H., Abraham, S., Noegel, A. A., et al. (2005). The inner nuclear membrane protein Sun1 mediates the anchorage of Nesprin-2 to the nuclear envelope. *Journal of Cell Science, 118*, 3419–3430.

Parvinen, M., & Söderström, K. O. (1976). Chromosome rotation and formation of synapsis. *Nature, 260*, 534–535.

Patterson, K., Molofsky, A. B., Robinson, C., Acosta, S., Cater, C., & Fischer, J. A. (2004). The functions of Klarsicht and nuclear lamin in developmentally regulated nuclear migrations of photoreceptor cells in the Drosophila eye. *Molecular Biology of the Cell, 15*, 600–610.

Penkner, A. M., Fridkin, A., Gloggnitzer, J., Baudrimont, A., Machacek, T., Woglar, A., et al. (2009). Meiotic chromosome homology search involves modifications of the nuclear envelope protein Matefin/SUN-1. *Cell, 139*, 920–933.

Penkner, A., Tang, L., Novatchkova, M., Ladurner, M., Fridkin, A., Gruenbaum, Y., et al. (2007). The nuclear envelope protein Matefin/SUN-1 is required for homologous pairing in C. elegans meiosis. *Developmental Cell, 12*, 873–885.

Pierre, V., Martinez, G., Coutton, C., Delaroche, J., Yassine, S., Novella, C., et al. (2012). Absence of Dpy19l2, a new inner nuclear membrane protein, causes globozoospermia in mice by preventing the anchoring of the acrosome to the nucleus. *Development, 139*, 2955–2965.

Platner, G. (1885). Ueber die Entstehung des Nebenkerns und seine Beziehung zur Kerntheilung. *Archiv für Mikroskopische Anatomie, 26*, 343–369.

Rasmussen, S. W. (1986). Chromosome interlocking during synapsis—A transient disorder. *The Tokai Journal of Experimental and Clinical Medicine, 11*, 437–451.

Razafsky, D., & Hodzic, D. (2009). Bringing KASH under the SUN: The many faces of nucleo-cytoskeletal connections. *The Journal of Cell Biology, 186*, 461–472.

Rockmill, B., & Roeder, G. S. (1998). Telomere-mediated chromosome pairing during meiosis in budding yeast. *Genes & Development, 12*, 2574–2586.

Roux, K. J., Crisp, M. L., Liu, Q., Kim, D., Kozlov, S., Stewart, C. L., et al. (2009). Nesprin 4 is an outer nuclear membrane protein that can induce kinesin-mediated cell polarization. *Proceedings of the National Academy of Sciences of the United States of America, 106*, 2194–2199.

Russell, L. D., Russell, J. A., MacGregor, G. R., & Meistrich, M. L. (1991). Linkage of manchette microtubules to the nuclear envelope and observations of the role of the manchette in nuclear shaping during spermiogenesis in rodents. *The American Journal of Anatomy, 192*, 97–120.

Sato, A., Isaac, B., Phillips, C. M., Rillo, R., Carlton, P. M., Wynne, D. J., et al. (2009). Cytoskeletal forces span the nuclear envelope to coordinate meiotic chromosome pairing and synapsis. *Cell, 139*, 907–919.

Scherthan, H. (2001). A bouquet makes ends meet. *Nature Reviews. Molecular Cell Biology, 2*, 621–627.

Scherthan, H., Bahler, J., & Kohli, J. (1994). Dynamics of chromosome organization and pairing during meiotic prophase in fission yeast. *The Journal of Cell Biology, 127*, 273–285.

Scherthan, H., Wang, H., Adelfalk, C., White, E. J., Cowan, C., Cande, W. Z., et al. (2007). Chromosome mobility during meiotic prophase in *Saccharomyces cerevisiae*. *Proceedings of the National Academy of Sciences of the United States of America, 104*, 16934–16939.

Scherthan, H., Weich, S., Schwegler, H., Heyting, C., Harle, M., & Cremer, T. (1996). Centromere and telomere movements during early meiotic prophase of mouse and man are associated with the onset of chromosome pairing. *The Journal of Cell Biology, 134*, 1109–1125.

Schmitt, J., Benavente, R., Hodzic, D., Höög, C., Stewart, C. L., & Alsheimer, M. (2007). Transmembrane protein Sun2 is involved in tethering mammalian meiotic telomeres to the nuclear envelope. *Proceedings of the National Academy of Sciences of the United States of America, 104*, 7426–7431.

Schütz, W., Alsheimer, M., Öllinger, R., & Benavente, R. (2005a). Nuclear envelope remodelling during mouse spermiogenesis: Postmeiotic expression and redistribution of germline lamin B3. *Experimental Cell Research, 307*, 285–291.

Schütz, W., Benavente, R., & Alsheimer, M. (2005b). Dynamic properties of germline-specific lamin B3: The role of the shortened rod domain. *European Journal of Cell Biology, 84*, 649–662.

Seifert, G. J., Lawson, D., & Wiche, G. (1992). Immunolocalization of the intermediate filament-associated protein plectin at focal contacts and actin stress fibers. *European Journal of Cell Biology, 59*, 138–147.

Shao, X., Tarnasky, H. A., Lee, J. P., Oko, R., & van der Hoorn, F. A. (1999). Spag4, a novel sperm protein, binds outer dense-fiber protein Odf1 and localizes to microtubules of manchette and axoneme. *Developmental Biology, 211*, 109–123.

Sheehan, M. J., & Pawlowski, W. P. (2009). Live imaging of rapid chromosome movements in meiotic prophase I in maize. *Proceedings of the National Academy of Sciences of the United States of America, 106*, 20989–20994.

Shimanuki, M., Miki, F., Ding, D. Q., Chikashige, Y., Hiraoka, Y., Horio, T., et al. (1997). A novel fission yeast gene, *kms1+*, is required for the formation of meiotic prophase-specific nuclear architecture. *Molecular & General Genetics*, *254*, 238–249.

Sonntag Brown, M., Zanders, S., & Alani, E. (2011). Sustained and rapid chromosome movements are critical for chromosome pairing and meiotic progression in budding yeast. *Genetics*, *188*, 21–32.

Sosa, B. A., Rothballer, A., Kutay, U., & Schwartz, T. U. (2012). LINC complexes form by binding of three KASH peptides to domain interfaces of trimeric SUN proteins. *Cell*, *149*, 1035–1047.

Starr, D. A. (2009). A nuclear-envelope bridge positions nuclei and moves chromosomes. *Journal of Cell Science*, *122*, 577–586.

Starr, D. A., & Fischer, J. A. (2005). KASH 'n Karry: The KASH domain family of cargo-specific cytoskeletal adaptor proteins. *Bioessays*, *27*, 1136–1146.

Starr, D. A., & Fridolfsson, H. N. (2010). Interactions between nuclei and the cytoskeleton are mediated by SUN-KASH nuclear-envelope bridges. *Annual Review of Cell and Developmental Biology*, *26*, 421–444.

Starr, D. A., & Han, M. (2002). Role of ANC-1 in tethering nuclei to the actin cytoskeleton. *Science*, *298*, 406–409.

Starr, D. A., Hermann, G. J., Malone, C. J., Fixsen, W., Priess, J. R., Horvitz, H. R., et al. (2001). unc-83 encodes a novel component of the nuclear envelope and is essential for proper nuclear migration. *Development*, *128*, 5039–5050.

Stewart-Hutchinson, P. J., Hale, C. M., Wirtz, D., & Hodzic, D. (2008). Structural requirements for the assembly of LINC complexes and their function in cellular mechanical stiffness. *Experimental Cell Research*, *314*, 1892–1905.

Tang, X., Jin, Y., & Cande, W. Z. (2006). Bqt2p is essential for initiating telomere clustering upon pheromone sensing in fission yeast. *The Journal of Cell Biology*, *173*, 845–851.

Technau, M., & Roth, S. (2008). The Drosophila KASH domain proteins Msp-300 and Klarsicht and the SUN domain protein Klaroid have no essential function during oogenesis. *Fly*, *2*, 82–91.

Texada, M. J., Simonette, R. A., Johnson, C. B., Deery, W. J., & Beckingham, K. M. (2008). Yuri gagarin is required for actin, tubulin and basal body functions in Drosophila spermatogenesis. *Journal of Cell Science*, *121*, 1926–1936.

Tokuyasu, K. T. (1974). Dynamics of spermiogenesis in *Drosophila melanogaster*. IV. Nuclear transformation. *Journal of Ultrastructure Research*, *48*, 284–303.

Tokuyasu, K. T. (1975). Dynamics of spermiogenesis in *Drosophila melanogaster*. V. Head–tail alignment. *Journal of Ultrastructure Research*, *50*, 117–129.

Tomita, K., & Cooper, J. P. (2006). The meiotic chromosomal bouquet: SUN collects flowers. *Cell*, *125*, 19–21.

Toshimori, K., & Ito, C. (2003). Formation and organization of the mammalian sperm head. *Archives of Histology and Cytology*, *66*, 383–396.

Trelles-Sticken, E., Adelfalk, C., Loidl, J., & Scherthan, H. (2005). Meiotic telomere clustering requires actin for its formation and cohesin for its resolution. *The Journal of Cell Biology*, *170*, 213–223.

Trelles-Sticken, E., Dresser, M. E., & Scherthan, H. (2000). Meiotic telomere protein Ndj1p is required for meiosis-specific telomere distribution, bouquet formation and efficient homologue pairing. *The Journal of Cell Biology*, *151*, 95–106.

Tsai, J. H., & McKee, B. D. (2011). Homologous pairing and the role of pairing centers in meiosis. *Journal of Cell Science*, *124*, 1955–1963.

Tzur, Y. B., Margalit, A., Melamed-Book, N., & Gruenbaum, Y. (2006a). Matefin/SUN-1 is a nuclear envelope receptor for CED-4 during *Caenorhabditis elegans* apoptosis. *Proceedings of the National Academy of Sciences of the United States of America*, *103*, 13397–13402.

Tzur, Y. B., Wilson, K. L., & Gruenbaum, Y. (2006b). SUN-domain proteins: 'Velcro' that links the nucleoskeleton to the cytoskeleton. *Nature Reviews. Molecular Cell Biology*, *7*, 782–788.

Vester, B., Smith, A., Krohne, G., & Benavente, R. (1993). Presence of a nuclear lamina in pachytene spermatocytes of the rat. *Journal of Cell Science*, *104*, 557–563.

Wanat, J. J., Kim, K. P., Koszul, R., Zanders, S., Weiner, B., Kleckner, N., et al. (2008). Csm4, in collaboration with Ndj1, mediates telomere-led chromosome dynamics and recombination during yeast meiosis. *PLoS Genetics*, *4*, e1000188.

Wang, W., Zhu, J. Q., Yu, H. M., Tan, F. Q., & Yang, W. X. (2010). KIFC1-like motor protein associates with the cephalopod manchette and participates in sperm nuclear morphogenesis in *Octopus tankahkeei*. *PLoS One*, *5*, e15616.

Wilcox, E. V. (1895). Spermatogenesis of caloptenus femurrubrum and cicada tibicen. *Bulletin of the Museum of Comparative Zoology at Harvard College*, *27*, 1–32.

Wilhelmsen, K., Litjens, S. H., Kuikman, I., Tshimbalanga, N., Janssen, H., van den Bout, I., et al. (2005). Nesprin-3, a novel outer nuclear membrane protein, associates with the cytoskeletal linker protein plectin. *The Journal of Cell Biology*, *171*, 799–810.

Xie, X., & Fischer, J. A. (2008). On the roles of the Drosophila KASH domain proteins Msp-300 and Klarsicht. *Fly*, *2*, 74–81.

Xiong, H., Rivero, F., Euteneuer, U., Mondal, S., Mana-Capelli, S., Larochelle, D., et al. (2008). Dictyostelium Sun-1 connects the centrosome to chromatin and ensures genome stability. *Traffic*, *9*, 708–724.

Yamamoto, A., West, R. R., McIntosh, J. R., & Hiraoka, Y. (1999). A cytoplasmic dynein heavy chain is required for oscillatory nuclear movement of meiotic prophase and efficient meiotic recombination in fission yeast. *The Journal of Cell Biology*, *145*, 1233–1249.

Yan, W. (2009). Male infertility caused by spermiogenic defects: Lessons from gene knockouts. *Molecular and Cellular Endocrinology*, *306*, 24–32.

Yang, W. X., Jefferson, H., & Sperry, A. O. (2006). The molecular motor KIFC1 associates with a complex containing nucleoporin NUP62 that is regulated during development and by the small GTPase RAN. *Biology of Reproduction*, *74*, 684–690.

Yang, W. X., & Sperry, A. O. (2003). C-terminal kinesin motor KIFC1 participates in acrosome biogenesis and vesicle transport. *Biology of Reproduction*, *69*, 1719–1729.

Yu, J., Starr, D. A., Wu, X., Parkhurst, S. M., Zhuang, Y., Xu, T., et al. (2006). The KASH domain protein MSP-300 plays an essential role in nuclear anchoring during *Drosophila* oogenesis. *Developmental Biology*, *289*, 336–345.

Zhang, Q., Skepper, J. N., Yang, F., Davies, J. D., Hegyi, L., Roberts, R. G., et al. (2001). Nesprins: A novel family of spectrin-repeat-containing proteins that localize to the nuclear membrane in multiple tissues. *Journal of Cell Science*, *114*, 4485–4498.

Zhen, Y. Y., Libotte, T., Munck, M., Noegel, A. A., & Korenbaum, E. (2002). NUANCE, a giant protein connecting the nucleus and actin cytoskeleton. *Journal of Cell Science*, *115*, 3207–3222.

Zhou, X., Graumann, K., Evans, D. E., & Meier, I. (2012). Novel plant SUN–KASH bridges are involved in RanGAP anchoring and nuclear shape determination. *The Journal of Cell Biology*, *196*, 203–211.

Zhou, K., Rolls, M. M., Hall, D. H., Malone, C. J., & Hanna-Rose, W. (2009). A ZYG-12-dynein interaction at the nuclear envelope defines cytoskeletal architecture in the *C. elegans* gonad. *The Journal of Cell Biology*, *186*, 229–241.

Zickler, D., & Kleckner, N. (1998). The leptotene–zygotene transition of meiosis. *Annual Review of Genetics*, *32*, 619–697.

CHAPTER SIX

Small RNAs in Germline Development

Matthew S. Cook[*,†,‡,1], **Robert Blelloch**[*,†,‡]
[*]Department of Urology, University of California, San Francisco, California, USA
[†]Eli and Edythe Broad Center of Regeneration Medicine and Stem Cell Research, University of California, San Francisco, California, USA
[‡]Center for Reproductive Sciences, University of California, San Francisco, California, USA
[1]Corresponding author: e-mail address: Matthew.Cook@ucsf.edu

Contents

Abstract

One of the most important and evolutionarily conserved strategies to control gene expression in higher metazoa is posttranscriptional regulation via small regulatory RNAs such as microRNAs (miRNAs), endogenous small interfering RNAs (endo-siRNAs), and piwi-interacting RNAs (piRNAs). Primordial germ cells, which are defined by their totipotent potential and noted for their dependence on posttranscriptional regulation by RNA-binding proteins, rely on these small regulatory RNAs for virtually every aspect of their development, including specification, migration, and differentiation into competent gametes. Here, we review current knowledge of the roles miRNAs, endo-siRNAs, and piRNAs play at all stages of germline development in various organisms, focusing on studies in the mouse.

Current Topics in Developmental Biology, Volume 102
ISSN 0070-2153
http://dx.doi.org/10.1016/B978-0-12-416024-8.00006-4

1. INTRODUCTION: GERMLINE DEVELOPMENT AND SMALL REGULATORY RNAs

Germ cells represent the immortal line: they are guardians of a totipotent genome and are essential for the genetic perpetuation of an individual organism and, ultimately, a species. An error at any stage in development (specification, migration, colonization, differentiation, adult maintenance) can lead to disastrous outcomes, including germ cell death or unchecked proliferation leading to tumorigenesis. Tight regulation of germ cell development occurs at both genetic and epigenetic levels and is essential to prevent the untimely manifestation of latent totipotency at any stage other than fertilization when the gametes fuse to complete the cycle of reproduction.

Small RNAs have proven to be robust regulators of pluripotency and cell fate in many contexts, including germ cells. Three major classes of small regulatory RNAs have been discovered: microRNAs (miRNAs), endogenous small interfering RNAs (endo-siRNAs), and piwi-interacting RNAs (piRNAs). Though each small RNA pathway is functionally distinct, related proteins are required for small RNA biogenesis and function in all pathways. All three types of small RNAs are present in germ cells and are required for critical events in germline development including primordial germ cell (PGC) specification, migration, gonad colonization, sex-specific differentiation into competent gametes, and fertilization, resulting in the early development and specification of the next founding PGC population. Using examples from vertebrate and invertebrate species, this chapter will review the origin and function of each type of small RNA during germline development, from PGC to gamete and back again.

1.1. Germ cell life cycle and posttranscriptional regulation of gene expression

Germ cell specification, development, and differentiation involve complex regulatory networks at the transcriptional as well as translational levels. In many species, germ cells contain a cytoplasmic RNA-rich regulatory center known generally as a germ granule, indicating a central role for translational regulation in the germline (Seydoux & Braun, 2006). In fact, at particular stages in germ cell development, transient transcriptional quiescence highlights the importance of posttranscriptional regulatory mechanisms involving small RNAs and RNA-binding proteins (RBPs). Regulation of translation involves choreographed interactions of the messenger RNA (mRNA) with both RBPs and various classes of small regulatory RNAs, including miRNAs, endo-siRNAs, and piRNAs.

The miRNAs function by destabilizing and suppressing the translation of target transcripts. Targeting occurs through partial complementation between the miRNA and the target mRNA. The mechanism by which endo-siRNAs and piRNAs function is less clear, but interactions with complementary RNAs or chromatin-modifying complexes are likely to be involved. RBPs can either promote or antagonize small RNA activity, depending on the molecular milieu and cellular context (Kedde & Agami, 2008). In this chapter, we will review the role of small RNAs at different stages of germ cell development, focusing largely on work reported from mice, but also including related mechanisms in other major genetic model organisms such as the fly, nematode, and zebrafish. The following sections will provide a very brief overview of mouse germline development. Figure 6.1 summarizes the life cycle of germ cells in the mouse and provides a timeline of major events. For a more comprehensive review of the different stages of germ cell development in mouse and other metazoa, please see the following references: for PGC specification (Durcova-Hills & Capel, 2008; Raz, 2003; Saitou & Yamaji, 2010; Santos & Lehmann, 2004), PGC migration (Kimble & Crittenden, 2005; Kunwar, Siekhaus, & Lehmann, 2006; Tarbashevich & Raz, 2010), and PGC gonad colonization and differentiation (Fuller & Spradling, 2007; Kimble, 2011; Western, 2009).

1.1.1 PGC specification and migration

An interesting feature of animal germ cells is that they are specified very early in development, long before the formation of the gonad, which is the site where germ cells will later reside and carry out their function of gametogenesis (Fig. 6.1A). This means that PGCs are molecularly distinguished from their somatic neighbors at a very early stage in the embryo and must then migrate through developing tissues to reach the nascent gonad, maintaining a germ cell identity as they transit through a multitude of somatic cell types and associated molecular signals.

Depending on the species, PGCs are specified through a process of either "preformation" or "induction." Invertebrate species such as *Drosophila melanogaster* (fly) and *Caenorhabditis elegans* (nematode), as well as vertebrate species like *Danio rerio* (zebrafish) and *Xenopus laevis* (frog) utilize "preformation," or the inheritance of maternally deposited germ plasm components, to specify PGCs (Strome & Lehmann, 2007). In other words, RNAs and/or

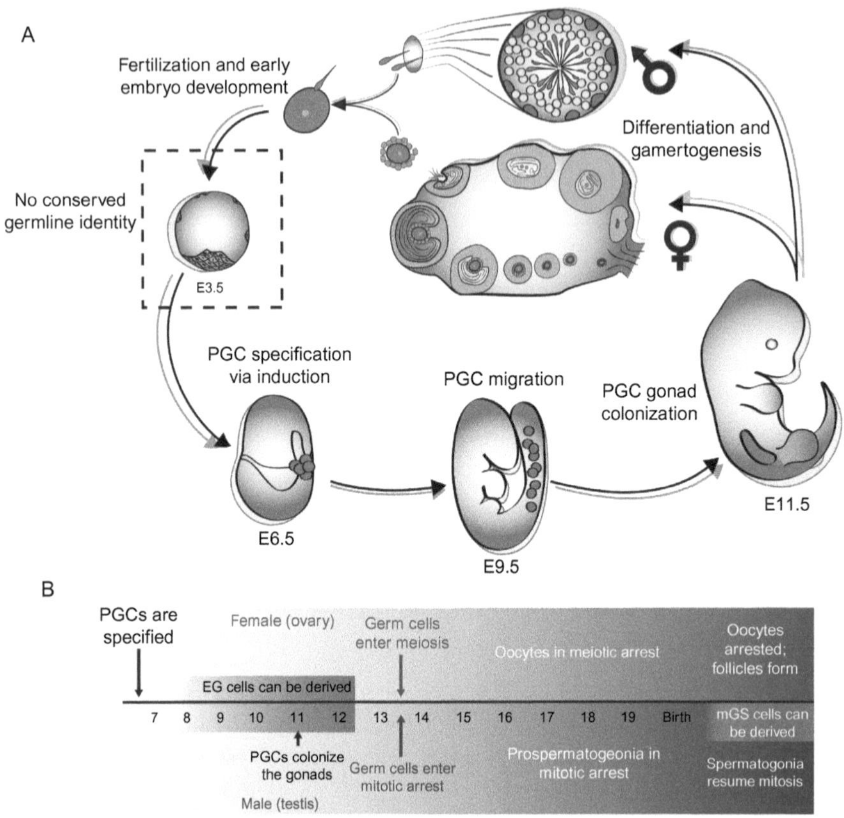

Figure 6.1 The life cycle of a mouse germ cell and a timeline of events. (A) Primordial germ cells (PGCs) are specified (green) from somatic progenitors in the developing mouse embryo around embryonic day 6.5 (E6.5). At this point, there is no gonad organ to populate and begin gametogenesis. Instead, PGCs migrate through the developing hindgut (E9.5) and enter the genital ridges to colonize the gonad by E11.5. After sex-specific differentiation into either mature sperm or an egg, haploid gametes fuse during fertilization and generate a totipotent zygote that gives rise to a blastocyst at E3.5. The dashed box indicates the stage during mouse development where no recognizable germline cell or conserved germ plasm component is expressed. This feature distinguishes the mouse from other vertebrate (zebrafish and frog) and invertebrate (fly and nematode) models, where the germ cell lineage is continuous and PGCs are induced via inheritance of germ plasm components maternally deposited in the mature oocyte. (B) A timeline of important events in the life cycle of mouse germ cells. PGCs are specified around E6.5. During the time that PGCs migrate and colonize the gonads, they can be explanted to form embryonic germ cells (EGCs), which resemble embryonic stem cells (ESCs) in many aspects. After gonad colonization, PGCs commit to their sex-specific fate: prospermatogonia in the male and oogonia in the female. Male germ cells enter mitotic arrest, and female germ cells immediately enter meiosis. After birth, male germ cells resume mitosis and maintain a spermatogonial stem cell that can be cultured to form multipotent germline stem (mGS) cells. Female germ cells associate with somatic granulosa cells to form follicles that will be induced to mature by hormones. (For interpretation of the references to color in this figure legend, the reader is referred to the online version of this chapter.)

proteins that define the germ cell fate are localized to a region of the early embryo such that the cells forming around those components "inherit" those properties and adopt the germ cell fate. Alternatively, in mammalian species, most notably the mouse (*Mus musculus*), specific signaling molecules instruct a small population of cells to adopt a PGC fate in a process known as "induction." Most evidence suggests that BMP4 induces mouse PGCs to form from somatic progenitors of the epiblast at embryonic day 6.5 (E6.5) amidst larger patterning events along the border of the extraembryonic ectoderm (Lawson et al., 1999; Ohinata et al., 2009; Tam & Zhou, 1996). The lack of expression of conserved germline genes in any cells of the early embryo preceding the induction of PGCs (Fig. 6.1A, dashed box) is a marked difference from species whose "preformation" strategy ensures an unbroken cellular lineage with the germline fate. In mouse, this method of induction is consistent with the highly regulative nature of early mammalian development. For example, early cell divisions appear to be symmetrical, with each cell maintaining equal potency to make all cell types. Distinct cell fates are only determined once inside–outside asymmetry occurs, just prior to the formation of the inner cell mass and trophectoderm at E2.5; PGCs are induced following establishment of anterior–posterior asymmetry 4 days later (E6.5).

The molecular basis of PGC specification in mice (Fig. 6.1A, E6.5) involves the silencing of genes associated with somatic cell differentiation (e.g., Hox genes). Suppression of somatic differentiation occurs through the action of the transcription factor BLIMP1 and associated chromatin-modifying complexes (Ancelin et al., 2006; Saitou, 2009), maintenance of the pluripotency transcriptional network (*Oct4*, *Nanog*, *Sox2,* and *Stella*), and activation of germ cell-specific genes (e.g., *Dnd1*, *Tdrd1*, and *Nanos3)*, many of which are RBPs involved in translational regulation (Ohinata et al., 2006; Saitou, Barton, & Surani, 2002; Yabuta, Kurimoto, Ohinata, Seki, & Saitou, 2006).

In *Drosophila*, PGCs form via preformation at the posterior pole where they inherit the germ plasm. Genes required for specifying PGC fate include *oskar*, *nanos*, *germ cell-less* , and *polar granule component*. Oskar is critical for defining the posterior region and localizing germ plasm components (Lehmann & Nusslein-Volhard, 1991), while Pgc inhibits phosphorylation of RNA polymerase II, thus suppressing transcription (Strome & Lehmann, 2007). The evolutionarily conserved RBP Nanos block somatic differentiation and is critical for germ cell specification and maintenance in many species from planaria to mouse (Forbes & Lehmann, 1998; Koprunner, Thisse,

Thisse, & Raz, 2001; Lehmann & Nusslein-Volhard, 1991; Subramaniam, 1999; Suzuki, Tsuda, & Saga, 2007; Tsuda et al., 2003; Wang, Zayas, Guo, & Newmark, 2007). A gradient of *nanos* in the fly embryo is necessary for posterior patterning and germ cell specification; this gene acts as a translational repressor of *hunchback* as well as other target transcripts (Irish, Lehmann, & Akam, 1989). These coordinated events restrict germ plasm transcripts to the pole cells of the fly embryo, thus specifying the PGC fate.

Once specified, PGCs in all species migrate to the site of the developing gonad via passive and active mechanisms. In *Drosophila*, migration is initially passive as the blastoderm folds and PGCs transition to the posterior midgut (Santos & Lehmann, 2004). The situation is similar in zebrafish and mouse where initial tissue growth and reorganization repositions the founding PGCs as they become motile. Whereas fly and mouse PGCs develop as one founding population, there are four founding populations of PGCs in zebrafish (totaling about 40–45 germ cells) that eventually make their way to the gonads (Raz, 2003; Weidinger, Wolke, Koprunner, Klinger, & Raz, 1999; Weidinger et al., 2002). Although the specification pattern of the founding PGC population differs in zebrafish, many of the same specification and migration cues are conserved, including Nanos (Koprunner et al., 2001) and Cxcr4 (Knaut, Werz, Geisler, & Nusslein-Volhard, 2003), respectively.

Mouse PGCs enter the hindgut and migrate along the dorsal aorta (Fig. 6.1A, E9.5) until they reach the genital ridge. During this time, they proliferate approximately every 16 h (Buehr, McLaren, Bartley, & Darling, 1993) and require specific survival factors to prevent apoptosis and promote cell division. Several factors known to be required for survival *in vivo* include *c-kit*, *kit-l*, TIAR, *Fgfr2-IIIb*, *Nanos3*, *Dnd1*, *Mvh*, *Ror2*, and *Wnt5a* (Beck, Miller, Anderson, & Streuli, 1998; Laird, Altshuler-Keylin, Kissner, Zhou, & Anderson, 2011; Sakurai, Iguchi, Moriwaki, & Noguchi, 1995; Suzuki, Tsuda, Kiso, & Saga, 2008; Takeuchi, Molyneaux, Runyan, Schaible, & Wylie, 2005; Tanaka et al., 2000; Zhao & Garbers, 2002). During active migration, guidance cues such as SDF-1 and its receptor CXCR4 (Molyneaux et al., 2003) as well as *c-kit* and *kit-l* (Buehr et al., 1993; Mintz & Russell, 1957) are necessary to keep PGCs from traveling off course. PGCs that migrate to ectopic locations are eliminated by BAX-mediated apoptosis (Stallock, Molyneaux, Schaible, Knudson, & Wylie, 2003). From a founding population of only a few dozen PGCs, thousands will reach the genital ridges to populate the nascent gonads (Godin, Wylie, & Heasman, 1990).

1.1.2 Gonad colonization and early differentiation

Once PGCs arrive in the gonad, they begin to commit to their sex-specific fate and are highly regulated by their somatic environment. In *Drosophila*, there are both male and female germline stem cells, and their respective gonadal niches closely regulate their cell cycle and differentiation. Alternatively, in mouse, only males have germline stem cells; females are instead endowed with a limited number of immature oocytes that will mature with each female ovulatory cycle until the original supply is depleted.

Mouse PGCs commit to their sex-specific fate after arriving in the somatic gonad at E11.5 (Fig. 6.1A) and will differentiate into sperm or egg depending on whether they develop in a testis or an ovary. PGCs in a testis will differentiate into prospermatogonia, later giving rise to mature sperm; PGCs in an ovary will differentiate as oogonia, later giving rise to mature eggs. The somatic environment directly controls the differentiation of germ cells into either sperm or eggs. Survival and commitment of germ cells to the male fate is dependent on expression of *Fgf9* in somatic cells (DiNapoli, Batchvarov, & Capel, 2006) and *Nanos2* in germ cells (Suzuki & Saga, 2008). Null mutations of these genes, or other downstream genes involved in blocking the activity of retinoic acid, like *Cyp26b1* (Bowles et al., 2006), lead to disruptions of male development. Commitment of germ cells to the female fate is, in turn, marked by immediate entry into meiosis (McLaren, 1984) and upregulation of genes associated with meiotic recombination (Menke, Koubova, & Page, 2003).

During gonad colonization, major differences arise in the gene expression and cell cycle profiles of male and female PGCs (Fig. 6.1B). Male germ cells enter mitotic arrest in G0 shortly after they reach the gonad and remain arrested until after birth (Matsui, 1998; McLaren, 1984; Western, Miles, van den Bergen, Burton, & Sinclair, 2008). Genes associated with the initiation of mitotic arrest include the cell cycle regulators *p27* (*Cdkn1b*), *p21* (*Cdkn1a*), and *p16* (*Cdkn2a*) (Western et al., 2008). Male germ cells do not resume mitosis until after birth, at which point prospermatogonia move to the periphery of testis cords, resume mitosis, and establish the male germline stem cell population of the adult testis (referred to as spermatogonial stem cells, SSCs). SSCs retain the potential to renew their own population but are normally restricted to give rise only to sperm throughout the lifetime of the male.

In the ovary, retinoic acid signaling and germ cell-specific *Dazl* expression are required for female germ cells to enter meiosis at E13.5, where they arrest at the end of the first prophase (Bowles et al., 2006; Bowles &

Koopman, 2010; Gill, Hu, Lin, & Page, 2011). The oocytes associate with *Foxl2*-expressing granulosa cells (Mork et al., 2012) to form primordial follicles that remain arrested at prophase I until the females reach sexual maturity. Oocyte maturation occurs as a result of signaling by cyclic waves of hormones, including follicle stimulating (FSH) and lutenizing hormone (LH). Once the ovary has been depleted of follicles, there is no female germline stem cell to replenish the pool of competent oocytes for fertilization. This is in stark contrast to females of many other invertebrate and vertebrate species such as fly and zebrafish, where oocytes continue to be produced throughout much of the lifetime of the organism.

1.1.3 Gametogenesis and fertilization

In mammals, male gametogenesis begins in the male gonad with the establishment of a SSC, which can self-renew and/or produce daughter cells that differentiate (de Rooij & Grootegoed, 1998). *In vitro*, these SSCs can be explanted to culture where they form multipotent germline stem (mGS) cells (Kanatsu-Shinohara et al., 2004). *In vivo*, these differentiating spermatogonia give rise to several intermediate cell types before forming type B spermatogonia. Type B cells develop into primary spermatocytes, enter meiosis I to produce secondary spermatocytes, and finally complete meiosis II to form early and late spermatids (Bellve et al., 1977).

Spermatogenesis and oogenesis in mammals share the common thread of meiosis, but each is quite distinct from the other in several aspects (Fig. 6.1A, top; Matova & Cooley, 2001). Whereas male meiosis will produce four gametes from a single meiotic division, only one egg is produced from a female meiotic division. Once females reach sexual maturity, oocytes arrested at prophase I of meiosis are recruited in small numbers by hormones (FSH and LH) to produce mature follicles. Maturing follicles contain oocytes intimately associated with somatic support cells that regulate meiotic reentry, leading ultimately to germinal vesicle breakdown in the oocyte, completion of meiosis I, arrest at metaphase II, and ovulation into the fallopian tube, where they await fertilization (for a review of this process, see Russell & Robker, 2007).

Fertilization, the fusion of mature gametes, is a critical reprogramming event where a terminally differentiated oocyte transitions to a totipotent zygote. The fusion of haploid genomes and reorganization of chromatin at this critical developmental stage is accompanied by transcriptional quiescence, emphasizing the importance of posttranscriptional regulation of gene expression. The storage and localization of maternal transcripts in the mature oocyte is essential for fertilization and early development in many species. Thus, small RNAs and RBPs are critical at this stage to ensure that normal translation programs facilitate the oocyte-to-zygote transition.

1.2. Small RNAs

The different classes of small regulatory noncoding RNAs include miRNAs, endo-siRNAs, and piRNAs. These small RNAs range in size from 18 to 32 nucleotides (nts), and they differ in both their biogenesis and function. The miRNAs function largely by suppressing translation of target transcripts (Bazzini, Lee, & Giraldez, 2012; Djuranovic, Nahvi, & Green, 2012). While little is known about mammalian endo-siRNA function, their exogenously synthesized counterparts function by cleavage and hence direct destabilization of target mRNA (Montgomery & Fire, 1998; Novina & Sharp, 2004). In fission yeast, endo-siRNAs are involved in the spreading of heterochromatin (Cam et al., 2005; Verdel et al., 2004), though there is little evidence to suggest such a role in mammals. In contrast, there is good evidence that piRNAs function, at least in part, at the transcriptional level via epigenetic modifications to the genome (Kuramochi-Miyagawa et al., 2008).

The burgeoning field of small RNA biology began with the discovery of a genetic mutation in a region of the genome coding for small RNAs that are complementary to the transcript of another gene (Lee, Feinbaum, & Ambros, 1993; Wightman, Ha, & Ruvkun, 1993) and the subsequent evidence that double-stranded RNA (dsRNA) could alter gene expression through a process known as RNA interference (RNAi) (Fire et al., 1998; Montgomery, Xu, & Fire, 1998). Further research has revealed that these small regulatory RNAs exist in large numbers across many species from yeast to plants to animals. The following sections will briefly introduce the three classes of small RNAs and what is known in terms of their biogenesis and function in mammals (see Fig. 6.2 for a summary).

1.2.1 miRNA and siRNA biogenesis and function

The miRNAs and endo-siRNAs can be distinguished from each other based on either their biogenesis or function; however, exceptions arise for both methods of definition. In general, miRNAs bind target mRNAs to inhibit translation and secondarily destabilize the transcript, whereas endo-siRNAs are thought to bind to target mRNAs and facilitate direct cleavage. This functional difference is due to the type of physical interaction between the small RNA and the target RNA: although complementary binding between the small RNA and mRNA target is involved in both cases, miRNAs

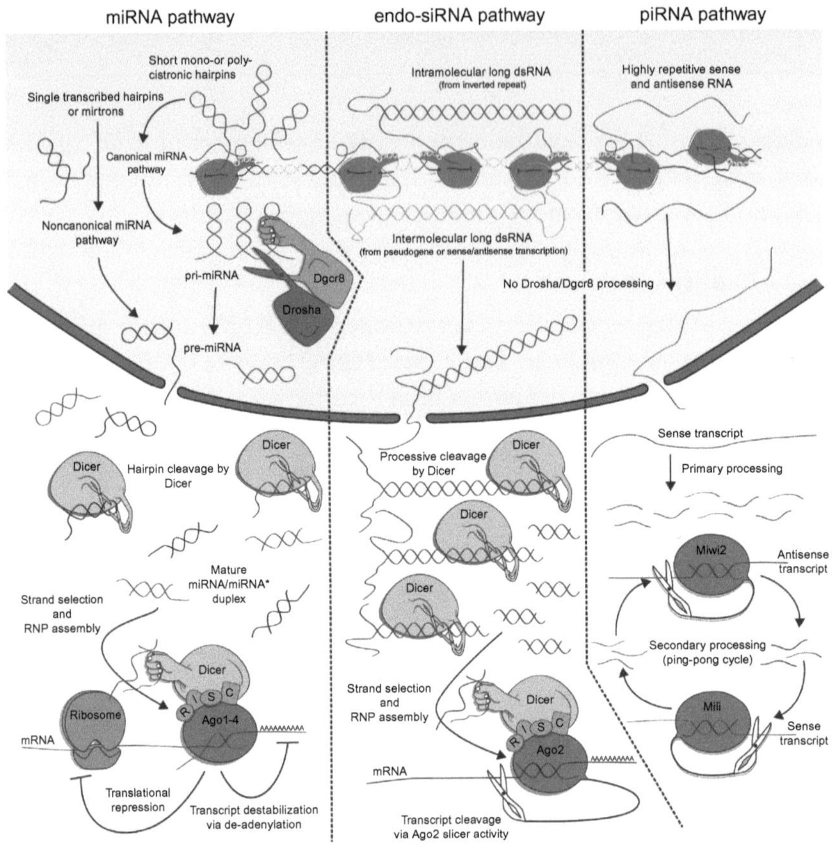

Figure 6.2 Schematic representation of small RNA biogenesis and function in mammals. (*miRNA pathway*) Canonical miRNAs are transcribed by RNA polymerase II as mono- or poly-cistronic hairpins known as primary miRNA (pri-miRNA) transcripts, which are subsequently cleaved by the microprocessor (Drosha/Dgcr8 complex) into pre-miRNAs. Noncanonical miRNAs are transcribed directly as single short hairpins, or they can be derived from intronic regions of a host mRNA, by-passing Microprocessor cleavage or other unknown endonuclease activity. After export to the cytoplasm, the RNase III-containing enzyme Dicer cleaves the hairpin, producing the mature miRNA/miRNA* duplex. The Dicer/miRNA complex then associates with Argonaute, where it guides one strand of the miRNA/miRNA* duplex into RISC with the complementary mRNA based on its seed sequence match. Target repression is generally achieved by inhibition of the translational machinery and/or destabilization of the transcript via deadenylation of the mRNA. (*endo-siRNA pathway*) Long double-stranded RNAs (dsRNAs) are the precursors to siRNAs. Whether long dsRNAs form from intramolecular or intermolecular complementary binding, they are exported from the nucleus without processing. In the cytoplasm, Dicer binds to and cleaves the dsRNA in a processive fashion, generating short dsRNA duplexes 21 nucleotides (21-nt) in length. The Dicer/siRNA complex guides one strand of the siRNA duplex into the Ago2-containing RISC based on perfect complementarity with the appropriate mRNA. Target repression is thought to be achieved by cleavage of the target transcript by Ago2, but little is known about endo-siRNA/target interactions in mammals. (*piRNA pathway*) Highly repetitive regions of the genome are transcribed to produce long sense and antisense RNAs. The sense transcript is exported, and a poorly understood primary processing event produces ~27-nt fragments that are then incorporated into a secondary processing event. Miwi2 forms a complex with the short sense RNAs bound to a long complementary antisense transcript, which is then cleaved into short RNAs that then

do not have a perfect match to their target transcript, whereas almost all siRNAs do. This results in a different mechanism by which target repression occurs: (1) an imperfect match (miRNA) results in a bulge in the duplex between the miRNA and its mRNA, leading to translational inhibition and protection from direct cleavage by endonucleases. Additionally, miRNA targeting promotes mRNA destabilization via decapping and shortening of the poly(A) tail (Fig. 6.2, left column); (2) a perfect match (siRNA) results in cleavage of the target transcript (Fig. 6.2, middle column; Doench, Petersen, & Sharp, 2003; Fabian, Sonenberg, & Filipowicz, 2010; Zeng, Yi, & Cullen, 2003). It should be noted that there are rare exceptions to this definition: miRNAs with a perfect match to their mRNA transcript are known to cleave targets in the same manner that siRNAs usually function (Yekta, Shih, & Bartel, 2004). It is also possible for siRNAs to bind imperfectly to their target transcripts, resulting in a bulge that prevents cleavage. For the purpose of this chapter, we will refer to miRNA function as an imperfect match with target transcripts resulting in translational inhibition and siRNA function as a perfect match with target transcripts and direct cleavage.

To better understand how miRNAs and siRNAs differ, it is important to distinguish their origins. There are two types of miRNAs, canonical and noncanonical. Both types are derived from either coding or noncoding RNA transcripts that contain hairpins (in coding RNAs, the miRNA component usually resides in the intron) (Cai, Hagedorn, & Cullen, 2004; Lee, Kim, et al., 2004). Canonical miRNAs derive from a primary miRNA (pri-miRNA), a long RNA that contains hairpins approximately 60–75 nt in length (Fig. 6.2, top left). The hairpins are recognized and cleaved at their base by a protein complex called the Microprocessor, which includes the RBP DGCR8 and the RNase enzyme DROSHA. The released hairpin is called the pre-miRNA, and it is exported to the cytoplasm by Exportin-5 for further processing (Bohnsack, Czaplinski, & Gorlich, 2004; Denli, Tops, Plasterk, Ketting, & Hannon, 2004; Gregory et al., 2004). Noncanonical miRNA production circumvents DROSHA-DGCR8 processing; instead, pre-miRNAs are produced from cleavage by other endonucleases or generated directly from the transcription of a single, short hairpin (Babiarz, Ruby, Wang, Bartel, & Blelloch, 2008; Okamura, Hagen, Duan, Tyler, & Lai, 2007; Ruby, Jan, & Bartel, 2007). Regardless of canonical or noncanonical origin, pre-miRNAs are exported to the

associate with Mili to cleave more sense transcripts, producing a positive feedback loop known as the ping-pong cycle. The ultimate functional mechanism of piRNAs is poorly understood in mammals, but it is possible that PIWI-related proteins associate with piRNAs and recruit chromatin-modifying machinery to specific regions of the genome to promote epigenetic changes that repress expression of mobile genetic elements. (See Color Insert.)

cytoplasm where the RNase III enzyme, DICER, cleaves them into a single 18–25 nt dsRNA (Bernstein, Caudy, Hammond, & Hannon, 2001; Ketting et al., 2001). At this point, one strand of the duplex miRNA will associate with a host of proteins known as the RNA-induced silencing complex (RISC, described below) (Fig. 6.2, bottom left). Correct strand specificity can be influenced by DICER association with TRBP (Chendrimada et al., 2005; Gregory, Chendrimada, Cooch, & Shiekhattar, 2005). The extent to which TRBP is necessary for strand selection is unclear, however, as the knockout phenotype is mild (Zhong, Peters, Lee, & Braun, 1999) relative to the embryonic lethality of the *Dicer* knockout (Bernstein et al., 2003).

In contrast to miRNAs, endo-siRNAs arise from long dsRNAs (Fig. 6.2, top middle), either as long hairpins or long sense/antisense RNA pairs (duplexes). In metazoa, DICER binds at one end of the long dsRNAs and then cuts processively at approximately 21-nt intervals along the length of the dsRNA, producing multiple siRNAs (Chung, Okamura, Martin, & Lai, 2008; Czech et al., 2008; Ghildiyal et al., 2008; Kawamura et al., 2008; Okamura, Balla, Martin, Liu, & Lai, 2008). Therefore, DICER is necessary for the final cleavage event in the production of mature siRNAs, canonical miRNAs, and noncanonical miRNAs.

Similar to the case with miRNAs, one strand of the siRNA duplex associates with an Argonaute protein to form an active RISC. At this point, the RISC-loaded siRNA or miRNA binds to and regulates the translation and stability of its target mRNAs. For miRNAs, almost all target identification is facilitated by a 7-nt "seed" sequence found at positions 2 through 8 of the mature miRNA as well as by other less well-characterized features downstream of the seed (Grimson et al., 2007; Lewis, Shih, Jones-Rhoades, Bartel, & Burge, 2003). For almost all siRNAs, there is perfect complementation of the siRNA to the target transcript. Most miRNA target sites are found within the 3′UTR of mRNAs, and RBPs can bind to and alter secondary structure within the 3′UTR to enhance or inhibit the effects of small RNAs on target transcripts (Kedde et al., 2007, 2010); this often has profound biological significance. For example, in the case of zebrafish germ cells, the presence of the RBP Dazl is essential to ensure translation of germ cell-specific genes that are otherwise suppressed by miRNAs (Takeda, Mishima, Fujiwara, Sakamoto, & Inoue, 2009).

In mouse, there are four related Argonaute proteins (Ago1–4). Only AGO2 can lead to cleavage of a perfect target (Liu et al., 2004). Thus, siRNA target cleavage is dependent on AGO2. However, all four Argonaute proteins are highly redundant in terms of miRNA activity—loss of all four in ES cells is required to ablate miRNA activity, but reintroduction of any one of the four can fully rescue miRNA activity (Su, Trombly, Chen, & Wang, 2009). This is in contrast to the situation in the fly where there is largely a division of labor among Ago proteins: Ago1 and Loquacious facilitate miRNA activity, while Ago2 and its partner R2D2 drive siRNA function (Ghildiyal & Zamore, 2009).

1.2.2 piRNA biogenesis and function

Three specific features distinguish piRNAs from miRNAs and endo-siRNAs: piRNAs are slightly longer (25–32 nt), do not require DICER for their processing (Fig. 6.2, right), and are expressed predominantly in the germline (Aravin et al., 2006; Houwing et al., 2007; Vagin et al., 2006; Watanabe et al., 2006). The complete mechanism by which piRNAs are produced and function remains unclear; however, most map to clusters in repeat sequences of the genome and are thought to defend against transposable elements (Klattenhoff & Theurkauf, 2008).

Much of what is known about piRNA biogenesis has been uncovered in the fly. The piRNA-interacting protein for which piRNAs were named, Piwi, was first identified in *Drosophila* as a mutation that caused defects in spermatogenesis (Lin & Spradling, 1997). PIWI stands for "P-element-induced wimpy testis." Since then, additional PIWI-like proteins have been discovered in *Drosophila*, including Aubergine (Aub) and Argonaute 3 (Ago3) (not to be confused with the mammalian Ago3, which functions in miRNA activity). All three of these proteins are highly expressed in the germline and associate with piRNAs (Carmell, Xuan, Zhang, & Hannon, 2002; Cox, Chao, & Lin, 2000; Harris & Macdonald, 2001; Parker & Barford, 2006). They are distantly related to the other Ago proteins and therefore form a subclade in the Ago family.

The biogenesis of piRNAs is poorly understood in mammals, although preliminary insights from the fly have been gained in recent years. Primary piRNA production in *Drosophila* involves the proteins Armitage, Zuchinni, and Yb (Saito et al., 2010; Siomi, Miyoshi, & Siomi, 2010). Long, single-stranded RNA transcripts are cleaved by these proteins and loaded onto Piwi/Aubergine to direct cleavage of transposon targets—these Aubergine-bound piRNAs represent class I piRNAs (Li et al., 2009; Malone et al., 2009; Saito et al., 2010). Piwi-mediated cleavage of the transposon target RNA generates class II piRNAs, which then associate with Ago3 and target more primary piRNA transcripts, thus creating a feedback loop known as the ping-pong model (Aravin, Sachidanandam, Girard, Fejes-Toth, & Hannon, 2007; Brennecke et al., 2007). These ping-pong-derived piRNAs are thought to be only one type of piRNA. More research is needed to understand how other piRNAs are generated from complex intergenic regions.

Similar to *Drosophila* Piwi, the mammalian orthologs *Mili*, *Miwi*, and *Miwi2* are required for spermatogenesis (Carmell et al., 2007; Deng & Lin, 2002; Kuramochi-Miyagawa et al., 2001). Mammalian piRNAs can be divided into the pachytene and prepachytene subtypes (Aravin et al., 2007). Prepachytene piRNAs (26–28 nt) are expressed in spermatogonia prior to meiosis and associate with the mouse PIWI-family proteins MILI or MIWI-2 before decreasing in levels at mid-pachytene. Pachytene piRNAs (29–31 nt) are expressed in the pachytene stage of meiosis and associate with MIWI until the cells reach the round spermatid stage (Aravin et al., 2007). It is thought that mouse pachytene piRNAs are generated by the primary processing pathway discovered in fly (Fig. 6.2, right), though their specific function remains unclear. Prepachytene piRNAs, in turn, are thought to defend against transposable elements because they are derived from repeat elements (Brennecke et al., 2007); their biogenesis and mechanism of action remain unknown.

An additional class of Piwi-associated small RNAs, known as the 21U-RNAs, has been discovered in *C. elegans* (Ruby et al., 2006). These RNAs also silence transposon activity and bind to piwi-like proteins but are synthesized via a different mechanism (Batista et al., 2008; Ruby et al., 2006; Wang & Reinke, 2008). Interestingly, 21U-RNA sequencing and transposon profiling data do not suggest direct silencing of transposons by 21U-RNAs. However, several recent reports have provided the first evidence that the 21U-RNAs direct the biogenesis of a second pool of small RNAs, termed 22G-RNAs. These are considered the "effector" components (which map to expressed transposons in *C. elegans*) and are thought to direct gene silencing, presumably through heterochromatin formation around the genomic loci that produce these transposons, sustaining a transgenerational silencing effect (Ashe et al., 2012; Bagijn et al., 2012; Lee et al., 2012; Shirayama et al., 2012).

2. SMALL RNAs IN GERM CELLS

Much of what has been learned about the functional roles of small regulatory RNAs in mammals has come from studies of loss-of-function mutants blocking the biogenesis or function of the different classes. In particular, mutants for *Drosha*, *Dgcr8*, *Dicer*, the Argonaute proteins, and the Piwi-related proteins have uncovered numerous roles for small RNAs during development. Table 6.1 provides a summary of knowledge gained from genetic models where individual small RNAs or the proteins important for their biogenesis and/or function have been depleted and analyzed in

Table 6.1 Summary of the genetic evidence for a role of small RNAs in germ cell development

Model organism	Mutation	small RNAs affected	Phenotype	Reference
Mus musculus (Mouse)	Dicer1 (TNAP-Cre)	miRNAs and siRNAs	Gonad colonization: slower proliferation during early differentiation	Hayashi et al. (2008) and Maatouk, Loveland, McManus, Moore, and Harfe (2008)
			Spermatogenesis: early proliferation and later sperm morphology and motility defects	Hayashi et al. (2008) and Maatouk et al. (2008)
	Dicer (Amh2-Cre)	miRNAs and siRNAs	Spermatogenesis: complete testicular degeneration, defects starting five days after birth	Papaioannou et al. (2009) and Kim et al. (2010)
	Dicer (Amhr2-Cre)	miRNAs and siRNAs	Oogenesis: larger pool of primordial follicles, premature folliculogenesis, oocyte migration defects from oviduct to uterus, shorter uterine horns and oviductal cysts, female infertility	Hong, Luense, McGinnis, Nothnick, and Christenson (2008), Nagaraja et al. (2008), Gonzalez and Behringer (2009) and Lei, Jin, Gonzalez, Behringer, and Woodruff (2010)
	Dicer1 (Zp3-Cre)	miRNAs and siRNAs	Oogenesis: arrest at meiosis I with a disorganized spindle and chromosome segregation defects	Murchison et al. (2007)
	Dgcr8 (Zp3-Cre)	miRNAs	No phenotype[a]	Suh et al. (2010)
	Drosha (Stra8-Cre)	miRNAs	Spermatogenesis: depletion of spermatocytes and spermatids	Wu et al. (2012)
	LIN-28	let-7 gain-of-function	PGC specification: fewer stella-positive PGCs are specified, rescued by Blimp-1 overexpression	West et al. (2009)

Continued

Table 6.1 Summary of the genetic evidence for a role of small RNAs in germ cell development—cont'd

Model organism	Mutation	small RNAs affected	Phenotype	Reference
	miR-290 cluster	miR-290–295	PGC migration: PGCs migrate off course	Medeiros et al. (2011)
			Oogenesis: the few PGCs that colonize the gonad die off, making females sterile	Medeiros et al. (2011)
	miR-17–92 cluster	miR-17–92	No phenotype[a]	Ventura et al. (2008)
	Ago2 (TNAP-Cre)	siRNAs	No phenotype[a]	Hayashi et al. (2008)
	Ago2 (Zp3-Cre)	siRNAs	Oogenesis: mature oocytes with disorganized spindles and chromosome segregation defects	Kaneda, Tang, O'Carroll, Lao, and Surani (2009)
	Miwi	piRNAs	Spermatogenesis: spermatogenic arrest at the beginning of the round spermatid stage (meiosis II) and sterile males; females are fertile	Deng & Lin (2002)
	Mili	piRNAs	Spermatogenesis: spermatogenic arrest at zygotene to early pacytene (meiosis I), sterile males; females are fertile	Kuramochi-Miyagawa et al. (2004)
	Miwi2	piRNAs	Spermatogenesis: spermatogenic arrest in early prohase (meiosis I), sterile males; females are fertile	Carmell et al. (2007)
Drosophila melanogaster (Fly)	Dicer1	miRNAs	PGC specification: Vasa-positive PGCs do not form in mutants	Megosh, Cox, Campbell, and Lin (2006)
	Loquacious	miRNAs	Oogenesis: necessary for GSC maintenance	Forstemann et al. (2005) and Park, Liu, Strauss, McKearin, and Liu (2007)
	Ago1	miRNAs	Oogenesis: necessary for GSC maintenance	Yang, Chen, et al. (2007), Yang, Duan, et al. (2007) and Neumuller et al. (2008)

	R2D2	miRNAs[b]	Oogenesis: necessary for stalk and follicle cell formation and, thus, normal oogenesis	Kalidas et al. (2008)
	Dicer2	siRNAs	No phenotype[a]	Hatfield et al. (2005) and Megosh et al. (2006)
	Ago2	siRNAs	No phenotype[a]	Okamura, Ishizuka, Siomi, and Siomi (2004)
	Piwi	piRNAs	PGC specification: Vasa-positive PGCs do not form in mutants	Megosh et al. (2006)
			Gametogenesis: necessary for GSCs of both sexes to be maintained, but not for further differentiation	Lin and Spradling (1997), Cox et al. (1998), and Szakmary, Cox, Wang, and Lin (2005)
	Aubergine	piRNAs	PGC specification: pole cells are lost in mutants	Schmidt et al. (1999), Harris and Macdonald (2001), and Thomson, Liu, Arkov, Lehmann, and Lasko (2008)
	Ago3	piRNAs	Oogenesis: females are sterile	Li et al. (2009)
			Spermatogenesis: subfertile due to misregulation of GSC mainteance	Li et al. (2009)
Danio rerio (Zebrafish)	Dicer1	miRNAs and siRNAs	No phenotype[a]	Giraldez et al. (2005)
	Ziwi	piRNAs	Early gametogenesis: loss of germ cells due to apoptosis	Houwing et al. (2007)
	Zili	piRNAs	Early gametogenesis: loss of germ cells due to apoptosis	Houwing, Berezikov, and Ketting (2008)

Continued

Table 6.1 Summary of the genetic evidence for a role of small RNAs in germ cell development—cont'd

Model organism	Mutation	small RNAs affected	Phenotype	Reference
Caenorhabditis elegans (Nematode)	Dicer1	miRNAs and siRNAs	Gametogenesis: abnormal oocyte morphology and migration through the spermatheca	Knight and Bass (2001)
	Ego-1	siRNAs	Gametogenesis: defects in mitotisis, premature, meiosis, errors in meiotic recombination	Smardon et al. (2000) and Knight and Bass (2001)
	Prg-1	21U-RNAs (piRNAs)	Spermatogenesis: mutant sperm exhibit extensive defects in activation and fertilization	Batista et al. (2008) and Wang and Reinke (2008)

[a]There is no germline phenotype as reported by the correlating reference; there may be a somatic phenotype.
[b]R2D2 normally associates with Ago2 in the siRNA pathway, but depletion causes a phenotype in the germline due to effects on the miRNA pathway.
All mutations are whole organism loss-of-function experiments, unless a conditional deletion was analyzed (indicated by the type of Cre recombinase used in parentheses).

the germline. Follow-up studies involving small RNA profiling and individual small RNA knockdown or introduction have further refined our knowledge of the role these small molecules play in critical germ cell fate decisions. These following sections will explore our current understanding of small RNAs at each stage of germ cell development: specification and migration, gonad colonization and differentiation, and gametogenesis.

2.1. Small RNAs in PGC specification and migration

In mouse, mutations deleting *Dicer1* are embryonic lethal around the time PGCs are specified, revealing a role for miRNAs and/or siRNAs in early development but leaving the question of PGC specification unresolved due to the timing of the defect (Bernstein et al., 2003). To better understand the role of small RNAs at the stage of PGC specification, various groups have utilized expression studies, directed miRNA or RBP knockouts, and generated *in vitro* models to uncover a role for miRNAs and the RBPs that regulate them.

One prominent miRNA, let-7, has been implicated in PGC specification (West et al., 2009). Mouse PGCs are induced by BMP4 from somatic cells to express *Prdm1* (*Blimp1*) and *Prdm14*, which suppress Hox genes and facilitate the reacquisition of pluripotency (Ohinata et al., 2005; Robertson et al., 2007; Vincent et al., 2005). *Prdm1* is suppressed in other tissues by the widely expressed miRNA let-7 via binding sites in its 3′UTR (Nie et al., 2008). Let-7 itself is antagonized by the activity of the RNA-binding protein LIN28, which is expressed in PGCs. LIN28 can bind to pre-let-7 and recruit TUT-4 to uridylate the 3′ end, blocking further processing by DICER (Heo et al., 2008; Newman, Thomson, & Hammond, 2008; Piskounova et al., 2008; Rybak et al., 2008; Viswanathan, Daley, & Gregory, 2008). The knockdown of LIN28 decreases the number of PGCs (West et al., 2009). Moreover, loss of LIN28 in mouse ES cells differentiated into PGC-like cells results in loss of *Prdm1* expression and a failure to maintain a germ cell-like state. Alternatively, overexpression of LIN28 results in more *in vitro* PGCs and more Stella-positive PGCs in chimeric embryos (West et al., 2009). Thus, LIN28 plays a critical role in PGC specification by preventing let-7 from suppressing *Prdm1* expression. However, it should be noted that LIN28 has additional targets (Jin et al., 2011; Peng et al., 2011), as does let-7 (Melton, Judson, & Blelloch, 2010), which are likely to play additional downstream roles in PGC specification.

LIN28 is not the only RBP with likely roles in regulating miRNA function in PGCs. The RBP *Dead end 1* (*Dnd1*) is essential in early PGC development (Youngren et al., 2005). One known function of DND1 is to block miRNA function via binding mRNA transcripts and blocking miRNA access (Kedde et al., 2007). The exact targets of DND1 at the time of PGC specification are unknown. Loss of *Dnd1* does not specifically affect PGC specification or the path of migration but does affect the ability of PGCs to proliferate and survive during migration such that only a small number colonize the gonad (Cook, Coveney, Batchvarov, Nadeau, & Capel, 2009; Sakurai et al., 1995). These data suggest a critical role for DND1 in blocking the activity of specific miRNAs that target transcripts promoting growth and survival.

Profiling experiments have revealed high expression of certain miRNAs during PGC migration, including the miR-17–92 and miR-290–295 clusters, which promote cell cycle and maintain embryonic stem cell pluripotency (Hayashi et al., 2008; Wang et al., 2008). Deletion of the miR-17–92 cluster does not appear to affect germ cell development, although it was not carefully evaluated as mutants die soon after birth (Ventura et al., 2008). In contrast, deletion of the miR-290–295 cluster results in migration defects such that a significantly reduced number of PGCs colonize the gonads (Medeiros et al., 2011). Males remain fertile, suggesting no essential role for the miR-290–295 cluster during adult spermatogenesis. Females, however, fail to compensate for the loss of PGCs after specification and migration, and they are sterile secondary to premature ovarian failure (Medeiros et al., 2011).

Strikingly, in zebrafish, deletion of *dicer* does not result in any germ cell phenotype (Giraldez et al., 2005). However, miRNAs are still important in zebrafish for regulating maternally deposited mRNAs, especially those required for PGC specification (Giraldez et al., 2005, 2006). In particular, miR-430, one of the most highly expressed miRNAs, is necessary to prevent germ plasm RNAs from being expressed in somatic cells. Transcripts that specify the PGC fate in zebrafish are not limited to the germ plasm but are also expressed in somatic cells. Many of these transcripts (*nanos*, *tdrd7*, *dead end*) are miR-430 targets and are thus destabilized and translationally repressed in the somatic compartment (Blaser et al., 2005; Giraldez et al., 2006; Koprunner et al., 2001; Mishima et al., 2006). These transcripts are ectopically expressed in somatic cells where *dicer* is mutated and miR-430 levels are diminished (Mishima et al., 2006). Surprisingly, miR-430 is also expressed in germ cells. How is miR-430 kept from inhibiting expression of these critical germ plasm determinants in the cells that become PGCs? The answer involves RBPs. The zebrafish *Dnd1* ortholog binds to the 3′UTR of germ plasm transcripts and blocks miR-430 activity (Kedde et al., 2007; Mickoleit, Banisch, & Raz, 2011; Slanchev et al., 2009), similar to what is suspected to occur in the mouse. In the absence of *dnd* in zebrafish, PGCs never become motile: they do not polarize or form pseudopodia, and they begin to undergo apoptosis (Weidinger et al., 2003). Zebrafish Dazl also plays a role in preventing the action of miR-430 in PGCs but through a different mechanism than Dnd. Dazl counteracts miR-430 activity by promoting translation of target transcripts via poly(A) tail elongation (Takeda et al., 2009).

Together, these findings highlight the role RBPs play in mitigating the function of miRNAs and emphasize the need to understand the cellular context when determining whether small RNAs will significantly affect target expression. In fact, a recent report attempting to define the minimal criteria for normal PGC morphology and motility during PGC migration in zebrafish found that a delicate balance between miR-430 and the RBP Dnd is necessary to generate a threshold of protein expression of target transcripts that encode factors regulating myosin contractility, cell adhesion, and cell cortex properties (Goudarzi et al., 2012).

C. elegans mutants lacking EGO-1 and Dicer1 (proteins necessary for endo-siRNA synthesis) only showed defects after germline specification (Knight & Bass, 2001; Smardon et al., 2000). Similarly, in fly, Dicer2 and Ago2 (both required for siRNA activity) are dispensable for germline specification and maintenance, indicating that endo-siRNA activity is not necessary (Hatfield et al., 2005; Lee, Nakahara, et al., 2004; Megosh et al., 2006; Okamura et al., 2004). However, loss of Piwi, dFmrp (a component of the RISC complex), and Dicer1 (required for miRNA activity) leads to pole plasm defects and reduced PGC numbers (Megosh et al., 2006). Interestingly, overexpression of Piwi increases Oskar and Vasa expression leading to a corresponding increase in PGC number, suggesting a central role for piRNAs (Megosh et al., 2006). Deletion of Aubergine, another PIWI homolog associated with polar granules, is also necessary for the normal translation of Oskar and ultimately the formation of PGCs (Harris & Macdonald, 2001; Thomson et al., 2008). Together, these findings indicate roles for piRNAs and miRNAs in PGC specification.

How piRNAs function in the early germline is unclear, though there are some reports that piRNAs may target transcripts rather than transposons. Normal *nanos* deadenylation in the anterior end of the embryo is facilitated by piRNA activity in the *nanos* 3′-UTR (Rouget et al., 2010). When piRNA-induced regulation is impaired by Aubergine depletion, *nanos* deadenylation decreases, and anterior defects are observed (Rouget et al., 2010). In addition to *nanos*, piRNAs might also target *vasa* transcripts, which encode another RBP essential for early germ cell development (Nishida et al., 2007). The only evidence supporting a role for piRNAs in PGC specification and migration in vertebrates comes from a recent report characterizing a deletion of the medaka (Japanese killifish) homolog of Piwi, *Opiwi*. Loss of *Opiwi* results in fewer PGCs being specified and major migration defects (Li, Hong, Gui, & Hong, 2012).

2.2. Small RNAs in PGC gonad colonization and early differentiation

Across metazoa, PGCs migrate to the gonad and then begin to associate with somatic cells as they commit to a sex-specific fate. Conditional deletion of critical components of small RNA biogenesis and function has been used to dissect the role of these pathways at this stage of germ cell development. *TNAP-Cre* deletion of *Dicer1* in mice at a time when PGCs are just arriving in the gonad results in reduced proliferation and differentiation defects in male, but not female, mice (Hayashi et al., 2008). These defects are likely a result of diminished miRNA activity, as the *TNAP-Cre Ago2* conditional deletion shows no phenotype (Hayashi et al., 2008). That is, deletion of *Ago2* is expected to ablate all siRNA function, while the three other Ago proteins, if expressed, can maintain miRNA function.

Sex differentiation of gametes occurs upon their arrival at the gonad. miRNA expression profiling of PGCs colonizing the gonad reveals differential expression patterns of miRNAs between the sexes: let-7, miR-125a, and miR-9 are upregulated in males (Hayashi et al., 2008), possibly contributing to their sex-specific differentiation and exit from the cell cycle; female germ cells express miR-29b (Takada, Berezikov, Choi, Yamashita, & Mano, 2009), which may be important for meiotic entry. It is believed that miR-29b targets *Dnmt3a* and *Dnmt3b* to facilitate changes in the methylation status of the female germline (Takada et al., 2009). The miR-290–295 cluster, which is important in PGC migration, is dispensable for male germ cell commitment and progression, but females lacking this cluster suffer germ cell loss and premature ovarian failure (Medeiros et al., 2011). It remains unclear whether female sterility is secondary to the premature ovarian failure or whether the miR-290–295 cluster plays a direct role in female germ cell development.

RBPs that regulate miRNA function in early development also have a role in PGC colonization of the gonad. In mouse, male germ cells enter mitotic arrest and downregulate markers of pluripotency as they upregulate cell cycle inhibitors of the G1/S transition such as *p21* (*Cdkn1a*) and *p27* (*Cdkn1b*) (Western et al., 2008). The RBP DND1 binds the 3′UTR of these transcripts and protects them from miRNA-mediated inhibition by miR-221/222 such that mitotic arrest can occur—mutants of *Dnd1* lose expression of *p21* and *p27*, maintain markers of cell cycle and pluripotency, and develop teratomas on certain genetic backgrounds (Cook, Munger, Nadeau, & Capel, 2011; Kedde et al., 2007; Youngren et al., 2005). Alternatively, the RBP PUM1, expressed in many tissues including germ cells, can alter the secondary structure of target transcripts to enhance the binding of miR-221/222 and the subsequent inhibition of translation (Kedde et al., 2010), though it is unclear whether this particular role is essential for germ cell development. These examples reveal that the coordinate expression (and timing) of multiple RBPs is essential to promote or antagonize miRNA activity of target transcripts that define a complex cell fate/identity.

In the fly, PGCs, together with somatic cells, form a GSC niche where the GSCs divide asymmetrically to produce a daughter stem cell and a progenitor cell that proceeds through gametogenesis. The loss of Dicer-1 results in a delay of the G1/S checkpoint in GSCs, in part, due to an increase in the cell cycle inhibitor Dacapo (a p21/p27 homolog) (Hatfield et al., 2005). Further analysis of mutants revealed that insulin signaling drives expression of miR-7, miR-278 m and miR-309, which target *dacapo* and are essential for normal GSC division (Yu et al., 2009). In *dicer-1* mutants, GSCs eventually differentiate and are lost (Jin & Xie, 2007). Similar GSC phenotypes were observed in other miRNA pathway mutants including *loquacious*, *ago1*, and *dfmrp* (Forstemann et al., 2005; Neumuller et al., 2008; Park et al., 2007; Pek, Lim, & Kai, 2009; Yang, Chen, et al., 2007; Yang, Duan, Chen, Wang, & Jin, 2007). For example, dFMRP is an RBP that binds to the miRNA *bantam*, and this interaction is essential to maintain GSC self-renewal (Yang, Duan, et al., 2007; Yang et al., 2009). These studies support the idea that miRNAs play a crucial role in maintaining the GSC niche of the fly, regulating cell cycle and differentiation genes that are critical for early gametogenesis.

Another example of small RNAs regulating early differentiation of the GSC niche in the fruit fly is miR-184 in the female germline. miRNA-184 is expressed in female GSCs and regulates differentiation by controlling the level of expression of the Saxophone receptor (Iovino, Pane, & Gaul, 2009). Dpp (Decapentaplegic) from the niche signals through the Saxophone receptor to suppress Bam, a protein essential for initiating gametogenesis (Xie & Spradling, 1998); thus, miR-184 suppresses Dpp signaling, resulting in more Bam and promoting differentiation (Iovino et al., 2009). Consequently, mutant miR-184 GSCs fail to differentiate because they never activate Bam signaling. Thus, the exquisite timing of Bam-mediated differentiation in the *Drosophila* germarium is dependent on the expression of miR-184 to prevent the niche from generating only GSCs.

Some miRNAs inhibit, rather than promote, normal GSC niche maintenance. In fly, the Piwi-associated nuage component, Maelstrom, drives GSC maintenance and self-renewal by blocking miRNA activity (Findley, Tamanaha, Clegg, & Ruohola-Baker, 2003; Soper et al., 2008). Maelstrom represses the transcription of miR-7, which targets *Bam*—loss of Maelstrom results in miR-7 repression of *Bam* and a failure to enter meiosis (Pek et al., 2009). Some proteins involved in GSC development prevent miRNA activity globally. For example, the RBP Vasa, which associates with the nuage, acts as an inhibitor of miRNA activity, albeit indirectly: Vasa binds to and promotes the expression of the *mei-P26* transcript, which in turn binds to and inhibits Ago1, resulting in downregulation of many miRNAs (Liu, Han, & Lasko, 2009; Neumuller et al., 2008). Importantly, upregulation of Mei-P26 occurs in 16 cell cysts and not GSCs themselves (Neumuller et al., 2008), thus theoretically preserving miRNA activity in GSCs. Therefore, whether through direct interaction with individual miRNAs or indirect action on the proteins regulating miRNA biogenesis and/or function, RBPs and other proteins cooperate with miRNAs in normal GSC maintenance and differentiation in the fly.

The requirement for miRNAs and siRNAs in early germline development differs across systems. In zebrafish, for example, loss of Dicer does not affect the ability of animals to produce mature sperm and eggs (Giraldez et al., 2005). However, *C. elegans* mutants lacking *dicer-1* show severe defects in spermatogenesis (Knight & Bass, 2001). Consistent with a role for siRNAs in spermatogenesis, nematodes lacking the RNA-dependent RNA Polymerase (RdRP) homolog EGO-1 (required for siRNA processing and RNAi) display severe defects in both oogenesis and spermatogenesis that result in infertility due to defects in mitosis, premature meiosis, and errors in meiotic recombination (Smardon et al., 2000). As an RdRP family member, EGO-1 is believed to be important for the secondary step of siRNA synthesis in *C. elegans* (Gent et al., 2009).

Drosophila is one of the few model organisms for which a role of piRNAs has been determined in early germline development. In the fly, piRNAs have been reported to have a role in PGC colonization of the gonad and early differentiation. Somatic and germline cells of both sexes express Piwi proteins, which are necessary for piRNA activity. GSCs are lost in *piwi* mutant flies of both sexes, and Piwi overexpression in female somatic cells results in an increased rate of GSC division and number (Cox et al., 1998; Szakmary et al., 2005). Aubergine and Ago3 are also required for fertility in both sexes (Li et al., 2009; Lin & Spradling, 1997; Schmidt et al., 1999). Piwi proteins and piRNAs also play a critical role in mouse germ cells, but at a later stage of differentiation, as described below.

2.3. Small RNAs in spermatogenesis

In mouse, the miRNA and piRNA pathways are essential for spermatogenesis (Carmell et al., 2007; Hayashi et al., 2008; Maatouk et al., 2008; Unhavaithaya et al., 2009). The proteins DICER, AGO1, and DROSHA, as well as individual miRNAs, localize to the chromatoid body, a center for RNA processing that is specific to male germ cells during spermatogenesis (Gonzalez-Gonzalez, Lopez-Casas, & del Mazo, 2008; Kotaja et al., 2006; Kotaja & Sassone-Corsi, 2007).

Loss of *Dicer* in mouse PGCs results in early male germ cell proliferation defects, impaired spermatogenesis, and defective sperm morphology and motility (Hayashi et al., 2008; Maatouk et al., 2008). These defects are due to problems in miRNA activity because *Ago2* deletion (which ablates siRNA function while leaving miRNA activity intact) is indistinguishable from wild type with regard to germ cell proliferation, differentiation, and meiotic progression (Hayashi et al., 2008). Further, deletion of Drosha (required for canonical miRNA, but not endo-siRNA, biogenesis) in spermatocytes using a *Stra8-Cre* results in loss of spermatocytes and spermatids and eventual male sterility (Wu et al., 2012).

Roles for some individual miRNAs in spermatogenesis have been characterized. The miR-17–92 cluster, which targets *E2F1*, protects meiotic cells from apoptosis (Novotny et al., 2007). miR-122a represses *Tnp2* in postmeiotic germ cells (Yu, Raabe, & Hecht, 2005), while TRANSLIN, another RBP, binds to miR-122a and promotes its stability in late-stage spermatids (Yu & Hecht, 2008). Thus, miR-122a is critical as part of a coordinated effort to repress *Tnp2* in differentiating spermatocytes. miRNA-34c is highly expressed in male germ cells during late spermatogenesis, and its overexpression in an *in vitro* model results in upregulation of germ cell-specific genes (Bouhallier et al., 2010). Alternatively, miR-18 of the miR-17–92 cluster targets *Hsf2*, a germ cell transcription factor that is essential for normal regulation of spermatogenesis (Bjork, Sandqvist, Elsing,

Kotaja, & Sistonen, 2010). Another clue that miRNAs are important in spermatogenesis comes from miRNAs encoded by the X-chromosome. Although all X-linked mRNA genes studied to date have been shown to be transcriptionally silent during spermatogenesis, many X-linked miRNAs circumvent this transcriptional quiescence (Song et al., 2009). It remains unclear what specific function these miRNAs perform and whether they are essential for normal spermatogenesis.

Some RBPs promote the action of miRNAs by binding target transcripts and altering secondary structure to permit the binding of complementary miRNAs. For example, the RBP Mouse Vasa Homolog interacts with Dicer and is thought to be the germ cell-specific helicase for some miRNA pathways during spermatogenesis (Kotaja et al., 2006). While promoting miRNA activity is essential at some stages of spermatogenesis, suppression of miRNA activity at other stages is also critical. Of particular interest is the RBP LIN28, known for its inhibition of the let-7 family of miRNAs; LIN28 is expressed in human and nonhuman primate SSC populations (Aeckerle et al., 2012), suggesting that blocking let-7 may be essential to maintain an adult population of SSCs.

Expression of miRNAs in somatic cells is also important for normal regulation of spermatogenesis. In mouse, deletion of *Dicer1* in Sertoli cells (one of the main supporting cell lineages of the testis) results in major gene expression changes within the testis (Papaioannou et al., 2011) and eventually complete testicular degeneration, with defects manifesting about 5 days after birth (Kim et al., 2010; Papaioannou et al., 2009). This suggests a critical role for miRNAs in regulating Sertoli cell influence on germ cells, most likely through effects on paracrine signaling. It is possible that miRNAs regulate hormonal signaling, which is also important for spermatogenesis. Hormone suppression during spermiation results in an increase in over 20 miRNAs in Sertoli cells (Nicholls et al., 2011). This suggests that long-range hormonal control of male germ cell differentiation might be facilitated in part by miRNA activity.

Another great example of miRNA function in somatic cells regulating spermatogenesis comes from a recent report in the fly (Toledano, D'Alterio, Czech, Levine, & Jones, 2012). In *Drosophila*, male GSCs exhaust over time due to loss of signaling from the somatic hub cells. Specifically, loss of the self-renewing factor Unpaired in hub cells results in decreased JAK-STAT signaling and eventual loss of GSCs (Boyle, Wong, Rocha, & Jones, 2007). Normally, the RBP Imp binds to and promotes translation of Unpaired, sustaining normal JAK-STAT signaling for GSC maintenance.

However, the *Imp* transcript has binding sites for the miRNA let-7 in its 3′UTR. Over time, expression of let-7 increases in hub cells and represses expression of *Imp*, causing a subsequent decrease in Unpaired protein and GSC exhaustion (Toledano et al., 2012). The aging effect can be rescued by maintaining expression of Imp in hub cells over time, demonstrating the critical balance between miRNAs and RBPs that regulate GSC development in the aging *Drosophila* niche.

While many studies have shown roles for miRNAs in mouse spermatogenesis, endo-siRNAs appear to be dispensable (Hayashi et al., 2008). However, in nematodes, several mutations in components of the RNAi pathway have implicated roles for siRNAs in spermatogenesis (Gent et al., 2009; Han et al., 2009; Ketting, Haverkamp, van Luenen, & Plasterk, 1999; Pavelec, Lachowiec, Duchaine, Smith, & Kennedy, 2009; Rocheleau et al., 2008; She, Xu, Fedotov, Kelly, & Maine, 2009; Smardon et al., 2000; Tabara et al., 1999). In particular, the 26G and 22G RNAs (primary and secondary endo-siRNAs produced specifically in *C. elegans*) are necessary to prevent defects in chromosome segregation during both mitosis and meiosis (Han et al., 2009). The 22G RNAs recruit chromatin-modifying factors that promote local histone modifications that influence the binding and organization of Condensins and Cohesins necessary for normal chromosome segregation (a distinct role for this special class of dsRNA), and 26G RNAs regulate gene expression via association with the AGO proteins Alg3 and Alg4 (Claycomb et al., 2009; Conine et al., 2010; Gu et al., 2009; Han et al., 2009; van Wolfswinkel et al., 2009). Why siRNAs appear to play a critical role in *C. elegans* spermatogenesis and not in mouse remains to be determined.

In contrast to endo-siRNAs, piRNAs play an essential role in mouse spermatogenesis. Deletion of *Mili*, *Miwi*, and *Miwi-2* (mammalian Piwi homologs) results in activation of retrotransposons in male germ cells, meiotic arrest, and eventual male sterility (Aravin et al., 2007; Carmell et al., 2007; Deng & Lin, 2002; Kuramochi-Miyagawa et al., 2004, 2008). These same reports suggest that piRISC (piRNA-RNA-induced silencing complex) may also influence the *de novo* methylation machinery that silences transposons, though this interaction may be independent of slicer activity because a *piwi* mutant lacking slicer activity in the fly can also direct methylation of transposons (Saito et al., 2010). Supporting the idea that Piwi proteins could be driving epigenetic changes in mice, loss of *Mili* or *Miwi2* results in defective DNA methylation of regulatory regions of retrotransposons, suggesting a specific role in *de novo* DNA methylation of transposable elements in male germ cells (Aravin et al., 2007, 2008; Carmell et al., 2007;

Kuramochi-Miyagawa et al., 2008). Further, Piwi proteins can recruit heterochromatin protein 1A to genomic loci (Brower-Toland et al., 2007; Pal-Bhadra et al., 2004). More research will be necessary to determine the full extent to which small RNAs are directing epigenetic changes in mouse spermatogenesis.

In mice, the activation of mobile genetic elements coinciding with the loss of components of the piRNA machinery results in genomic instability. For example, in mouse *Miwi-2* mutants, most cells arrest at the leptotene stage of meiosis, which correlates with defects in dsDNA break repair, consistent with a role for MIWI2 in normal meiotic recombination (Carmell et al., 2007; Kuramochi-Miyagawa et al., 2004). *Miwi* mutants suffer spermatogenic arrest at the beginning of the round spermatid stage (Deng & Lin, 2002), and loss of *Mili* leads to a block at the zygotene or early pachytene stages (Kuramochi-Miyagawa et al., 2004). All of these defects are consistent with the timing of expression for each of these proteins: *Mili* and *Miwi-2* mutants showing arrest in meiosis I (primary spermatocyte stage) and deletion of *Miwi* results in arrest following meiosis II (round spermatid stage) (Deng & Lin, 2002).

The piRNAs also play a role in spermatogenesis in other vertebrate and invertebrate model organisms. In zebrafish, loss of piwi homologs, Ziwi and Zili, results in activation of transposable elements and male sterility (Houwing et al., 2007). Ziwi mutant germ cells undergo apoptosis, leading to male infertility, and loss of Zili results in a failure of germ cells to differentiate to mature oocytes or sperm. In the fly, piRNAs target repetitive sequences of the protein-coding gene *stellate* to maintain male fertility (Aravin et al., 2001). In nematodes, loss of prg-1, a piwi-like protein that is localized to P-granules, results in a reduction in the expression of 21U-RNAs (a *C. elegans*-specific class of piwi-interacting small RNAs that are shorter than other piRNAs) during spermatogenesis, which correlates with multiple defects in sperm activation and fertilization (Batista et al., 2008; Wang & Reinke, 2008). These results clearly demonstrate a critical role for piRNAs and Piwi proteins in spermatogenesis. However, the full mechanism of action for the piRNA pathway in each context has yet to be fully elucidated.

2.4. Small RNAs in oogenesis

Gene regulation in maturing oocytes is critical to ensure reprogramming to a totipotent zygote. In the female mouse germline, all three types of small RNAs have been shown to be present by deep sequencing analysis (Tam et al., 2008; Watanabe et al., 2006, 2008). A role for small RNAs in oogenesis was demonstrated based on knockout studies of *Dicer1* and *Ago2*, where loss of these proteins resulted in meiotic and spindle organization defects (Kaneda et al., 2009; Murchison et al., 2007; Tang et al., 2007; Watanabe et al., 2008). It was believed that both endo-siRNAs and miRNAs were playing a role in the *Dicer* knockout; so, it was a surprise to learn that loss of *Dgcr8* (which is only necessary for miRNA, not siRNA, biogenesis; Fig. 6.2, top) presented no defects in oocyte maturation, fertilization, or early development (Ma et al., 2010; Suh et al., 2010; Wang, Medvid, Melton, Jaenisch, & Blelloch, 2007). Together with reporter studies showing siRNA (but not miRNA) activity in oocytes (Ma et al., 2010), these findings suggest that endo-siRNAs are largely responsible for the *Dicer1* phenotype, and that although miRNAs are expressed, their activity is suppressed in maturing oocytes. Additionally, analyses of the small RNAs expressed in oocytes and the transcripts upregulated in *Dicer1* mutants have revealed expressed pseudogenes as a possible source of dsRNA for production of the endo-siRNAs (Tam et al., 2008; Watanabe et al., 2008).

What molecular events occur in the oocyte to block miRNA activity while leaving endo-siRNA activity intact? It is possible that the miRNA loss of function in oocytes and early embryos is secondary to the loss of processing bodies (P-bodies) that occurs at the GV stage and does not recover until the blastocyst stage (Flemr, Ma, Schultz, & Svoboda, 2010; Swetloff et al., 2009). However, it is equally likely that the loss of P-bodies, known sites of miRNA activity, is secondary to a primary block in miRNA function (Parker & Sheth, 2007). After fertilization, it is believed that miRNAs and siRNAs do not play a role in preimplantation development because zygotic deletions of *Dgcr8*, *Dicer*, and *Ago2* do not have a phenotype until after embryo implantation (Bernstein et al., 2003; Morita et al., 2007; Suh & Blelloch, 2011). Moreover, maternal deletion of *Dgcr8* has no phenotype until postimplantation, suggesting no essential role for maternal miRNAs throughout oocyte maturation and all of preimplantation development (Suh et al., 2010).

Similar to miRNAs, piRNAs are expressed in mouse oocytes but do not appear to be essential as deletion of the Piwi proteins does not result in an oocyte phenotype (Carmell et al., 2007; Deng & Lin, 2002; Kuramochi-Miyagawa et al., 2004). In flies, a lack of maternally deposited piRNAs that target transposons carried by the paternal genome can result in "hybrid dysgenesis," where sterile progeny result from DNA damage caused by transposon mobilization (Bourc'his & Voinnet, 2010; Brennecke et al., 2008). Fly piRNAs also control Fasciclin 3 (Fas3) levels in follicle cells of the ovaries to regulate oogenesis, and loss of Piwi results in a failure of the somatic cells to intermingle with germ cells (Saito et al., 2009). In zebrafish, loss of Zili results in a failure of germ cells to differentiate to mature oocytes due to defects in chromosome segregation and early meiotic processes (Houwing et al., 2008; Houwing et al., 2007). These studies also revealed a crucial function for Zili independent of its function in transposon defense: in the presence of hypomorphic alleles of Zili where transposon transcripts were not ectopically upregulated, females remained sterile due to arrest in meiosis I (Houwing et al., 2008). More research will be needed to elucidate the mechanism by which the piRNA pathway affects female germline development and why these components are not essential for mouse oogenesis.

Oocyte development and maturation in mouse require intimate cellular interactions between the oocyte and soma. In mice, somatic cell loss of small RNAs leads to defects in oogenesis. Specifically, depletion of *Dicer1* in ovarian granulosa cells as well as the mesenchyme-derived cells of the oviduct and uterus causes female infertility (Gonzalez & Behringer, 2009; Hong et al., 2008; Lei et al., 2010; Nagaraja et al., 2008). Loss of *Dicer1* in granulosa cells, the cells surrounding the oocyte during development (utilizing the *Amhr2-Cre*), results in the establishment of a larger pool of primordial follicles in the neonatal ovary (Lei et al., 2010). Despite this larger pool of primordial follicles, these *Dicer1* mutants are infertile due to premature follicle maturation and loss of oocytes resulting from defects in granulosa cells. It is believed that the defects are due in part to a loss of miR-503, which normally targets Cyclin D2 (an important regulator of the cell cycle) as well as other genes important for granulosa cell proliferation and luteinization (Lei et al., 2010). The *Amhr2-Cre* also removes *Dicer1* from all Mullerian duct mesenchyme-derived tissues. Loss of *Dicer1* in these tissues leads to defects in ovulation due to a failure of the oocytes to migrate from the oviduct to the uterus (Hong et al., 2008). This defect may be secondary to structural defects, as the uterine horns are shorter and oviductal cysts develop (Gonzalez & Behringer, 2009; Nagaraja et al., 2008), presumably due to derepression of targets detrimental to normal reproductive tract development.

Like Sertoli cells of the testis, miRNAs present in mural granulosa cells also regulate the hormonal signals important for oogenesis. miR-132 and miR-212 are stimulated by human chorionic gonadotropin (which mimics endogenous LH activity) and are necessary for the downstream cAMP-dependent signaling events that result in normal meiotic progression (Fiedler, Carletti, Hong, & Christenson, 2008). Indeed, the link between miRNAs and hormone signaling has also been validated in human primary granulosa cell cultures, where a screen for miRNAs that affect expression and release of testosterone and estradiol found that overexpression of more than 50 different miRNAs had a significant effect (Sirotkin, Ovcharenko, Grossmann, Laukova, & Mlyncek, 2009). Follow-up studies found that some miRNAs directly affect proliferation and apoptosis (Sirotkin, Laukova, Ovcharenko, Brenaut, & Mlyncek, 2010), though the direct targets through which these miRNAs act remain to be determined. Discovering the mechanism by which miRNAs mediate hormone signals *in vivo* will require more research.

Small RNA regulation of the soma is also crucial in fly oogenesis. *Drosophila* somatic stem cells (SomaSCs) in the ovary lacking Dicer1 failed to self-renew, which resulted in defective follicle cell growth and proliferation (Jin & Xie, 2007). Moreover, SomaSCs lacking Loquacious and Ago1 (affecting the miRNA pathway) resulted in an abnormally developed germarium (Forstemann et al., 2005; Park et al., 2007; Yang, Chen, et al., 2007). Further supporting a role for miRNAs in SomaSCs, SomaSC deletion of *belle*, another important component for RNAi in Drosophila S2 cells, resulted in GSC cell cycle and maturation defects due to misregulated timing of Notch activity in follicle cells (Deng, Althauser, & Ruohola-Baker, 2001; Lopez-Schier & St Johnston, 2001; Poulton et al., 2011). Specifically, miRNA targeting of the Notch ligand Delta regulates the timing of Notch expression in follicle cells, which then promotes normal GSC development (Poulton et al., 2011). Another role for miRNAs in the follicle cells of the fly ovary has also been uncovered in R2D2 mutants. R2D2, which is normally involved in siRNA production through an interaction with Dicer2, has also been found to bind Dicer1, indicating a role for this protein in the miRNA pathway. A loss of R2D2 results in follicle and stalk cell formation defects, leading to abnormal oogenesis and lower fertility (Kalidas et al., 2008). Together, these studies have revealed a profound impact of miRNAs in oogenesis and reproductive function through small RNA expression in somatic tissues and subsequent regulation of ovulation and early embryo development.

3. SMALL RNAS IN GERM CELL TUMOR FORMATION

Misregulation of germ cell development at any stage (specification, migration, gonad colonization, differentiation) can have disastrous outcomes, including abnormal differentiation and proliferation resulting in tumor formation. The latent potential of germ cells to produce tumors is evident by their ability to form embryonic germ cells (EGCs) when migratory PGCs are explanted to culture and placed under the influence of a specific cocktail of signals (Fig. 6.1B; Cheng et al., 1994; Donovan & de Miguel, 2003; Labosky, Barlow, & Hogan, 1994). EGCs are very similar to embryonic stem cells (ESCs) in their developmental potential, and the PGCs from which EGCs are derived demonstrate this potential when they develop into various types of germ cell tumors (GCTs). Small RNAs and the RBPs that regulate them in normal germline development also have critical roles in the formation of GCTs. The following section will briefly highlight known roles of small RNAs in both controlling and promoting the development of GCTs.

3.1. miRNAs, siRNAs, and piRNAs in GCTs

Many of the processes that small RNAs regulate during the germ cell life cycle are linked to tumorigenesis. In the fly, Mei-P26 binds Ago1 and reduces miRNA levels in ovarian GSCs, resulting in restricted growth and proliferation (Neumuller et al., 2008). Loss of *mei-P26* results in expansion of the transit amplifying cell pool, unrestricted proliferation, and eventually ovarian tumor formation. The RBP Vasa normally promotes translation of *mei-P26* in the germline (Liu et al., 2009), suggesting that aberrant function of Vasa may also cause tumor formation. Thus, balancing levels of miRNA activity in the *Drosophila* female germline is essential to prevent the unchecked growth and proliferation of a tumor.

Similarly, in mouse, miRNA activity can promote tumor development. The RBP DND1 mitigates miRNA activity and regulates germ cell development at multiple stages (Cook et al., 2009, 2011; Kedde et al., 2007; Youngren et al., 2005). Loss of *Dnd1* in mouse results in miRNA suppression of cell cycle inhibitors that normally facilitate differentiation of male PGCs such that uncontrolled growth and proliferation gives rise to full blown teratomas in a substrain of mice (Cook et al., 2011; Stevens, 1973; Youngren et al., 2005). A screen for mutations in *Dnd1* in human testicular germ cell tumors (TGCTs), however, revealed only one patient with a detrimental mutation out of more than 260 TGCT samples, indicating alternative pathways of cell cycle derepression that potentially act through miRNA misregulation (Linger et al., 2008).

Other studies have linked small RNAs and RBPs regulating small RNAs with human GCT models. High levels of miR-372/373 are associated with human teratocarcinomas (Voorhoeve et al., 2007). The human miR-371–373 cluster is homologous to the mouse miR-290–295 (Houbaviy, Murray, & Sharp, 2003) and miR-17–92 clusters, which are highly upregulated in migrating PGCs and ESCs (Hayashi et al., 2008; Wang et al., 2008). These human miRNAs are known to target *Lats2*, which is expressed in germ cells and is important for maintaining genomic stability; intriguingly, miRNA suppression of *Lats2* in humans can be relieved by the RBP DND1 (Kedde et al., 2007). This suggests that miRNAs and the RBPs that regulate them control sets of transcripts that work within common molecular pathways affecting broad cellular processes (Keene, 2007). To determine what other miRNAs and pathways might be involved in GCT development, a high-throughput miRNA qPCR screen in human samples confirmed elevated levels of the miR-371 cluster in nonseminomatous GCTs and revealed a differential expression pattern for 156 other miRNAs in various testicular tumors (Gillis et al., 2007).

Some RBPs that directly suppress miRNA activity are associated with tumor progression. For example, LIN28, which directly inhibits the activity of the let-7 family of miRNAs and thereby promotes pluripotency and self-renewal in ESCs, is overexpressed in malignant human GCTs, but not benign tumors or normal testicular tissue (West et al., 2009). Additionally, LIN28 is a strong marker for ovarian primitive GCTs (Xue, Peng, Wang, Allan, & Cao, 2011). Inhibition of LIN28 activity in human epithelial ovarian cancer cells resulted in significant reductions in cell growth and survival, providing corroborative evidence that LIN28 acts directly as an oncogene (Peng, Maihle, & Huang, 2010).

The piRNA pathway may also be involved in testicular tumors. Another TGCT sample study in human tissues revealed overexpression of the piRNA-binding protein HIWI (homolog of Piwi) in a majority of seminomas, but not nonseminomas or spermatocytic seminomas (Qiao, Zeeman, Deng, Looijenga, & Lin, 2002). The extent to which piRNA misregulation plays a role in GCT development remains unclear.

4. CONCLUSION

Small RNAs have emerged as critical regulators of cell cycle, pluripotency, and cell fate transitions in many cell types, especially germ cells. The complex nature of the germ cell life cycle has made it difficult to discern roles for each type of small RNA at every developmental stage. This review has summarized the broad scope of research that has been undertaken to begin to understand the role small RNAs are playing in this very important cell type. Recent technological advances and newer genetic models are paving the way for an even better understanding of how these small molecules regulate significant germ cell fate transitions. What small

RNAs are essential for germline sex-specific development? How might piRNAs direct epigenetic changes, and what is the full mechanism of their biogenesis? How is miRNA activity, but not siRNA activity, suppressed in the vertebrate egg? What are the small RNAs and pathways they regulate that are clinically relevant for GCT development and treatment? Further research will answer these questions and provide insight into how small RNAs shape our germline and, thus, the survival of our species.

ACKNOWLEDGMENTS

We thank Joseph Wilson and members of the Blelloch lab for their critical review of this chapter. We also thank Lindsey Mork for her critical reading and her artistic inspiration of Fig. 6.1. We are supported by grants from NIH (R01 NS057221, R01 GM101180, U54 HD055764) and CIRM (RN2-00906).

REFERENCES

Aeckerle, N., Eildermann, K., Drummer, C., Ehmcke, J., Schweyer, S., Lerchl, A., et al. (2012). The pluripotency factor LIN28 in monkey and human testis: A marker for spermatogonial stem cells? *Molecular Human Reproduction, 18*, 477–488.

Ancelin, K., Lange, U. C., Hajkova, P., Schneider, R., Bannister, A. J., Kouzarides, T., et al. (2006). Blimp1 associates with Prmt5 and directs histone arginine methylation in mouse germ cells. *Nature Cell Biology, 8*(6), 623–630.

Aravin, A., Gaidatzis, D., Pfeffer, S., Lagos-Quintana, M., Landgraf, P., Iovino, N., et al. (2006). A novel class of small RNAs bind to MILI protein in mouse testes. *Nature, 442*(7099), 203–207.

Aravin, A. A., Naumova, N. M., Tulin, A. V., Vagin, V. V., Rozovsky, Y. M., & Gvozdev, V. A. (2001). Double-stranded RNA-mediated silencing of genomic tandem repeats and transposable elements in the D. melanogaster germline. *Current Biology, 11* (13), 1017–1027.

Aravin, A. A., Sachidanandam, R., Bourc'his, D., Schaefer, C., Pezic, D., Toth, K. F., et al. (2008). A piRNA pathway primed by individual transposons is linked to de novo DNA methylation in mice. *Molecular Cell, 31*(6), 785–799.

Aravin, A. A., Sachidanandam, R., Girard, A., Fejes-Toth, K., & Hannon, G. J. (2007). Developmentally regulated piRNA clusters implicate MILI in transposon control. *Science, 316*(5825), 744–747.

Ashe, A., Sapetschnig, A., Weick, E. M., Mitchell, J., Bagijn, M. P., Cording, A. C., et al. (2012). piRNAs can trigger a multigenerational epigenetic memory in the germline of C. elegans. *Cell, 150*(1), 88–99.

Babiarz, J. E., Ruby, J. G., Wang, Y., Bartel, D. P., & Blelloch, R. (2008). Mouse ES cells express endogenous shRNAs, siRNAs, and other Microprocessor-independent, Dicer-dependent small RNAs. *Genes & Development, 22*(20), 2773–2785.

Bagijn, M. P., Goldstein, L. D., Sapetschnig, A., Weick, E. M., Bouasker, S., Lehrbach, N. J., et al. (2012). Function, targets, and evolution of Caenorhabditis elegans piRNAs. *Science, 337*, 574–578.

Batista, P. J., Ruby, J. G., Claycomb, J. M., Chiang, R., Fahlgren, N., Kasschau, K. D., et al. (2008). PRG-1 and 21U-RNAs interact to form the piRNA complex required for fertility in C. elegans. *Molecular Cell, 31*(1), 67–78.

Bazzini, A. A., Lee, M. T., & Giraldez, A. J. (2012). Ribosome profiling shows that miR-430 reduces translation before causing mRNA decay in zebrafish. *Science, 336*(6078), 233–237.

Beck, A. R. P., Miller, I. J., Anderson, P., & Streuli, M. (1998). RNA-binding protein TIAR is essential for primordial germ cell development. *Proceedings of the National Academy of Sciences, 95*(5), 2331–2336.

Bellve, A. R., Cavicchia, J. C., Millette, C. F., O'Brien, D. A., Bhatnagar, Y. M., & Dym, M. (1977). Spermatogenic cells of the prepuberal mouse. Isolation and morphological characterization. *The Journal of Cell Biology, 74*(1), 68–85.

Bernstein, E., Caudy, A. A., Hammond, S. M., & Hannon, G. J. (2001). Role for a bidentate ribonuclease in the initiation step of RNA interference. *Nature, 409*(6818), 363–366.

Bernstein, E., Kim, S. Y., Carmell, M. A., Murchison, E. P., Alcorn, H., Li, M. Z., et al. (2003). Dicer is essential for mouse development. *Nature Genetics, 35*(3), 215–217.

Bjork, J. K., Sandqvist, A., Elsing, A. N., Kotaja, N., & Sistonen, L. (2010). miR-18, a member of Oncomir-1, targets heat shock transcription factor 2 in spermatogenesis. *Development, 137*(19), 3177–3184.

Blaser, H., Eisenbeiss, S., Neumann, M., Reichman-Fried, M., Thisse, B., Thisse, C., et al. (2005). Transition from non-motile behaviour to directed migration during early PGC development in zebrafish. *Journal of Cell Science, 118*(Pt 17), 4027–4038.

Bohnsack, M. T., Czaplinski, K., & Gorlich, D. (2004). Exportin 5 is a RanGTP-dependent dsRNA-binding protein that mediates nuclear export of pre-miRNAs. *RNA, 10*(2), 185–191.

Bouhallier, F., Allioli, N., Lavial, F., Chalmel, F., Perrard, M. H., Durand, P., et al. (2010). Role of miR-34c microRNA in the late steps of spermatogenesis. *RNA, 16*(4), 720–731.

Bourc'his, D., & Voinnet, O. (2010). A small-RNA perspective on gametogenesis, fertilization, and early zygotic development. *Science, 330*(6004), 617–622.

Bowles, J., Knight, D., Smith, C., Wilhelm, D., Richman, J., Mamiya, S., et al. (2006). Retinoid signaling determines germ cell fate in mice. *Science, 312*(5773), 596–600.

Bowles, J., & Koopman, P. (2010). Sex determination in mammalian germ cells: Extrinsic verses intrinsic factors. *Reproduction, 139*, 943–958.

Boyle, M., Wong, C., Rocha, M., & Jones, D. L. (2007). Decline in self-renewal factors contributes to aging of the stem cell niche in the Drosophila testis. *Cell Stem Cell, 1*(4), 470–478.

Brennecke, J., Aravin, A. A., Stark, A., Dus, M., Kellis, M., Sachidanandam, R., et al. (2007). Discrete small RNA-generating loci as master regulators of transposon activity in Drosophila. *Cell, 128*(6), 1089–1103.

Brennecke, J., Malone, C. D., Aravin, A. A., Sachidanandam, R., Stark, A., & Hannon, G. J. (2008). An epigenetic role for maternally inherited piRNAs in transposon silencing. *Science, 322*(5906), 1387–1392.

Brower-Toland, B., Findley, S. D., Jiang, L., Liu, L., Yin, H., Dus, M., et al. (2007). Drosophila PIWI associates with chromatin and interacts directly with HP1a. *Genes & Development, 21*(18), 2300–2311.

Buehr, M., McLaren, A., Bartley, A., & Darling, S. (1993). Proliferation and migration of primordial germ cells in We/We mouse embryos. *Developmental Dynamics, 198*(3), 182–189 issn: 1058-8388.

Cai, X., Hagedorn, C. H., & Cullen, B. R. (2004). Human microRNAs are processed from capped, polyadenylated transcripts that can also function as mRNAs. *RNA, 10*(12), 1957–1966.

Cam, H. P., Sugiyama, T., Chen, E. S., Chen, X., FitzGerald, P. C., & Grewal, S. I. (2005). Comprehensive analysis of heterochromatin- and RNAi-mediated epigenetic control of the fission yeast genome. *Nature Genetics, 37*(8), 809–819.

Carmell, M. A., Girard, A., van de Kant, H. J., Bourc'his, D., Bestor, T. H., de Rooij, D. G., et al. (2007). MIWI2 is essential for spermatogenesis and repression of transposons in the mouse male germline. *Developmental Cell, 12*(4), 503–514.

Carmell, M. A., Xuan, Z., Zhang, M. Q., & Hannon, G. J. (2002). The Argonaute family: Tentacles that reach into RNAi, developmental control, stem cell maintenance, and tumorigenesis. *Genes & Development, 16*(21), 2733–2742.

Chendrimada, T. P., Gregory, R. I., Kumaraswamy, E., Norman, J., Cooch, N., Nishikura, K., et al. (2005). TRBP recruits the Dicer complex to Ago2 for microRNA processing and gene silencing. *Nature, 436*(7051), 740–744.

Cheng, L., Gearing, D. P., White, L. S., Compton, D. L., Schooley, K., & Donovan, P. J. (1994). Role of leukemia inhibitory factor and its receptor in mouse primordial germ cell growth. *Development, 120*(11), 3145–3153.

Chung, W. J., Okamura, K., Martin, R., & Lai, E. C. (2008). Endogenous RNA interference provides a somatic defense against Drosophila transposons. *Current Biology, 18*(11), 795–802.

Claycomb, J. M., Batista, P. J., Pang, K. M., Gu, W., Vasale, J. J., van Wolfswinkel, J. C., et al. (2009). The Argonaute CSR-1 and its 22G-RNA cofactors are required for holocentric chromosome segregation. *Cell, 139*(1), 123–134.

Conine, C. C., Batista, P. J., Gu, W., Claycomb, J. M., Chaves, D. A., Shirayama, M., et al. (2010). Argonautes ALG-3 and ALG-4 are required for spermatogenesis-specific 26G-RNAs and thermotolerant sperm in Caenorhabditis elegans. *Proceedings of the National Academy of Sciences of the United States of America, 107*(8), 3588–3593.

Cook, M. S., Coveney, D., Batchvarov, I., Nadeau, J. H., & Capel, B. (2009). BAX-mediated cell death affects early germ cell loss and incidence of testicular teratomas in Dnd1Ter/Ter mice. *Developmental Biology, 328*(2), 377–383.

Cook, M. S., Munger, S. C., Nadeau, J. H., & Capel, B. (2011). Regulation of male germ cell cycle arrest and differentiation by DND1 is modulated by genetic background. *Development, 138*(1), 23–32.

Cox, D. N., Chao, A., Baker, J., Chang, L., Qiao, D., & Lin, H. (1998). A novel class of evolutionarily conserved genes defined by piwi are essential for stem cell self-renewal. *Genes & Development, 12*(23), 3715–3727.

Cox, D. N., Chao, A., & Lin, H. (2000). piwi encodes a nucleoplasmic factor whose activity modulates the number and division rate of germline stem cells. *Development, 127*(3), 503–514.

Czech, B., Malone, C. D., Zhou, R., Stark, A., Schlingeheyde, C., Dus, M., et al. (2008). An endogenous small interfering RNA pathway in Drosophila. *Nature, 453*(7196), 798–802.

Deng, W. M., Althauser, C., & Ruohola-Baker, H. (2001). Notch-Delta signaling induces a transition from mitotic cell cycle to endocycle in Drosophila follicle cells. *Development, 128*(23), 4737–4746.

Deng, W., & Lin, H. (2002). miwi, a murine homolog of piwi, encodes a cytoplasmic protein essential for spermatogenesis. *Developmental Cell, 2*(6), 819–830.

Denli, A. M., Tops, B. B., Plasterk, R. H., Ketting, R. F., & Hannon, G. J. (2004). Processing of primary microRNAs by the Microprocessor complex. *Nature, 432*(7014), 231–235.

de Rooij, D. G., & Grootegoed, J. A. (1998). Spermatogonial stem cells. *Current Opinion in Cell Biology, 10*(6), 694–701.

DiNapoli, L., Batchvarov, J., & Capel, B. (2006). FGF9 promotes survival of germ cells in the fetal testis. *Development, 133*(8), 1519–1527.

Djuranovic, S., Nahvi, A., & Green, R. (2012). miRNA-mediated gene silencing by translational repression followed by mRNA deadenylation and decay. *Science, 336*(6078), 237–240.

Doench, J. G., Petersen, C. P., & Sharp, P. A. (2003). siRNAs can function as miRNAs. *Genes & Development, 17*(4), 438–442.

Donovan, P. J., & de Miguel, M. P. (2003). Turning germ cells into stem cells. *Current Opinion in Genetics & Development, 13*(5), 463–471.

Durcova-Hills, G., & Capel, B. (2008). Chapter 6 development of germ cells in the mouse. *Current Topics in Developmental Biology, 83*, 185–212.

Fabian, M. R., Sonenberg, N., & Filipowicz, W. (2010). Regulation of mRNA translation and stability by microRNAs. *Annual Review of Biochemistry, 79*, 351–379.

Fiedler, S. D., Carletti, M. Z., Hong, X., & Christenson, L. K. (2008). Hormonal regulation of MicroRNA expression in periovulatory mouse mural granulosa cells. *Biology of Reproduction, 79*(6), 1030–1037.

Findley, S. D., Tamanaha, M., Clegg, N. J., & Ruohola-Baker, H. (2003). Maelstrom, a Drosophila spindle-class gene, encodes a protein that colocalizes with Vasa and RDE1/AGO1 homolog, Aubergine, in nuage. *Development, 130*(5), 859–871.

Fire, A., Xu, S., Montgomery, M. K., Kostas, S. A., Driver, S. E., & Mello, C. C. (1998). Potent and specific genetic interference by double-stranded RNA in Caenorhabditis elegans. *Nature, 391*(6669), 806–811.

Flemr, M., Ma, J., Schultz, R. M., & Svoboda, P. (2010). P-body loss is concomitant with formation of a messenger RNA storage domain in mouse oocytes. *Biology of Reproduction, 82*(5), 1008–1017.

Forbes, A., & Lehmann, R. (1998). Nanos and Pumilio have critical roles in the development and function of Drosophila germline stem cells. *Development, 125*(4), 679–690.

Forstemann, K., Tomari, Y., Du, T., Vagin, V. V., Denli, A. M., Bratu, D. P., et al. (2005). Normal microRNA maturation and germ-line stem cell maintenance requires Loquacious, a double-stranded RNA-binding domain protein. *PLoS Biology, 3*(7), e236.

Fuller, M. T., & Spradling, A. C. (2007). Male and female Drosophila germline stem cells: Two versions of immortality. *Science, 316*(5823), 402–404.

Gent, J. I., Schvarzstein, M., Villeneuve, A. M., Gu, S. G., Jantsch, V., Fire, A. Z., et al. (2009). A Caenorhabditis elegans RNA-directed RNA polymerase in sperm development and endogenous RNA interference. *Genetics, 183*(4), 1297–1314.

Ghildiyal, M., Seitz, H., Horwich, M. D., Li, C., Du, T., Lee, S., et al. (2008). Endogenous siRNAs derived from transposons and mRNAs in Drosophila somatic cells. *Science, 320* (5879), 1077–1081.

Ghildiyal, M., & Zamore, P. D. (2009). Small silencing RNAs: An expanding universe. *Nature Reviews. Genetics, 10*(2), 94–108.

Gill, M. E., Hu, Y. C., Lin, Y., & Page, D. C. (2011). Licensing of gametogenesis, dependent on RNA binding protein DAZL, as a gateway to sexual differentiation of fetal germ cells. *Proceedings of the National Academy of Sciences of the United States of America, 108*, 7443–7448.

Gillis, A. J., Stoop, H. J., Hersmus, R., Oosterhuis, J. W., Sun, Y., Chen, C., et al. (2007). High-throughput microRNAome analysis in human germ cell tumours. *The Journal of Pathology, 213*(3), 319–328.

Giraldez, A. J., Cinalli, R. M., Glasner, M. E., Enright, A. J., Thomson, J. M., Baskerville, S., et al. (2005). MicroRNAs regulate brain morphogenesis in zebrafish. *Science, 308*(5723), 833–838.

Giraldez, A. J., Mishima, Y., Rihel, J., Grocock, R. J., Van Dongen, S., Inoue, K., et al. (2006). Zebrafish MiR-430 promotes deadenylation and clearance of maternal mRNAs. *Science, 312*(5770), 75–79.

Godin, I., Wylie, C., & Heasman, J. (1990). Genital ridges exert long-range effects on mouse primordial germ cell numbers and direction of migration in culture. *Development, 108*(2), 357–363.

Gonzalez, G., & Behringer, R. R. (2009). Dicer is required for female reproductive tract development and fertility in the mouse. *Molecular Reproduction and Development, 76*(7), 678–688.

Gonzalez-Gonzalez, E., Lopez-Casas, P. P., & del Mazo, J. (2008). The expression patterns of genes involved in the RNAi pathways are tissue-dependent and differ in the germ and somatic cells of mouse testis. *Biochimica et Biophysica Acta, 1779*(5), 306–311.

Goudarzi, M., Banisch, T. U., Mobin, M. B., Maghelli, N., Tarbashevich, K., Strate, I., et al. (2012). Identification and regulation of a molecular module for bleb-based cell motility. *Developmental Cell, 23*, 210–218.

Gregory, R. I., Chendrimada, T. P., Cooch, N., & Shiekhattar, R. (2005). Human RISC couples microRNA biogenesis and posttranscriptional gene silencing. *Cell, 123*(4), 631–640.

Gregory, R. I., Yan, K. P., Amuthan, G., Chendrimada, T., Doratotaj, B., Cooch, N., et al. (2004). The Microprocessor complex mediates the genesis of microRNAs. *Nature, 432* (7014), 235–240.

Grimson, A., Farh, K. K., Johnston, W. K., Garrett-Engele, P., Lim, L. P., & Bartel, D. P. (2007). MicroRNA targeting specificity in mammals: Determinants beyond seed pairing. *Molecular Cell, 27*(1), 91–105.

Gu, W., Shirayama, M., Conte, D., Jr., Vasale, J., Batista, P. J., Claycomb, J. M., et al. (2009). Distinct argonaute-mediated 22G-RNA pathways direct genome surveillance in the C. elegans germline. *Molecular Cell, 36*(2), 231–244.

Han, T., Manoharan, A. P., Harkins, T. T., Bouffard, P., Fitzpatrick, C., Chu, D. S., et al. (2009). 26G endo-siRNAs regulate spermatogenic and zygotic gene expression in Caenorhabditis elegans. *Proceedings of the National Academy of Sciences of the United States of America, 106*(44), 18674–18679.

Harris, A. N., & Macdonald, P. M. (2001). Aubergine encodes a Drosophila polar granule component required for pole cell formation and related to eIF2C. *Development, 128*(14), 2823–2832.

Hatfield, S. D., Shcherbata, H. R., Fischer, K. A., Nakahara, K., Carthew, R. W., & Ruohola-Baker, H. (2005). Stem cell division is regulated by the microRNA pathway. *Nature, 435*(7044), 974–978.

Hayashi, K., de Sousa, Chuva, Lopes, S. M., Kaneda, M., Tang, F., Hajkova, P., et al. (2008). MicroRNA biogenesis is required for mouse primordial germ cell development and spermatogenesis. *PLoS One, 3*(3), e1738.

Heo, I., Joo, C., Cho, J., Ha, M., Han, J., & Kim, V. N. (2008). Lin28 mediates the terminal uridylation of let-7 precursor MicroRNA. *Molecular Cell, 32*(2), 276–284.

Hong, X., Luense, L. J., McGinnis, L. K., Nothnick, W. B., & Christenson, L. K. (2008). Dicer1 is essential for female fertility and normal development of the female reproductive system. *Endocrinology, 149*(12), 6207–6212.

Houbaviy, H. B., Murray, M. F., & Sharp, P. A. (2003). Embryonic stem cell-specific MicroRNAs. *Developmental Cell, 5*(2), 351–358.

Houwing, S., Berezikov, E., & Ketting, R. F. (2008). Zili is required for germ cell differentiation and meiosis in zebrafish. *The EMBO Journal, 27*(20), 2702–2711.

Houwing, S., Kamminga, L. M., Berezikov, E., Cronembold, D., Girard, A., van den Elst, H., et al. (2007). A role for Piwi and piRNAs in germ cell maintenance and transposon silencing in Zebrafish. *Cell, 129*(1), 69–82.

Iovino, N., Pane, A., & Gaul, U. (2009). miR-184 has multiple roles in Drosophila female germline development. *Developmental Cell, 17*(1), 123–133.

Irish, V., Lehmann, R., & Akam, M. (1989). The Drosophila posterior-group gene nanos functions by repressing hunchback activity. *Nature, 338*(6217), 646–648.

Jin, J., Jing, W., Lei, X. X., Feng, C., Peng, S., Boris-Lawrie, K., et al. (2011). Evidence that Lin28 stimulates translation by recruiting RNA helicase A to polysomes. *Nucleic Acids Research, 39*(9), 3724–3734.

Jin, Z., & Xie, T. (2007). Dcr-1 maintains Drosophila ovarian stem cells. *Current Biology, 17* (6), 539–544.

Kalidas, S., Sanders, C., Ye, X., Strauss, T., Kuhn, M., Liu, Q., et al. (2008). Drosophila R2D2 mediates follicle formation in somatic tissues through interactions with Dicer-1. *Mechanisms of Development, 125*(5–6), 475–485.

Kanatsu-Shinohara, M., Inoue, K., Lee, J., Yoshimoto, M., Ogonuki, N., Miki, H., et al. (2004). Generation of pluripotent stem cells from neonatal mouse testis. *Cell, 119*(7), 1001–1012.

Kaneda, M., Tang, F., O'Carroll, D., Lao, K., & Surani, M. A. (2009). Essential role for Argonaute2 protein in mouse oogenesis. *Epigenetics & Chromatin, 2*(1), 9.

Kawamura, Y., Saito, K., Kin, T., Ono, Y., Asai, K., Sunohara, T., et al. (2008). Drosophila endogenous small RNAs bind to Argonaute 2 in somatic cells. *Nature, 453*(7196), 793–797.

Kedde, M., & Agami, R. (2008). Interplay between microRNAs and RNA-binding proteins determines developmental processes. *Cell Cycle, 7*(7), 899–903.

Kedde, M., Strasser, M. J., Boldajipour, B., Oude Vrielink, J. A., Slanchev, K., le Sage, C., et al. (2007). RNA-binding protein Dnd1 inhibits microRNA access to target mRNA. *Cell, 131*(7), 1273–1286.

Kedde, M., van Kouwenhove, M., Zwart, W., Oude Vrielink, J. A., Elkon, R., & Agami, R. (2010). A Pumilio-induced RNA structure switch in p27-3' UTR controls miR-221 and miR-222 accessibility. *Nature Cell Biology, 12*(10), 1014–1020.

Keene, J. D. (2007). RNA regulons: Coordination of post-transcriptional events. *Nature Reviews. Genetics, 8*(7), 533–543.

Ketting, R. F., Fischer, S. E., Bernstein, E., Sijen, T., Hannon, G. J., & Plasterk, R. H. (2001). Dicer functions in RNA interference and in synthesis of small RNA involved in developmental timing in C. elegans. *Genes & Development, 15*(20), 2654–2659.

Ketting, R. F., Haverkamp, T. H., van Luenen, H. G., & Plasterk, R. H. (1999). Mut-7 of C. elegans, required for transposon silencing and RNA interference, is a homolog of Werner syndrome helicase and RNaseD. *Cell, 99*(2), 133–141.

Kim, G. J., Georg, I., Scherthan, H., Merkenschlager, M., Guillou, F., Scherer, G., et al. (2010). Dicer is required for Sertoli cell function and survival. *The International Journal of Developmental Biology, 54*(5), 867–875.

Kimble, J. (2011). Molecular regulation of the mitosis/meiosis decision in multicellular organisms. *Cold Spring Harbor Perspectives in Biology, 3*(8), a002683.

Kimble, J., & Crittenden, S. L. (2005). Germline proliferation and its control. *WormBook*, 1–14.

Klattenhoff, C., & Theurkauf, W. (2008). Biogenesis and germline functions of piRNAs. *Development, 135*(1), 3–9.

Knaut, H., Werz, C., Geisler, R., & Nusslein-Volhard, C. (2003). A zebrafish homologue of the chemokine receptor Cxcr4 is a germ-cell guidance receptor. *Nature, 421*(6920), 279–282.

Knight, S. W., & Bass, B. L. (2001). A role for the RNase III enzyme DCR-1 in RNA interference and germ line development in Caenorhabditis elegans. *Science, 293*(5538), 2269–2271.

Koprunner, M., Thisse, C., Thisse, B., & Raz, E. (2001). A zebrafish nanos-related gene is essential for the development of primordial germ cells. *Genes & Development, 15*(21), 2877–2885.

Kotaja, N., Bhattacharyya, S. N., Jaskiewicz, L., Kimmins, S., Parvinen, M., Filipowicz, W., et al. (2006). The chromatoid body of male germ cells: Similarity with processing bodies and presence of Dicer and microRNA pathway components. *Proceedings of the National Academy of Sciences of the United States of America, 103*(8), 2647–2652.

Kotaja, N., & Sassone-Corsi, P. (2007). The chromatoid body: A germ-cell-specific RNA-processing centre. *Nature Reviews. Molecular Cell Biology, 8*(1), 85–90.

Kunwar, P. S., Siekhaus, D. E., & Lehmann, R. (2006). In vivo migration: A germ cell perspective. *Annual Review of Cell and Developmental Biology, 22*, 237–265.

Kuramochi-Miyagawa, S., Kimura, T., Ijiri, T. W., Isobe, T., Asada, N., Fujita, Y., et al. (2004). Mili, a mammalian member of piwi family gene, is essential for spermatogenesis. *Development, 131*(4), 839–849.

Kuramochi-Miyagawa, S., Kimura, T., Yomogida, K., Kuroiwa, A., Tadokoro, Y., Fujita, Y., et al. (2001). Two mouse piwi-related genes: Miwi and mili. *Mechanisms of Development, 108*(1–2), 121–133.

Kuramochi-Miyagawa, S., Watanabe, T., Gotoh, K., Totoki, Y., Toyoda, A., Ikawa, M., et al. (2008). DNA methylation of retrotransposon genes is regulated by Piwi family members MILI and MIWI2 in murine fetal testes. *Genes & Development, 22*(7), 908–917.

Labosky, P. A., Barlow, D. P., & Hogan, B. L. (1994). Embryonic germ cell lines and their derivation from mouse primordial germ cells. *Ciba Foundation Symposium, 182*, 157–168.

Laird, D. J., Altshuler-Keylin, S., Kissner, M. D., Zhou, X., & Anderson, K. V. (2011). Ror2 enhances polarity and directional migration of primordial germ cells. *PLoS Genetics*, 7(12), e1002428.

Lawson, K. A., Dunn, N. R., Roelen, B. A., Zeinstra, L. M., Davis, A. M., Wright, C. V., et al. (1999). Bmp4 is required for the generation of primordial germ cells in the mouse embryo. *Genes & Development, 13*(4), 424–436.

Lee, R. C., Feinbaum, R. L., & Ambros, V. (1993). The C. elegans heterochronic gene lin-4 encodes small RNAs with antisense complementarity to lin-14. *Cell, 75*(5), 843–854.

Lee, H. C., Gu, W., Shirayama, M., Youngman, E., Conte, D., Jr., & Mello, C. C. (2012). C. elegans piRNAs mediate the genome-wide surveillance of germline transcripts. *Cell, 150*(1), 78–87.

Lee, Y., Kim, M., Han, J., Yeom, K. H., Lee, S., Baek, S. H., et al. (2004). MicroRNA genes are transcribed by RNA polymerase II. *The EMBO Journal, 23*(20), 4051–4060.

Lee, Y. S., Nakahara, K., Pham, J. W., Kim, K., He, Z., Sontheimer, E. J., et al. (2004). Distinct roles for Drosophila Dicer-1 and Dicer-2 in the siRNA/miRNA silencing pathways. *Cell, 117*(1), 69–81.

Lehmann, R., & Nusslein-Volhard, C. (1991). The maternal gene nanos has a central role in posterior pattern formation of the Drosophila embryo. *Development, 112*(3), 679–691.

Lei, L., Jin, S., Gonzalez, G., Behringer, R. R., & Woodruff, T. K. (2010). The regulatory role of Dicer in folliculogenesis in mice. *Molecular and Cellular Endocrinology, 315*(1–2), 63–73.

Lewis, B. P., Shih, I. H., Jones-Rhoades, M. W., Bartel, D. P., & Burge, C. B. (2003). Prediction of mammalian microRNA targets. *Cell, 115*(7), 787–798.

Li, M., Hong, N., Gui, J., & Hong, Y. (2012). Medaka piwi is essential for primordial germ cell migration. *Current Molecular Medicine, 12*, 1040–1049.

Li, C., Vagin, V. V., Lee, S., Xu, J., Ma, S., Xi, H., et al. (2009). Collapse of germline piRNAs in the absence of Argonaute3 reveals somatic piRNAs in flies. *Cell, 137*(3), 509–521.

Lin, H., & Spradling, A. C. (1997). A novel group of pumilio mutations affects the asymmetric division of germline stem cells in the Drosophila ovary. *Development, 124*(12), 2463–2476.

Linger, R., Dudakia, D., Huddart, R., Tucker, K., Friedlander, M., Phillips, K. A., et al. (2008). Analysis of the DND1 gene in men with sporadic and familial testicular germ cell tumors. *Genes, Chromosomes & Cancer, 47*(3), 247–252.

Liu, J., Carmell, M. A., Rivas, F. V., Marsden, C. G., Thomson, J. M., Song, J. J., et al. (2004). Argonaute2 is the catalytic engine of mammalian RNAi. *Science, 305*(5689), 1437–1441.

Liu, N., Han, H., & Lasko, P. (2009). Vasa promotes Drosophila germline stem cell differentiation by activating mei-P26 translation by directly interacting with a (U)-rich motif in its 3' UTR. *Genes & Development, 23*(23), 2742–2752.

Lopez-Schier, H., & St Johnston, D. (2001). Delta signaling from the germ line controls the proliferation and differentiation of the somatic follicle cells during Drosophila oogenesis. *Genes & Development, 15*(11), 1393–1405.

Ma, J., Flemr, M., Stein, P., Berninger, P., Malik, R., Zavolan, M., et al. (2010). MicroRNA activity is suppressed in mouse oocytes. *Current Biology, 20*(3), 265–270.

Maatouk, D. M., Loveland, K. L., McManus, M. T., Moore, K., & Harfe, B. D. (2008). Dicer1 is required for differentiation of the mouse male germline. *Biology of Reproduction, 79*(4), 696–703.

Malone, C. D., Brennecke, J., Dus, M., Stark, A., McCombie, W. R., Sachidanandam, R., et al. (2009). Specialized piRNA pathways act in germline and somatic tissues of the Drosophila ovary. *Cell, 137*(3), 522–535.

Matova, N., & Cooley, L. (2001). Comparative aspects of animal oogenesis. *Developmental Biology, 231*(2), 291–320.

Matsui, Y. (1998). Developmental fates of the mouse germ cell line. *The International Journal of Developmental Biology, 42*(7), 1037.

McLaren, A. (1984). Meiosis and differentiation of mouse germ cells. *Symposia of the Society for Experimental Biology, 38*, 7–23.

Medeiros, L. A., Dennis, L. M., Gill, M. E., Houbaviy, H., Markoulaki, S., Fu, D., et al. (2011). Mir-290-295 deficiency in mice results in partially penetrant embryonic lethality and germ cell defects. *Proceedings of the National Academy of Sciences of the United States of America, 108*(34), 14163–14168.

Megosh, H. B., Cox, D. N., Campbell, C., & Lin, H. (2006). The role of PIWI and the miRNA machinery in Drosophila germline determination. *Current Biology, 16*(19), 1884–1894.

Melton, C., Judson, R. L., & Blelloch, R. (2010). Opposing microRNA families regulate self-renewal in mouse embryonic stem cells. *Nature, 463*(7281), 621–626.

Menke, D. B., Koubova, J., & Page, D. C. (2003). Sexual differentiation of germ cells in XX mouse gonads occurs in an anterior-to-posterior wave. *Developmental Biology, 262*(2), 303–312.

Mickoleit, M., Banisch, T. U., & Raz, E. (2011). Regulation of hub mRNA stability and translation by miR430 and the dead end protein promotes preferential expression in zebrafish primordial germ cells. *Developmental Dynamics, 240*(3), 695–703.

Mintz, B., & Russell, E. S. (1957). Gene-induced embryological modifications of primordial germ cells in the mouse. *The Journal of Experimental Zoology, 134*(2), 207–237.

Mishima, Y., Giraldez, A. J., Takeda, Y., Fujiwara, T., Sakamoto, H., Schier, A. F., et al. (2006). Differential regulation of germline mRNAs in soma and germ cells by zebrafish miR-430. *Current Biology, 16*(21), 2135–2142.

Molyneaux, K. A., Zinszner, H., Kunwar, P. S., Schaible, K., Stebler, J., Sunshine, M. J., et al. (2003). The chemokine SDF1/CXCL12 and its receptor CXCR4 regulate mouse germ cell migration and survival. *Development, 130*(18), 4279–4286.

Montgomery, M. K., & Fire, A. (1998). Double-stranded RNA as a mediator in sequence-specific genetic silencing and co-suppression. *Trends in Genetics, 14*(7), 255–258.

Montgomery, M. K., Xu, S., & Fire, A. (1998). RNA as a target of double-stranded RNA-mediated genetic interference in Caenorhabditis elegans. *Proceedings of the National Academy of Sciences of the United States of America, 95*(26), 15502–15507.

Morita, S., Horii, T., Kimura, M., Goto, Y., Ochiya, T., & Hatada, I. (2007). One Argonaute family member, Eif2c2 (Ago2), is essential for development and appears not to be involved in DNA methylation. *Genomics, 89*(6), 687–696.

Mork, L., Maatouk, D. M., McMahon, J. A., Guo, J. J., Zhang, P., McMahon, A. P., et al. (2012). Temporal differences in granulosa cell specification in the ovary reflect distinct follicle fates in mice. *Biology of Reproduction, 86*(2), 37.

Murchison, E. P., Stein, P., Xuan, Z., Pan, H., Zhang, M. Q., Schultz, R. M., et al. (2007). Critical roles for Dicer in the female germline. *Genes & Development, 21*(6), 682–693.

Nagaraja, A. K., Andreu-Vieyra, C., Franco, H. L., Ma, L., Chen, R., Han, D. Y., et al. (2008). Deletion of Dicer in somatic cells of the female reproductive tract causes sterility. *Molecular Endocrinology, 22*(10), 2336–2352.

Neumuller, R. A., Betschinger, J., Fischer, A., Bushati, N., Poernbacher, I., Mechtler, K., et al. (2008). Mei-P26 regulates microRNAs and cell growth in the Drosophila ovarian stem cell lineage. *Nature, 454*(7201), 241–245.

Newman, M. A., Thomson, J. M., & Hammond, S. M. (2008). Lin-28 interaction with the Let-7 precursor loop mediates regulated microRNA processing. *RNA, 14*(8), 1539–1549.

Nicholls, P. K., Harrison, C. A., Walton, K. L., McLachlan, R. I., O'Donnell, L., & Stanton, P. G. (2011). Hormonal regulation of sertoli cell micro-RNAs at spermiation. *Endocrinology, 152*(4), 1670–1683.

Nie, K., Gomez, M., Landgraf, P., Garcia, J. F., Liu, Y., Tan, L. H., et al. (2008). MicroRNA-mediated down-regulation of PRDM1/Blimp-1 in Hodgkin/Reed-Sternberg cells: A potential pathogenetic lesion in Hodgkin lymphomas. *The American Journal of Pathology, 173*(1), 242–252.

Nishida, K. M., Saito, K., Mori, T., Kawamura, Y., Nagami-Okada, T., Inagaki, S., et al. (2007). Gene silencing mechanisms mediated by Aubergine piRNA complexes in Drosophila male gonad. *RNA, 13*(11), 1911–1922.

Novina, C. D., & Sharp, P. A. (2004). The RNAi revolution. *Nature, 430*(6996), 161–164.

Novotny, G. W., Sonne, S. B., Nielsen, J. E., Jonstrup, S. P., Hansen, M. A., Skakkebaek, N. E., et al. (2007). Translational repression of E2F1 mRNA in carcinoma in situ and normal testis correlates with expression of the miR-17-92 cluster. *Cell Death and Differentiation, 14*(4), 879–882.

Ohinata, Y., Ohta, H., Shigeta, M., Yamanaka, K., Wakayama, T., & Saitou, M. (2009). A signaling principle for the specification of the germ cell lineage in mice. *Cell, 137*(3), 571–584.

Ohinata, Y., Payer, B., O'Carroll, D., Ancelin, K., Ono, Y., Sano, M., et al. (2005). Blimp1 is a critical determinant of the germ cell lineage in mice. *Nature, 436*(7048), 207–213.

Ohinata, Y., Seki, Y., Payer, B., O'Carroll, D., Surani, M. A., & Saitou, M. (2006). Germline recruitment in mice: A genetic program for epigenetic reprogramming. *Ernst Schering Research Foundation Workshop*, (60), 143–174.

Okamura, K., Balla, S., Martin, R., Liu, N., & Lai, E. C. (2008). Two distinct mechanisms generate endogenous siRNAs from bidirectional transcription in Drosophila melanogaster. *Nature Structural & Molecular Biology, 15*(9), 998.

Okamura, K., Hagen, J. W., Duan, H., Tyler, D. M., & Lai, E. C. (2007). The mirtron pathway generates microRNA-class regulatory RNAs in Drosophila. *Cell, 130*(1), 89–100.

Okamura, K., Ishizuka, A., Siomi, H., & Siomi, M. C. (2004). Distinct roles for Argonaute proteins in small RNA-directed RNA cleavage pathways. *Genes & Development, 18*(14), 1655–1666.

Pal-Bhadra, M., Leibovitch, B. A., Gandhi, S. G., Rao, M., Bhadra, U., Birchler, J. A., et al. (2004). Heterochromatic silencing and HP1 localization in Drosophila are dependent on the RNAi machinery. *Science, 303*(5658), 669–672.

Papaioannou, M. D., Lagarrigue, M., Vejnar, C. E., Rolland, A. D., Kuhne, F., Aubry, F., et al. (2011). Loss of Dicer in Sertoli cells has a major impact on the testicular proteome of mice. *Molecular & Cellular Proteomics, 10*(4), M900587MCP200.

Papaioannou, M. D., Pitetti, J. L., Ro, S., Park, C., Aubry, F., Schaad, O., et al. (2009). Sertoli cell Dicer is essential for spermatogenesis in mice. *Developmental Biology, 326*(1), 250–259.

Park, J. K., Liu, X., Strauss, T. J., McKearin, D. M., & Liu, Q. (2007). The miRNA pathway intrinsically controls self-renewal of Drosophila germline stem cells. *Current Biology, 17* (6), 533–538.

Parker, J. S., & Barford, D. (2006). Argonaute: A scaffold for the function of short regulatory RNAs. *Trends in Biochemical Sciences, 31*(11), 622–630.

Parker, R., & Sheth, U. (2007). P bodies and the control of mRNA translation and degradation. *Molecular Cell, 25*(5), 635–646.

Pavelec, D. M., Lachowiec, J., Duchaine, T. F., Smith, H. E., & Kennedy, S. (2009). Requirement for the ERI/DICER complex in endogenous RNA interference and sperm development in Caenorhabditis elegans. *Genetics, 183*(4), 1283–1295.

Pek, J. W., Lim, A. K., & Kai, T. (2009). Drosophila maelstrom ensures proper germline stem cell lineage differentiation by repressing microRNA-7. *Developmental Cell, 17*(3), 417–424.

Peng, S., Chen, L. L., Lei, X. X., Yang, L., Lin, H., Carmichael, G. G., et al. (2011). Genome-wide studies reveal that Lin28 enhances the translation of genes important for growth and survival of human embryonic stem cells. *Stem Cells, 29*(3), 496–504.

Peng, S., Maihle, N. J., & Huang, Y. (2010). Pluripotency factors Lin28 and Oct4 identify a sub-population of stem cell-like cells in ovarian cancer. *Oncogene, 29*(14), 2153–2159.

Piskounova, E., Viswanathan, S. R., Janas, M., LaPierre, R. J., Daley, G. Q., Sliz, P., et al. (2008). Determinants of microRNA processing inhibition by the developmentally regulated RNA-binding protein Lin28. *The Journal of Biological Chemistry, 283*(31), 21310–21314.

Poulton, J. S., Huang, Y. C., Smith, L., Sun, J., Leake, N., Schleede, J., et al. (2011). The microRNA pathway regulates the temporal pattern of Notch signaling in Drosophila follicle cells. *Development, 138*(9), 1737–1745.

Qiao, D., Zeeman, A. M., Deng, W., Looijenga, L. H., & Lin, H. (2002). Molecular characterization of hiwi, a human member of the piwi gene family whose overexpression is correlated to seminomas. *Oncogene, 21*(25), 3988–3999.

Raz, E. (2003). Primordial germ-cell development: The zebrafish perspective. *Nature Reviews. Genetics, 4*(9), 690–700.

Robertson, E. J., Charatsi, I., Joyner, C. J., Koonce, C. H., Morgan, M., Islam, A., et al. (2007). Blimp1 regulates development of the posterior forelimb, caudal pharyngeal arches, heart and sensory vibrissae in mice. *Development, 134*(24), 4335–4345.

Rocheleau, C. E., Cullison, K., Huang, K., Bernstein, Y., Spilker, A. C., & Sundaram, M. V. (2008). The Caenorhabditis elegans ekl (enhancer of ksr-1 lethality) genes include putative components of a germline small RNA pathway. *Genetics, 178*(3), 1431–1443.

Rouget, C., Papin, C., Boureux, A., Meunier, A. C., Franco, B., Robine, N., et al. (2010). Maternal mRNA deadenylation and decay by the piRNA pathway in the early Drosophila embryo. *Nature, 467*(7319), 1128–1132.

Ruby, J. G., Jan, C. H., & Bartel, D. P. (2007). Intronic microRNA precursors that bypass Drosha processing. *Nature, 448*(7149), 83–86.

Ruby, J. G., Jan, C., Player, C., Axtell, M. J., Lee, W., Nusbaum, C., et al. (2006). Large-scale sequencing reveals 21U-RNAs and additional microRNAs and endogenous siRNAs in C. elegans. *Cell, 127*(6), 1193–1207.

Russell, D. L., & Robker, R. L. (2007). Molecular mechanisms of ovulation: Co-ordination through the cumulus complex. *Human Reproduction Update, 13*(3), 289–312.

Rybak, A., Fuchs, H., Smirnova, L., Brandt, C., Pohl, E. E., Nitsch, R., et al. (2008). A feedback loop comprising lin-28 and let-7 controls pre-let-7 maturation during neural stem-cell commitment. *Nature Cell Biology, 10*(8), 987–993.

Saito, K., Inagaki, S., Mituyama, T., Kawamura, Y., Ono, Y., Sakota, E., et al. (2009). A regulatory circuit for piwi by the large Maf gene traffic jam in Drosophila. *Nature, 461*(7268), 1296–1299.

Saito, K., Ishizu, H., Komai, M., Kotani, H., Kawamura, Y., Nishida, K. M., et al. (2010). Roles for the Yb body components Armitage and Yb in primary piRNA biogenesis in Drosophila. *Genes & Development, 24*(22), 2493–2498.
Saitou, M. (2009). Germ cell specification in mice. *Current Opinion in Genetics & Development, 19*, 386–395.
Saitou, M., Barton, S. C., & Surani, M. A. (2002). A molecular programme for the specification of germ cell fate in mice. *Nature, 418*(6895), 293–300.
Saitou, M., & Yamaji, M. (2010). Germ cell specification in mice: Signaling, transcription regulation, and epigenetic consequences. *Reproduction, 139*(6), 931–942.
Sakurai, T., Iguchi, T., Moriwaki, K., & Noguchi, M. (1995). The ter mutation first causes primordial germ cell deficiency in ter/ter mouse embryos at 8 days of gestation. *Development, Growth & Differentiation, 37*(3), 293–302.
Santos, A. C., & Lehmann, R. (2004). Germ cell specification and migration in Drosophila and beyond. *Current Biology, 14*(14), R578–R589.
Schmidt, A., Palumbo, G., Bozzetti, M. P., Tritto, P., Pimpinelli, S., & Schafer, U. (1999). Genetic and molecular characterization of sting, a gene involved in crystal formation and meiotic drive in the male germ line of Drosophila melanogaster. *Genetics, 151*(2), 749–760.
Seydoux, G., & Braun, R. E. (2006). Pathway to totipotency: Lessons from germ cells. *Cell, 127*(5), 891–904.
She, X., Xu, X., Fedotov, A., Kelly, W. G., & Maine, E. M. (2009). Regulation of heterochromatin assembly on unpaired chromosomes during Caenorhabditis elegans meiosis by components of a small RNA-mediated pathway. *PLoS Genetics, 5*(8), e1000624.
Shirayama, M., Seth, M., Lee, H. C., Gu, W., Ishidate, T., Conte, D., Jr., et al. (2012). piRNAs initiate an epigenetic memory of nonself RNA in the C. elegans germline. *Cell, 150*(1), 65–77.
Siomi, M. C., Miyoshi, T., & Siomi, H. (2010). piRNA-mediated silencing in Drosophila germlines. *Seminars in Cell & Developmental Biology, 21*(7), 754–759.
Sirotkin, A. V., Laukova, M., Ovcharenko, D., Brenaut, P., & Mlyncek, M. (2010). Identification of microRNAs controlling human ovarian cell proliferation and apoptosis. *Journal of Cellular Physiology, 223*(1), 49–56.
Sirotkin, A. V., Ovcharenko, D., Grossmann, R., Laukova, M., & Mlyncek, M. (2009). Identification of microRNAs controlling human ovarian cell steroidogenesis via a genome-scale screen. *Journal of Cellular Physiology, 219*(2), 415–420.
Slanchev, K., Stebler, J., Goudarzi, M., Cojocaru, V., Weidinger, G., & Raz, E. (2009). Control of Dead end localization and activity—Implications for the function of the protein in antagonizing miRNA function. *Mechanisms of Development, 126*(3–4), 270–277.
Smardon, A., Spoerke, J. M., Stacey, S. C., Klein, M. E., Mackin, N., & Maine, E. M. (2000). EGO-1 is related to RNA-directed RNA polymerase and functions in germ-line development and RNA interference in C. elegans. *Current Biology, 10*(4), 169–178.
Song, R., Ro, S., Michaels, J. D., Park, C., McCarrey, J. R., & Yan, W. (2009). Many X-linked microRNAs escape meiotic sex chromosome inactivation. *Nature Genetics, 41*(4), 488–493.
Soper, S. F., van der Heijden, G. W., Hardiman, T. C., Goodheart, M., Martin, S. L., de Boer, P., et al. (2008). Mouse maelstrom, a component of nuage, is essential for spermatogenesis and transposon repression in meiosis. *Developmental Cell, 15*(2), 285–297.
Stallock, J., Molyneaux, K., Schaible, K., Knudson, C. M., & Wylie, C. (2003). The pro-apoptotic gene Bax is required for the death of ectopic primordial germ cells during their migration in the mouse embryo. *Development, 130*(26), 6589–6597.
Stevens, L. C. (1973). A new inbred subline of mice (129-terSv) with a high incidence of spontaneous congenital testicular teratomas. *Journal of the National Cancer Institute, 50* (1), 235–242 issn: 0027-8874.

Strome, S., & Lehmann, R. (2007). Germ versus soma decisions: Lessons from flies and worms. *Science, 316*(5823), 392–393.

Su, H., Trombly, M. I., Chen, J., & Wang, X. (2009). Essential and overlapping functions for mammalian Argonautes in microRNA silencing. *Genes & Development, 23*(3), 304–317.

Subramaniam, K. S. G. (1999). nos-1 and nos-2, two genes related to Drosophila nanos, regulate primordial germ cell development and survival in Caenorhabditis elegans. *Development, 126*(21), 4861.

Suh, N., Baehner, L., Moltzahn, F., Melton, C., Shenoy, A., Chen, J., et al. (2010). MicroRNA function is globally suppressed in mouse oocytes and early embryos. *Current Biology, 20*(3), 271–277.

Suh, N., & Blelloch, R. (2011). Small RNAs in early mammalian development: From gametes to gastrulation. *Development, 138*(9), 1653–1661.

Suzuki, A., & Saga, Y. (2008). Nanos2 suppresses meiosis and promotes male germ cell differentiation. *Genes & Development, 22*(4), 430–435.

Suzuki, H., Tsuda, M., Kiso, M., & Saga, Y. (2008). Nanos3 maintains the germ cell lineage in the mouse by suppressing both Bax-dependent and -independent apoptotic pathways. *Developmental Biology, 318*(1), 133–142.

Suzuki, A., Tsuda, M., & Saga, Y. (2007). Functional redundancy among Nanos proteins and a distinct role of Nanos2 during male germ cell development. *Development, 134*(1), 77–83.

Swetloff, A., Conne, B., Huarte, J., Pitetti, J. L., Nef, S., & Vassalli, J. D. (2009). Dcp1-bodies in mouse oocytes. *Molecular Biology of the Cell, 20*(23), 4951–4961.

Szakmary, A., Cox, D. N., Wang, Z., & Lin, H. (2005). Regulatory relationship among piwi, pumilio, and bag-of-marbles in Drosophila germline stem cell self-renewal and differentiation. *Current Biology, 15*(2), 171–178.

Tabara, H., Sarkissian, M., Kelly, W. G., Fleenor, J., Grishok, A., Timmons, L., et al. (1999). The rde-1 gene, RNA interference, and transposon silencing in C. elegans. *Cell, 99*(2), 123–132.

Takada, S., Berezikov, E., Choi, Y. L., Yamashita, Y., & Mano, H. (2009). Potential role of miR-29b in modulation of Dnmt3a and Dnmt3b expression in primordial germ cells of female mouse embryos. *RNA, 15*(8), 1507–1514.

Takeda, Y., Mishima, Y., Fujiwara, T., Sakamoto, H., & Inoue, K. (2009). DAZL relieves miRNA-mediated repression of germline mRNAs by controlling poly(A) tail length in zebrafish. *PLoS One, 4*(10), e7513.

Takeuchi, Y., Molyneaux, K., Runyan, C., Schaible, K., & Wylie, C. (2005). The roles of FGF signaling in germ cell migration in the mouse. *Development, 132*(24), 5399–5409.

Tam, O. H., Aravin, A. A., Stein, P., Girard, A., Murchison, E. P., Cheloufi, S., et al. (2008). Pseudogene-derived small interfering RNAs regulate gene expression in mouse oocytes. *Nature, 453*(7194), 534–538.

Tam, P. P., & Zhou, S. X. (1996). The allocation of epiblast cells to ectodermal and germline lineages is influenced by the position of the cells in the gastrulating mouse embryo. *Developmental Biology, 178*(1), 124–132.

Tanaka, S. S., Toyooka, Y., Akasu, R., Katoh-Fukui, Y., Nakahara, Y., Suzuki, R., et al. (2000). The mouse homolog of Drosophila Vasa is required for the development of male germ cells. *Genes & Development, 14*(7), 841–853.

Tang, F., Kaneda, M., O'Carroll, D., Hajkova, P., Barton, S. C., Sun, Y. A., et al. (2007). Maternal microRNAs are essential for mouse zygotic development. *Genes & Development, 21*(6), 644–648.

Tarbashevich, K., & Raz, E. (2010). The nuts and bolts of germ-cell migration. *Current Opinion in Cell Biology, 22*(6), 715–721.

Thomson, T., Liu, N., Arkov, A., Lehmann, R., & Lasko, P. (2008). Isolation of new polar granule components in Drosophila reveals P body and ER associated proteins. *Mechanisms of Development, 125*(9–10), 865–873.

Toledano, H., D'Alterio, C., Czech, B., Levine, E., & Jones, D. L. (2012). The let-7-Imp axis regulates ageing of the Drosophila testis stem-cell niche. *Nature*, *485*(7400), 605–610.

Tsuda, M., Sasaoka, Y., Kiso, M., Abe, K., Haraguchi, S., Kobayashi, S., et al. (2003). Conserved role of nanos proteins in germ cell development. *Science*, *301*(5637), 1239–1241.

Unhavaithaya, Y., Hao, Y., Beyret, E., Yin, H., Kuramochi-Miyagawa, S., Nakano, T., et al. (2009). MILI, a PIWI-interacting RNA-binding protein, is required for germ line stem cell self-renewal and appears to positively regulate translation. *The Journal of Biological Chemistry*, *284*(10), 6507–6519.

Vagin, V. V., Sigova, A., Li, C., Seitz, H., Gvozdev, V., & Zamore, P. D. (2006). A distinct small RNA pathway silences selfish genetic elements in the germline. *Science*, *313*(5785), 320–324.

van Wolfswinkel, J. C., Claycomb, J. M., Batista, P. J., Mello, C. C., Berezikov, E., & Ketting, R. F. (2009). CDE-1 affects chromosome segregation through uridylation of CSR-1-bound siRNAs. *Cell*, *139*(1), 135–148.

Ventura, A., Young, A. G., Winslow, M. M., Lintault, L., Meissner, A., Erkeland, S. J., et al. (2008). Targeted deletion reveals essential and overlapping functions of the miR-17 through 92 family of miRNA clusters. *Cell*, *132*(5), 875–886.

Verdel, A., Jia, S., Gerber, S., Sugiyama, T., Gygi, S., Grewal, S. I., et al. (2004). RNAi-mediated targeting of heterochromatin by the RITS complex. *Science*, *303*(5658), 672–676.

Vincent, S. D., Dunn, N. R., Sciammas, R., Shapiro-Shalef, M., Davis, M. M., Calame, K., et al. (2005). The zinc finger transcriptional repressor Blimp1/Prdm1 is dispensable for early axis formation but is required for specification of primordial germ cells in the mouse. *Development*, *132*(6), 1315–1325.

Viswanathan, S. R., Daley, G. Q., & Gregory, R. I. (2008). Selective blockade of microRNA processing by Lin28. *Science*, *320*(5872), 97–100.

Voorhoeve, P. M., le Sage, C., Schrier, M., Gillis, A. J., Stoop, H., Nagel, R., et al. (2007). A genetic screen implicates miRNA-372 and miRNA-373 as oncogenes in testicular germ cell tumors. *Advances in Experimental Medicine and Biology*, *604*, 17–46.

Wang, Y., Baskerville, S., Shenoy, A., Babiarz, J. E., Baehner, L., & Blelloch, R. (2008). Embryonic stem cell-specific microRNAs regulate the G1-S transition and promote rapid proliferation. *Nature Genetics*, *40*(12), 1478–1483.

Wang, Y., Medvid, R., Melton, C., Jaenisch, R., & Blelloch, R. (2007). DGCR8 is essential for microRNA biogenesis and silencing of embryonic stem cell self-renewal. *Nature Genetics*, *39*(3), 380–385.

Wang, G., & Reinke, V. (2008). A C. elegans Piwi, PRG-1, regulates 21U-RNAs during spermatogenesis. *Current Biology*, *18*(12), 861–867.

Wang, Y., Zayas, R. M., Guo, T., & Newmark, P. A. (2007). nanos function is essential for development and regeneration of planarian germ cells. *Proceedings of the National Academy of Sciences of the United States of America*, *104*(14), 5901–5906.

Watanabe, T., Takeda, A., Tsukiyama, T., Mise, K., Okuno, T., Sasaki, H., et al. (2006). Identification and characterization of two novel classes of small RNAs in the mouse germline: Retrotransposon-derived siRNAs in oocytes and germline small RNAs in testes. *Genes & Development*, *20*(13), 1732–1743.

Watanabe, T., Totoki, Y., Toyoda, A., Kaneda, M., Kuramochi-Miyagawa, S., Obata, Y., et al. (2008). Endogenous siRNAs from naturally formed dsRNAs regulate transcripts in mouse oocytes. *Nature*, *453*(7194), 539–543.

Weidinger, G., Stebler, J., Slanchev, K., Dumstrei, K., Wise, C., Lovell-Badge, R., et al. (2003). dead end, a novel vertebrate germ plasm component, is required for zebrafish primordial germ cell migration and survival. *Current Biology*, *13*(16), 1429–1434.

Weidinger, G., Wolke, U., Koprunner, M., Klinger, M., & Raz, E. (1999). Identification of tissues and patterning events required for distinct steps in early migration of zebrafish primordial germ cells. *Development*, *126*(23), 5295–5307.

Weidinger, G., Wolke, U., Koprunner, M., Thisse, C., Thisse, B., & Raz, E. (2002). Regulation of zebrafish primordial germ cell migration by attraction towards an intermediate target. *Development, 129*(1), 25–36.
West, J. A., Viswanathan, S. R., Yabuuchi, A., Cunniff, K., Takeuchi, A., Park, I. H., et al. (2009). A role for Lin28 in primordial germ-cell development and germ-cell malignancy. *Nature, 460*(7257), 909–913.
Western, P. (2009). Foetal germ cells: Striking the balance between pluripotency and differentiation. *The International Journal of Developmental Biology, 53*(2–3), 393–409.
Western, P. S., Miles, D. C., van den Bergen, J. A., Burton, M., & Sinclair, A. H. (2008). Dynamic regulation of mitotic arrest in fetal male germ cells. *Stem Cells, 26*(2), 339–347.
Wightman, B., Ha, I., & Ruvkun, G. (1993). Posttranscriptional regulation of the heterochronic gene lin-14 by lin-4 mediates temporal pattern formation in C. elegans. *Cell, 75*(5), 855–862.
Wu, Q., Song, R., Ortogero, N., Zheng, H., Evanoff, R., Small, C. L., et al. (2012). The RNase III enzyme DROSHA is essential for microRNA production and spermatogenesis. *The Journal of Biological Chemistry, 287*, 25173–25190.
Xie, T., & Spradling, A. C. (1998). decapentaplegic is essential for the maintenance and division of germline stem cells in the Drosophila ovary. *Cell, 94*(2), 251–260.
Xue, D., Peng, Y., Wang, F., Allan, R. W., & Cao, D. (2011). RNA-binding protein LIN28 is a sensitive marker of ovarian primitive germ cell tumours. *Histopathology, 59*(3), 452–459.
Yabuta, Y., Kurimoto, K., Ohinata, Y., Seki, Y., & Saitou, M. (2006). Gene expression dynamics during germline specification in mice identified by quantitative single-cell gene expression profiling. *Biology of Reproduction, 75*(5), 705–716.
Yang, L., Chen, D., Duan, R., Xia, L., Wang, J., Qurashi, A., et al. (2007). Argonaute 1 regulates the fate of germline stem cells in Drosophila. *Development, 134*(23), 4265–4272.
Yang, L., Duan, R., Chen, D., Wang, J., & Jin, P. (2007). Fragile X mental retardation protein modulates the fate of germline stem cells in Drosophila. *Human Molecular Genetics, 16* (15), 1814–1820.
Yang, Y., Xu, S., Xia, L., Wang, J., Wen, S., Jin, P., et al. (2009). The bantam microRNA is associated with drosophila fragile X mental retardation protein and regulates the fate of germline stem cells. *PLoS Genetics, 5*(4), e1000444.
Yekta, S., Shih, I. H., & Bartel, D. P. (2004). MicroRNA-directed cleavage of HOXB8 mRNA. *Science, 304*(5670), 594–596.
Youngren, K. K., Coveney, D., Peng, X., Bhattacharya, C., Schmidt, L. S., Nickerson, M. L., et al. (2005). The Ter mutation in the dead end gene causes germ cell loss and testicular germ cell tumours. *Nature, 435*(7040), 360–364.
Yu, Z., & Hecht, N. B. (2008). The DNA/RNA-binding protein, translin, binds microRNA122a and increases its in vivo stability. *Journal of Andrology, 29*(5), 572–579.
Yu, Z., Raabe, T., & Hecht, N. B. (2005). MicroRNA Mirn122a reduces expression of the posttranscriptionally regulated germ cell transition protein 2 (Tnp2) messenger RNA (mRNA) by mRNA cleavage. *Biology of Reproduction, 73*(3), 427–433.
Yu, J. Y., Reynolds, S. H., Hatfield, S. D., Shcherbata, H. R., Fischer, K. A., Ward, E. J., et al. (2009). Dicer-1-dependent Dacapo suppression acts downstream of Insulin receptor in regulating cell division of Drosophila germline stem cells. *Development, 136*(9), 1497–1507.
Zeng, Y., Yi, R., & Cullen, B. R. (2003). MicroRNAs and small interfering RNAs can inhibit mRNA expression by similar mechanisms. *Proceedings of the National Academy of Sciences of the United States of America, 100*(17), 9779–9784.
Zhao, G. Q., & Garbers, D. L. (2002). Male germ cell specification and differentiation. *Developmental Cell, 2*(5), 537–547.
Zhong, J., Peters, A. H., Lee, K., & Braun, R. E. (1999). A double-stranded RNA binding protein required for activation of repressed messages in mammalian germ cells. *Nature Genetics, 22*(2), 171–174.

CHAPTER SEVEN

The Control of Meiotic Maturation in Mammalian Oocytes

Janet E. Holt[*,1], **Simon I.R. Lane**[*,†], **Keith T. Jones**[*,†,1]
[*]Center for Reproductive Sciences & School of Biomedical Sciences, University of Newcastle, Callaghan, New South Wales, Australia
[†]Centre for Biological Sciences, Southampton University, SO17 1BJ, UK.
[1]Corresponding authors: e-mail address: janet.holt@newcastle.edu.au; k.t.jones@soton.ac.uk

Contents

Abstract

Mammalian oocytes spend the majority of their lives in a dormant state, residing in primordial follicles. This arrest, most analogous to the G2 stage of the mitotic cell cycle division, is only broken in the hours preceding ovulation, when a hormonal rise induces meiotic resumption and entry into the first meiotic division. At a molecular level, this event is triggered by CDK1 activity, and here, we examine how CDK1 is suppressed during meiotic arrest and raised for oocyte maturation. We focus on signaling: intercellular signaling between the oocyte and the somatic cells of the follicle, and spatial signaling involving the anaphase-promoting complex (APC) within the oocyte. Meiotic arrest is achieved through APC^{FZR1}-mediated cyclin B1 degradation. Once meiotic resumption resumes, CDK1 levels rise, but its activity eventually needs to be suppressed for completion of the first meiotic division. This is achieved by APC^{CDC20}, whose activity is critically regulated by the spindle assembly checkpoint, and which induces both a loss in CDK1 activity as well as the cohesive ties holding chromosomes together.

Current Topics in Developmental Biology, Volume 102
ISSN 0070-2153
http://dx.doi.org/10.1016/B978-0-12-416024-8.00007-6

1. INTRODUCTION

This chapter examines the control of chromosome segregation in the first meiotic division of the mammalian oocyte. The primary focus will be on the mouse because most is known about this species, and the inference will be that the processes described are more or less conserved amongst different mammalian species to make the mouse a valid model system. However, if this is known not to be the case, we will highlight such differences.

Our focus will be on the timing of meiosis and principally how meiotic events are coordinated to ensure faithful segregation of chromosomes. We will not be concerned with how oocytes are created in fetal life from primordial germ cells. The only point to note here is that the newborn female animal is endowed with oocytes that have to last her entire reproductive lifespan; in mouse, 5000–6000 oocytes per ovary are present at birth. The fate of the vast majority of these oocytes is to undergo a process of cell death termed atresia, which is still poorly understood. The remainder gets periodically recruited into the growing follicular pool, which results either in their eventual ovulation or demise within nonovulated atretic follicles. For the most part, we are only concerned with the meiotic events taking place in the one or small number of oocytes that are destined to be ovulated in each female reproductive cycle.

The theme of the chapter will commence from the time of meiotic arrest in the fully grown antral follicles, which have therefore been recruited to grow, and are now destined to be ovulated. It will conclude with the completion of the first meiotic division. A detailed account of how the second meiotic division is controlled and broken by the sperm at fertilization can be found elsewhere (Jones, 2007; Madgwick & Jones, 2007). Here, our entire focus will be on the segregation of chromosomes, a process often described as nuclear maturation, which delineates it from events going on in the cytoplasm (cytoplasmic maturation) to produce a competent mature egg readily capable of forming an embryo at fertilization.

A second important theme, tied with the stops and starts of meiosis, is that principally the events of meiosis are controlled by just one kinase: CDK1 (also known as p34, cdc2) (Adhikari et al., 2012). Historically CDK1 activity has been called "maturation-promoting factor" and here we use CDK1 to be more precise, given this is the molecular entity that physiologically fulfils the epithet. Obviously, the picture is more complicated, often with intricate interactions of multiple proteins involved in regulating a particular meiotic event. However, in principle, it aids comprehension to take a reductionist view that all of meiosis is controlled through the activity of just one kinase: CDK1 (Fig. 7.1).

The chapter lends itself to the examination of meiosis at two important stops/starts, whose junctions are governed by dramatic changes in CDK1 activity. Firstly, the process of prophase I arrest, or so-called germinal vesicle, or GV arrest, because of the very prominent single nucleolus contained within the nucleus of such oocytes. CDK1 activity is low at this time and the important job of luteinizing hormone, which physiologically cues meiotic maturation and ovulation, is to trigger a dramatic rise in this kinase. Secondly, the nuclear events associated with the first meiotic division

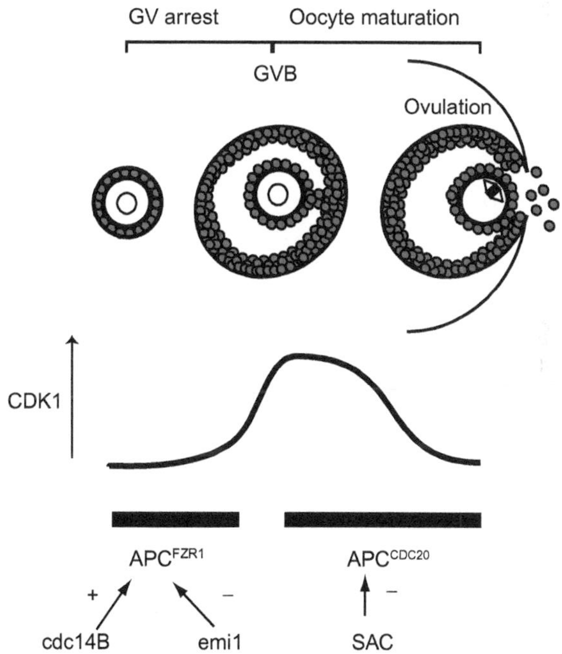

Figure 7.1 CDK1 levels during GV arrest and meiosis I. During GV arrest the follicle grows in size but CDK1 activity remains low. At GVB an LH-signaling cascade switches on CDK1, and this is an essential step for meiotic resumption. Meiosis I exit requires CDK1 activity to fall. Many signaling processes are involved in driving these CDK1 changes. However, illustrated here is the involvement of the APC within the oocyte. It tends to decrease or suppress CDK1 activity by promoting cyclin B1 degradation. APC activity is controlled by levels of Emi1 and CDC14B, and is negatively regulated by the SAC. The APC needs one of two coactivators to be active, and its association with these two proteins, FZR1 and CDC20, changes during meiosis.

(MI), in which the homologous chromosomes (also called bivalents) are segregated. During this time CDK1 activity rises and then precipitously falls at the end of MI.

2. CONTROL OF GV ARREST

2.1. Prolonged GV arrest of the mammalian oocyte

A unique feature of the mammalian oocyte is its capacity to maintain a protracted prophase I arrest that extends from fetal life until adulthood. For the majority of this period, the oocyte exists within the ovary as a primordial follicle surrounded by quiescent granulosa cells. This cell cycle arrest is maintained even as the primordial follicle enters the growing follicle pool and goes on to increase in volume by ~150-fold. Fully grown at a size of around 80 μm diameter in the mouse, the oocyte by this stage has accumulated all the transcripts required for subsequent meiotic progression and fertilization and transcription ceases (Moore & Lintern-Moore, 1978; Sorensen & Wassarman, 1976). Now regarded as "meiotically competent," the oocyte responds to the LH surge at ovulation with a burst of CDK1 activity that precipitates the events of meiotic resumption, including chromatin condensation and nuclear membrane breakdown (Doree & Hunt, 2002; Jones, 2004).

Despite the fact that small nongrowing primordial oocytes and fully grown GV oocytes are similarly arrested in prophase I, they differ in their capacity to resume meiosis. It is not until the oocyte reaches approximately 80% of its final size that it becomes capable of this feat (Sorensen & Wassarman, 1976). The inability of small follicles to undergo GVB is likely due to insufficient CDK1 activity. By their definition, all CDKs require the binding of a cyclin subunit, and as such limited levels of CDK1 and/or the cyclin will prevent the kinase activity. Similar to mitosis, cyclin B1 is the key binding partner of CDK1 during meiosis, with both proteins accumulating and developing a greater capacity to associate with one another as the oocyte grows (de Vantery, Stutz, Vassalli, & Schorderet-Slatkine, 1997). However, CDK1 is likely to be the rate limiting factor, since cyclin B1 always appears to be in excess of CDK1 (Chesnel & Eppig, 1995; de Vantery et al., 1997; Kanatsu-Shinohara, Schultz, & Kopf, 2000). Limited CDK1 and cyclin B1 availabilities appear to play a particularly essential role in maintaining the GV arrest of larger mammalian species (i.e., not rodents), where protein translation even in the fully grown oocyte is an essential prerequisite for GVB (Fulka, Flechon, Motlik, & Fulka, 1988; Mattioli, Galeati, Bacci, & Barboni, 1991; Tatemoto & Horiuchi, 1995).

2.2. Cyclic adenosine 3′,5′-monophosphate-mediated pathway for CDK1 regulation in the GV oocyte

Once the oocyte is fully grown and meiotically component, a complex regulatory system involving both the oocyte and the somatic cells of the follicle ensures GV arrest is maintained until ovulation. Here again, the goal is to keep CDK1 activity low, but in the presence of sufficient CDK1 and cyclin B1 protein, the oocyte now employs additional posttranslational regulatory mechanisms that allow coordination of meiotic resumption with the follicular environment. Cyclic adenosine 3′,5′-monophosphate (cAMP) plays a central role in this process and levels in the oocyte are tightly controlled by the surrounding granulosa cells (Bornslaeger, Mattei, & Schultz, 1986; Cho, Stern, & Biggers, 1974; Magnusson & Hillensjo, 1977). Removal of the mature GV oocyte from its follicular environment results in spontaneous meiotic resumption triggered by a rapid drop in oocyte cAMP levels, underscoring the need for oocyte–granulosa cell communication in this process (Conti, Andersen, Richard, Shitsukawa, & Tsafriri, 1998; Eppig, 1989).

cAMP is produced in the oocyte through the constitutive action of an oolemmal G protein-coupled receptor (GPR3 in mouse) that stimulates G protein (Gs) to activate adenylyl cyclase (Hinckley, Vaccari, Horner, Chen, & Conti, 2005; Horner et al., 2003; Mehlmann, Jones, & Jaffe, 2002; Mehlmann et al., 2004; Norris et al., 2007). Importantly, the oocyte also possesses the capacity to hydrolyse cAMP through the action of phosphodiesterases, specifically PDE3A (Masciarelli et al., 2004). Although the cumulus cells of the follicle also produce cAMP that can be transferred via transzonal gap junctions (Bornslaeger & Schultz, 1985; Dekel, Lawrence, Gilula, & Beers, 1981), studies using the PDE3 and GPR3 knockout mice revealed that oocyte mediated regulation of cAMP is sufficient and necessary for meiotic reentry (Hinckley et al., 2005; Ledent et al., 2005; Mehlmann et al., 2004). The granulosa cells do, however, play an indirect role in elevating the cAMP to levels sufficient to maintain the meiotic arrest by transferring a PDE3 inhibitor to the oocyte in the form of another cyclic nucleotide, namely cGMP. Cumulus cells possess a guanylyl cyclase, NPR2, which is responsible for the cGMP production in this cohort of granulosa cells directly surrounding the oocyte (Zhang, Su, Sugiura, Xia, & Eppig, 2010). Expression of the receptor is maintained by estradiol while its activity is promoted by the binding of ligand NPPC (Zhang et al., 2011). In a paracrine paradigm that is unique to the ovary, NPPC is produced in the outermost mural granulosa such that the GV oocyte arrest is intrinsically linked to all the somatic cells that make up the interior of the follicle (Fig. 2).

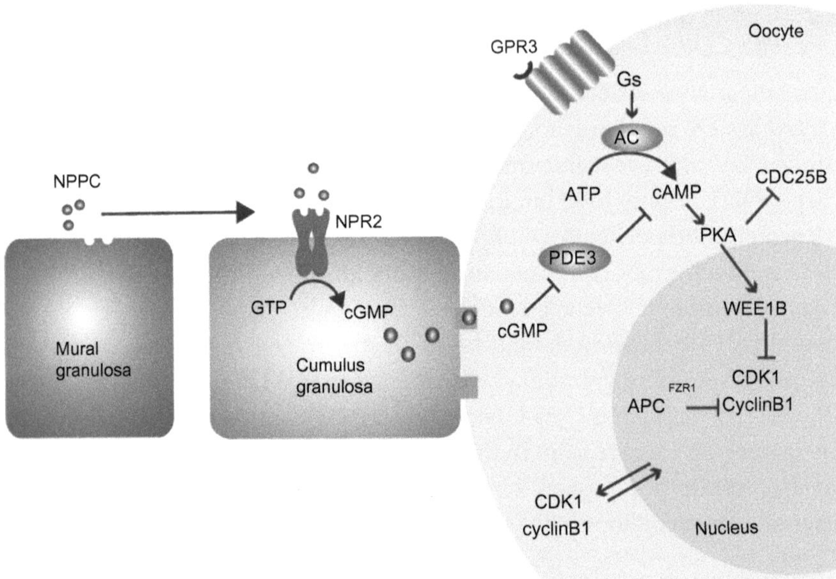

Figure 7.2 Pathways involved in maintenance of Prophase I arrest in the fully grown mammalian oocyte. A G-protein (Gs) linked GPR membrane receptor on the oocyte membrane stimulates adenylate cyclase (AC) to produce cAMP which promotes the activity of PKA. Loss of cAMP is prevented by cumulus cell contribution of cGMP produced by guanylyl cyclase NPR2 in response NPPC produced by the mural granulosa cells. High PKA activity 1) stimulates WEE1B kinase activity which ensures inhibitory phosphorylation of CDK1, and 2) inhibits the activity of CDC25B, the phosphatase responsible for removing this inhibitory phosphorylation. APC^{FZR1} activity ensures the degradation of cyclin B1, the regulatory subunit of CDK1. Both WEE1B and APC^{FZR1} localize predominantly to the nucleus during prophase I arrest with cyclin B1 and CDC25B sequestered primarily to the cytoplasm.

In response to the LH surge at ovulation, cGMP influx into the oocyte is reduced concomitant with a decrease in NPR2 activity and ligand availability (Norris et al., 2009; Robinson et al., 2012; Vaccari, Weeks, Hsieh, Menniti, & Conti, 2009). Closure of granulosa cell junctions also occurs at this time, which is mediated at least in part by an EGF receptor signaling pathway (Norris et al., 2009; Vaccari et al., 2009). The resulting increase in PDE activity promotes a rapid drop in oocyte cAMP, initiating a downstream signaling cascade that culminates in meiotic resumption. The link between cAMP and CDK1 activation appears to be provided by a single intermediary serine/threonine kinase, Protein Kinase A (PKA) in the oocyte. Upon cAMP binding, the catalytic subunits of this tetramer kinase are released allowing the phosphorylation of target proteins (Kirschner, Yin, Jones, & Mahoney, 2009). Key known PKA substrates in the oocyte are members of the WEE/MYT kinase and CDC25 phosphatase families that act directly upon CDK1 to provide an additional level of kinase regulation distinct from cyclin B1 binding. WEE/MYT kinases hold CDK1 in an inactive state by inhibitory phosphorylation at Threonine 14 and Tyrosine 15, changes which are counterbalanced by CDC25 phosphatases (Krek & Nigg, 1991a, 1991b; Morgan, 1995; Norbury, Blow, & Nurse, 1991; Sebastian, Kakizuka, & Hunter, 1993). Oocytes possess a unique isoform of WEE, WEE1B (also known as WEE2), that is positively regulated by PKA phosphorylation at Serine 15 and subsequently leads to further autophosphorylation (Han, Chen, Paronetto, & Conti, 2005). MYT1 seems to act synergistically with WEE1B since antisense knockdown studies reveal that maximal precocious meiotic reentry occurs when there is concurrent loss of WEE1B and MYT1 (Oh, Han, & Conti, 2010).

Acting in opposition to the WEE/MYT kinases, CDC25B is the essential isoform of the CDC25 phosphatase family that is solely responsible for CDK1 activation in oocytes. This has been demonstrated by the failure of CDC25B knockout oocytes to resume meiosis despite the remaining presence of CDC25A and CDC25C (Lincoln et al., 2002). During meiotic arrest, CDC25B is rendered inactive by PKA-mediated phosphorylation at Serine 321, which promotes binding to 14-3-3 protein and cytoplasmic sequestering (Duckworth, Weaver, & Ruderman, 2002; Pirino, Wescott, & Donovan, 2009; Zhang et al., 2008). Therefore, when oocyte levels of cAMP are high, WEE1B activity and CDC25B inhibition is favored thereby maintaining inactive CDK1 and so GV arrest (Oh et al., 2010; Pirino et al., 2009; Zhang et al., 2008). At the onset of GVB, WEE1B is inactivated around the same time as CDC25B is activated thereby reenforcing the oocytes commitment to full, irreversible CDK1 activation.

The potential also exists for direct activating phosphorylation of CDK1 itself. In somatic cells and oocytes of nonmammalian species, active CDK1 is associated with phosphorylation of Threonine 161 by CDK-activating kinase (CAK), a complex of CDK7/cyclin H/MAT1 (Lorca et al., 1992; Solomon, Lee, & Kirschner, 1992). Recent evidence suggests that porcine oocyte CDK1 undergoes CAK-mediated T161 phosphorylation that is associated with GVB (Fujii, Nishimura, Kano, Sugiura, & Naito, 2011); however, it remains to be established if this is also the case in other mammalian species.

2.3. Regulation of cyclin B1 levels also influences CDK1 activity in the GV oocyte

Prevention of precocious meiotic reentry is fundamental for the oocyte and so it is not surprising that the oocyte possesses a secondary mechanism for maintaining GV arrest by regulating cyclin B1 protein levels. Like the somatic cell cycle that is driven by a balance between cyclin degradation and resynthesis, the fully grown oocyte also utilizes proteasomal protein cyclin B1 degradation to limit CDK1 activity until required. Regulation of cyclin B1 levels in fact becomes an essential means of CDK1 regulation for the subsequent stops and starts of MI and MII.

Protein degradation during the mitotic metaphase–anaphase and G1–S transitions is driven by the anaphase-promoting complex (APC), a multisubunit E3 ligase responsible for ubiquitin tagging of protein substrates that promotes their destruction by the 26 proteasome (Peters, 2006; Sullivan & Morgan, 2007). The APC requires the binding of a coactivator protein, either CDC20 or FZR1 (also known CDH1), which confers a degree of substrate specificity to the enzyme. APC bound to CDC20 (APC^{CDC20}) allows mitotic and meiotic anaphase progression by the degradation of substrates securin and cyclin B1, but it is replaced by FZR1 at the time of mitotic exit (Jones, 2011; Peters, 2006). Interestingly, in oocytes, it is FZR1 that is important in regulating the APC. Oocytes lacking FZR1 undergo precocious GVB in response to accumulating cyclin B1 *in vivo* and *in vitro*, even in the presence of PDE inhibitors (Holt et al., 2011; Jones, 2011; Qiao, Zhang, Gamper, Fujita, & Wan, 2010; Reis, Chang, Levasseur, & Jones, 2006; Yamamuro, Kano, & Naito, 2008). The need to ensure low cyclin B1 levels during this period is also underscored by studies where cyclin B1 is overexpressed in oocytes resulting in a similar increased propensity to undergo GVB (Holt, Weaver, & Jones, 2010; Ledan, Polanski, Terret, & Maro, 2001; Marangos & Carroll, 2004; Reis et al., 2006).

Clearly, there is a need for the oocyte to achieve a balance between high cyclin B1 levels that may precipitate the GVB onset and very low cyclin B1 levels that could make the oocyte refractory to GVB. Consistent with this, APC^{FZR1} is itself subject to regulation in the fully grown GV oocyte. The phosphatase CDC14B promotes the activity of FZR1 in the mouse oocyte, but it is currently unknown whether this is due to direct dephosphorylation as occurs during activation of the G2 DNA damage checkpoint in somatic cells (Schindler & Schultz, 2009). Early mitotic inhibitor (Emi1), identified as an S-phase negative regulator of FZR1 in somatic cells appears to be active in GV oocytes until the onset of GVB at which time it is rapidly degraded (Marangos, Verschuren, Chen, Jackson, & Carroll, 2007). Therefore, GV oocytes contain both a known activator and a known inhibitor of FZR1.

The ability of the oocyte to finely tune cyclin B1 levels is likely to be important in the timely coordination of events at GVB. Recent studies using somatic cell extracts reveal that CDK1 activation occurs in a stepwise manner in response to gradually rising levels of cyclin B1, with different CDK1 thresholds triggering specific mitotic events (Deibler & Kirschner, 2010; Gavet & Pines, 2010b). Low levels of cyclin B1 and therefore CDK1 activity actually promotes WEE phosphorylation and so maintains a negative feedback loop. At higher cyclin B1/CDK1 activity, the substrate preference turns instead to CDC25 thus promoting a positive feedback and full CDK1 activation (Deibler & Kirschner, 2010). It is possible the stepwise activation of CDK1 substrate specificity also occurs in oocytes that may operate alongside PKA-mediated phosphorylation events; however, this remains to be investigated.

2.4. Spatial regulation of proteins is important for control of CDK1 activity in oocytes

The GV oocyte therefore possesses two interrelated mechanisms for controlling the onset of GVB, in the form of a cAMP and a cyclin B1-mediated pathway. A reoccurring theme between them is the spatial separation of proteins that either promote or inhibit the CDK1 activity. The key CDK1 substrates include the nuclear lamins and DNA histone proteins, and so the nucleus would seem to be the important site of action for CDK1 immediately prior to GVB. Studies of cyclin B1 dynamics suggest that the CDK1–cyclin B1 complex continually shuttles between the cytoplasm and the nucleus of mammalian cells and oocytes (Hagting, Jackman, Simpson, & Pines, 1999; Hagting, Karlsson, Clute, Jackman, & Pines, 1998; Marangos & Carroll, 2004). In the GV oocyte, the complex appears mostly cytoplasmic, probably due to the propensity for APC^{FZR1} to degrade it in its nuclear location (Holt et al., 2010; Oh et al., 2010) and also perhaps like the S/G2 somatic cell, rates of nuclear export exceed those of import (Hagting et al., 1998; Toyoshima, Moriguchi, Wada, Fukuda, & Nishida, 1998; Yang et al., 1998). Low levels of nuclear cyclin B1–CDK1 may be kept in check by WEE1B and FZR1, both of which possess nuclear localization signals and reside predominantly in the nucleus of the GV oocyte (Oh et al., 2010; Zhou, Ching, Chun, & Jin, 2003). In contrast, the CDK1 activator CDC25B is restricted to the cytoplasm by sequestration to 14-3-3 protein (Pirino et al., 2009)(Fig 2).

In the minutes prior to GVB, a series of dynamic changes in subcellular localization of these components occurs which likely drives the activation of CDK1. Firstly, there is a rapid influx of cyclin B1–CDK1 into the nucleus (Marangos & Carroll, 2004; Mitra & Schultz, 1996). In somatic cells, this movement is thought to be driven by changes in the nuclear import and export machinery, perhaps in response to rising CDK1 itself (Gavet & Pines, 2010a; Hagting et al., 1999). Such high nuclear levels of cyclin B1 are likely to overwhelm APC^{FZR1}-mediated degradation and ensure that the activating subunit of CDK1 is locally available and possibly in excess. Secondly, the cAMP-mediated drop in PKA activity results in the loss of WEE1B activity and concurrent movement out of the nucleus and potentially out of reach of CDK1 (Oh et al., 2010). Finally, loss of PKA activity also allows release of CDC25B from 14-3-3 protein in the cytoplasm, and by virtue of an NLS, it can move into the nucleus (Oh et al., 2010; Zhang et al., 2008) where its substrate CDK1 is now accumulating. Hence the oocyte is able to coordinate the GVB onset by appropriately relocating the CDK1 regulators between the nucleus and cytoplasm in the minutes immediately prior to nuclear membrane dissolution.

3. CONTROL OF MEIOSIS I

The goal of meiosis I is the faithful reductional division of the homologous chromosomes termed bivalents, such that only one copy of each chromosome remains in the oocyte, the other being discarded into the first polar body. The rise in CDK1 activity associated with the break from GV arrest brings about the dissolution of the nuclear lamins, inducing GVB, and condensation of chromatin. Both events are believed to be due to direct phosphorylation by CDK1 (Adhikari et al., 2012; Enserink & Kolodner, 2010).

The regulation of CDK1 activity through its binding partner cyclin B1 is important during maturation, as it is during prophase arrest. During early prometaphase, it continues to rise, as cyclin B1 synthesis outweighs destruction by the APC. Additionally, the high CDK1 activity inhibits those proteins, which previously restrained it during the GV arrest such as the MYT1 and WEE1 kinases (Oh et al., 2010; Ruiz, Vilar, & Nebreda, 2010).

Ultimately, increasing levels of cyclin B1 and CDK1 activity are brought into check by the activity of the APC in coordination with the spindle assembly checkpoint (SAC), a critical component of the division process which will be discussed later. However, during the first several hours of maturation, high CDK1 activity persists and drives the formation of the first meiotic spindle.

3.1. Setting up the meiotic spindle

In somatic cells, spindle formation is guided by the presence of two centrioles, which are gamma-tubulin nucleating centers responsible for promoting the rapid formation of a bipolar spindle, at whose equator the chromosomes congregate, align, and faithfully separate at anaphase-onset (Gadde & Heald, 2004). Oocytes lack such structures and instead rely on the intrinsic properties of microtubules and their associated motor proteins. Firstly, microtubule organizing centers (MTOCs) form *de novo* in great numbers in the cytoplasm (Schuh & Ellenberg, 2007). Over the course of 3–4 h, these MTOCs coalesce into a microtubule ball centered around the chromosomes, and emit multiple poles. These protrusions gradually form into two dominant poles which separate from each other, elongating the spindle. These poles are poorly focused, giving the meiotic spindle its characteristic barrel shape. The process is long in duration when compared to somatic cells, which complete the spindle formation in a matter of minutes (Kapoor, Mayer, Coughlin, & Mitchison, 2000), but nonetheless the bipolar spindle is formed with the chromosomes in its midst, setting the scene for the next stage in the division process.

3.2. Bivalent biorientation

Unique to meiosis I is the configuration of the bivalent, which has to be divided in such a way that each of the two sister chromatid pairs stays intact, with only the two pairs of chromatids segregating, in a so-called reductional division. The bivalent is probably able to move toward the spindle equator under the indiscriminate pushing of microtubules emanating from both the poles. However, once at or near the equator, more control has to be placed on establishing biorientation because if this goes wrong, the chromosomes could missegregate, generating an aneuploid embryo. It is important to appreciate that the two sister kinetochore pairs contained within each bivalent act as a single functional unit. Thus the two kinetochores belonging to the sister chromatid pair, which together form half of the complete bivalent structure, will both bind microtubules emanating from the same pole (Fig. 7.3A). The other kinetochore pair of the bivalent will establish microtubule connections with the other pole and some studies suggest bivalents have a geometry predisposing them to such biorientation (Sakuno, Tada, & Watanabe, 2009). In addition to providing docking sites for microtubules, the kinetochore also creates an important platform for a regulatory cascade called the SAC, which ensures faithful biorientation.

The two halves of the bivalent remain tethered to each other during the process of biorientation because of homologous recombination events that took place in the early fetal life created crossover events (Page & Hawley, 2004). Multisubunit protein rings, termed cohesins, are required to bind the

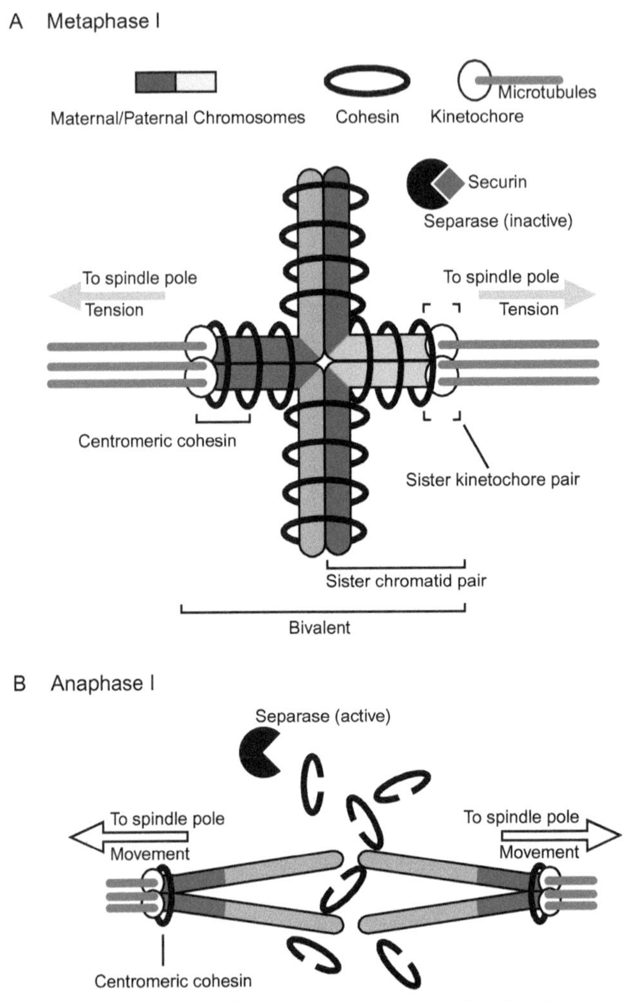

Figure 7.3 The structure of a bivalent during meiosis I. (A) A bivalent in metaphase I under tension due to its biorientation from the two spindle poles. Both kinetochores of each pair of sister chromatids attach to the same spindle pole. The bivalent is able to resist the pulling forces of the microtubules because the chromosome arms are held together by cohesin. (B) Following the activation of separase at anaphase I loss of cohesin on the chromosome arms causes the bivalent to segregate, with one sister chromatid pair moving toward each pole. During this time centromeric cohesin is protected from cleavage.

sister chromatids together (Buonomo et al., 2000; Fig. 7.3A), and at anaphase-onset, the cohesin rings are removed from the chromosome arms due to cleavage by the enzyme separase (Fig. 7.3B). At anaphase-onset, the cohesin rings are removed from the chromosome arms due to cleavage by the enzyme separase (Kudo et al., 2009; Fig. 7.3B). However, centromeric cohesins are protected from this enzymatic cleavage, probably by their association with Sgo2 and the protein phosphatase 2A, which may reverse cleavage promoting phosphorylation (Lee et al., 2008). Cohesin removal is irreversible and is a prerequisite for the chromosome segregation at the anaphase. For faithful segregation, anaphase must not be attempted before all bivalents have achieved biorientation, as such cohesin cleavage is highly regulated and premature activation of separase is normally prevented. This function is fulfilled by the SAC.

3.3. The SAC in oocytes

Initially, it was suggested that oocytes may not have a functional SAC. This was because rates of aneuploidy are high in oocytes when compared to somatic cells, and oocytes appeared unable to respond to the presence of univalents, which often segregate intact in meiosis I, suggesting they had not biorientated (LeMaire-Adkins, Radke, & Hunt, 1997). However, subsequent studies of the SAC in oocytes have revealed that it is both present and functional. Oocytes respond to spindle poisons with a robust arrest as would be expected of cells with a SAC. Furthermore, each of the SAC proteins in turn has been investigated and their knockdown or knockout invariably leads to a shortening of the duration of meiosis I and to increased aneuploidy in the mature egg (Mps1: Hached et al., 2011; Bub1: McGuinness et al., 2009; BubR1: Tsurumi, Hoffmann, Geley, Graeser, & Polanski, 2004; Mad2:Waters, Chen, Murray, & Salmon, 1998; Aurora Kinase: Yang et al., 2010; Mad1: Zhang et al., 2005).

More recent investigations have started to examine why oocytes, which apparently have the full repertoire of functional SAC proteins, are susceptible to aneuploidy. It seems that bivalents are prone to forming erroneous attachments to the spindle, with about three attempts required to correctly biorientate (Kitajima, Ohsugi, & Ellenberg, 2011). It is only when this biorientation is achieved that chromosomes are likely to be segregated faithfully. In addition, it also appears that oocytes do not respond to low numbers of attachment errors. For instance, in unperturbed oocytes, some bivalents fail to establish biorientation and consequently end up close to one spindle pole, instead of under tension near the spindle equator. These bivalents do not generate a sufficient SAC signal to inhibit the APC, and so unless they recorrect their erroneous attachments in time, will likely missegregate at anaphase (Lane, Yun, & Jones, 2012). Artificially, increasing the number of nonaligned bivalents, without disrupting the SAC, by means of mild spindle disruption, or knockdown of key motor proteins also results in an increase in aneuploidy (Gui & Homer, 2012; Kolano, Brunet, Silk, Cleveland, & Verlhac, 2012). Together these finding suggest that in spite of a functional SAC, oocytes fail to arrest in response to low numbers of attachment errors.

3.4. CDK1 and exit from meiosis I

Upon entry into prometaphase, rising CDK1 activity drives the formation of the spindle. However, once the bivalents become sufficiently attached to the spindle to activate the APC, a role reversal occurs and the spindle then controls CDK1 activity. During GV arrest, it is APC^{FZR1} that is used to control cyclin B1 levels (Section 2.3), but soon after GVB, FZR1 levels drop and APC^{FZR1} activity wanes (Reis et al., 2007). From somatic cell studies, it has been established that FZR1 is negatively regulated by CDK1 activity, thus after GVB, we would predict such a decrease in APC^{FZR1}. However, the rising CDK1 activity actually promotes CDC20 as a coactivator of the APC, and as such it is APC^{CDC20} that is tasked with degrading both cyclin B1 and securin, so permitting meiotic exit (Jin et al., 2010; Reis et al., 2007). However, as described above, this activity is initially held in check by the SAC.

The satisfaction of the SAC and the consequent increase in APC^{CDC20} activity starts to lower the levels of cyclin B1 in the oocyte causing a drop in CDK1 activity (Hampl & Eppig, 1995). This, and the equivalent drop in securin permit the activation of separase, which by cleaving the cohesin rings allows the disjunction of the sister chromatid pairs making up each bivalent (Herbert et al., 2003). Due to their opposing poleward orientation, the two chromatid pairs are pulled away from each other along the lengthening spindle. In mitosis, this process is dependent on low CDK1 activity as an artificial anaphase in the presence of high CDK1 activity (in metaphase) results in a very ineffective anaphase (Oliveira, Hamilton, Pauli, Davis, & Nasmyth, 2010).

Unlike dividing somatic cells, which follow division with DNA replication, oocytes enter directly into meiosis II. There is only a brief, partial decondensation of the chromatin, the nuclear envelope does not reform, and there is no DNA synthesis. This may be because during this brief period separating the two meiotic divisions, termed interkinesis, CDK1 activity does not decline completely (Hampl & Eppig, 1995), allowing rapid entry into meiosis II. It is interesting to note that in mouse oocytes, this rapid reactivation of CDK1 activity may be helped, at least in part, by not fully degrading all of the oocyte's cyclin B1. This would allow rapid CDK1 activation by virtue of it being independent of *de novo* cyclin B1 synthesis. However, this creates a problem, in that the residual CDK1 activity, caused by persistent cyclin B1, could block meiotic exit. It seems likely that this is resolved by free separase, which can bind CDK1 and in so doing inhibit the activity of this kinase (Gorr, Boos, & Stemmann, 2005). In mouse oocytes, when the interaction of free separase with CDK1 is blocked using an antibody raised against the region in separase that binds CDK1, polar body extrusion is inhibited although this does not interfere with the ability of separase to cleave cohesins (Gorr et al., 2006).

4. CONCLUDING REMARKS

There has been a greater understanding of the control of GV arrest and meiosis I in recent years, uncovered by molecular and knockout approaches on mice. In the coming years, it is expected that further breakthroughs will be made in our knowledge of the signal transduction process linking a rise of luteinizing hormone with meiotic resumption, especially with respect to a better appreciation of the spatial aspects of CDK1 control. Still to be established is how APC^{FZR1} activity is delicately regulated in oocytes, such that there is not too much activity, which would make the oocyte refractory to hormonal stimulation by there being too little cyclin B1, nor too little activity, which would induce precocious GVB. This balance appears important, but at present, our knowledge is limited to the fact that potential negative and positive regulators of FZR1 are present in GV oocytes. Following GVB, our knowledge of what regulates the SAC activity, which is important in preventing premature APC^{CDC20} induced securin and cyclin B1 degradation, is still limited. What the SAC responds to, and what satisfies this checkpoint is critical in preventing aneuploidy, yet, oocytes seem to be different to many other cells in how this is achieved. Answers to this question have a clinical dimension because the end result of the APC dysregulation at this time is missegregated chromosomes that go on to form aneuploid embryos. Interesting times!

REFERENCES

Adhikari, D., Zheng, W., Shen, Y., Gorre, N., Ning, Y., Halet, G., et al. (2012). Cdk1, but not Cdk2, is the sole Cdk that is essential and sufficient to drive resumption of meiosis in mouse oocytes. *Human Molecular Genetics*, *21*, 2476–2484.

Bornslaeger, E. A., Mattei, P., & Schultz, R. M. (1986). Involvement of cAMP-dependent protein kinase and protein phosphorylation in regulation of mouse oocyte maturation. *Developmental Biology*, *114*, 453–462.

Bornslaeger, E. A., & Schultz, R. M. (1985). Regulation of mouse oocyte maturation: Effect of elevating cumulus cell cAMP on oocyte cAMP levels. *Biology of Reproduction*, *33*, 698–704.

Buonomo, S. B., Clyne, R. K., Fuchs, J., Loidl, J., Uhlmann, F., & Nasmyth, K. (2000). Disjunction of homologous chromosomes in meiosis I depends on proteolytic cleavage of the meiotic cohesin Rec8 by separin. *Cell*, *103*, 387–398.

Chesnel, F., & Eppig, J. J. (1995). Synthesis and accumulation of p34cdc2 and cyclin B in mouse oocytes during acquisition of competence to resume meiosis. *Molecular Reproduction and Development*, *40*, 503–508.

Cho, W. K., Stern, S., & Biggers, J. D. (1974). Inhibitory effect of dibutyryl cAMP on mouse oocyte maturation in vitro. *The Journal of Experimental Zoology*, *187*, 383–386.

Conti, M., Andersen, C. B., Richard, F. J., Shitsukawa, K., & Tsafriri, A. (1998). Role of cyclic nucleotide phosphodiesterases in resumption of meiosis. *Molecular and Cellular Endocrinology*, *145*, 9–14.

Deibler, R. W., & Kirschner, M. W. (2010). Quantitative reconstitution of mitotic CDK1 activation in somatic cell extracts. *Molecular Cell*, *37*, 753–767.

Dekel, N., Lawrence, T. S., Gilula, N. B., & Beers, W. H. (1981). Modulation of cell-to-cell communication in the cumulus-oocyte complex and the regulation of oocyte maturation by LH. *Developmental Biology*, *86*, 356–362.

de Vantery, C., Stutz, A., Vassalli, J. D., & Schorderet-Slatkine, S. (1997). Acquisition of meiotic competence in growing mouse oocytes is controlled at both translational and posttranslational levels. *Developmental Biology*, *187*, 43–54.

Doree, M., & Hunt, T. (2002). From Cdc2 to Cdk1: When did the cell cycle kinase join its cyclin partner? *Journal of Cell Science*, *115*, 2461–2464.

Duckworth, B. C., Weaver, J. S., & Ruderman, J. V. (2002). G2 arrest in Xenopus oocytes depends on phosphorylation of cdc25 by protein kinase A. *Proceedings of the National Academy of Sciences of the United States of America*, *99*, 16794–16799.

Enserink, J. M., & Kolodner, R. D. (2010). An overview of Cdk1-controlled targets and processes. *Cell Division*, *5*, 11.

Eppig, J. J. (1989). The participation of cyclic adenosine monophosphate (cAMP) in the regulation of meiotic maturation of oocytes in the laboratory mouse. *Journal of Reproduction and Fertility. Supplement*, *38*, 3–8.

Fujii, W., Nishimura, T., Kano, K., Sugiura, K., & Naito, K. (2011). CDK7 and CCNH are components of CDK-activating kinase and are required for meiotic progression of pig oocytes. *Biology of Reproduction*, *85*, 1124–1132.

Fulka, J., Jr., Flechon, J. E., Motlik, J., & Fulka, J. (1988). Does autocatalytic amplification of maturation-promoting factor (MPF) exist in mammalian oocytes? *Gamete Research*, *21*, 185–192.

Gadde, S., & Heald, R. (2004). Mechanisms and molecules of the mitotic spindle. *Current Biology*, *14*, R797–R805.

Gavet, O., & Pines, J. (2010a). Activation of cyclin B1-Cdk1 synchronizes events in the nucleus and the cytoplasm at mitosis. *The Journal of Cell Biology*, *189*, 247–259.

Gavet, O., & Pines, J. (2010b). Progressive activation of CyclinB1-Cdk1 coordinates entry to mitosis. *Developmental Cell*, *18*, 533–543.

Gorr, I. H., Boos, D., & Stemmann, O. (2005). Mutual inhibition of separase and Cdk1 by two-step complex formation. *Molecular Cell*, *19*, 135–141.

Gorr, I. H., Reis, A., Boos, D., Wuhr, M., Madgwick, S., Jones, K. T., et al. (2006). Essential CDK1-inhibitory role for separase during meiosis I in vertebrate oocytes. *Nature Cell Biology*, *8*, 1035–1037.

Gui, L., & Homer, H. (2012). Spindle assembly checkpoint signalling is uncoupled from chromosomal position in mouse oocytes. *Development*, *139*, 1941–1946.

Hached, K., Xie, S. Z., Buffin, E., Cladiere, D., Rachez, C., Sacras, M., et al. (2011). Mps1 at kinetochores is essential for female mouse meiosis I. *Development*, *138*, 2261–2271.

Hagting, A., Jackman, M., Simpson, K., & Pines, J. (1999). Translocation of cyclin B1 to the nucleus at prophase requires a phosphorylation-dependent nuclear import signal. *Current Biology*, *9*, 680–689.

Hagting, A., Karlsson, C., Clute, P., Jackman, M., & Pines, J. (1998). MPF localization is controlled by nuclear export. *The EMBO Journal*, *17*, 4127–4138.

Hampl, A., & Eppig, J. J. (1995). Analysis of the mechanism(s) of metaphase I arrest in maturing mouse oocytes. *Development*, *121*, 925–933.

Han, S. J., Chen, R., Paronetto, M. P., & Conti, M. (2005). Wee1B is an oocyte-specific kinase involved in the control of meiotic arrest in the mouse. *Current Biology*, *15*, 1670–1676.

Herbert, M., Levasseur, M., Homer, H., Yallop, K., Murdoch, A., & McDougall, A. (2003). Homologue disjunction in mouse oocytes requires proteolysis of securin and cyclin B1. *Nature Cell Biology*, *5*, 1023–1025.

Hinckley, M., Vaccari, S., Horner, K., Chen, R., & Conti, M. (2005). The G-protein-coupled receptors GPR3 and GPR12 are involved in cAMP signaling and maintenance of meiotic arrest in rodent oocytes. *Developmental Biology*, *287*, 249–261.

Holt, J. E., Tran, S. M., Stewart, J. L., Minahan, K., Garcia-Higuera, I., Moreno, S., et al. (2011). The APC/C activator FZR1 coordinates the timing of meiotic resumption during prophase I arrest in mammalian oocytes. *Development*, *138*, 905–913.

Holt, J. E., Weaver, J., & Jones, K. T. (2010). Spatial regulation of APCCdh1-induced cyclin B1 degradation maintains G2 arrest in mouse oocytes. *Development*, *137*, 1297–1304.

Horner, K., Livera, G., Hinckley, M., Trinh, K., Storm, D., & Conti, M. (2003). Rodent oocytes express an active adenylyl cyclase required for meiotic arrest. *Developmental Biology*, *258*, 385–396.

Jin, F., Hamada, M., Malureanu, L., Jeganathan, K. B., Zhou, W., Morbeck, D. E., et al. (2010). Cdc20 is critical for meiosis I and fertility of female mice. *PLoS Genetics*, *6*, e1001147.

Jones, K. T. (2004). Turning it on and off: M-phase promoting factor during meiotic maturation and fertilization. *Molecular Human Reproduction*, *10*, 1–5.

Jones, K. T. (2007). Intracellular calcium in the fertilization and development of mammalian eggs. *Clinical and Experimental Pharmacology & Physiology*, *34*, 1084–1089.

Jones, K. T. (2011). Anaphase-promoting complex control in female mouse meiosis. *Results and Problems in Cell Differentiation*, *53*, 343–363.

Kanatsu-Shinohara, M., Schultz, R. M., & Kopf, G. S. (2000). Acquisition of meiotic competence in mouse oocytes: Absolute amounts of p34(cdc2), cyclin B1, cdc25C, and wee1 in meiotically incompetent and competent oocytes. *Biology of Reproduction*, *63*, 1610–1616.

Kapoor, T. M., Mayer, T. U., Coughlin, M. L., & Mitchison, T. J. (2000). Probing spindle assembly mechanisms with monastrol, a small molecule inhibitor of the mitotic kinesin, Eg5. *The Journal of Cell Biology*, *150*, 975–988.

Kirschner, L. S., Yin, Z., Jones, G. N., & Mahoney, E. (2009). Mouse models of altered protein kinase A signaling. *Endocrine-Related Cancer*, *16*, 773–793.

Kitajima, T. S., Ohsugi, M., & Ellenberg, J. (2011). Complete kinetochore tracking reveals error-prone homologous chromosome biorientation in mammalian oocytes. *Cell*, *146*, 568–581.

Kolano, A., Brunet, S., Silk, A. D., Cleveland, D. W., & Verlhac, M. H. (2012). Error-prone mammalian female meiosis from silencing the spindle assembly checkpoint without normal interkinetochore tension. *Proceedings of the National Academy of Sciences of the United States of America*, *109*, E1858–E1867.

Krek, W., & Nigg, E. A. (1991a). Differential phosphorylation of vertebrate p34cdc2 kinase at the G1/S and G2/M transitions of the cell cycle: Identification of major phosphorylation sites. *The EMBO Journal*, *10*, 305–316.

Krek, W., & Nigg, E. A. (1991b). Mutations of p34cdc2 phosphorylation sites induce premature mitotic events in HeLa cells: Evidence for a double block to p34cdc2 kinase activation in vertebrates. *The EMBO Journal*, *10*, 3331–3341.

Kudo, N. R., Anger, M., Peters, A. H., Stemmann, O., Theussl, H. C., Helmhart, W., et al. (2009). Role of cleavage by separase of the Rec8 kleisin subunit of cohesin during mammalian meiosis I. *Journal of Cell Science*, *122*, 2686–2698.

Lane, S. I., Yun, Y., & Jones, K. T. (2012). Timing of anaphase-promoting complex activation in mouse oocytes is predicted by microtubule-kinetochore attachment but not by bivalent alignment or tension. *Development*, *139*, 1947–1955.

Ledan, E., Polanski, Z., Terret, M. E., & Maro, B. (2001). Meiotic maturation of the mouse oocyte requires an equilibrium between cyclin B synthesis and degradation. *Developmental Biology*, *232*, 400–413.

Ledent, C., Demeestere, I., Blum, D., Petermans, J., Hamalainen, T., Smits, G., et al. (2005). Premature ovarian aging in mice deficient for Gpr3. *Proceedings of the National Academy of Sciences of the United States of America*, *102*, 8922–8926.

Lee, J., Kitajima, T. S., Tanno, Y., Yoshida, K., Morita, T., Miyano, T., et al. (2008). Unified mode of centromeric protection by shugoshin in mammalian oocytes and somatic cells. *Nature Cell Biology*, *10*, 42–52.

LeMaire-Adkins, R., Radke, K., & Hunt, P. A. (1997). Lack of checkpoint control at the metaphase/anaphase transition: A mechanism of meiotic nondisjunction in mammalian females. *The Journal of Cell Biology*, *139*, 1611–1619.

Lincoln, A. J., Wickramasinghe, D., Stein, P., Schultz, R. M., Palko, M. E., De Miguel, M. P., et al. (2002). Cdc25b phosphatase is required for resumption of meiosis during oocyte maturation. *Nature Genetics*, *30*, 446–449.

Lorca, T., Labbe, J. C., Devault, A., Fesquet, D., Capony, J. P., Cavadore, J. C., et al. (1992). Dephosphorylation of cdc2 on threonine 161 is required for cdc2 kinase inactivation and normal anaphase. *The EMBO Journal*, *11*, 2381–2390.

Madgwick, S., & Jones, K. T. (2007). How eggs arrest at metaphase II: MPF stabilisation plus APC/C inhibition equals Cytostatic Factor. *Cell Division*, *2*, 4.

Magnusson, C., & Hillensjo, T. (1977). Inhibition of maturation and metabolism in rat oocytes by cyclic amp. *The Journal of Experimental Zoology*, *201*, 139–147.

Marangos, P., & Carroll, J. (2004). The dynamics of cyclin B1 distribution during meiosis I in mouse oocytes. *Reproduction*, *128*, 153–162.

Marangos, P., Verschuren, E. W., Chen, R., Jackson, P. K., & Carroll, J. (2007). Prophase I arrest and progression to metaphase I in mouse oocytes are controlled by Emi1-dependent regulation of APC(Cdh1). *The Journal of Cell Biology*, *176*, 65–75.

Masciarelli, S., Horner, K., Liu, C., Park, S. H., Hinckley, M., Hockman, S., et al. (2004). Cyclic nucleotide phosphodiesterase 3A-deficient mice as a model of female infertility. *The Journal of Clinical Investigation*, *114*, 196–205.

Mattioli, M., Galeati, G., Bacci, M. L., & Barboni, B. (1991). Changes in maturation-promoting activity in the cytoplasm of pig oocytes throughout maturation. *Molecular Reproduction and Development*, *30*, 119–125.

McGuinness, B. E., Anger, M., Kouznetsova, A., Gil-Bernabe, A. M., Helmhart, W., Kudo, N. R., et al. (2009). Regulation of APC/C activity in oocytes by a Bub1-dependent spindle assembly checkpoint. *Current Biology*, *19*, 369–380.

Mehlmann, L. M., Jones, T. L., & Jaffe, L. A. (2002). Meiotic arrest in the mouse follicle maintained by a Gs protein in the oocyte. *Science, 297*, 1343–1345.
Mehlmann, L. M., Saeki, Y., Tanaka, S., Brennan, T. J., Evsikov, A. V., Pendola, F. L., et al. (2004). The Gs-linked receptor GPR3 maintains meiotic arrest in mammalian oocytes. *Science, 306*, 1947–1950.
Mitra, J., & Schultz, R. M. (1996). Regulation of the acquisition of meiotic competence in the mouse: Changes in the subcellular localization of cdc2, cyclin B1, cdc25C and wee1, and in the concentration of these proteins and their transcripts. *Journal of Cell Science, 109* (Pt 9), 2407–2415.
Moore, G. P., & Lintern-Moore, S. (1978). Transcription of the mouse oocyte genome. *Biology of Reproduction, 18*, 865–870.
Morgan, D. O. (1995). Principles of CDK regulation. *Nature, 374*, 131–134.
Norbury, C., Blow, J., & Nurse, P. (1991). Regulatory phosphorylation of the p34cdc2 protein kinase in vertebrates. *The EMBO Journal, 10*, 3321–3329.
Norris, R. P., Freudzon, L., Freudzon, M., Hand, A. R., Mehlmann, L. M., & Jaffe, L. A. (2007). A G(s)-linked receptor maintains meiotic arrest in mouse oocytes, but luteinizing hormone does not cause meiotic resumption by terminating receptor-G(s) signaling. *Developmental Biology, 310*, 240–249.
Norris, R. P., Ratzan, W. J., Freudzon, M., Mehlmann, L. M., Krall, J., Movsesian, M. A., et al. (2009). Cyclic GMP from the surrounding somatic cells regulates cyclic AMP and meiosis in the mouse oocyte. *Development, 136*, 1869–1878.
Oh, J. S., Han, S. J., & Conti, M. (2010). Wee1B, Myt1, and Cdc25 function in distinct compartments of the mouse oocyte to control meiotic resumption. *The Journal of Cell Biology, 188*, 199–207.
Oliveira, R. A., Hamilton, R. S., Pauli, A., Davis, I., & Nasmyth, K. (2010). Cohesin cleavage and Cdk inhibition trigger formation of daughter nuclei. *Nature Cell Biology, 12*, 185–192.
Page, S. L., & Hawley, R. S. (2004). The genetics and molecular biology of the synaptonemal complex. *Annual Review of Cell and Developmental Biology, 20*, 525–558.
Peters, J. M. (2006). The anaphase promoting complex/cyclosome: A machine designed to destroy. *Nature Reviews. Molecular Cell Biology, 7*, 644–656.
Pirino, G., Wescott, M. P., & Donovan, P. J. (2009). Protein kinase A regulates resumption of meiosis by phosphorylation of Cdc25B in mammalian oocytes. *Cell Cycle, 8*, 665–670.
Qiao, X., Zhang, L., Gamper, A. M., Fujita, T., & Wan, Y. (2010). APC/C-Cdh1: From cell cycle to cellular differentiation and genomic integrity. *Cell Cycle, 9*, 3904–3912.
Reis, A., Chang, H. Y., Levasseur, M., & Jones, K. T. (2006). APCcdh1 activity in mouse oocytes prevents entry into the first meiotic division. *Nature Cell Biology, 8*, 539–540.
Reis, A., Madgwick, S., Chang, H. Y., Nabti, I., Levasseur, M., & Jones, K. T. (2007). Prometaphase APCcdh1 activity prevents non-disjunction in mammalian oocytes. *Nature Cell Biology, 9*, 1192–1198.
Robinson, J. W., Zhang, M., Shuhaibar, L. C., Norris, R. P., Geerts, A., Wunder, F., et al. (2012). Luteinizing hormone reduces the activity of the NPR2 guanylyl cyclase in mouse ovarian follicles, contributing to the cyclic GMP decrease that promotes resumption of meiosis in oocytes. *Developmental Biology, 366*, 308–316.
Ruiz, E. J., Vilar, M., & Nebreda, A. R. (2010). A two-step inactivation mechanism of Myt1 ensures CDK1/cyclin B activation and meiosis I entry. *Current Biology, 20*, 717–723.
Sakuno, T., Tada, K., & Watanabe, Y. (2009). Kinetochore geometry defined by cohesion within the centromere. *Nature, 458*, 852–858.
Schindler, K., & Schultz, R. M. (2009). CDC14B acts through FZR1 (CDH1) to prevent meiotic maturation of mouse oocytes. *Biology of Reproduction, 80*, 795–803.

Schuh, M., & Ellenberg, J. (2007). Self-organization of MTOCs replaces centrosome function during acentrosomal spindle assembly in live mouse oocytes. *Cell, 130*, 484–498.
Sebastian, B., Kakizuka, A., & Hunter, T. (1993). Cdc25M2 activation of cyclin-dependent kinases by dephosphorylation of threonine-14 and tyrosine-15. *Proceedings of the National Academy of Sciences of the United States of America, 90*, 3521–3524.
Solomon, M. J., Lee, T., & Kirschner, M. W. (1992). Role of phosphorylation in p34cdc2 activation: Identification of an activating kinase. *Molecular Biology of the Cell, 3*, 13–27.
Sorensen, R. A., & Wassarman, P. M. (1976). Relationship between growth and meiotic maturation of the mouse oocyte. *Developmental Biology, 50*, 531–536.
Sullivan, M., & Morgan, D. O. (2007). Finishing mitosis, one step at a time. *Nature Reviews. Molecular Cell Biology, 8*, 894–903.
Tatemoto, H., & Horiuchi, T. (1995). Requirement for protein synthesis during the onset of meiosis in bovine oocytes and its involvement in the autocatalytic amplification of maturation-promoting factor. *Molecular Reproduction and Development, 41*, 47–53.
Toyoshima, F., Moriguchi, T., Wada, A., Fukuda, M., & Nishida, E. (1998). Nuclear export of cyclin B1 and its possible role in the DNA damage-induced G2 checkpoint. *The EMBO Journal, 17*, 2728–2735.
Tsurumi, C., Hoffmann, S., Geley, S., Graeser, R., & Polanski, Z. (2004). The spindle assembly checkpoint is not essential for CSF arrest of mouse oocytes. *The Journal of Cell Biology, 167*, 1037–1050.
Vaccari, S., Weeks, J. L., 2nd, Hsieh, M., Menniti, F. S., & Conti, M. (2009). Cyclic GMP signaling is involved in the luteinizing hormone-dependent meiotic maturation of mouse oocytes. *Biology of Reproduction, 81*, 595–604.
Waters, J. C., Chen, R. H., Murray, A. W., & Salmon, E. D. (1998). Localization of Mad2 to kinetochores depends on microtubule attachment, not tension. *The Journal of Cell Biology, 141*, 1181–1191.
Yamamuro, T., Kano, K., & Naito, K. (2008). Functions of FZR1 and CDC20, activators of the anaphase-promoting complex, during meiotic maturation of swine oocytes. *Biology of Reproduction, 79*, 1202–1209.
Yang, J., Bardes, E. S., Moore, J. D., Brennan, J., Powers, M. A., & Kornbluth, S. (1998). Control of cyclin B1 localization through regulated binding of the nuclear export factor CRM1. *Genes & Development, 12*, 2131–2143.
Yang, K. T., Li, S. K., Chang, C. C., Tang, C. J., Lin, Y. N., Lee, S. C., et al. (2010). Aurora-C kinase deficiency causes cytokinesis failure in meiosis I and production of large polyploid oocytes in mouse. *Molecular Biology of the Cell, 21*, 2371–2383.
Zhang, D., Li, M., Ma, W., Hou, Y., Li, Y. H., Li, S. W., et al. (2005). Localization of mitotic arrest deficient 1 (MAD1) in mouse oocytes during the first meiosis and its functions as a spindle checkpoint protein. *Biology of Reproduction, 72*, 58–68.
Zhang, M., Su, Y. Q., Sugiura, K., Wigglesworth, K., Xia, G., & Eppig, J. J. (2011). Estradiol promotes and maintains cumulus cell expression of natriuretic peptide receptor 2 (NPR2) and meiotic arrest in mouse oocytes in vitro. *Endocrinology, 152*, 4377–4385.
Zhang, M., Su, Y. Q., Sugiura, K., Xia, G., & Eppig, J. J. (2010). Granulosa cell ligand NPPC and its receptor NPR2 maintain meiotic arrest in mouse oocytes. *Science, 330*, 366–369.
Zhang, Y., Zhang, Z., Xu, X. Y., Li, X. S., Yu, M., Yu, A. M., et al. (2008). Protein kinase A modulates Cdc25B activity during meiotic resumption of mouse oocytes. *Developmental Dynamics, 237*, 3777–3786.
Zhou, Y., Ching, Y. P., Chun, A. C., & Jin, D. Y. (2003). Nuclear localization of the cell cycle regulator CDH1 and its regulation by phosphorylation. *The Journal of Biological Chemistry, 278*, 12530–12536.

CHAPTER EIGHT

Signaling Modalities During Oogenesis in Mammals

Lynda K. McGinnis, S. Darlene Limback, David F. Albertini[1]
Department of Molecular and Integrative Physiology, Kansas University Medical Center, Kansas City, Missouri, USA
[1]Corresponding author: e-mail address: dalbertini@kumc.edu

Contents

Abstract

During oogenesis in mammals, the germ line interacts with ovarian somatic cells as follicles assemble, grow, ovulate, and die. As a result, these interactions with granulosa cells determine germ cell fate as the oocyte undergoes hypertrophy, the final stages of meiosis, and preparations required for successful fertilization. Over the past 15 years, investigators using a range of experimental approaches have uncovered the existence of multiple modalities for signaling between the oocyte and companion granulosa cells that play essential and sometimes overlapping roles during the growth and maturative phases of oogenesis. Five modalities of intercellular signaling are considered in the context of regulating oocyte gene expression, metabolism, spatial patterning, and the cell cycle. While some forms of signaling predominate at specific stages of oogenesis, such as during the assembly of primordial follicles, it is apparent that combinations of modalities work in concert to control events associated with ovulation when both nuclear and cytoplasmic maturation occur. A final key feature of the signaling platform underscoring the protracted process of oogenesis is the existence of negative and positive feedback loops designed to coordinate the tempo of oogenesis and folliculogenesis at key developmental transitions.

Current Topics in Developmental Biology, Volume 102
ISSN 0070-2153
http://dx.doi.org/10.1016/B978-0-12-416024-8.00008-8

1. INTRODUCTION

The mammalian ovary contains a finite reserve of primordial follicles that supports the reproductive lifespan by ensuring the availability of mature oocytes for fertilization at ovulation. Yielding single or multiple oocytes during each reproductive cycle is the result of a complex and protracted process known as oogenesis. From the population of primordial follicles existing at the onset of reproductive maturity, diplotene stage oocytes initiate and sustain a developmental progression entailing a remarkable growth phase of cellular hypertrophy followed by the resumption and completion of meiosis (Rodrigues, Limback, McGinnis, Plancha, & Albertini, 2008). The growth phase of oogenesis has long been appreciated to provide sufficient time to acquire and accumulate substantive alterations in organellar and macromolecular components essential to the early stages of embryonic development. This is followed by the maturative phase of oogenesis, when completion of the meiotic cell cycle occurs at ovulation and after fertilization; the maturative phase of oogenesis consists of biosynthetic and secretory events that are compressed into a matter of hours following the ovulation-inducing LH surge. In contrast, the growth phase of oogenesis is weeks to months in duration, and in either case, the oocyte is actively involved in a signaling syncitium within the follicle through heterocellular interactions with the so-called granulosa or follicle cells. It is this population of granulosa cells that maintains direct physical contact with the oocyte once primordial follicle formation has occurred. There is an increasing body of evidence indicating that coordinated interactions between somatic and germ cell compartments of the ovarian follicle entail multiple signaling modalities without which oogenesis fails and fertility is compromised.

From a signaling perspective, the advancement from growth to maturation can be viewed as a series of switches that relay information between oocytes and granulosa in a metabolically efficient manner. Exchanging information between these two distinct cell types provides an opportunity to adjust differentiation and cell cycle status and would assume different characters depending on the temporal dimensions of the process being anticipated. For example, some switches may be turned on or off over a time course of relatively short or acute dimensions such as those triggered in anticipation of ovulation, which impact both key cell cycle transitions and mRNA processing at transcriptional and posttranscriptional levels in oocytes. In contrast, the growth phase of oogenesis is characterized by the deployment of largely negative feedback mechanisms, involving the sustained release of oocyte-derived factors such as growth differentiation factor 9 (GDF9) that temper the rate of oocyte hypertrophy with that of granulosa cell hyperplasia (Hutt & Albertini, 2007). Achieving synchrony throughout the course of oogenesis, not surprisingly, invokes multiple modalities of cell signaling, the topic of this chapter.

Over the past 20 years, five modalities of signaling have been recognized to operate at different stages of oogenesis in mammals. As shown in Fig. 8.1, these include (a) contact-mediated signals that do not involve intercellular junctions, (b) classical ligand–receptor interactions, (c) paracrine and autocrine signaling pathways, (d) junctional contact via transzonal projections (TZPs), and (e) the most recently discovered mechanism that invokes selective activation of receptor tyrosine kinases (RTKs) through remodeling of the oocyte membrane and cortical cytoplasm. Some of these pathways operate individually at discrete stages of oogenesis via first and second messenger relay once the donor cell stimulus has been received by the target cell, either oocyte or granulosa cell. However, other developmental transitions require integration of two or more pathways to establish positive or negative feedback mechanisms at either the level of ligand/receptor expression or because of overlapping signal propagation within the target cell. While recognizing that at one level it is possible to describe and deconstruct signaling

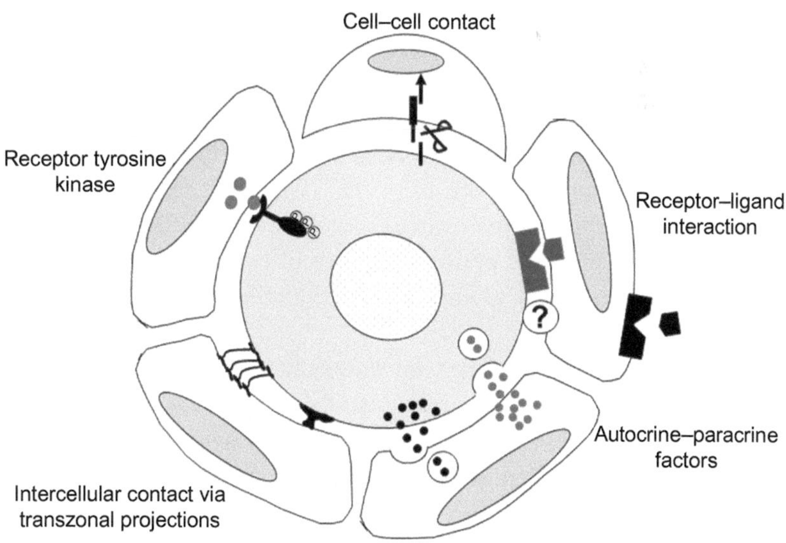

Figure 8.1 Schematic depicting the major signaling modaltities that have been implicated in the regulation of oogenesis in mammals. Note that types of interactions between the oocyte and surrounding granulosa cells are expected to vary according to the stage of follicle development.

pathways as operationally independent, as is likely the case of follicle activation, we are mindful that a more realistic perspective is that the key developmental transitions typically encountered during oogenesis rely upon two or more pathways that converge during signal propagation and amplification. It is our intention to provide a framework that will inform and guide future studies into the fundamental developmental properties of oogenesis in mammals that remain obscure and await further resolution.

2. FRAMING AND QUALIFYING THE COMPLEXITY OF OOGENESIS

From the earlier discussion, it is evident that reducing the complexity of signaling during oogenesis requires a framework encompassing multiple signaling modalities. Oogenesis does not proceed autonomously. Rather, oogenesis is inextricably coupled to the concurrent process of folliculogenesis in mammals and other metazoans. For this reason, it is useful to distinguish five stages of follicle development in order to designate the spatial and temporal alterations at key steps in oogenesis relative to the follicular environment within which gametogenesis proceeds. Since the ovarian follicle is a dynamic structure, the sequential steps most relevant to this discussion include: (1) follicle assembly which typically occurs in the fetal or newborn ovary, (2) establishment and maintenance of quiescence in the primordial follicle pool, (3) activation of cohorts of primordial follicles, (4) growth of the follicle during gonadotropin-independent preantral and gonadotropin-dependent antral stages of development, and (5) ovulation of the follicle in response to the LH surge. Qualifiers for this framework include the stages at which the ovarian vasculature has yet to be developed, such as during the pre- and perinatal periods of ovarian development, and restrictions imposed by the architecture of the follicle at different stages of folliculogenesis.

Most notable among the latter are extracellular matrix barriers that separate granulosa cells from vascular and stromal cell types at the follicle basement membrane. Moreover, the zona pellucida is assembled at the interface of the oocyte and granulosa, thereby posing a definitive structural impasse for cell contact-mediated signaling events or soluble ligand diffusion. As we will see, these architectural principles directly impact the manner in which intercellular communication can be obtained and reinforce the importance of why different signaling modalities may be required to support sequential stages of oogenesis Fig 8.2.

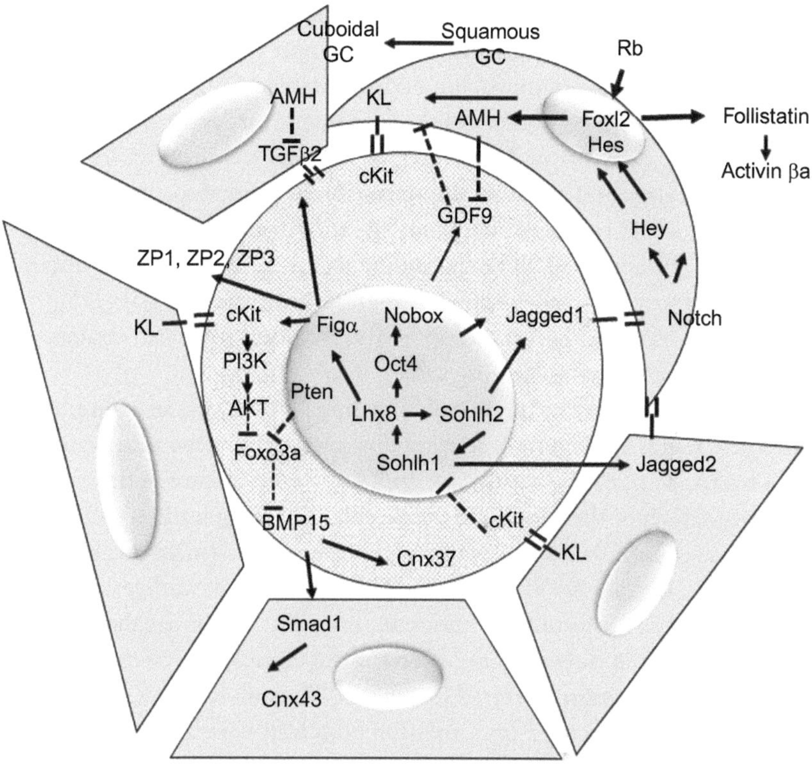

Figure 8.2 Primary signaling interactions at the oocyte–granulosa interface and their downstream effects on gene expression patterns are shown by the arrows; inhibitory effects indicated by dotted lines and stimulatory effects are depicted by the solid lines. Note that the cKit/KL pathway is implicated in several developmentally distinct downstream responses at the level of gene expression.

3. CELL CONTACT-MEDIATED SIGNALING

The physical approximation of ovarian somatic cells with germ cells in the developing ovary constitutes the initial and most long-lived modality for signaling during oogenesis (McLaren, 1988). Cell–cell contact between dissimilar cell types is widely recognized to be a major determinant of cell fate specification in a variety of developmental systems where ligand–receptor interactions are based upon a donor- or signal-generating cell binding to a receptor-bearing cell under conditions favorable for direct cell contact (Borggrefe & Liefke, 2012). One of the most studied examples of this form of signaling that is widely implicated in cell fate decisions during development is the case of Notch (Gridley, 2010).

The basic mechanism by which Notch signaling is effected has been analyzed in detail (Borggrefe & Liefke, 2012). Upon Notch receptor interaction with ligand, the receptor is cleaved by an extracellular gamma secretase allowing the intracellular domain of the receptor to target the nucleus where it activates gene expression. This modality bypasses the need for generation of a second message and results in the activation of a broad spectrum of target genes and is subject to regulation by a variety of posttranslational modifications in the intracellular cleavage product that influence both nuclear-cytoplasmic transport and protein stability. Importantly, triggering the cascade of transcriptional activation requires direct physical contact between Notch- and ligand-bearing cells.

In the case of the ovary, the Notch signaling pathway has been implicated in the early stages of oogenesis that involve assembly of the primordial follicle (Tingen, Kim, & Woodruff, 2009). A consistent feature of this pathway is the designation of the Notch-bearing cell as the pregranulosa cell and the presentation of ligands such as Jagged-1 by the oocyte (Johnson, Espinoza, McGaughey, Rawls, & Wilson-Rawls, 2001). Initial studies implicated Notch as essential for fertility in knockout mice but attributed the major effect to later stages in oogenesis required for the completion of meiotic maturation (Hahn, Johnson, Beres, Howard, & Wilson-Rawls, 2005). More recent studies have focused on a role for Notch in resolution of germ cell cysts at the time of primordial follicle formation in the newborn mouse ovary. That Notch is required for cyst breakdown has been shown by following temporal patterns of protein expression and demonstrating sensitivity to gamma-secretase inhibitors. A recent study by Guo et al. (2012) implicates changes in progesterone levels prior to and at the time of birth in the regulation of Notch expression in the developing ovary. Because cyst breakdown is restricted to the perinatal period, when maternal progesterone levels have declined, *in vitro* ovary culture experiments were designed to test the idea of steroid involvement via the Notch pathway. This work shows that follicle formation is inhibited by both progesterone or Notch inhibitors and importantly demonstrates that the levels of mRNA and protein for Notch, Jagged-2, and Hey-2 are increased as maternal progesterone levels decline.

While these studies suggest a link between endocrine and Notch signaling, the details of this pathway have yet to be worked out. Moreover, a level of control that is receiving much attention of late in terms of the Notch signaling mechanism in other developmental systems is the role of primary cilia in Notch presentation and the involvement of exosomes or endosomes (Bray, 2006; Gridley, 2010; Sheldon et al., 2010). How these new elements of the Notch signaling pathway pertain to the assembly of the primordial follicle will be an important area for future study especially in the context of germ cell survival given the major loss of follicles associated with the transition from a prepubertal to post pubertal state.

4. RECEPTOR–LIGAND INTERACTIONS

The transition from primordial to primary follicle involves a complex interplay of signaling between the immature oocyte and its companion granulosa cells. Autocrine and paracrine factors control oocyte growth, the cell cycle and differentiation (Rodrigues et al., 2008). Among the many factors currently shown to influence these events are basic fibroblast growth factor (FGF2), bone morphogenic proteins (BMP-4 and -15), kit ligand (KL), keratinocyte growth factor (KGF), insulin, tumor necrosis factor (TNFα), leukemia inhibitory factor (LIF), and Mullerian inhibitory substance (MIS) as reviewed by Skinner (Skinner, 2005). While most of these are dealt with in subsequent sections, the advances made in understanding the epidermal growth factor (EGF) receptor system during ovulation prompt particular attention in this regard.

EGF signaling in the ovarian follicle represents a classic case of receptor–ligand signaling that has recently been deciphered, owing largely to seminal contributions from the Conti laboratory. EGF-family members known to be expressed in cumulus cells include amphiregulin (AREG), epiregulin (EREG), and beta-cellulin (BTC) (Conti, Hsieh, Park, & Su, 2006). EGF-family members are transmembrane proteins released from the cell surface by proteolytic cleavage. Once released, they function as effectors of cell activation signaling, which in this case attributes the role of target cell to the oocyte. EGF receptors also encompass a subfamily of closely related RTKs including ErbB1, ErbB2, ErbB3, and ErbB4. As EGF receptors are expressed on cumulus cells and are upregulated in the cumulus by oocyte-derived GDF9 and BMP-15, it is safe to conclude that both germline and somatic cells invoke classical forms of ligand–receptor activation and TGFβ signaling (see next section) concurrently. It is generally appreciated, however, that EGF signaling induced in the cumulus cell following the LH surge contributes to the normal process of ovulation and the production of healthy, mature, and fully grown oocytes (Conti et al., 2006; Su et al., 2010).

With this background in hand, it is interesting to note that at the cellular level, few if any of the processes typically elicited by EGF signaling have been studied in the context of oogenesis, including endocytosis, receptor recycling, and overlap with other transduction pathways.

5. AUTOCRINE–PARACRINE SIGNALING PATHWAYS

A subject that has gained widespread attention in terms of signaling in the mammalian ovary is that of autocrine–paracrine modes of cell communication. There are several reasons this is the case. The first is that many members of the TGFβ and growth factor family have been shown to have an oocyte-specific pattern of gene expression (Choi & Rajkovic, 2006; Eppig, 2001) Second, molecular genetic studies demonstrate essential roles for certain TGFβ family members as exhibited in the dramatic phenotypes shown in TGFB and growth factor family knockout animals (Dong et al., 1996; Matzuk, Burns, Viveiros, & Eppig, 2002; McNatty et al., 2005). Third, paracrine regulators of ovarian function have been established for many mammalian species (Knight & Glister, 2006). Together, the role for intraovarian paracrine–autocrine regulators is well established and in many cases has been identified as functioning at the oocyte–granulosa cell interface.

As in other TGFβ signaling model systems, details regarding the proteolytic processing of ligands, the types of receptors involved in signal propagation, and the signal transduction pathway employed have been the focus of most studies (Mazerbourg et al., 2004; Moore, Otsuka, & Shimasaki, 2003; Otsuka, McTavish, & Shimasaki, 2011). From the ligand point of view, the best studied of these are GDF9 and BMP-15. Genetic ablation of GDF9 in mice results in female sterility due to an arrest of follicles at the primary stage of development (Dong et al., 1996). Detailed phenotype characterizations on these animals revealed two notable consequences of GDF9 loss in oocytes. Carabatsos, Elvin, Matzuk, and Albertini (1998) showed that in GDF9 null follicles, somatic cells are unable to establish contact with oocytes and as a result these oocytes grow to abnormally large dimensions and fail to acquire full meiotic competence. Further, the impact on oocyte growth has also been shown to be due to an overexpression of KL by granulosa cells demonstrating the importance of negative feedback control during the growth phase of oogenesis (Miyoshi et al., 2012). Thus, in the case of GDF9, a well-characterized paracrine signaling pathway has been documented as essential for the earliest stages of oogenesis. This is also true for BMP-15 signaling in murine and ovine model systems and has prompted much interest in the field of infertility treatments

(Li, Rajanahally, Edson, & Matzuk, 2009; McNatty et al., 2005; Mottershead, Ritter, & Gilchrist, 2012).

Beyond these TGFβ family members, other growth factors have been implicated in the communication pathways that control follicle formation and growth and, again, discrete cell-specific patterns of gene expression reinforce the importance of this type of signaling modality. For example, oocyte-specific growth factors include FGF8 (Miyoshi et al., 2012; Valve, Penttila, Paranko, & Harkonen, 1997) whereas indirect roles have been implicated for CTGF (Schindler, Nilsson, & Skinner, 2010), GDNF (Dole, Nilsson, & Skinner, 2008), and a range of neurotropins (Childs et al., 2010; Choi & Rajkovic, 2006; Kerr, Garcia-Rudaz, Dorfman, Paredes, & Ojeda, 2009; Nilsson, Dole, & Skinner, 2009). It remains unknown as to how these factors affect their actions at the level of receptor and signal transduction pathways but the experimental studies offered so far clearly suggest additional important pathways of autocrine or paracrine nature are at play.

6. HETEROCELLULAR CONTACT VIA TZPs

For most of oogenesis in mammals, somatic granulosa cells are in contact with the oocyte cell surface (Rodrigues et al., 2008). Zones of cell–cell adhesion are evident in primordial follicles before the synthesis and elaboration of the zona pellucida takes place. Following activation of the primordial follicle, the zona pellucida is assembled within the intercellular gap that separates granulosa cells from the oocyte as both oocyte and follicle growth proceeds in a gonadotropin-independent manner. How the physical interactions are maintained and amplified during oocyte growth is poorly understood but the structural basis for continued communication of the junctional variety is clearly a byproduct of TZPs.

Initially identified by electron microscopy, TZPs are prominent and abundant cellular extensions emanating from granulosa cells and forming a variety of junctional specializations with the oolemma (Albertini, Combelles, Benecchi, & Carabatsos, 2001). These include connexin-based gap junctions, desmosome-like junctions, and broad zonula adherens (Mora et al., 2012). To date, the only clear evidence implicating signaling in the regulation of TZPs comes from the studies of Combelles, Carabatsos, Kumar, Matzuk, and Albertini (2004) who showed that the formation and stability of TZPs was maximal at preantral stages of follicle development. Importantly, these studies established a role for FSH in that TZP stability was maximal in FSHβ knockout mice but rapid turnover and retraction occurred upon treatment of animals with FSH. While hormonal regulation of the formation and turnover of TZPs remains an important aspect of oogenesis and has been implicated during *in vitro* oocyte maturation in the bovine (Allworth & Albertini, 1993), the fact that TZPs are required for the formation of gap junctions at the oocyte–granulosa cell interface signifies a critical aspect of their role in signaling.

Metabolic cooperation, whereby neighboring cells are capable of substrate exchange through gap junctions, is a central feature of the oocyte–granulosa cell complex (Sugiura & Eppig, 2005). The metabolic demands of the oocyte during the growth phase of oogenesis have been thought to be served by the combined efforts of both somatic and germ cell compartments of the follicle. A spate of recent studies is changing this viewpoint to one where the oocyte itself deploys a signaling system that enhances and optimizes the extent of metabolic cooperation during the growth and probably maturative phases of oogenesis. Leading this effort has been a series of elegant studies from the Eppig and Matzuk laboratories. For example, oocytes were shown to secrete products that enhanced granulosa cell glycolysis to the benefit of the oocyte as a substrate recipient (Sugiura, Pendola, & Eppig, 2005). More recently, these studies have been extended to cholesterol metabolism where gene expression profiling suggested that oocytes, and not granulosa cells, were unlikely to be able to generate cholesterol through the *de novo* biosynthetic pathway. In fact, a series of experiments convincingly demonstrated that granulosa cells are the major source of oocyte cholesterol and that cell contact was necessary for efficient sterol exchange (Su et al., 2008). Most interestingly, the efficiency of cholesterol transport from somatic to germ cells was modulated by both GDF9 and BMP-15, factors already discussed as oocyte-specific members of the TGFβ family of signaling molecules.

This line of investigation takes on added significance when viewed in the context of TZP-based signaling modalities. Because metabolic cooperation entails formation of functional gap junctions at the ends of TZPs, factors that impact the expression of TZPs would predictably amplify the extent of metabolic cooperation observed. Earlier studies showed that GDF9 null mice fail to form TZPs (Carabatsos et al., 1998), implying that the benefits realized by oocytes for cholesterol uptake from granulosa could, in part, be due to the more general influence oocyte secreted factors have on establishing and maintaining TZPs. The combination of signaling modalities deployed to achieve metabolic cooperation is but one example where multiple, and possibly redundant, mechanisms are called into play at critical junctures of oogenesis.

This is not surprising and is likely to be the case during oocyte maturation as well. In fact, due to glutathione biosynthetic deficiencies manifest in bovine oocytes, it is again left up to the biosynthetic performance of somatic granulosa cells to provide the oocyte with the levels of GSH required for male pronuclear formation (de Matos, Furnus, & Moses, 1997). While not directly demonstrated, this form of metabolic cooperation is also likely to be mediated by TZPs (Allworth & Albertini, 1993).

7. RTKs AND CORTICAL MATURATION

Broadly speaking, RTKs and their cognate ligands have been implicated at several stages of oogenesis. Many studies have implicated growth factors that exert their effects on target cells through RTKs. The most well studied of these is the KL–cKit system that was described above (Hutt, McLaughlin, & Holland, 2006). Of the approximately 20 distinct classes of RTKs, 6 have been definitively identified and implicated in various aspects of follicle development that impact the course of oogenesis. These include class I (EGF/Erbβ), class II (Insulin receptor), class III (PDGF receptor), class IV (FGF receptor), class V (VEGF receptor), and class VII (Trk receptor). Whereas the VEGF-mediated events are associated with development of the follicle vasculature (Gridley, 2010), and beyond the scope of this chapter, the class VII Trk receptors appear to be directly involved in neural-like signaling modalities that have been revealed in the studies on rodents and humans (Childs et al., 2010; Choi & Rajkovic, 2006; Kerr et al., 2009). We emphasize below, those RTK subtypes that have been studied in detail.

Insulin/insulin-like growth factors: Insulin growth factor receptors are expressed in the oocyte and cumulus cells and the levels of expression are increased following gonadotropin stimulation (Acevedo, Ding, & Smith, 2007; Purcell, Chi, & Moley, 2012). Although *in vitro* exposure of oocytes to insulin during meiotic maturation causes aberrations in chromatin condensation and reduced embryonic developmental capacity, the nature of these defects remains to be resolved. Current studies on this class of growth factors are aimed at understanding the influence of insulin-like growth factors and their receptors in diabetic animal models.

Platlet-derived growth factor: PDGF is produced by the oocyte and appears to be critical for the primordial to primary follicle transition. This role was shown by studies with cultured ovaries from 4-day-old rats where PDGF supplementation lead to an increase in the number of primordial follicles transitioned to primary follicles. Because PDGF inhibitors reverse this process causing a net increase in the number of follicles still at the primordial stage, this form of RTK-mediated signaling appears to be an initial and prominent element in follicle activation (Nilsson, Detzel, & Skinner, 2006). Whether this pathway is positioned to contribute to the PI3 kinase mechanism proposed by the studies of Liu et al. (2006) is not yet established.

Fibroblast growth factor: Basic fibroblast growth factor (FGF2) and three of its receptors (FGFr2, FGFr3, and FGFr4) are expressed in follicular granulosa cells of women and girls but not in fetal ovaries (Ben-Haroush et al., 2005). FGF8 is among oocyte-derived factors such as BMP-15 that work in concert to induce glycolysis and metabolism in granulosa cells exemplifying yet another case where multiple signaling pathways converge to exert their actions on oogenesis (Sugiura et al., 2007). Another example of cross pathway integration that involves FGFs affects the KL/cKit signaling system. Evidence for this includes the demonstration that FGF7 produced by pretheca cells (Kezele, Ague, Nilsson, & Skinner, 2005) and FGF2 from oocytes (Nilsson & Skinner, 2004) increase levels of KL in pregranulosa and affect the primordial to primary follicle transition.

While RTKs appear to be central to processes associated with follicle activation and growth, a new role for these enzymes is emerging during the final, maturative phase of oogenesis. The primary evidence for this is the recent demonstration of tyrosine phosphorylation of cortical proteins in mouse oocytes undergoing *in vitro* maturation. While low levels of phosphorylation are apparent in the oocyte cortex of immature GV stage oocytes, progress through the meiotic cell cycle is associated with a significant increase in metaphase-II oocytes (McGinnis & Albertini, 2010) Interestingly, cortical phosphorylation was influenced by culture conditions since addition of EGF and insulin, physiological ligands for RTKs, causes increased tyrosine phosphorylation in the actin-rich oocyte cortex (McGinnis & Albertini, 2010). Several studies point to the Src-Family Kinase (SFK) member c-fyn as being responsible for these dramatic biochemical changes in the oocyte cortex and whether FYN or other SFKs are involved with remodeling of the oocyte in preparation for fertilization is actively under study (Luo, McGinnis, & Kinsey, 2009; McGinnis & Albertini, 2010; McGinnis, Carroll, & Kinsey, 2011).

8. SUMMARY AND CONCLUSIONS

Integrating oogenesis within the context of the ovarian follicle is a fundamental objective for achieving successful reproduction in female mammals. While higher order and systemic control of ovarian function has long been appreciated through the involvement of hypothalamic and pituitary axis, only recently has the delicate and precise contribution within the ovary been reduced to aspects of intercellular signaling within the follicle. In a cursory, but hopefully stimulating way for future research, we have assembled a framework for signaling modalities between the germ line and somatic compartments of the ovarian follicle so as to bring light to mechanistic underpinnings that are likely to be at the heart of the diagnosis and treatment of human infertility. Whether the imperative is clinical or fundamental basic science, defining the signaling mechanisms that play a central role in the development of the mammalian oocyte will continue to be the subject of much enquiry. Achieving this goal will require technological advances as efforts proceed beyond the traditional and tractable murine model system. Whether the foundations arrived at, using in most cases the biology of the mouse oocyte, will be translatable to humans and agriculturally important species is not yet clear. However, further attempts to manipulate human reproduction for restoration and preservation of fertility, or for dealing with the matter of endangered species preservation, will benefit from harnessing the complexities of signaling within the ovarian follicle that ensure the embryonic destiny of the ovulated oocyte.

ACKNOWLEDGMENTS

The authors recognize the valuable input and ideas generated from members of the Albertini laboratory over the years and regret that many of the key findings in the literature could not be included due to space constraints. Funding from the NIH, the ESHE Fund, and the Hall family Foundation is graciously acknowledged.

REFERENCES

Acevedo, N., Ding, J., & Smith, G. D. (2007). Insulin signaling in mouse oocytes. *Biology of Reproduction*, 77, 872–879.

Albertini, D. F., Combelles, C. M., Benecchi, E., & Carabatsos, M. J. (2001). Cellular basis for paracrine regulation of ovarian follicle development. *Reproduction*, *121*, 647–653.

Allworth, A. E., & Albertini, D. F. (1993). Meiotic maturation in cultured bovine oocytes is accompanied by remodeling of the cumulus cell cytoskeleton. *Developmental Biology*, *158*, 101–112.

Ben-Haroush, A., Abir, R., Ao, A., Jin, S., Kessler-Icekson, G., Feldberg, D., et al. (2005). Expression of basic fibroblast growth factor and its receptors in human ovarian follicles from adults and fetuses. *Fertility and Sterility*, *84*(Suppl. 2), 1257–1268.

Borggrefe, T., & Liefke, R. (2012). Fine-tuning of the intracellular canonical Notch signaling pathway. *Cell Cycle*, *11*, 264–276.

Bray, S. J. (2006). Notch signalling: A simple pathway becomes complex. *Nature Reviews Molecular Cell Biology*, 7, 678–689.

Carabatsos, M. J., Elvin, J., Matzuk, M. M., & Albertini, D. F. (1998). Characterization of oocyte and follicle development in growth differentiation factor-9-deficient mice. *Developmental Biology*, *204*, 373–384.

Childs, A. J., Bayne, R. A., Murray, A. A., Martins Da Silva, S. J., Collins, C. S., Spears, N., et al. (2010). Differential expression and regulation by activin of the neurotrophins BDNF and NT4 during human and mouse ovarian development. *Developmental Dynamics*, *239*, 1211–1219.

Choi, Y., & Rajkovic, A. (2006). Genetics of early mammalian folliculogenesis. *Cellular and Molecular Life Sciences: CMLS*, *63*, 579–590.

Combelles, C. M. H., Carabatsos, M. J., Kumar, T. R., Matzuk, M. M., & Albertini, D. F. (2004). Hormonal control of somatic cell oocyte interactions during ovarian follicle development. *Molecular Reproduction and Development*, *69*, 347–355.

Conti, M., Hsieh, M., Park, J. Y., & Su, Y. Q. (2006). Role of the epidermal growth factor network in ovarian follicles. *Molecular Endocrinology*, *20*, 715–723.

de Matos, D. G., Furnus, C. C., & Moses, D. F. (1997). Glutathione synthesis during in vitro maturation of bovine oocytes: Role of cumulus cells. *Biology of Reproduction*, *57*, 1420–1425.

Dole, G., Nilsson, E. E., & Skinner, M. K. (2008). Glial-derived neurotrophic factor promotes ovarian primordial follicle development and cell-cell interactions during folliculogenesis. *Reproduction*, *135*, 671–682.

Dong, J., Albertini, D. F., Nishimori, K., Kumar, T. R., Lu, N., & Matzuk, M. M. (1996). Growth differentiation factor-9 is required during early ovarian folliculogenesis. *Nature*, *383*, 531.

Eppig, J. J. (2001). Oocyte control of ovarian follicular development and function in mammals. *Reproduction*, *122*, 829–838.

Gridley, T. (2010). Notch signaling in the vasculature. *Current Topics in Developmental Biology*, *92*, 277–309.

Guo, M., Zhang, H., Bian, F., Li, G., Mu, X., Wen, J., et al. (2012). P4 down-regulates Jagged2 and Notch1 expression during primordial folliculogenesis. *Frontiers in Bioscience (Elite Edition)*, *4*, 2731–2744.

Hahn, K. L., Johnson, J., Beres, B. J., Howard, S., & Wilson-Rawls, J. (2005). Lunatic fringe null female mice are infertile due to defects in meiotic maturation. *Development (Cambridge, England)*, *132*, 817–828.

Hutt, K. J., & Albertini, D. F. (2007). An oocentric view of folliculogenesis and embryogenesis. *Reproductive Biomedicine Online*, *14*, 758–764.

Hutt, K. J., McLaughlin, E. A., & Holland, M. K. (2006). KIT/KIT Ligand in mammalian oogenesis and folliculogenesis: Roles in rabbit and murine ovarian follicle activation and oocyte growth. *Biology of Reproduction*, *75*, 421–433.

Johnson, J., Espinoza, T., McGaughey, R. W., Rawls, A., & Wilson-Rawls, J. (2001). Notch pathway genes are expressed in mammalian ovarian follicles. *Mechanisms of Development*, *109*, 355–361.

Kerr, B., Garcia-Rudaz, C., Dorfman, M., Paredes, A., & Ojeda, S. R. (2009). NTRK1 and NTRK2 receptors facilitate follicle assembly and early follicular development in the mouse ovary. *Reproduction*, *138*, 131–140.

Kezele, P. R., Ague, J. M., Nilsson, E., & Skinner, M. K. (2005). Alterations in the ovarian transcriptome during primordial follicle assembly and development. *Biology of Reproduction*, *72*, 241–255.

Knight, P. G., & Glister, C. (2006). TGF-{beta} superfamily members and ovarian follicle development. *Reproduction*, *132*, 191–206.

Li, Q., Rajanahally, S., Edson, M. A., & Matzuk, M. M. (2009). Stable expression and characterization of N-terminal tagged recombinant human bone morphogenetic protein 15. *Molecular Human Reproduction*, *15*, 779–788.

Liu, K., Rajareddy, S., Liu, L., Jagarlamudi, K., Boman, K., Selstam, G., et al. (2006). Control of mammalian oocyte growth and early follicular development by the oocyte PI3 kinase pathway: New roles for an old timer. *Developmental Biology, 299*, 1–11.
Luo, J., McGinnis, L. K., & Kinsey, W. H. (2009). Fyn kinase activity is required for normal organization and functional polarity of the mouse oocyte cortex. *Molecular Reproduction and Development, 76*, 819–831.
Matzuk, M. M., Burns, K. H., Viveiros, M. M., & Eppig, J. J. (2002). Intercellular communication in the mammalian ovary: Oocytes carry the conversation. *Science, 296*, 2178–2180.
Mazerbourg, S., Klein, C., Roh, J., Kaivo-Oja, N., Mottershead, D. G., Korchynskyi, O., et al. (2004). Growth differentiation factor-9 signaling is mediated by the type I receptor, activin receptor-like kinase 5. *Molecular Endocrinology, 18*, 653–665.
McGinnis, L. K., & Albertini, D. F. (2010). Dynamics of protein phosphorylation during meiotic maturation. *Journal of Assisted Reproduction and Genetics, 27*, 169–182.
McGinnis, L. K., Carroll, D. J., & Kinsey, W. H. (2011). Protein tyrosine kinase signaling during oocyte maturation and fertilization. *Molecular Reproduction and Development, 78*, 831–845.
McLaren, A. (1988). The developmental history of female germ cells in mammals. *Oxford Reviews of Reproductive Biology, 10*, 162–179.
McNatty, K. P., Juengel, J. L., Reader, K. L., Lun, S., Myllymaa, S., Lawrence, S. B., et al. (2005). Bone morphogenetic protein 15 and growth differentiation factor 9 co-operate to regulate granulosa cell function in ruminants. *Reproduction, 129*, 481–487.
Miyoshi, T., Otsuka, F., Nakamura, E., Inagaki, K., Ogura-Ochi, K., Tsukamoto, N., et al. (2012). Regulatory role of kit ligand-c-kit interaction and oocyte factors in steroidogenesis by rat granulosa cells. *Molecular and Cellular Endocrinology, 358*, 18–26.
Moore, R. K., Otsuka, F., & Shimasaki, S. (2003). Molecular basis of bone morphogenetic protein-15 signaling in granulosa cells. *The Journal of Biological Chemistry, 278*, 304–310.
Mora, J. M., Fenwick, M. A., Castle, L., Baithun, M., Ryder, T. A., Mobberley, M., et al. (2012). Characterization and significance of adhesion and junction-related proteins in mouse ovarian follicles. *Biology of Reproduction, 86*, 153.
Mottershead, D. G., Ritter, L. J., & Gilchrist, R. B. (2012). Signalling pathways mediating specific synergistic interactions between GDF9 and BMP15. *Molecular Human Reproduction, 18*, 121–128.
Nilsson, E. E., Detzel, C., & Skinner, M. K. (2006). Platelet-derived growth factor modulates the primordial to primary follicle transition. *Reproduction, 131*, 1007–1015.
Nilsson, E., Dole, G., & Skinner, M. K. (2009). Neurotrophin NT3 promotes ovarian primordial to primary follicle transition. *Reproduction, 138*, 697–707.
Nilsson, E. E., & Skinner, M. K. (2004). Kit ligand and basic fibroblast growth factor interactions in the induction of ovarian primordial to primary follicle transition. *Molecular and Cellular Endocrinology, 214*, 19–25.
Otsuka, F., McTavish, K. J., & Shimasaki, S. (2011). Integral role of GDF-9 and BMP-15 in ovarian function. *Molecular Reproduction and Development, 78*, 9–21.
Purcell, S. H., Chi, M. M., & Moley, K. H. (2012). Insulin-stimulated glucose uptake occurs in specialized cells within the cumulus oocyte complex. *Endocrinology, 153*, 2444–2454.
Rodrigues, P., Limback, D., McGinnis, L. K., Plancha, C. E., & Albertini, D. F. (2008). Oogenesis: Prospects and challenges for the future. *Journal of Cellular Physiology, 216*, 355–365.
Schindler, R., Nilsson, E., & Skinner, M. K. (2010). Induction of ovarian primordial follicle assembly by connective tissue growth factor CTGF. *PLoS One, 5*, e12979.
Sheldon, H., Heikamp, E., Turley, H., Dragovic, R., Thomas, P., Oon, C. E., et al. (2010). New mechanism for Notch signaling to endothelium at a distance by Delta-like 4 incorporation into exosomes. *Blood, 116*, 2385–2394.

Skinner, M. K. (2005). Regulation of primordial follicle assembly and development. *Human Reproduction Update, 11*, 461–471.

Su, Y. Q., Sugiura, K., Li, Q., Wigglesworth, K., Matzuk, M. M., & Eppig, J. J. (2010). Mouse oocytes enable LH-induced maturation of the cumulus-oocyte complex via promoting EGF receptor-dependent signaling. *Molecular Endocrinology, 24*, 1230–1239.

Su, Y. Q., Sugiura, K., Wigglesworth, K., O'Brien, M. J., Affourtit, J. P., Pangas, S. A., et al. (2008). Oocyte regulation of metabolic cooperativity between mouse cumulus cells and oocytes: BMP15 and GDF9 control cholesterol biosynthesis in cumulus cells. *Development (Cambridge, England), 135*, 111–121.

Sugiura, K., & Eppig, J. J. (2005). Society for Reproductive Biology Founders Lecture 2005. Control of metabolic cooperativity between oocytes and their companion granulosa cells by mouse oocytes. *Reproduction, Fertility, and Development, 17*, 667–674.

Sugiura, K., Pendola, F. L., & Eppig, J. J. (2005). Oocyte control of metabolic cooperativity between oocytes and companion granulosa cells: Energy metabolism. *Developmental Biology, 279*, 20–30.

Sugiura, K., Su, Y.-Q., Diaz, F. J., Pangas, S. A., Sharma, S., Wigglesworth, K., et al. (2007). Oocyte-derived BMP15 and FGFs cooperate to promote glycolysis in cumulus cells. *Development (Cambridge, England), 134*, 2593–2603.

Tingen, C., Kim, A., & Woodruff, T. K. (2009). The primordial pool of follicles and nest breakdown in mammalian ovaries. *Molecular Human Reproduction, 15*, 795–803.

Valve, E., Penttila, T. L., Paranko, J., & Harkonen, P. (1997). FGF-8 is expressed during specific phases of rodent oocyte and spermatogonium development. *Biochemical and Biophysical Research Communications, 232*, 173–177.

CHAPTER NINE

Biogenesis of the Mouse Egg's Extracellular Coat, the Zona Pellucida

Paul M. Wassarman[1], Eveline S. Litscher
Department of Developmental and Regenerative Biology, Mount Sinai School of Medicine, New York, USA
[1]Corresponding author: e-mail address: paul.wassarman@mssm.edu

Contents

Abstract

Biogenesis of the zona pellucida (ZP), the extracellular coat that surrounds all mammalian eggs, is a universal and essential feature of mammalian oogenesis and reproduction. The mouse egg's ZP consists of only three glycoproteins, called ZP1–3, that are synthesized, secreted, and assembled into an extracellular coat exclusively by growing oocytes during late stages of oogenesis while oocytes are arrested in meiosis. Expression of ZP genes and synthesis of ZP1–3 are gender-specific. Nascent ZP1–3 are synthesized by oocytes as precursor polypeptides that possess several elements necessary for their secretion and assembly into a matrix of long fibrils outside of growing oocytes. Failure to synthesize either ZP2 or ZP3 by homozygous null female mice precludes formation of

Current Topics in Developmental Biology, Volume 102
ISSN 0070-2153
http://dx.doi.org/10.1016/B978-0-12-416024-8.00009-X

a ZP during oocyte growth and, due to faulty folliculogenesis and a paucity of ovulated eggs, results in infertility. High-resolution structural analyses suggest that ZP glycoproteins consist largely of immunoglobulin (Ig)-like folds and that the glycoproteins probably arose by duplication of a common Ig-like domain. Mouse ZP1–3 share many features, particularly a ZP domain, with extracellular coat glycoproteins of eggs from other vertebrate and invertebrate animals whose origins date back more than 600 million years. These and other aspects of ZP biogenesis are discussed in this review.

1. INTRODUCTION TO THE MAMMALIAN EGG'S ZONA PELLUCIDA

The zona pellucida (ZP) is a relatively thick extracellular coat that surrounds all mammalian oocytes, ovulated eggs, and preimplantation embryos (Dietl, 1989; Fig. 9.1). It is a highly specialized extracellular coat that carries out functions specific to mammalian reproduction. The ZP ranges in thickness from less than 1 μm to more than 25 μm (~1–30 ng protein) for eggs from different mammals and consists of a network of interconnected fibrils, with the innermost fibrils oriented radially to the oolemma and closely packed and the outermost fibrils oriented tangentially to the oolemma and more loosely packed (Keefe, Tran, Pellegrini, & Oldenbourg, 1997). The ZP is known to be quite porous and permeable to relatively large

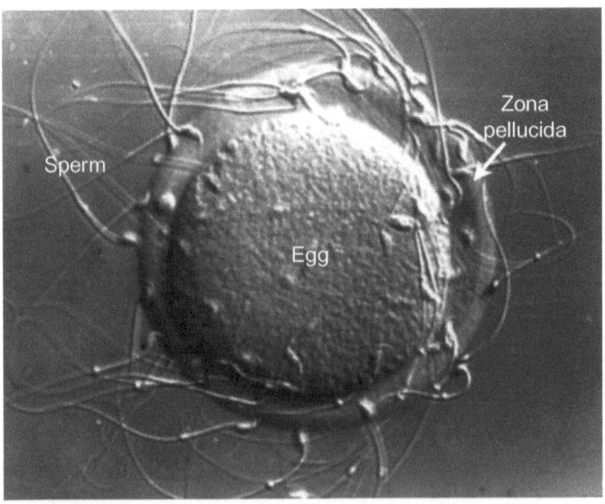

Figure 9.1 Binding of free-swimming mouse sperm to the zona pellucida of an ovulated mouse egg. The light micrograph was taken using Nomarski differential interference contrast microscopy.

macromolecules, including many viruses. Microvilli emanating from growing oocytes and long processes emanating from surrounding follicle cells traverse the ZP and establish regions of intercellular contact (gap junctions) through which small molecules can pass. Expanded blastocysts hatch from the ZP by using a specific protease(s) just prior to implantation into the uterus.

The ZP is a vital participant in mammalian reproduction, playing important roles during oogenesis, fertilization, and preimplantation development. For example, during fertilization, sperm must bind to and penetrate the ZP in order to reach and fuse with the egg's plasma membrane. In this context, the ZP prevents fusion of sperm from most, but not all heterologous species with the oolemma of ovulated eggs (i.e., supports species-restricted fertilization) and prevents supernumerary sperm from binding to and penetrating eggs following fusion of an egg with a single sperm (i.e., assists in preventing polyspermic fertilization) (Florman & Ducibella, 2006; Wassarman, 1999; Yanagimachi, 1994). Because of the ZP's vital role in mammalian reproduction, many aspects of ZP biogenesis in mice have been investigated during the past 30 years and several aspects of these investigations are presented here.

2. APPEARANCE OF THE MOUSE EGG'S ZP DURING OOGENESIS

Unfertilized eggs are the end-product of oogenesis, a process that is initiated during fetal development and ends months-to-years later in the sexually mature adult (Austin & Short, 1982; Wassarman & Albertini, 1994). Oogenesis in mice begins with the appearance of primordial germ cells (fetal days 7–9) which become the oogonia that populate fetal ovaries (fetal days 11–12). Oogonia in turn become nongrowing oocytes (fetal days 12–14) that populate ovaries of neonatal mice. The transition from oogonia to oocytes involves a change from mitotic to meiotic cells. Progression through the first meiotic prophase (leptotene, zygotene, pachytene, and diplotene), with pairing of homologous chromosomes, crossing-over, and recombination, takes 4–5 days. Shortly after birth (postcoitus day-21), nearly all oocytes are arrested in late diplotene (so-called, dictyate stage), where they remain until stimulated to resume meiotic progression at the time of ovulation. This pool of small, nongrowing oocytes is the source of unfertilized eggs in the sexually mature mouse (~6 weeks-of-age). Nongrowing oocytes do not have a ZP.

In sexually mature mice, each ovary contains ~8000 nongrowing oocytes. Each nongrowing oocyte (~12 μm in diameter) is contained within a cellular follicle that grows concomitantly with the oocyte, from a single layer of a few epithelial-like cells to three layers of cuboidal granulosa cells (~900 cells; ~125 μm in diameter) by the time the oocyte has completed its growth (~80 μm in diameter). It is during the ~2-week growth phase, while oocytes are continually arrested at the dictyate stage of the first meiotic prophase, that they acquire a ZP. As a growing oocyte increases in diameter, its ZP increases in thickness, achieving a final width of 6.2 ± 1.9 μm. It should be noted that the dictyate stage of meiosis is characterized by very diffuse chromosomes and oocyte growth is characterized by relatively high rates of transcription and translation (from ~1 to 40 pg protein synthesized per hour per oocyte during growth of mouse oocytes; Davidson, 1986).

3. NATURE OF MOUSE ZP GLYCOPROTEINS

Each mouse egg ZP contains ~3.5 ng of protein (~10% of total egg protein) and only three glycoproteins, called ZP1 (~200 kDa; 1246 aa; dimer), ZP2 (~120 kDa; 713 aa; monomer), and ZP3 (~83 kDa; 424 aa; monomer) account for the protein content (Bleil & Wassarman, 1980a; Shimizu, Tsuji, & Dean, 1983; Wassarman, 1988, 2008; Fig. 9.2). ZP2

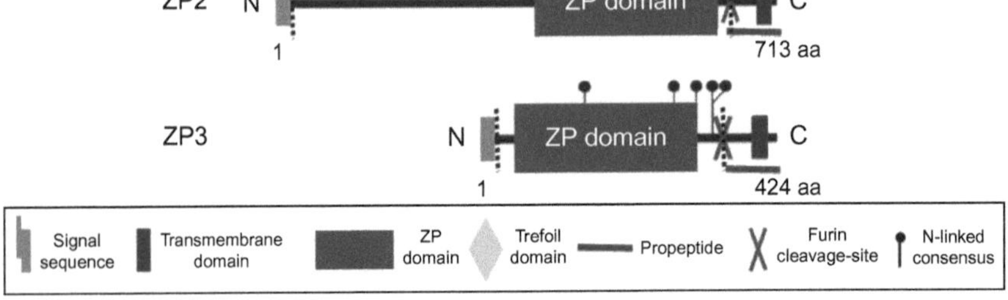

Figure 9.2 Schematic representation of the overall architecture of mouse ZP glycoproteins mZP1, mZP2, and mZP3. The polypeptide of each ZP glycoprotein is drawn to scale with the N- and C-termini indicated. Shown are the signal peptide (green), trefoil domain (yellow), ZP domain (red), consensus furin cleavage-site (red X), transmembrane domain (purple), and C-terminal propeptide region (blue bar). Putative N-linked glycosylation sites are shown (black lollipops). The number of amino acids in each polypeptide is indicated. (See Color Insert.)

and ZP3 are present in the ZP in approximately equimolar amounts and represent ~80% of total ZP protein; whereas ZP1 represents only ~20% of total ZP protein. ZP polypeptides are heterogeneously glycosylated with complex-type, asparagine-linked (N-) and serine/threonine-linked (O-) oligosaccharides that causes them to migrate as broad bands on SDS-PAGE. ZP1–3 possess 4, 6, and 5 N-linked oligosaccharides/monomer (Boja, Hoodbhoy, Fales, & Dean, 2003). The oligosaccharides are both sialylated and sulfated such that native ZP1–3 have rather low isoelectric points. The heterogeneity and acidic nature of ZP glycoproteins is due entirely to their N- and O-linked oligosaccharides and not to their polypeptides (Wassarman, 1988).

ZP polypeptides from different mammals are well conserved; ZP2 and ZP3 exhibit ~65–98% identity and ZP1 ~40% identity. For example, the sequences of mouse and human ZP2 and ZP3, two species separated by more than 100 million years of evolution, are 56% and 67% identical, respectively. ZP polypeptides have several regions in common, suggesting that they may be derived from a common ancestral gene. Nascent ZP1–3 are synthesized as precursor polypeptides that have an N-terminal signal-sequence that directs the glycoproteins to the secretory pathway, a C-terminal propeptide (CTP) that has a consensus furin cleavage-site (CFCS), a transmembrane domain (TMD) ~40 aa downstream of the CFCS, and a short cytoplasmic tail (CT); in addition, ZP1 has a P or trefoil domain (Wassarman, 2008; Fig. 9.2).

ZP1–3 also have an ~260 aa sequence, called a ZP domain (ZPD) (Bork & Sander, 1992), that contains eight conserved Cys residues present as intramolecular disulfides (Fig. 9.2). ZPDs are found in hundreds of proteins of diverse functions from a wide variety of tissues in all multicellular eukaryotes and are often glycosylated modular structures consisting of multiple types of domains (Jovine, Darie, Litscher, & Wassarman, 2005). Among these many proteins are α- and β-tectorin, uromodulin, endoglin, β-glycan, vomeroglandin, cuticlin, dumpy, NompA, and DYF-7. Several of these proteins are known to assemble into fibrils and to play a role in organizing and shaping specialized apical structures in epithelial cells (Plaza, Chanut-Delalande, Fernandes, Wassarman, & Payre, 2010). Mutations in genes encoding ZPD proteins can result in severe human pathologies, including infertility, deafness, and cancer. ZPDs are bipartite structures consisting of an N-terminal subdomain (NTS) with four conserved Cys residues and a C-terminal subdomain (CTS) separated by a short, protease-sensitive, linker region (Han et al., 2010; Jovine, Qi, Williams, Litscher, & Wassarman, 2004; Llorca, Trujillo, Blanco, & Bernabeu, 2007; Fig. 9.3). It is of interest

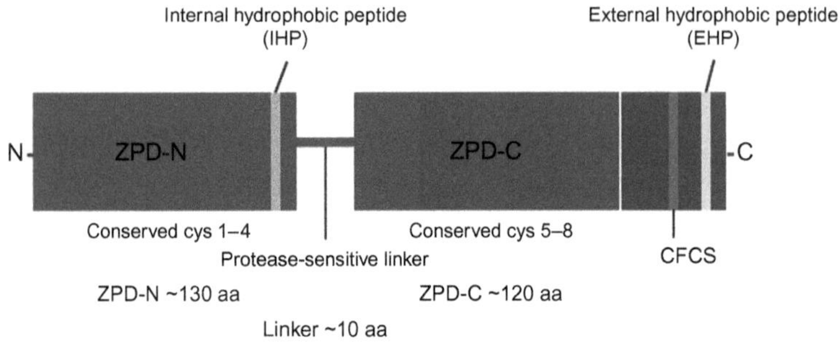

Figure 9.3 Schematic representation of the ZP domain (ZPD) of mZP3. Shown are the N-terminal region of the ZPD (ZPD-N; ~130 residues) with four conserved Cys residues, the C-terminal region of the ZPD (ZPD-C; ~120 residues) with four conserved Cys residues, the protease-sensitive linker between the ZPD-N and ZPD-C (~10 residues), the positions of the internal hydrophobic peptide (IHP) and external hydrophobic peptide (EHP), and the position of the consensus furin cleavage-site (CFCS) in the C-terminal propeptide. (For color version of this figure, the reader is referred to the online version of this chapter.)

that several proteins have been identified in flies, mice, and man that consist of only the NTS of the ZPD; none have been found that consist of only the CTS (Jovine, Janssen, Litscher, & Wassarman, 2006).

Isolated ZPDs polymerize into long fibrils on their own and it is their NTS that is responsible for polymerization (Jovine, Qi, Williams, Litscher, & Wassarman, 2002). This is of particular interest since ultrastructural analyses of solubilized ZP revealed their fibrillar nature, with long fibrils interconnected by ZP1 (a dimer held together by intermolecular disulfides), and exhibiting a structural repeat (~140 Å) attributed to ZP2–ZP3 dimers (Greve & Wassarman, 1985; Wassarman, Liu, & Litscher, 1996; Wassarman & Mortillo, 1991). In this context, the three-dimensional structure of chicken ZP3 (53% identical to human ZP3) has been determined at high resolution (~2.0 Å) by using X-ray crystallographic methods (Han et al., 2010; Monné, Han, Schwend, Burendahl, & Jovine, 2008; Monné & Jovine, 2011). Both the NTS and CTS of the ZPD adopt an immunoglobulin (Ig)-like fold, suggesting that ZP polypeptides may have arisen by duplication of a common Ig-like domain. Furthermore, within crystals, ZP3 forms antiparallel dimers held together by interactions between the NTS and CTS of opposing ZP3 molecules. Although relevant high-resolution X-ray crystallographic data are not yet available, it is possible that ZP2–ZP3 dimers adopt a similar orientation within ZP fibrils giving rise to the aforementioned ~140 Å structural repeat.

4. GENES ENCODING MOUSE ZP GLYCOPROTEINS

ZP1–3 polypeptides are encoded by single-copy genes located on three different chromosomes (ZP1–3, chromosomes 19, 7, and 5, respectively) (Epifano, Liang, & Dean, 1995; Lunsford et al., 1990). Genes encoding ZP1–3 consist of 12, 18, and 8 exons, respectively. Although the intron size of ZP1–3 genes varies extensively between species, the number and length of exons is highly conserved. The genes share TATAA boxes ~30 bp upstream of transcription start-sites and exhibit conserved organization, with exon/intron boundaries defining distinct domains. E-boxes (CANNTG) located at −200 bp are involved in expression of ZP polypeptides upon binding E12/FIGα heterodimers (Liang et al., 1997; Soyal, Amieh, & Dean, 2000), and two ovary-specific DNA-binding proteins, ZAP-1 (Millar, Lader, & Dean, 1993; Millar, Lader, Liang, & Dean, 1991) and OSP-1 (Schickler, Lira, Kinloch, & Wassarman, 1992), have been identified that bind either upstream of (at the sequence 5′-CAC(G/C)TG-3′) or within (at the sequence 5′-TGATAA-3′) ZP gene promoters. Both ZP2 and ZP3 genes possess 8–12 bp sequences within 250 bp of their transcription start-sites that bear ~85% identity (Liang, Chamow, & Dean, 1990). The similarity between mouse and human ZP genes, especially the promoters, enables the human promoter to utilize the heterologous transcription machinery in mouse oocytes (Liang & Dean, 1993).

5. SYNTHESIS OF ZP GLYCOPROTEINS BY MOUSE OOCYTES

Eggs from all invertebrates and vertebrates are surrounded by one or more extracellular coats (Dumont & Brummett, 1985; Monné, Han, & Jovine, 2006). Among mammalian and nonmammalian eggs, the extracellular coat is called the ZP and vitelline envelope (VE), respectively. The source of proteins that make up the coats depends on the animal species. Among mammals the proteins are synthesized solely by the ovary (oocytes and/or follicle cells), whereas among nonmammals they are synthesized by either the liver or the ovary, or by both organs. For example, among fish (Litscher & Wassarman, 2007) and birds (Bausek, Waclawek, Schneider, & Wohlrab, 2000; Takeuchi et al., 1999), the proteins are synthesized by the ovary and/or liver (e.g., trout-liver, zebrafish-ovary, medaka-liver

Table 9.1 Source of ZP glycoproteins

Source	**Site(s) of ZP glycoprotein synthesis**	
	Ovary[a]	**Liver**
Mammals (ZP)[b]	+	–
Bird (VE)[c]	+	+
Amphibians (VE)[b]	+	–
Fish (VE)[d]	+	+

[a]Oocytes and/or follicle cells.
[b]For example, mice/humans/*Xenopus*-oocytes.
[c]For example, chickens-liver and ovary.
[d]For example, trout-liver, zebrafish-ovary, medaka-liver, and ovary.

and ovary, chicken-liver and ovary), whereas among amphibians (Kubo et al., 1997; Yang & Hedrick, 1997), they are synthesized solely by the ovary (Table 9.1). Regardless of their source, ZP1–3 polypeptides share many common features with VE polypeptides from birds, frogs, fish, ascidians, and molluscs—an evolutionary time-span of more than 600 million years.

What is the source of ZP1–3 in mice? Several lines of evidence presented below strongly suggest that mouse ZP1–3 are synthesized only by the ovary and exclusively by growing oocytes, not by surrounding follicle cells.

To investigate the origin of the mouse egg's ZP, ovarian thick sections obtained from mice injected with radiolabeled fucose or *N*-acetylglucosamine were subjected to autoradiography. The autoradiograms revealed a progression of grains from the inside to the outside of the ZP as a function of time after injection, consistent with growing oocytes being the source of the radiolabeled material (Haddad & Nagai, 1977).

By 1980, it was known that the mouse egg's ZP is composed of ZP1–3 (Bleil & Wassarman, 1980a). To determine the site of synthesis of mouse ZP1–3, denuded and follicle-enclosed oocytes isolated at several stages of growth were cultured *in vitro* in the presence of either radiolabeled methionine or fucose (Bleil & Wassarman, 1980b). Electrophoretic analyses of radiolabel incorporated into oocyte ZP and follicle cell proteins revealed that growing oocytes, but not follicle cells, synthesized and secreted ZP1–3. Subsequently, results of analogous *in vitro* experiments with isolated growing oocytes provided further support for the conclusion that only growing mouse oocytes synthesized ZP1–3 (Greve, Salzmann, Roller, & Wassarman, 1982; Roller & Wassarman, 1983; Salzmann, Greve, Roller, & Wassarman, 1983; Shimizu et al., 1983).

cDNA and genomic cloning of mouse ZP3 (Kinloch, Roller, Fimiani, Wassarman, & Wassarman, 1988; Kinloch & Wassarman, 1989; Ringuette, Chamberlin, Baur, Sobieski, & Dean, 1988; Ringuette, Sobieski, Chamow, & Dean, 1986), followed by cloning of ZP2 (Liang et al., 1990) and ZP1 (Epifano, Liang, & Dean, 1995, Epifano, Liang, Familiari, Moos, & Dean, 1995), permitted several alternative experimental approaches to be taken to investigate further the site of synthesis of ZP glycoproteins. Accordingly, results of Northern blotting (Epifano, Liang, Familiari, Moos, & Dean, 1995; Philpott, Ringuette, & Dean, 1987; Roller, Kinloch, Hiraoka, Li, & Wassarman, 1989), RNase protection (Epifano, Liang, Familiari, Moos, & Dean, 1995; Roller et al., 1989; Tong, Nelson, & Dean, 1995), *in situ* hybridization (Epifano, Liang, Familiari, Moos, & Dean, 1995; Philpott et al., 1987; Ringuette et al., 1986; Roller et al., 1989), and antisense oligonucleotide injection (Tong et al., 1995) experiments once again confirmed that ZP1–3 are synthesized exclusively by growing mouse oocytes (Fig. 9.4).

With the advent of transgenic mice, it became possible to test whether a particular transgene would be expressed exclusively in growing mouse oocytes when placed under the control of ZP gene 5′-flanking region possessing *cis*-acting sequences. A transgene was constructed in which 6.5 kb of ZP3 gene 5′-flanking sequence was fused to the coding region of the firefly luciferase gene (ZP3/6.5-LUC) and was used to generate several independent lines of transgenic mice (Lira, Kinloch, Mortillo, & Wassarman, 1990). Evaluation of LUC activity in a dozen tissues revealed that the reporter gene was expressed only in the ovary (Table 9.2) and that there were ~2400 units of luciferase activity per growing oocyte; as little as 153 nt of the ZP3 5′-flanking region was sufficient to target expression of luciferase to growing oocytes. As expected, the levels of luciferase increased dramatically during the first 21 days *postpartum*, and exhibited a pattern consistent with that of ZP3 expression during mouse development. Furthermore, *in situ* hybridization assays of luciferase transcripts, as well as assays of the luciferase activity of isolated oocytes and follicles from transgenic mice, revealed that luciferase was localized exclusively to growing oocytes and not to follicle cells (Fig. 9.4).

Collectively, results of these experiments demonstrate conclusively that expression of ZP genes in mice is restricted to females and to growing oocytes. In addition, isolation of the oocyte-specific promoter of ZP3 provided the means by which to express foreign genes in mouse oocytes, reduce expression of mouse genes by using antisense RNAs, or delete mouse genes by using Cre/loxP technology (e.g., Lan, Xu, & Cooney, 2004;

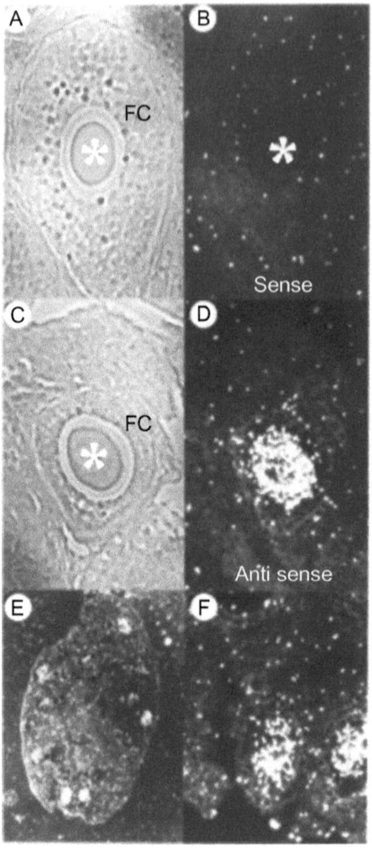

Figure 9.4 *In situ* hybridization analysis of the site of luciferase gene expression in ovaries from transgenic mice. Shown are bright (A) and (C) and dark (B), (D)–(F) -field images of ovarian sections probed with ^{35}S-labeled sense (A) and (B) and antisense (C)–(F) RNA probes specific for luciferase mRNA. In panels (A)–(C), an asterisk marks the position of the growing oocyte. FC, follicle cells.

Lewandoski, Wassarman, & Martin, 1997; Richards, Carroll, Kinloch, Wassarman, & Strickland, 1993; Sun, Liu, & Kikuchi, 2008).

6. MOLECULAR ASPECTS OF SYNTHESIS OF MOUSE ZP GLYCOPROTEINS

Expression of each of the three ZP genes is restricted to mouse oocytes and synthesis of ZP1–3 takes place concomitantly during growth of oocytes. During oocyte growth, the number of copies of ZP transcripts increases dramatically. For example, as the diameter of growing oocytes increases from

Table 9.2 Luciferase activity in tissues of transgenic mice

Tissue	Mean luciferase activity
Adrenal	56
Brain	92
Epididymus	66
Gut	53
Heart	56
Kidney	51
Liver	65
Lung	54
Ovary	**398,000**
Spleen	50
Testis	89
Thymus	54

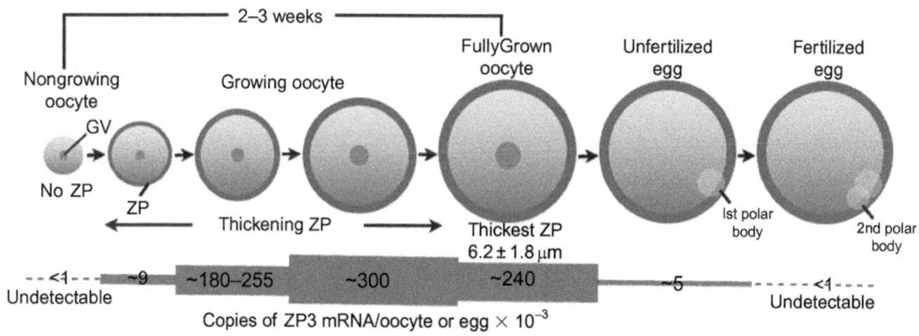

Figure 9.5 Schematic representation of ZP3 mRNA accumulation during oocyte growth, ovulation of eggs, and fertilization of eggs in the mouse. ZP3 mRNA is undetectable in non-growing oocytes, appears just as oocytes begin to grow, accumulates to high levels during the final stages of oocyte growth, decreases dramatically during ovulation of eggs, and is undetectable in fertilized eggs. (See Color Insert.)

10–20 to 60–70 μm, the level of ZP3 mRNA increases from ~9000 to ~300,000 copies/oocyte (Roller et al., 1989; Fig. 9.5). During this period, ZP3 synthesis represents from 1.5% to 2.5% of total protein synthesis. In mid-sized oocytes, 50–60 μm in diameter, ZP transcripts represent ~1.5% (~1% ZP2 and ~0.4% ZP3) of total poly(A)$^+$-RNA, whereas in ovulated eggs the transcripts are present at less than 5% of their maximal levels (Epifano, Liang, Familiari, Moos, & Dean, 1995; Liang et al., 1990). For example, during ovulation, the number of copies of ZP3 mRNA declines by ~98%, from ~240,000 copies/fully grown oocyte (~0.27% of total poly $(A)^+$-RNA and 1.8% of polysomal poly$(A)^+$-RNA) to only ~5000 copies/ovulated egg (Roller et al., 1989). It is known that more than one-half of the poly$(A)^+$-RNA present in fully grown mouse oocytes is either deadenylated or degraded during the period when the oocytes convert to ovulated eggs (Paynton, Rempel, & Bachvarova, 1988). ZP3 mRNA is undetectable in fertilized eggs (<1000 copies) and preimplantation stage embryos.

As discussed earlier, mice bearing transgene ZP3/6.5-LUC express luciferase specifically in growing oocytes (Lira et al., 1990). These experiments were extended to include mice harboring transgenes consisting of either 153 or 470 nt of ZP3 5′-flanking sequence fused to the luciferase gene (ZP3/153-LUC and ZP3/470-LUC, respectively) (Lira, Schickler, & Wassarman, 1993). It was found that as little as 153 nt of ZP3 5′-flanking sequence was sufficient to target expression of luciferase to growing oocytes. On the other hand, the level of luciferase expression was much lower, from 18- to 30-fold less, than observed with larger segments (470–6500 nt) of ZP3 5′-flanking sequence. Comparisons of levels of expression suggest that sequences between −153 and −470 nt greatly affect the level of luciferase expression. As mentioned previously, there is a consensus sequence within this region, -GATAA- at −97 to −93 nt, to which the transcription factor OSP-1, a member of the GATA family of DNA-binding proteins (Ko & Engel, 1993), binds (Schickler et al., 1992). It is likely that other ZP3 promoter sequences between −153 and −470 nt act in the capacity of enhancers of transcription rather than in directing expression of ZP3 specifically in growing oocytes. Candidates for such elements are the sequences -CACGTG- located at −181 to −186 nt and -CTCATTCTGA- at −7 to −16 nt; both of these sequences bind proteins belonging to the helix-loop-helix family of transcription factors (Pabo & Sauer, 1992).

7. ELIMINATION OF SYNTHESIS OF MOUSE ZP GLYCOPROTEINS

ZP1–3 heterozygous and homozygous null mutant female mice have been produced by homologous recombination in embryonic stem cells. Mutant females heterozygous for ZP3 are fertile, although the egg's ZP is about one-half the width (~2.7 μm) of a wild-type ZP (~6.2 μm) and contains ~55% and ~44% of the ZP3 and ZP2, respectively, present in a

Table 9.3 Phenotypes of m ZP 1–3 null females

	Presence of a ZP			
	Early-growing oocyte	**Late-growing oocyte**	**Fully grown oocyte**	**Fertility**
Wild-type	+	+	+	+
$mZP1^{-/-}$	+[a]	+	+	+/–[d]
$mZP2^{-/-}$	+[b]	–	–	–[e]
$mZP3^{-/-}$	–	–	–	–[e]
$mZP3^{+/-}$	+[c]	+	+	+

[a]A loosely organized, abnormal ZP.
[b]A very thin, abnormal ZP.
[c]A thinner than normal ZP.
[d]Fewer than normal offspring.
[e]No offspring.

wild-type ZP (Wassarman, Qi, & Litscher, 1997). On the other hand, homozygous mutant females in which synthesis of either ZP2 or ZP3 is completely eliminated during oocyte growth, fail to lay down a ZP around oocytes and are infertile (Liu et al., 1996; Rankin et al., 1996, 2001) (Table 9.3). These results are consistent with those of antisense oligonucleotide injection experiments in which abolishment of either ZP2 or ZP3 synthesis prevented incorporation of the other protein into the ZP (Tong et al., 1995). This, as well as other evidence (Hoodbhoy et al., 2006; Qi & Wassarman, 1999), strongly suggests that ZP2 and ZP3 are synthesized independently, but are dependent on each other for assembly into the ZP. Furthermore, ZP assembly from nascent glycoproteins may be a stochastic process, as a significant portion of nascent ZP2 and ZP3 is not incorporated into the thickening ZP (Qi & Wassarman, 1999). Rather, the unincorporated glycoproteins are deposited inside and outside the follicle. Such behavior could account for the rather large range of values (4.3–8.1 μm; ave. 6.2 μm) for the thickness of the ZP around fully grown oocytes.

In the absence of synthesis of either ZP2 or ZP3, female mice possess fewer growing oocytes and Graafian follicles in their ovaries and ovulate fewer eggs (~10%) compared with wild-type females (Wassarman, Liu, Chen, Qi, & Litscher, 1998). Interestingly, human ZP2 (~75 kDa) and ZP3 (~55 kDa) can replace mouse ZP2 and ZP3 and restore a ZP to oocytes in homozygous null females and the rescued mice are fertile (Rankin et al., 2003). In the absence of ZP1 synthesis, oocytes and ovulated eggs from females have a ZP, but it is characterized by unusually large pores due to insufficient cross-linking of ZP fibrils by ZP1 (Rankin, Talbot, Lee, & Dean, 1999). ZP1 homozygous null females are less fertile than wild-type mice, giving birth to litters about one-half the size of wild-type litters; this difference is attributed to early embryonic loss by ZP1 null mice. These finding emphasize that both the successful completion of oogenesis and fertility of female mice rely on ZP glycoprotein synthesis and the formation of a ZP around growing oocytes.

8. SECRETION AND ASSEMBLY OF MOUSE ZP GLYCOPROTEINS

As discussed earlier (Section 7), several lines of evidence strongly suggest that ZP2 and ZP3 are synthesized and secreted independently by growing mouse oocytes, but are dependent on each other for incorporation into the ZP. Several additional studies described below have been carried out to gain further insight into elements of nascent ZP glycoproteins that affect their secretion and incorporation into the ZP. Many, but not all of these elements are associated with the CTP of nascent ZP glycoproteins (Jovine, Qi, Williams, Litscher, & Wassarman, 2007; Fig. 9.6).

8.1. Role of N-linked oligosaccharides of ZP glycoproteins

Two different experimental approaches were taken to assess whether N-linked oligosaccharides are required for secretion and assembly of ZP glycoproteins. Radiolabeled ZP2 and ZP3 synthesized by oocytes in the presence

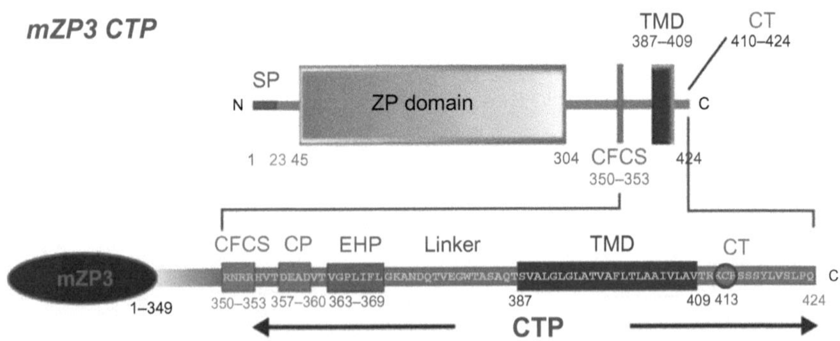

Figure 9.6 Schematic representation of the C-terminal propeptide (CTP) of mZP3. Polypeptide boundaries are marked by gray bars, with the signal peptide (SP) in red; the ZP domain, consensus furin cleavage-site (CFCS), charged peptide (CP), external hydrophobic peptide (EHP), and transmembrane domain (TMD) are depicted in pink, orange, green, cyan, and blue rectangles, respectively; conserved Cys_{413} is circled in red. (For interpretation of the references to color in this figure legend, the reader is referred to the online version of this chapter.)

of tunicamycin (Roller & Wassarman, 1983), an inhibitor of N-linked glycosylation, and a fluorescent, mutated ZP3, lacking N-linked glycosylation sites (-Asn-X-Ser/Thr-, where X is any amino acid other than Pro) and synthesized by oocytes of transgenic mice (Hoodbhoy et al., 2006), were examined. In both instances, nascent ZP3, either depleted of or completely lacking N-linked oligosaccharides, was secreted and incorporated at normal levels into the ZP. On the other hand, ZP2 synthesized by oocytes in the presence of tunicamycin accumulated intracellularly such that very little was incorporated into the ZP (Roller & Wassarman, 1983). These results indicate that ZP3 N-linked oligosaccharides are not required for secretion and assembly, but N-linked oligosaccharides are necessary for normal levels of ZP2 secretion and assembly into the ZP.

8.2. Intracellular localization of nascent ZP glycoproteins

To examine the intracellular location of nascent ZP glycoproteins in growing mouse oocytes, epitope-tagged (*Myc* and *Flag*) cDNAs for ZP2 and ZP3 were microinjected into the germinal vesicle (nucleus) of growing oocytes and recombinant ZP glycoproteins were detected by confocal microscopy using monoclonal antibodies directed against the specific epitopes (Qi, Williams, & Wassarman, 2002). Tagged glycoproteins were found in oocytes in large, membrane-bound, doughnut-shaped vesicles (~2–3 μm in diameter; secretory vesicles of somatic cells are ~0.1–0.2 μm in diameter) that originated from the *trans*-Golgi; the vesicles were labeled with a monoclonal antibody directed against vesicle associated membrane protein. Similarly, using enhanced green fluorescent protein-tagged ZP2 and ZP3, it was found that ZP glycoproteins were localized in growing oocytes to large (1–5 μm in diameter) multivesicular aggregates that originated from the Golgi (Hoodbhoy et al., 2006). By both approaches, ZP2 and ZP3 were found colocalized together in vesicular compartments derived from Golgi, but no interactions between the glycoproteins were detected. Interaction of ZP2 and ZP3 only occurred after release of the nascent ZP glycoproteins from the oocyte's plasma membrane (Jimenez-Movilla & Dean, 2011; Qi et al., 2002). Furthermore, by both approaches, nascent ZP2 and ZP3 were found associated with the innermost region of the thickening ZP, in close proximity to the oocyte's plasma membrane.

8.3. Role of the CFCS of nascent ZP glycoproteins

Prior to incorporation into the oocyte's ZP, nascent ZP2 and ZP3 polypeptides are processed at their CFCS (ZP2: -Arg-Ser-Lys-Arg-; ZP3: -Arg-Asn-Arg-Arg-), located near the C-terminus, by a member of the furin family of serine proteases (Litscher, Qi, & Wassarman, 1999). Human ZP polypeptides also undergo proteolytic processing at their CFCS (Kiefer & Saling, 2002). In the case of mouse ZP2 and ZP3, this results in removal of the C-terminal 78 and 71 aa, respectively, encoded by *ZP2* exon-18 and *ZP3* exon-8, prior to incorporation of the glycoproteins into the ZP.

Is cleavage at the CFCS essential for secretion and assembly of ZP glycoproteins? Two different experimental approaches were employed to address this question. First, embryonal carcinoma (EC) cells transfected with *Flag-ZP3* were cultured in the presence or absence of decanoylated peptidyl chloromethylketone (DPC), a modified peptide inhibitor of furin-like enzymes; an inhibitor that binds covalently to the substrate binding site of proprotein convertases (Feldman, Nichol, Klenk, Peters, & Sanchez, 1994). EC cells stably transfected with either *ZP3* or *Flag-ZP3* secreted large amounts of nascent ZP3 into culture medium (Kinloch, Mortillo, Stewart, & Wassarman, 1991). On the other hand, it was found that secretion of ZP3 from transfected EC cells was reduced more than 20-fold when cells were cultured in the presence of DPC (Williams and Wassarman, 2001). In the presence of DPC, uncleaved ZP3 accumulated in the endoplasmic reticulum (ER) and, as expected, possessed high-mannose-type (added cotranslationally in the ER) rather than complex-type (high-mannose-type modified posttranslationally) N-linked oligosaccharides.

Alternatively, *ZP2* and *ZP3* were subjected to site-specific mutagenesis at their CFCS to remove the cleavage-site and secretion of ZP2 and ZP3 was examined in oocytes from wild-type and transgenic mice, as well as in transfected cells (Williams and Wassarman, 2001; Qi et al., 2002; Zhao et al., 2002). For example, as in the case of transfected EC cells cultured in the presence of DPC (above), growing oocytes microinjected with *ZP3* mutated at its CFCS resulted in a slowdown of trafficking of nascent glycoprotein through the secretory pathway and an enhanced retention of ZP3 within oocytes (Qi et al., 2002). Although a small amount of mutated ZP3 escaped retention and was incorporated into the ZP, it had undergone proteolytic processing at its C-terminus, possibly by a protease other than furin. Very similar results were obtained with EC cells transfected with *ZP3* mutated at the CFCS that also failed to secrete nascent ZP3 (Williams and Wassarman, 2001). On the other hand, such effects on ZP3 secretion by mutation of the CFCS were not observed by others using transgenic mice; they invoked a role for membrane-anchored secretases in ZP glycoprotein processing as an alternative to furin cleavage (Zhao et al., 2002).

8.4. Role of the TMD of nascent ZP glycoproteins

The CTPs of nascent ZP1–3 have a TMD consisting of 26, 17, and 23 aa, respectively. Truncation of ZP2 and ZP3 polypeptides before the TMD did not prevent packaging of ZP glycoproteins into secretory vesicles, but it completely prevented their incorporation into the ZP (Jimenez-Movilla & Dean, 2011; Jovine et al., 2002). Unlike the situation with wild-type recombinant ZP glycoproteins, the truncated ZP glycoproteins that filled the lumen of secretory vesicles were not anchored in the lipid bilayer of vesicles. Although the TMD is not required for secretion, it is required for cleavage at the CFCS and incorporation of nascent ZP glycoproteins into the ZP (Jovine et al., 2004). Replacement of the TMD of ZP3 (aa 387–409) with the single-spanning TMD of human CD7 (aa 178–201; Schanberg, Fleenor, Kurtzberg, Haynes, & Kaufman, 1991) had no effect on secretion of ZP3 or incorporation of ZP3 into the ZP of oocytes (Jovine et al., 2004). These observations suggest that the conserved TMD is not involved in specific interactions, as reported for other biological systems (Harrison, 1996), but ensures proper localization and/or topological orientation of nascent ZP polypeptides so that assembly into the ZP can take place.

8.5. Role of the CT of nascent ZP glycoproteins

The CTPs of nascent ZP1–3 possess a short (9–15 aa in length), hydrophilic CT downstream of the TMD. Like the TMD, the CT is not required for secretion, but is required for incorporation of nascent ZP glycoproteins into the ZP (Jovine et al., 2004). Although there is a conserved Cys residue in the CT of ZP glycoproteins, it is not required for either secretion or incorporation of ZP glycoproteins into the ZP (Jovine et al., 2004). Furthermore, results of experiments using fluorescent complementation of recombinant ZP2 and ZP3 in Chinese hamster ovary cells and growing oocytes revealed that CTs are sufficient and necessary to prevent intracellular oligomerization of nascent ZP glycoproteins (Jimenez-Movilla & Dean, 2011). Of particular interest was the finding that β-tectorin bearing the TMD and CT of ZP3 was incorporated into the oocyte's ZP, perhaps indicating that specific recognition of the CT at the plasma membrane is required for proper cleavage at the CFCS and incorporation of nascent ZP glycoproteins. Consistent with this possibility, recombinant ZP2 and ZP3 bearing the same CT, derived from either ZP2 or ZP3, failed to be incorporated into the ZP. This suggests that CTs of ZP2 and ZP3 can be distinguished from one another once nascent ZP glycoproteins arrive at the oocyte's plasma membrane.

8.6. Role of the hydrophobic patches of nascent ZP glycoproteins

Several of the findings described earlier strongly suggest that elements required for secretion of ZP glycoproteins are located in the CTP between the CFCS and TMD. One of these, a short stretch rich in charged amino acids, called a charged patch (CP), was shown using mutated recombinant proteins not to be involved in secretion of ZP glycoproteins (Jovine et al., 2004). However, just C-terminal to the CP is a short conserved motif, consisting of an almost invariant GlyPro sequence immediately followed by 4–5 hydrophobic amino acids, called an external hydrophobic peptide (EHP) (Jovine et al., 2004; Zhao et al., 2003). Much of the evidence suggests that ZP glycoproteins must contain either an EHP or TMD to be secreted, and that both the EHP and TMD must be present for incorporation of nascent ZP glycoproteins into the ZP. Another sequence of hydrophobic amino acids, called an internal hydrophobic peptide (IHP), is present in the ZPD itself, following conserved Cys residues 1–4 (Jovine et al., 2004). Mutation of the IHP prevents incorporation of full-length ZP glycoproteins into the ZP without affecting secretion; similar to the effect of deletion of the EHP on incorporation. On the other hand, secretion is inhibited when either the EHP or IHP is mutated in a ZP glycoprotein truncated before the TMD.

8.7. A general mechanism for assembly of ZP glycoproteins

The presence of both an EHP and IHP within ZP glycoprotein precursors is thought to prevent the ZPD from participating in polymerization of nascent ZP glycoproteins within oocytes or transfected cells. The close proximity of the EHP and IHP to each other in ZP glycoprotein precursors is strongly supported by the high-resolution structure of full-length ZP3 (Han et al., 2010). When the CTP of ZP glycoprotein precursors is removed by proteolytic cleavage at the CFCS, the EHP can no longer interact with the IHP and polymerization ensues. This mechanism may apply to all ZPD proteins (e.g., uromodulin; Schaeffer, Santambrogio, Perucca, Casari, & Rampoldi, 2009) since it relies on sequence elements, EHP and IHP, and events, coupling between proteolytic processing and polymerization, conserved in all ZPD proteins (Jovine et al., 2004, 2005). In this context, proteolytic cleavage of inhibitory sequences of protein precursors, with concomitant exposure of polymerization elements, has been shown to control assembly of several other types of proteins (Jovine et al., 2007).

8.8. Summary

Nascent ZP glycoproteins of growing oocytes are packaged in unusually large secretory vesicles and subsequently incorporated into the innermost region of the thickening ZP. Different ZP glycoproteins traffic independently in oocytes, apparently due to their CTs, but are dependent on each other and on their ZPD for incorporation into the ZP. Proteolytic cleavage of nascent ZP glycoproteins at their CFCS permits normal levels of secretion, but failure to do so can result in accumulation of ZP glycoproteins in the oocyte's ER. The presence of either an EHP or TMD is required for secretion of ZP glycoproteins, but the presence of both is required for incorporation of ZP glycoproteins into the ZP. The presence of both an EHP (in the CTP) and an IHP (in the ZPD) prevents premature polymerization of ZP glycoproteins in the oocyte's cytoplasm. Polymerization of ZP glycoproteins takes place extracellularly as a result of proteolytic cleavage at the CFCS and release of the EHP.

ACKNOWLEDGMENTS

Many thanks to the postdoctoral fellows, graduate students, and research technicians who contributed to the research on ZP biogenesis carried out in our laboratory at Harvard Medical School, Roche Institute of Molecular Biology, and Mount Sinai Medical School. Our research was supported in part by the National Institutes of Health (NICHD) and by Hoffmann-La Roche Inc.

REFERENCES

Austin, C. R., & Short, R. V. (Eds.), (1982). (2nd ed.). In: *Reproduction in mammals: Germ cells and fertilization 1*, (pp. 1–177). Cambridge, England: Cambridge University Press.

Bausek, E. S., Waclawek, M., Schneider, W. J., & Wohlrab, F. (2000). The major chicken egg envelope protein ZP1 is different from ZPB and is synthesized in the liver. *The Journal of Biological Chemistry*, *275*, 28866–28872.

Bleil, J. D., & Wassarman, P. M. (1980a). Structure and function of the zona pellucida: Identification and characterization of the proteins of the mouse oocyte's zona pellucida. *Developmental Biology*, *76*, 185–202.

Bleil, J. D., & Wassarman, P. M. (1980b). Synthesis of zona pellucida proteins by denuded and follicle-enclosed mouse oocytes during culture *in vitro*. *Proceedings of the National Academy of Sciences of the United States of America*, 77, 1029–1033.

Boja, E. S., Hoodbhoy, T., Fales, H. M., & Dean, J. (2003). Structural characterization of native mouse zona pellucida proteins using mass spectrometry. *The Journal of Biological Chemistry*, *278*, 34189–34202.

Bork, P., & Sander, C. (1992). A large domain common to sperm receptors (ZP2 and ZP3) and TGF-b type III receptor. *FEBS Letters*, *300*, 237–240.

Davidson, E. H. (1986). *Gene activity in early development* (3rd ed). New York: Academic Press.

Dietl, J. (Ed.), (1989). *The mammalian egg coat: Structure and function*. Berlin: Springer-Verlag KG.

Dumont, J. N., & Brummett, A. R. (1985). Egg envelopes in vertebrates. In L. Browder (Ed.), *Developmental biology: A comprehensive synthesis 1*, (pp. 235–288). New York: Plenum Press.

Epifano, O., Liang, L.-F., & Dean, J. (1995). Mouse Zp1 encodes a zona pellucida protein homologous to egg envelope proteins in mammals and fish. *The Journal of Biological Chemistry, 270*, 27254–27258.

Epifano, O., Liang, L.-F., Familiari, M., Moos, M. C., & Dean, J. (1995). Coordinate expression of the three zona pellucida genes during mouse oogenesis. *Development, 121*, 1947–1956.

Feldman, H., Nichol, S. T., Klenk, H.-D., Peters, C. J., & Sanchez, A. (1994). Characterization of filoviruses based on differences in structure and antigenicity of the virion glycoprotein. *Virology, 199*, 469–473.

Florman, H. M., & Ducibella, T. (2006). Fertilization in mammals. In J. D. Neill (Ed.), *The physiology of reproduction1*, (pp. 55–112). San Diego: Elsevier.

Greve, J. M., Salzmann, G. S., Roller, R. J., & Wassarman, P. M. (1982). Biosynthesis of the major zona pellucida glycoprotein secreted by oocytes during mammalian oogenesis. *Cell, 31*, 749–759.

Greve, J. M., & Wassarman, P. M. (1985). Mouse egg extracellular coat is a matrix of interconnected filaments possessing a structural repeat. *Journal of Molecular Biology, 181*, 253–264.

Haddad, A., & Nagai, M. E. (1977). Radioautographic study of glycoprotein biosynthesis and renewal in the ovarian follicles of mice and the origin of the zona pellucida. *Cell and Tissue Research, 177*, 347–369.

Han, L., Monné, M., Okumura, H., Schwend, T., Cherry, A. L., Flot, D., et al. (2010). Insights into egg coat assembly and egg-sperm interaction from the X-ray structure of full-length ZP3. *Cell, 143*, 404–415.

Harrison, P. T. (1996). Protein:protein interactions in the lipid bilayer. *Molecular Membrane Biology, 13*, 67–79.

Hoodbhoy, T., Aviles, M., Baibakov, B., Epifano, O., Jimenez-Movilla, M., Gauthier, L., et al. (2006). ZP2 and ZP3 traffic independently within oocytes prior to assembly into the extracellular zona pellucida. *Molecular and Cellular Biology, 26*, 7991–7998.

Jimenez-Movilla, M., & Dean, J. (2011). ZP2 and ZP3 cytoplasmic tails prevent premature interactions and ensure incorporation into the zona pellucida. *Journal of Cell Science, 124*, 940–950.

Jovine, L., Darie, C. C., Litscher, E. S., & Wassarman, P. M. (2005). Zona pellucida domain proteins. *Annual Review of Biochemistry, 74*, 83–114.

Jovine, L., Janssen, W. G., Litscher, E. S., & Wassarman, P. M. (2006). The PLAC-1 homology region of the ZP domain is sufficient for protein polymerization. *BMC Biochemistry*, 7, 11–19.

Jovine, L., Qi, H., Williams, Z., Litscher, E. S., & Wassarman, P. M. (2002). The ZP domain is a conserved module for protein polymerization. *Nature Cell Biology, 4*, 457–461.

Jovine, L., Qi, H., Williams, Z., Litscher, E. S., & Wassarman, P. M. (2004). A duplicated motif controls assembly of zona pellucida domain proteins. *Proceedings of the National Academy of Sciences of the United States of America, 101*, 5922–5927.

Jovine, L., Qi, H., Williams, Z., Litscher, E. S., & Wassarman, P. M. (2007). Features that affect secretion and assembly of zona pellucida glycoproteins during mammalian oogenesis. In S. Gupta, K. Koyama & J. F. Murray (Eds.), *Gamete biology: Emerging frontiers in fertility and contraceptive development. Reproduction supplement*, Nottingham, UK: Nottingham University Press.

Keefe, D., Tran, P., Pellegrini, C., & Oldenbourg, R. (1997). Polarized light microscopy and digital imaging processing identify a multilaminar structre of the hamster zona pellucida. *Human Reproduction, 12*, 1250–1252.

Kiefer, S. M., & Saling, P. (2002). Proteolytic processing of human zona pellucida proteins. *Biology of Reproduction, 66*, 407–414.

Kinloch, R. A., Mortillo, S., Stewart, C. L., & Wassarman, P. M. (1991). Embryonal carcinoma cells transfected with ZP3 genes differentially glycosylate similar polypeptides and secrete active mouse sperm receptor. *The Journal of Cell Biology, 115*, 655–664.

Kinloch, R. A., Roller, R. J., Fimiani, C. M., Wassarman, D. A., & Wassarman, P. M. (1988). Primary structure of the mouse sperm receptor's polypeptide chain determined by genomic cloning. *Proceedings of the National Academy of Sciences of the United States of America*, *85*, 6409–6413.

Kinloch, R. A., & Wassarman, P. M. (1989). Nucleotide sequence of the gene encoding zona pellucida glycoprotein ZP3—The mouse sperm receptor. *Nucleic Acid Research*, *17*, 2861–2863.

Ko, L. J., & Engel, J. D. (1993). DNA-binding specificities of the GATA transcription factor family. *Molecular and Cellular Biology*, *13*, 4011–4022.

Kubo, H., Kawano, T., Tsubuki, S., Kawashima, S., Katagiri, C., & Suzuki, A. (1997). A major glycoprotein of Xenopus egg vitelline envelope, gp41, is a frog homolog of mammalian ZP3. *Development, Growth & Differentiation*, *39*, 405–417.

Lan, Z. J., Xu, X., & Cooney, A. J. (2004). Differential oocyte-specific expression of Cre recombinase activity in GDF-9-iCre, Zp3cre, and Msx2Cre transgenic mice. *Biology of Reproduction*, *71*, 1469–1474.

Lewandoski, M., Wassarman, K. M., & Martin, G. R. (1997). Zp3-cre, a transgenic mouse line for the activation or inactivation of loxP-flanked target genes specifically in the female germ line. *Current Biology*, *7*, 148–151.

Liang, L.-F., Chamow, S. M., & Dean, J. (1990). Oocyte-specific expression of mouse Zp-2: Developmental regulation of the zona pellucida genes. *Molecular and Cellular Biology*, *10*, 1507–1515.

Liang, L.-F., & Dean, J. (1993). Conservation of mammalian secondary sperm receptor genes enables the promoter of the human gene to function in mouse oocytes. *Developmental Biology*, *156*, 399–408.

Liang, L.-F., Soyal, S., & Dean, J. (1997). FIGα, a germ cell specific transcription factor involved in the coordinate expression of the zona pellucida genes. *Development*, *124*, 4939–4949.

Lira, S. A., Kinloch, R. A., Mortillo, S., & Wassarman, P. M. (1990). An upstream region of the mouse ZP3 gene directs expression of firefly luciferase specifically to growing oocytes in transgenic mice. *Proceedings of the National Academy of Sciences of the United States of America*, *87*, 7215–7219.

Lira, S. A., Schickler, M., & Wassarman, P. M. (1993). Cis-acting DNA elements involved in oocyte-specific expression of mouse sperm receptor gene mZP3 are located close to the gene's transcription start-site. *Molecular Reproduction and Development*, *36*, 494–499.

Litscher, E. S., Qi, H., & Wassarman, P. M. (1999). Mouse zona pellucida glycoproteins mZP2 and mZP3 undergo carboxy-terminal proteolytic processing in growing oocytes. *Biochemistry*, *38*, 12280–12287.

Litscher, E. S., & Wassarman, P. M. (2007). Egg extracellular coat proteins: From fish to mammals. *Histology and Histopathology*, *22*, 337–347.

Liu, C., Litscher, E. S., Mortillo, S., Sakai, Y., Kinloch, R. A., Stewart, C. L., et al. (1996). Targeted disruption of the mZP3 gene results in production of eggs lacking a zona pellucida and infertility in female mice. *Proceedings of the National Academy of Sciences of the United States of America*, *93*, 5431–5436.

Llorca, O., Trujillo, A., Blanco, F. J., & Bernabeu, C. (2007). Structural model of human endoglin, a transmembrane receptor responsible for hereditary hemorrhagic telangiectasia. *Journal of Molecular Biology*, *365*, 694–705.

Lunsford, R. D., Jenkins, N. A., Kozak, C. A., Liang, L. F., Silan, C. M., Copeland, N. G., et al. (1990). Genomic mapping of murine Zp-2 and Zp-3, two oocyte-specific loci encoding zona pellucida proteins. *Genomics*, *6*, 184–187.

Millar, S. E., Lader, E. S., & Dean, J. (1993). ZAP-1 DNA binding activity is first detected at the onset of zona pellucida gene expression in embryonic mouse oocytes. *Developmental Biology*, *158*, 410–413.

Millar, S. E., Lader, E., Liang, L. F., & Dean, J. (1991). Oocyte-specific factors bind a conserved upstream sequence required for mouse zona pellucida promoter activity. *Molecular and Cellular Biology, 11,* 6197–6204.

Monné, M., Han, L., & Jovine, L. (2006). Tracking down the ZP domain: From the mammalian zona pellucida to the molluscan vitelline envelope. *Seminars in Reproductive Medicine, 24,* 204–216.

Monné, M., Han, L., Schwend, T., Burendahl, S., & Jovine, L. (2008). Crystal structure of the ZP-N domain of ZP3 reveals the core fold of animal egg coats. *Nature, 456,* 653–657.

Monné, M., & Jovine, L. (2011). A structural view of egg coat architecture and function in fertilization. *Biology of Reproduction, 85,* 661–669.

Pabo, C. O., & Sauer, R. T. (1992). Transcription factors: Structural families and principles of DNA recognition. *Annual Review of Biochemistry, 61,* 1053–1095.

Paynton, B. V., Rempel, R., & Bachvarova, R. (1988). Changes in state of adenylation and time course of degradation of maternal mRNAs during oocyte maturation and early embryonic development in the mouse. *Developmental Biology, 129,* 304–314.

Philpott, C. C., Ringuette, M. J., & Dean, J. (1987). Oocyte-specific expression and developmental regulation of ZP3, the sperm receptor of the mouse zona pellucida. *Developmental Biology, 121,* 568–575.

Plaza, S., Chanut-Delalande, H., Fernandes, I., Wassarman, P. M., & Payre, F. (2010). From A to Z: Apical structures and zona pellucida-domain proteins. *Trends in Cell Biology, 20,* 524–532.

Qi, H., & Wassarman, P. M. (1999). Secretion of zona pellucida glycoprotein mZP2 by growing oocytes from *mZP3*$^{+/+}$ and *mZP3*$^{-/-}$ mice. *Developmental Genetics, 25,* 95–102.

Qi, H., Williams, Z., & Wassarman, P. M. (2002). Secretion and assembly of zona pellucida glycoproteins by growing mouse oocytes microinjected with epitope-tagged cDNAs for mZP2 and mZP3. *Molecular Biology of the Cell, 13,* 530–541.

Rankin, T. L., Coleman, J. S., Epifano, O., Hoodbhoy, T., Turner, S. G., Castle, P. E., et al. (2003). Fetility and taxon-specific sperm binding persist after replacement of mouse sperm receptors with human homologs. *Developmental Cell, 5,* 33–43.

Rankin, T. L., Familiari, M., Lee, E., Ginsberg, A., Dwyer, N., Blanchette-Mackie, J., et al. (1996). Mice homozygous for an insertional mutation in the ZP3 gene lack a zona pellucida and are infertile. *Development, 122,* 2903–2910.

Rankin, T. L., O'Brien, M., Lee, E., Wigglesworth, K., Eppig, J., & Dean, J. (2001). Defective zonae pellucidae in ZP2-null mice disrupt folliculogenesis, fertility, and development. *Development, 128,* 1119–1126.

Rankin, T. L., Talbot, P., Lee, E., & Dean, J. (1999). Abnormal zonae pellucidae in mice lacking ZP1 result in early embryonic loss. *Development, 126,* 3847–3855.

Richards, W. G., Carroll, P. M., Kinloch, R. A., Wassarman, P. M., & Strickland, S. (1993). Creating maternal effect mutations in transgenic mice: Antisense inhibition of a specific oocyte gene product. *Developmental Biology, 160,* 543–553.

Ringuette, M. J., Chamberlin, M. E., Baur, D. A., Sobieski, A., & Dean, J. (1988). Molecular analysis of cDNA coding for ZP3, a sperm binding protein of the mouse zona pellucida. *Developmental Biology, 127,* 287–295.

Ringuette, M. J., Sobieski, D. A., Chamow, S. M., & Dean, J. (1986). Oocyte-specific gene expression: Molecular characterization of a cDNA coding for ZP-3, the sperm receptor of the mouse zona pellucida. *Proceedings of the National Academy of Sciences of the United States of America, 83,* 4341–4345.

Roller, R. J., Kinloch, R. A., Hiraoka, B. Y., Li, S. S.-L., & Wassarman, P. M. (1989). Gene expression during mammalian oogenesis and early embryogenesis: Quantification of three messenger-RNAs abundant in fully-grown oocytes. *Development, 106,* 251–261.

Roller, R. J., & Wassarman, P. M. (1983). Role of asparagine-linked oligosaccharides in secretion of glycoproteins of the mouse egg's extracellular coat. *The Journal of Biological Chemistry, 258,* 13243–13249.

Salzmann, G. S., Greve, J. M., Roller, R. J., & Wassarman, P. M. (1983). Biosynthesis of the sperm receptor during oogenesis in the mouse. *The EMBO Journal, 2*, 1451–1456.

Schaeffer, C., Santambrogio, S., Perucca, S., Casari, G., & Rampoldi, L. (2009). Analysis of uromodulin polymerization provides new insights into the mechanisms regulating ZP domain-mediated protein assembly. *Molecular Biology of the Cell, 20*, 589–599.

Schanberg, L. E., Fleenor, D. E., Kurtzberg, J., Haynes, B. F., & Kaufman, R. E. (1991). Isolation and characterization of the genomic human *CD7* gene: Structural similarity with the murine Thy-1 gene. *Proceedings of the National Academy of Sciences of the United States of America, 88*, 603–607.

Schickler, M., Lira, S. A., Kinloch, R. A., & Wassarman, P. M. (1992). A mouse oocyte-specific protein that binds to a region of mZP3 promoter involved in regulating oocyte-specific expression of the mZP3 gene. *Molecular and Cellular Biology, 12*, 120–127.

Shimizu, S., Tsuji, M., & Dean, J. (1983). *In vitro* biosynthesis of three sulfated glycoproteins of murine zonae pellucidae by oocytes grown in follicle culture. *The Journal of Biological Chemistry, 258*, 5858–5863.

Soyal, S. M., Amieh, A., & Dean, J. (2000). FIGalpha, a germ cell-specific transcription factor required for ovarian follicle formation. *Development, 127*, 4645–4654.

Sun, Q. Y., Liu, K., & Kikuchi, K. (2008). Oocyte-specific knockout: A novel in vivo approach for studying gene functions during folliculogenesis, oocyte maturation, fertilization, and embryogenesis. *Biology of Reproduction, 79*, 1014–1020.

Takeuchi, Y., Nishimura, K., Aoki, N., Adachi, T., Sato, C., Kitajima, K., et al. (1999). A 42-kDa glycoprotein from chicken egg-envelope, an avian homolog of the ZPC family glycoproteins in mammalian zona pellucida, cDNA cloning and granulose cell-specific expression. *European Journal of Biochemistry, 260*, 736–742.

Tong, Z.-B., Nelson, L. M., & Dean, J. (1995). Inhibition of zona pellucida gene expression by antisense oligonucleotides injected into mouse oocytes. *The Journal of Biological Chemistry, 270*, 849–853.

Wassarman, P. M. (1988). Zona pellucida glycoproteins. *Annual Review of Biochemistry, 57*, 415–442.

Wassarman, P. M. (1999). Mammalian fertilization: Molecular aspects of gamete adhesion, exocytosis, and fusion. *Cell, 96*, 175–183.

Wassarman, P. M. (2008). Zona pellucida glycoproteins. *The Journal of Biological Chemistry, 283*, 24285–24289.

Wassarman, P. M., & Albertini, D. F. (1994). Cellular and molecular biology of the mammalian ovum. In E. Knobil & J. D. Neill (Eds.), *The physiology of reproduction1*, (pp. 79–122). New York: Raven Press.

Wassarman, P. M., Liu, C., Chen, J., Qi, H., & Litscher, E. S. (1998). Ovarian development in mice bearing homozygous and heterozygous null mutations in zona pellucida glycoprotein gene mZP3. *Histology and Histopathology, 13*, 293–300.

Wassarman, P. M., Liu, C., & Litscher, E. S. (1996). Constructing the mouse egg zona pellucida: Some new pieces of an old puzzle. *Journal of Cell Science, 109*, 2001–2004.

Wassarman, P. M., & Mortillo, S. (1991). Structure of the mouse egg extracellular coat, the zona pellucida. *International Review of Cytology, 130*, 85–110.

Wassarman, P. M., Qi, H., & Litscher, E. S. (1997). Mutant female mice carrying a single mZP3 allele produce eggs with a thin zona pellucida, but reproduce normally. *Proceedings of the Royal Society of London - Series B: Biological Sciences, 264*, 323–328.

Williams, Z., & Wassarman, P. M. (2001). Secretion of mouse ZP3, the sperm receptor, requires cleavage of its polypeptide at a consensus furin cleavage site. *Biochemistry, 40*, 929–937.

Yanagimachi, R. (1994). Mammalian fertilization. In E. Knobil & J. D. Neill (Eds.), *The physiology of reproduction1*, (pp. 189–317). New York: Raven Press.

Yang, J. C., & Hedrick, J. L. (1997). cDNA cloning and sequence analysis of the Xenopus laevis egg envelope glycoprotein gp43. *Development, Growth & Differentiation, 39*, 457–467.

Zhao, M., Gold, L., Ginsberg, A. M., Liang, L.-F., & Dean, J. (2002). Conserved furin cleavage site not essential for secretion and integration of ZP3 into the extracellular egg coat of transgenic mice. *Molecular Cellular Biology, 22*, 3111–3120.

Zhao, M., Gold, L., Dorward, H., Liang, L.-F., Hoodbhoy, T., Boja, E. S., Fales, H. M., & Dean, J. (2003). Mutation of a conserved hydrophobic patch prevents incorporation of ZP3 into the zona pellucida surrounding mouse eggs. *Molecular Cellular Biology, 23*, 8982–8991.

CHAPTER TEN

Molecular Changes During Egg Activation

Amber R. Krauchunas, Mariana F. Wolfner[1]
Department of Molecular Biology and Genetics, Cornell University, Ithaca, New York, USA
[1]Corresponding author: e-mail address: mfw5@cornell.edu

Contents

Abstract

Egg activation is the final transition that an oocyte goes through to become a developmentally competent egg. This transition is usually triggered by a calcium-based signal that is often, but not always, initiated by fertilization. Activation encompasses a number of changes within the egg. These include changes to the egg's membranes and outer coverings to prevent polyspermy and to support the developing embryo, as well as resumption and completion of the meiotic cell cycle, mRNA polyadenylation, translation of new proteins, and the degradation of specific maternal mRNAs and proteins. The transition from an arrested, highly differentiated cell, the oocyte, to a developmentally active, totipotent cell, the activated egg or embryo, represents a complete change in cellular state. This is accomplished by altering ion concentrations and by widespread changes in both the proteome and the suite of mRNAs present in the cell. Here, we review the role of calcium and zinc in the events of egg activation, and the importance of macromolecular changes during this transition. The latter include the degradation and translation of proteins, protein posttranslational regulation through phosphorylation, and the degradation, of maternal mRNAs.

Current Topics in Developmental Biology, Volume 102
ISSN 0070-2153
http://dx.doi.org/10.1016/B978-0-12-416024-8.00010-6

1. INTRODUCTION

At the end of oogenesis, mature oocytes are held in a developmentally quiescent, arrested state until an appropriate trigger initiates development. In these specialized cells, the cell cycle is paused and maternally provided mRNAs and proteins are stored until their time to function. If development is to be successful, the oocyte must undergo one final transition from this steady state to a new cellular state that can support embryogenesis. This transition is referred to as "egg activation." Without this step of oocyte development, fertilization cannot lead to the creation of an embryo.

The events of egg activation include cortical granule exocytosis (in many taxa) and the block to polyspermy, translation of new proteins, decreases in Mitogen-Activated Protein Kinase (MAPK) activity, and resumption of the cell cycle through inactivation of Maturation Promoting Factor (MPF; composed of Cdc2 kinase and cyclin B) and activation of the Anaphase Promoting Complex (APC/c). Without these changes, the egg cannot limit fertilization to a single sperm nor direct events of development such as mitosis and embryonic patterning. Egg activation as a whole process, as well as specific events such as meiotic cell cycle regulation, has been well reviewed elsewhere (Gadella & Evans, 2011; Horner & Wolfner, 2008b; Li, Zheng, & Dean, 2010; Liu, 2011; Marcello & Singson, 2010; Tsaadon, Eliyahu, Shtraizent, & Shalgi, 2006). Here, we present egg activation as it relates to a complete cellular/molecular transition through the regulation of ion concentrations, the suite of mRNAs present within the cell, and the composition of the proteome.

Understanding how an oocyte becomes developmentally competent through egg activation is an important aspect of the diagnosis and treatment of female infertility, especially as the field continues to work toward improved assisted reproductive technologies. In addition, egg activation provides a unique model system to study the transition from a differentiated cell to a totipotent cell. In some ways, egg activation can be viewed as the converse of the change that the daughter cell of a stem cell undergoes when it differentiates, and is similar in concept to the cellular change required to form induced pluripotent cells. Upon fertilization, the highly differentiated oocyte suddenly becomes a one-cell embryo, which will give rise to all the varied cells, tissues, and organs of a complete multicellular organism. This transition represents a large and rapid change in cellular state and requires a number of molecular changes that must be both highly regulated and properly coordinated.

2. ION LEVELS

2.1. Calcium

While the specific trigger of egg activation varies between organisms—from fertilization in vertebrates, to mechanical stimulation in some insects, to changes in pH or ionic environment for some marine invertebrates (reviewed in Horner & Wolfner, 2008b)—a rise in intracellular calcium appears to be the nearly universal signal that sets off the events of egg activation. As the regulation and role of this calcium signal have been extensively reviewed (Horner & Wolfner, 2008b; Nomikos, Swann, & Lai, 2012; Parrington, Davis, Galione, & Wessel, 2007; Wakai, Vanderheyden, & Fissore, 2011), we will describe it only briefly here.

In vertebrates, fertilization leads to a rise in cytosolic Ca^{2+} in the egg which is responsible for initiating all of the events of egg activation (Horner & Wolfner, 2008b; Wakai et al., 2011). In mammals, this Ca^{2+} rise is mediated by a sperm-specific phospholipase C (PLCζ) (Saunders et al., 2002; reviewed in Nomikos et al., 2012; Wakai et al., 2011). When a sperm enters the egg, PLCζ hydrolyzes phosphatidylinositol 4,5-bisphosphate into inositol 1,4,5-triphosphate (IP_3) and diacyl glycerol (DAG). IP_3 then binds its receptor (IP_3R1), which results in the release of Ca^{2+} from internal stores in the endoplasmic reticulum and an overall rise in cytosolic Ca^{2+} in the egg (Wakai et al., 2011). A recent report also showed that Ca^{2+} influx (from the external environment) is necessary for complete egg activation in the mouse, suggesting greater complexity to the regulation of the Ca^{2+} signal than previously presumed (Miao, Stein, Jefferson, Padilla-Banks, & Williams, 2012).

In some organisms, Ca^{2+} levels in the egg continue to rise and fall in an oscillatory pattern following fertilization. In mammals, these oscillations last for several hours (Stricker, 1999). The importance of the oscillatory nature of the Ca^{2+} signal is still a matter of discussion, though several theories have been proposed (reviewed in Wakai et al., 2011). For example, Ducibella et al. (2002) showed that, in mouse, different egg activation events are sensitive to different numbers of calcium oscillations. For instance, a single Ca^{2+} pulse is enough to initiate cortical granule exocytosis, while resumption of the cell cycle requires at least four Ca^{2+} pulses (Ducibella et al., 2002). In addition, the number of Ca^{2+} pulses required for completion of egg activation events is greater than the number required to initiate those events; eight Ca^{2+} pulses is not enough for polar body formation even though exit from metaphase arrest had been initiated (Ducibella et al., 2002). However, other evidence suggests that the Ca^{2+} oscillations are not strictly required for proper egg activation. A single large Ca^{2+} rise induced by ethanol can generate an equal number of embryos with a polar body as Ca^{2+} oscillations induced by Sr^{2+} (when observed 6 h after exposure) and produces an equivalent percentage of embryos that develop full term (Rogers et al., 2006; Suzuki, Yoshida, Suzuki, Okuda, & Perry, 2010). This has led to the proposal that it is the overall magnitude of the Ca^{2+} increase produced by the oscillations, rather than the actual oscillatory nature of the Ca^{2+} rise that is necessary for the events of egg activation to occur (Ducibella, Schultz, & Ozil, 2006; Ozil, Banrezes, Tóth, Pan, & Schultz, 2006).

The mechanisms of Ca^{2+} regulation at egg activation are less characterized in nonmammalian systems, though Ca^{2+} signaling itself appears to be a relatively conserved feature of the oocyte-to-embryo transition. A sperm PLC has not yet been identified outside of mammals, though a different PLC (PLCγ), along with Src-family kinases, has been implicated in the activation of the phosphoinositide pathway during fertilization in Xenopus (reviewed in Sato, Fukami, & Stith, 2006). In *Caenorhabditis elegans*, a single Ca^{2+} increase has been observed at the time of fertilization, but how this Ca^{2+} signal relates to the events of egg activation, or if it is dependent on a sperm-derived factor, is not yet known (Samuel, Murthy, & Hengartner, 2001). Supporting the idea that the Ca^{2+} rise seen in *C. elegans* plays a role analogous to that seen in other species, knocking out a *C. elegans* IP_3 receptor, *itr-1*, results in sterility, suggesting that the Ca^{2+} rise is necessary for development (Del Santo P, Logan, Chisholm, & Jorgensen, 1999; Samuel et al., 2001). A single Ca^{2+} rise also occurs in zebrafish during egg activation (Lee, Webb, & Miller, 1999). Here, the Ca^{2+} rise is dependent on the RNA-binding protein *brom bones*, which acts upstream of IP_3-mediated Ca^{2+} release (Mei, Lee, Marlow, Miller, & Mullins, 2009). In Drosophila, *in vitro* experiments show that removal of Ca^{2+} from the buffer, through the use of the calcium chelators BAPTA or EGTA, prevents egg activation from occurring (Horner & Wolfner, 2008a). Gadolinium, which inhibits calcium influx into the cell, also prevents egg activation in this system (Horner & Wolfner, 2008a). Thus, like other species, Drosophila requires Ca^{2+} for egg activation and some, or all, of the Ca^{2+} must come from the external environment.

A number of questions still remain regarding how well the mechanisms of Ca^{2+} regulation are conserved between invertebrates and mammals, and between species with sperm-dependent versus sperm-independent triggers of egg activation. What is clear is that for a complete cellular transition,

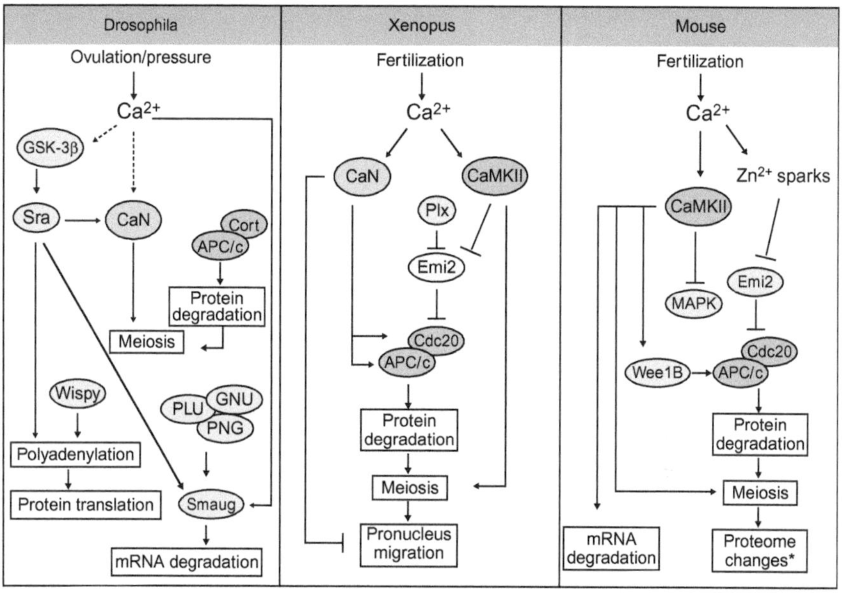

Figure 10.1 Egg activation pathways in Drosophila, Xenopus, and mouse showing only those aspects for which experimental evidence is reported in the literature. See text for references. *Knott et al. (2006) and Backs et al. (2010) show that the protein patterns observed by 2D- and 1D-gel electrophoresis in activated eggs are regulated by CaMKII and attribute these changes to maternal mRNA recruitment and protein synthesis. However, it cannot be ruled out that some of the observed changes are due to posttranslational modifications. As such, we have placed "proteome changes" in the pathway to encompass both of these possibilities. (See Color Insert.)

where coordination of all the changes is key, all of the events are initiated by a single signal. An increase in cytosolic Ca^{2+} is typically the signal that triggers egg activation, after which individual events then appear to be carried out by various Ca^{2+}-binding/Ca^{2+}-dependent proteins (Fig. 10.1).

2.2. Zinc

While a role for Ca^{2+} has long been established in egg activation, the regulation of Zn^{2+} levels during mammalian fertilization has gained attention in recent years. In contrast to Ca^{2+}, where increasing levels are required to activate egg activation pathways, Zn^{2+} levels decrease at egg activation. Upon fertilization or parthenogenetic activation in mouse, zinc is rapidly released from the egg in 1–5 exocytosis events termed "zinc sparks" (Kim et al., 2011). Zinc sparks are also observed in the two nonhuman primate species that have been tested (Kim et al., 2011). Each zinc spark is immediately preceded by a Ca^{2+} rise and none occur when Ca^{2+} rises are blocked by the calcium chelator BAPTA (Kim et al., 2011). This suggests that the release of Zn^{2+} may be dependent on the Ca^{2+} signal. If Zn^{2+} is artificially reduced in the mature oocyte by the zinc chelator TPEN, MAPK kinase activity decreases, cyclin B is degraded, and meiotic arrest is released (Kim et al., 2011; Suzuki, Yoshida, et al., 2010). These events occur in the absence of a Ca^{2+} increase in the oocyte and are unaffected by the addition of BAPTA, once again suggesting that the decrease of Zn^{2+} levels is downstream of the Ca^{2+} signal (Suzuki, Yoshida, et al., 2010). Zn^{2+} sequestration by TPEN appears to be sufficient to induce all the events of egg activation, as eggs fertilized with heat-inactivated sperm heads (that cannot produce a Ca^{2+} signal), and then activated with TPEN, are able to fully develop into viable mice (Suzuki, Yoshida, et al., 2010). Additionally, when zinc levels are increased with the zinc ionophore, ZnPT, mouse oocytes can no longer be parthenogenetically activated with $SrCl_2$ (Bernhardt, Kong, Kim, O'Halloran, & Woodruff, 2012). Thus, release of zinc is not only sufficient but also necessary for the events of egg activation.

Precisely, how Zn^{2+} release triggers exit from meiotic arrest has not yet been worked out. TPEN-induced resumption of the cell cycle is dependent on the APC/c activator Cdc20, suggesting that it requires APC/c activity (Suzuki, Yoshida, et al. 2010). However, TPEN does not induce degradation of the APC/c inhibitor Emi2 (Suzuki, Yoshida, et al., 2010). The Emi2 protein is required to maintain the metaphase II arrest in mature mouse oocytes by inhibiting the APC/c; Emi2 is typically degraded at egg activation to allow cell cycle progression (Schmidt et al., 2005; Shoji et al., 2006; discussed below). The fact that chelation of Zn^{2+} can induce resumption of the cell cycle without degradation of Emi2, along with the fact that Emi2 contains a putative zinc-binding region, has led to the model that zinc binding may directly modulate the activity of Emi2 (Bernhardt et al., 2012). This model is supported by the finding that a single mutation in the zinc-binding region of Emi2 completely inhibits its function (Schmidt et al., 2005). In addition, Zn^{2+} has been shown to affect the activity of the phosphatase Cdc25 and thus its ability to dephosphorylate Cdc2 (also known as Cdk1) (Sun, Chai, Hannigan, Bhogaraju, & Machaca, 2007). Therefore, it is possible that decreases in Zn^{2+} levels also contribute to progression of the cell cycle by reducing the efficiency of Cdc25 and allowing Cdc2 to become phosphorylated and inhibited. Further work is required to determine if these models are correct and to identify whether additional proteins are modulated by Zn^{2+} levels in the oocyte.

3. CHANGES IN PROTEOME COMPOSITION

Large-scale changes in the egg's proteome take place at egg activation that drastically alter the cellular properties from those of the mature oocyte. Because there is little or no transcription associated with the oocyte-to-embryo transition, changes in protein composition within the cell rely on posttranscriptional and posttranslational methods of regulation. Studies have been undertaken to characterize the proteome of oocytes and/or embryos in multiple organisms including mouse, frog, sea urchin, and Drosophila; these studies provide an overview of the dynamic changes that occur during egg activation. In the mouse, over 3500 proteins have been identified in metaphase II oocytes and approximately 2000 have been found in zygotes (Pfeiffer et al., 2011; Wang et al., 2010; reviewed in Yurttas, Morency, & Coonrod, 2010). This discrepancy in the number of proteins identified between oocytes and zygotes is suggestive of significant degradation of maternal proteins (Wang et al., 2010). A functional role for this degradation is seen in sea urchins, where inhibition of the proteasome blocks proper entry into the first mitosis following fertilization (Kawahara et al., 2000). In agreement with this result, two-dimensional gel analysis of unfertilized and fertilized sea urchin eggs shows a 23% decrease in the number of detectable spots 2 min postfertilization (Roux et al., 2006). This is likely due to a combination of protein degradation as well as posttranslational modifications as the same study showed that approximately 30% of the proteins in unfertilized eggs are phosphorylated (Roux et al., 2006). It is the overall combination of degradation and posttranslational modification of existing maternal proteins, and new protein translation from maternal mRNAs, that transitions an oocyte into an embryo.

3.1. Protein degradation

As mentioned above, substantial protein degradation is observed upon egg activation. One important role for this degradation is the regulation of the meiotic cell cycle by the APC/c, an E3 ubiquitin ligase that targets proteins for degradation by the 26S proteosome (reviewed in McLean, Chaix, Ohi, & Gould, 2011). Targets of the APC/c include securin and cyclins, which hold the mature oocyte in metaphase arrest until egg activation. As there are several recent reviews on the function and regulation of the APC/c (McLean et al., 2011; Pesin & Orr-Weaver, 2008; Verlhac, Terret, & Pintard, 2010), we will only mention the key points here.

The APC/c is regulated by the adaptor protein Cdc20, which acts as an activator of the APC/c and confers substrate specificity. While the role of Cdc20 in mitosis is well studied, its role in meiosis remains more elusive (reviewed in Pesin & Orr-Weaver, 2008). However, in Drosophila, a meiosis-specific Cdc20, Cortex, has been identified that is required for cyclin degradation and completion of meiosis during egg activation (Chu, Henrion, Haegeli, & Strickland, 2001; Page & Orr-Weaver, 1996; Pesin & Orr-Weaver, 2007; Swan & Schupbach, 2007). Cortex is also responsible for regulating its own degradation by the APC/c following egg activation, presumably to allow the canonical Cdc20 (Fizzy) to regulate the subsequent embryonic mitoses (Pesin & Orr-Weaver, 2007). Continuing studies in Drosophila will provide a unique model system for understanding how Cdc20 is regulated to activate the metaphase-to-anaphase transition during egg activation.

In Xenopus and mice, a novel mechanism for regulating the APC/c during egg activation was found through the identification of Emi2/XErp1 as a necessary protein for maintaining metaphase arrest (Schmidt et al., 2005; Shoji et al., 2006). Emi2 inhibits the APC/c in metaphase-arrested oocytes/egg extracts, evidenced by the fact that removal of Emi2 results in precocious metaphase exit, while overexpression of Emi2 prevents APC/c activation upon addition of $SrCl_2$ or Ca^{2+} (Schmidt et al., 2005; Shoji et al., 2006). Upon egg activation in Xenopus, Emi2 protein is degraded by the Skp1–Cullin–F-box (SCF) ubiquitin ligase complex, showing that at least one other degradation pathway besides the APC/c is active during egg activation (Schmidt et al., 2005). This degradation is believed to be dependent on phosphorylation by Polo-like kinase 1 (Plx1). Two Plx1 phosphorylation sites are present within the $DSGX_3S$ motif required for Emi2 degradation, and mutation of either of these sites reduces Emi2 degradation *in vivo* and Plx1 phosphorylation of Emi2 *in vitro* (Schmidt et al., 2005). While Emi2 is also degraded upon egg activation in the mouse, a similar mechanism of phosphorylation by Plx1, and degradation by the SCF, has not been demonstrated (Shoji et al., 2006). In addition, though Emi2 clearly inhibits the APC/c, the exact mechanism of this inhibition is not yet known. Emi2 has been proposed to interact with, and inhibit, both Cdc20 and the core APC/c (Shoji et al., 2006; Wu et al., 2007). Finally, it is possible that once activated, the APC/c can contribute to the inactivation of Emi2. Studies in Xenopus show that overexpression of the APC/c-cooperating E2 enzyme leads to the ubiquitylation of Emi2 and its disassociation from the APC/c (Hormanseder, Tischer, Heubes, Stemmann, & Mayer, 2011).

Although the APC/c clearly plays a critical role in protein degradation during egg activation, it is not known to what extent other pathways contribute to the global degradation that is observed. In Xenopus, action of the SCF is implicated by its role in Emi2 degradation, but whether it has any additional targets at this time remains to be investigated. Additionally, autophagy [a bulk degradation system of the cell (reviewed in Mizushima, 2007)] is highly induced in early mouse embryos and is required for preimplantation development (Tsukamoto, Kuma, & Mizushima, 2008; Tsukamoto, Murakami, et al., 2008). A full understanding of the role of protein degradation, and the ways in which individual pathways regulate unique or overlapping sets of targets, will require a more complete characterization of the proteome after egg activation and how inhibition of each pathway affects global changes in protein composition in the egg.

3.2. Protein phosphorylation

Multiple types of posttranslational modification are likely to act on the proteome during egg activation. For instance, maternal sumoylation is important for eggshell patterning and mitosis during Drosophila embryogenesis (Nie, Xie, Loo, & Courey, 2009). However, the main focus thus far has been on the role of phosphorylation and that is where we will also focus this review. In sea urchins, fertilization causes the number of phosphorylated proteins to double within 2 min, before returning to prefertilization levels 30 min later (Roux et al., 2006). In Xenopus, phosphoproteomic analysis has identified a combined total of 654 phosphoproteins in oocytes, unfertilized eggs, and embryos (McGivern, Swaney, Coon, & Sheets, 2009). And in Drosophila, we have identified 311 proteins that change in phosphorylation state between mature oocytes and unfertilized, activated eggs (Krauchunas, Horner, & Wolfner, 2012).

Phosphorylation can be a major contributor to a change in cellular state as it has the ability to rapidly modulate a large number of proteins and has a wide array of regulatory affects [i.e., altering protein activity (Cargnello & Roux, 2011) or localization (Jans & Hubner, 1996)]. The importance of phosphorylation during egg activation is evidenced by the dynamic phosphorylation changes that take place during the oocyte-to-embryo transition, and the finding that phosphorylation modulators such as Ca^{2+}/calmodulin-dependent protein kinase II (CaMKII) and calcineurin (discussed below) are critical for successful egg activation.

3.2.1 CaMKII

CaMKII is a serine/threonine protein kinase that is regulated by calcium and the calcium receptor molecule calmodulin (reviewed in Hudmon & Schulman, 2002). Because of the relationship between Ca^{2+} and CaMKII, and the importance of Ca^{2+} to initiate egg activation, CaMKII has been proposed to be one of the main effectors of the Ca^{2+} signal at egg activation. Numerous studies support this hypothesis, pointing primarily to a key role for CaMKII in the resumption of the cell cycle.

An increase in CaMKII activity occurs at the time of fertilization/egg activation that correlates with the observed increase in Ca^{2+} levels in both Xenopus and mouse. In Xenopus, where one Ca^{2+} spike occurs, CaMKII activity transiently spikes in the first 5 min after calcium addition (Liu & Maller, 2005; Nishiyama, Yoshizaki, Kishimoto, & Ohsumi, 2007). In mice, the activity of CaMKII is oscillatory, once again correlating with the pattern of Ca^{2+} rises in the egg (Johnson, Bierle, Gallicano, & Capco, 1998; Markoulaki, Matson, Abbott, & Ducibella, 2003; Markoulaki, Matson, & Ducibella, 2004; Tatone et al., 2002). When Ca^{2+} oscillations are produced by fertilization, or artificially induced, CaMKII activity is seen to peak at the time of Ca^{2+} peaks followed by a return to baseline activity levels; these increases in kinase activity can be observed for up to 1 h postinsemination (Markoulaki et al., 2003, 2004). In contrast, if mouse oocytes are activated by ethanol, which produces only a single Ca^{2+} rise, CaMKII activity increases only once, showing that the oscillations of CaMKII activity are dependent on the oscillations of Ca^{2+} levels in the egg (Tatone et al., 2002). When CaMKII is inhibited in Xenopus or mouse oocytes, cyclin B fails to be degraded, Cdc2 remains active, and the cell cycle remains arrested in metaphase (Johnson et al., 1998; Lorca et al., 1993; Madgwick, Levasseur, & Jones, 2005; Markoulaki et al., 2003; Morin, Abrieu, Lorca, Martin, & Dorée, 1994; Tatone, Iorio, Francione, Gioia, & Colonna, 1999). If a constitutively active form of CaMKII is added to the oocyte, cyclin B degradation is induced, Cdc2 and MAPK levels decrease, and meiosis resumes (Knott et al., 2006; Lorca et al., 1993; Madgwick et al., 2005; Morin et al., 1994).

CaMKII exists in four different isoforms—α, β, δ, and γ. In mouse oocytes, two splice forms of CaMKIIγ are predominantly expressed, while the other isoforms are present at low or undetectable levels (Backs et al., 2010; Chang, Minahan, Merriman, & Jones, 2009; Suzuki, Suzuki, et al., 2010). In CaMKIIγ conditional knockout mice, or when CaMKIIγ is knocked down by an antisense morpholino, eggs remain arrested in metaphase and MAPK and Cdc2 kinase levels remain high (Backs et al., 2010; Chang et al., 2009). Injection of CaMKII cDNA rescues this phenotype. Rescue is not isoform specific, suggesting that it is the general activity of CaMKII that is required for meiotic resumption rather than any CaMKIIγ-specific targets (Backs et al., 2010). In addition, characteristic proteome changes that occur at egg activation, observed by either 2D- or 1D-gel electrophoresis, are not seen in the CaMKIIγ mutants (Backs et al., 2010), though they can be triggered by addition of constitutively active CaMKII (Knott et al., 2006). However, these proteome changes can also be induced by a Ca^{2+} signal and direct inactivation of Cdc2 kinase, even when CaMKII is not present (Backs et al., 2010). Thus, these observed changes in proteome composition appear to rely on resumption of the cell cycle but are only indirectly regulated by CaMKII.

Two mechanisms for how CaMKII initiates the metaphase to anaphase transition have been elucidated in Xenopus. The first comes from the finding that CaMKII is both necessary and sufficient for Emi2 degradation in Xenopus egg extracts (Hansen, Tung, & Jackson, 2006; Rauh, Schmidt, Bormann, Nigg, & Mayer, 2005). As mentioned earlier, in Xenopus, Plx1 phosphorylates Emi2 to target it for degradation (Schmidt et al., 2005). Removing Plx1 prevents release from the metaphase arrest, even when constitutively active CaMKII is added (Liu & Maller, 2005). CaMKII is equally required for Plx1 to induce resumption of the cell cycle (Liu & Maller, 2005). Additionally, while expression of either CaMKII or Plx1 alone cannot overcome the arrest caused by Emi2 overexpression, a combination of both kinases can (Liu & Maller, 2005). This suggests that the kinases work together to regulate Emi2 degradation and the subsequent exit from meiotic arrest. *In vitro* experiments show that Emi2 is a good substrate for CaMKII and that Emi2 interacts with, and is phosphorylated by, Plx1 in a CaMKII-dependent manner (Hansen et al., 2006; Liu & Maller, 2005; Rauh et al., 2005). Thus, the current model is that when CaMKII is activated by the Ca^{2+} rise at egg activation, it phosphorylates Emi2, which primes Emi2 for subsequent phosphorylation by Plx1, eventually leading to its degradation (Hansen et al., 2006; Liu & Maller, 2005; Rauh et al., 2005). CaMKII also promotes microtubule destabilization at the onset of anaphase (Reber, Over, Kronja, & Gruss, 2008). This function is independent of APC/c activation, suggesting that CaMKII works at multiple levels to regulate progression of the cell cycle at egg activation.

Whether a similar mechanism for the role of CaMKII in Emi2 degradation or microtubule destabilization exists in mammals has not yet been established. However, in mouse, the kinase Wee1B has been shown to be a CaMKII target at egg activation. Knockdown of Wee1B in oocytes results in a failure to completely activate the APC/c and in high levels of MAPK in parthenogenetically activated eggs, preventing exit from metaphase. (Oh, Susor, & Conti, 2011). In addition, Wee1B knockdown reduces the ability of constitutively active CaMKII to induce polar body formation (Oh et al., 2011). *In vitro* experiments show that Wee1B can be phosphorylated by CaMKII, and that this phosphorylation increases Wee1B activity (Oh et al., 2011). Thus, in mouse, it has been shown that CaMKII regulates resumption of the cell cycle through phosphorylation of Wee1B, which in turn phosphorylates and inhibits Cdc2, leading to the inhibition of MPF activity and the transition from metaphase to anaphase.

3.2.2 Calcineurin

Calcineurin is a Ca^{2+}/calmodulin-dependent serine/threonine phosphatase composed of a catalytic A subunit (CnA) and regulatory B subunit (Rusnak & Mertz, 2000). A role for calcineurin in egg activation was first suggested by studies in Drosophila which showed that the calcineurin regulator, Sarah (Sra), is necessary for multiple aspects of egg activation (Horner et al., 2006; Takeo, Tsuda, Akahori, Matsuo, & Aigaki, 2006). Eggs laid by *sra* mutant females have defects in the polyadenylation and translation of maternal mRNAs and defects in protein degradation (Horner et al., 2006). In addition, while meiosis briefly resumes in eggs laid by *sra* mutants, it rearrests at anaphase I, perhaps due to the high levels of cyclin B in these eggs (Horner et al., 2006; Takeo et al., 2006).

The absence of these egg activation events in eggs laid by *sra* mutants is presumed to be due to misregulation of calcineurin when Sra is not present. This hypothesis is supported by both genetic and biochemical assays. Three calcineurin subunits are present in Drosophila oocytes and early embryos: two A subunits (Pp2B-14D and CanA-14F) and one B subunit (CanB2) (Takeo et al., 2006). Sra has been shown to interact with all three of these subunits, both in cell culture and *in vivo* by mass spectrometry experiments (Takeo et al., 2006, 2012). In addition, genetic interaction studies showed that the phenotype caused by expression of a dominant negative CnA is rescued by Sra overexpression and enhanced in *sra* heterozygotes (Takeo, Hawley, & Aigaki, 2010). Thus, it is proposed that Sra is a positive regulator of calcineurin function during egg activation.

A direct role for calcineurin in Drosophila egg activation has been shown through germline clonal analysis of both CanB2 mutants and CnA double mutants. In both cases, elimination of calcineurin function from the oocyte/embryo led to an anaphase I arrest identical to that seen in *sra* mutants (Takeo et al., 2010, 2012). As the CnA double mutants can be rescued by expression of wild-type Pp2B-14D, but not by a phosphatase-dead version of the protein, the phosphatase activity of calcineurin is required for its role in egg activation in Drosophila (Takeo et al., 2012). What remains to be determined is whether the defect in polyadenylation and translation seen in *sra* mutants also occurs in calcineurin mutants, or if *sra* has additional functions beyond the regulation of calcineurin. If calcineurin is indeed necessary for these events, it provides a link between the Ca^{2+} signal and maternal mRNA translation. However, the possible calcineurin target(s) regulating this event remain unidentified.

A role for calcineurin in egg activation has also been observed in Xenopus (Mochida & Hunt, 2007; Nishiyama et al., 2007). Addition of Ca^{2+} to Xenopus egg extracts causes a transient increase in calcineurin activity (Mochida & Hunt, 2007; Nishiyama et al., 2007). This transient activity occurs along the same time frame as the spike in CaMKII activity that is seen upon addition of calcium, and precedes the degradation of cyclin B (Nishiyama et al., 2007). When calcineurin is inhibited, Cdc2 activity remains high and eggs/extracts are not able to exit metaphase II (Nishiyama et al., 2007). Immunodepletion of calcineurin confirms these results, showing a failure to degrade both cyclin B and securin (Nishiyama et al., 2007). Similar to the results in Drosophila, adding back constitutively active calcineurin reverses this phenotype but a phosphatase-dead version of the protein does not (Nishiyama et al., 2007). Surprisingly, constitutively active calcineurin reduces the size of sperm asters formed around demembranated sperm added to Ca^{2+}-activated Xenopus egg extracts. This is in contrast to the situation in Drosophila, where sperm asters fail to form in eggs laid by *sra* mutants (Horner et al., 2006). Thus, at least in Xenopus, it appears that it is important that calcineurin activated only be transiently.

The necessity of calcineurin for cyclin B degradation and metaphase II exit leads to the hypothesis that calcineurin is involved in the regulation of the APC/c. In Xenopus, it appears that calcineurin does directly regulate components of the APC/c (Mochida & Hunt, 2007). Two APC/c subunits, Apc3 (a core subunit) and Fizzy (Cdc20), are phosphoregulated during the cell cycle (Chung & Chen, 2003; Peters, King, Hoog, & Kirschner, 1996), and both proteins undergo dephosphorylation when calcium is added to egg extracts (Mochida & Hunt, 2007). However, the calcineurin inhibitor Cyclosporin A prevents this dephosphorylation from occurring (Mochida & Hunt, 2007). The ability of calcineurin to dephosphorylate immunopurified Apc3 and Fizzy suggests that these proteins are direct targets of the phosphatase (Mochida & Hunt, 2007). Whether calcineurin also contributes to the regulation of the APC/c through Emi2 is less clear. While Nishiyama et al. (2007) report that inhibiting calcineurin reduces the degradation of Emi2, Mochida and Hunt (2007) find that Emi2 dephosphorylation and degradation is unaffected in the presence of Cyclosporin A. Almost all the Ca^{2+}-induced dephosphorylations that can be observed with an anti-(phospho-Ser-Pro) antibody are prevented by Cyclosporin A, suggesting that there are many more calcineurin targets that remain to be identified (Mochida & Hunt, 2007). However, how many of these dephosphorylations are direct targets and how many are indirect effects due to misregulation of the APC/c is unknown.

As with CaMKII, calcineurin appears to be a key regulator activated by the Ca^{2+} signal that triggers egg activation. The activation of calcineurin is independent of CaMKII activity, and vice versa (Nishiyama et al., 2007). However, the transient activity of both proteins is required for metaphase II exit in Xenopus (Nishiyama et al., 2007). This appears to contrast with the findings that addition of constitutively active CaMKII alone can induce resumption of the cell cycle. However, in those experiments, the activity of CaMKII is not transient, the way it is under fertilization or Ca^{2+}-induced activation. When constitutively active CaMKII is inactivated 25 min after its addition to Xenopus egg extracts, the inactivation of Cdc2 is no longer observed unless active calcineurin is also added (Nishiyama et al., 2007). In addition, the results from Xenopus show that calcineurin activity and the appropriately timed inhibition of calcineurin are both required for successful egg activation and the initiation of embryogenesis. In Drosophila, the regulation of calcineurin itself is more complex than simply direct activation by Ca^{2+}, as Sra is also required for calcineurin activity. Additionally, Sra itself is regulated through phosphorylation by glycogen synthase kinase 3β (GSK-3β) (Takeo et al., 2012). Whether similar regulators of calcineurin are also present in Xenopus oocytes, or if GSK-3β plays a role in egg activation of other organisms remains to be tested.

At present, there is no evidence to show that calcineurin plays an analogous role in mammalian egg activation. Suzuki, Suzuki, et al. (2010) failed to detect the CnA subunit in mouse oocytes, and although ICSI-induced metaphase II exit was slowed by the addition of calcineurin inhibitors, it was not prevented. More experimentation is required to conclusively rule out any contributions that calcineurin may make to egg activation in mammals, including testing Cyclosporin A and other inhibitors when eggs are activated by Ca^{2+} (rather than ICSI). Thorough investigation is also needed into what, if any, calcineurin subunits are present in mammalian oocytes, and the effects of knockout of those proteins.

3.2.3 Other phosphorylation regulators

While CaMKII and calcineurin have been identified as two of the most upstream regulators that transduce the Ca^{2+} signal and initiate multiple aspects of egg activation, other phosphorylation regulators have also been implicated in egg activation events and early embryogenesis. Polo-like kinase, GSK-3, Cdc25, and Wee1B have all been mentioned above. Pan Gu (PNG) is a serine/threonine kinase in Drosophila that complexes with two other proteins, Giant Nuclei (GNU) and Plutonium (PLU), to function in the oocyte-to-embryo transition (Lee, Van Hoewyk, & Orr-Weaver, 2003). Both GNU and PLU can be phosphorylated by the PNG kinase complex *in vitro* and GNU is dephosphorylated upon egg activation, suggesting that function of the complex is itself phosphoregulated (Lee et al., 2003; Renault et al., 2003). However, other phosphorylation targets of PNG and the importance of GNU dephosphorylation remain to be determined.

In *C. elegans*, one critical phosphorylation regulator is the kinase MBK-2, which regulates maternal protein degradation during the oocyte-to-embryo transition (Pellettieri, Reinke, Kim, & Seydoux, 2003). MBK-2 is regulated by the pseudophosphatases EGG-3, EGG-4, and EGG-5 (Cheng, Klancer, Singson, & Seydoux, 2009; Maruyama et al., 2007; Parry et al., 2009; reviewed in Parry & Singson, 2011). Some of the targets of MBK-2 have been identified and include major regulators of the oocyte-to-embryo transition (reviewed in Marcello & Singson, 2010). At least one role of MBK-2 phosphorylation appears to be providing a priming phosphorylation for subsequent modification by an additional kinase, as MBK-2 targets have been shown to be further phosphorylated by GSK-3 and polo kinases (Nishi & Lin, 2005; Nishi, Rogers, Robertson, & Lin, 2008).

Other kinases that have been implicated in the oocyte-to-embryo transition of various species include protein kinase C, myosin light chain kinase, and MAPKs. However, in many cases, the roles of these kinases at egg activation are less clear than the present understanding of those of CaMKII and calcineurin. The evidence for potential functions of these kinases as they relate to the events of egg activation are reviewed elsewhere (Ducibella & Fissore, 2008).

3.3. New protein translation

Some of the proteins necessary for the oocyte-to-embryo transition and early embryogenesis are not present in the mature oocyte. Instead, these proteins are translated from maternal mRNAs that had been loaded into the oocyte and held in a repressed state until egg activation. Analysis of the protein classes that are required in the early embryo, as well as the function of specific proteins encoded by the maternally deposited mRNAs, will inform us of the cellular events that are critical at this time and help identify potential regulators of those processes.

One way to determine the full set of proteins that are translated in the activated egg is by analyzing the transcripts that are associated with the polysome. This analysis has been carried out in mouse oocytes and in one-cell embryos to determine which maternal mRNAs are recruited to the polysome after egg activation (Potireddy, Vassena, Patel, & Latham, 2006). Nearly 2000 individual mRNAs were found to have greater signals in the one-cell polysomal RNA population when compared to the oocyte polysomal RNA population (Potireddy et al., 2006). These RNAs were enriched for functional classes related to metabolism, transcription, and cell cycle regulation, consistent with the idea that the proteins translated from maternal mRNAs are important for early embryonic mitosis and activation of the zygotic genome (Potireddy et al., 2006).

Proteomic studies comparing mouse oocytes and zygotes also provide information on protein translation at egg activation. While the overall number of proteins identified in zygotes was lower than the number of proteins in oocytes (Wang et al., 2010), a comparison of the identities of the proteins in each sample should reveal proteins that are present only in the zygote, indicating that they are not translated until egg activation. It will also be of interest to compare the polysomal mRNA and proteomic data sets to determine how well polysomal recruitment correlates with identification of proteins in the embryo and/or the level of saturation reached by either of these screens.

In addition to determining the identities of these proteins, another important aspect of protein translation at egg activation is understanding how maternal transcripts are regulated to ensure they are translated at the appropriate time. One mechanism by which the translation of these transcripts is regulated is through polyadenylation of the 3′-untranslated regions (UTR). During oocyte maturation in Xenopus, this process is regulated by opposing polymerase and deadenylase activities (Kim & Richter, 2006). The polymerase and deadenylase are both bound to the transcript through the cytoplasmic polyadenylation element binding protein (CPEB) and the cleavage and polyadenylation specificity factor (CPSF), which recognize *cis*-elements in the 3′-UTR (Kim & Richter, 2006). When maturation is signaled, CPEB is phosphorylated and the deadenylase subsequently disassociates from the transcript. As only the poly-A polymerase remains, the result is a lengthening of the poly-A tail. It is possible that a similar mechanism works during egg activation to regulate the polyadenylation and translation that occurs at this time.

In Drosophila, the maternal protein Wispy is a cytoplasmic poly-A polymerase that is responsible for the polyadenylation of numerous targets during both oocyte maturation and egg activation (Benoit, Papin, Kwak, Wickens, & Simonelig, 2008; Cui, Sackton, Horner, Kumar, & Wolfner, 2008). Targets of Wispy include Bicoid and Torso, which are important for patterning the early embryo and have been previously shown to be translated upon egg activation (Casanova & Struhl, 1989; Driever & Nusslein-Volhard, 1988). In *wispy* mutants, the *bicoid* and *torso* transcripts fail to be polyadenylated in activated eggs; for *bicoid*, this has been shown to result in a failure to translate the Bicoid protein (Benoit et al., 2008; Cui et al., 2008). However, lengthening of the poly-A tail alone is not sufficient for translation, as elongation of the *bicoid* poly-A tail by overexpression of a poly-A polymerase in the oocyte does not result in Bicoid protein expression in oocytes (Juge, Zaessinger, Temme, Wahle, & Simonelig, 2002). Thus, it appears that protein translation at egg activation not only requires the lengthening of maternal mRNA poly-A tails, but either an additional activational pathway or the repression of an inhibitory pathway that is present in the mature oocyte.

Wispy is not necessary for all protein translation during egg activation, as translation of the Smaug protein is unaffected in *wispy* mutants (Benoit et al., 2008; Cui et al., 2008). Smaug is important for degradation of maternal transcripts after egg activation (Tadros et al., 2007; discussed below). To ensure that maternal transcripts are protected from Smaug-mediated degradation prior to egg activation, Smaug protein is not present in the mature oocyte (Dahanukar, Walker, & Wharton, 1999; Smibert, Lie, Shillinglaw, Henzel, & Macdonald, 1999). Instead, maternal *smaug* mRNA is actively translated at egg activation; a process dependent on both *pan gu* and *sra* (Cui et al., 2008; Tadros et al., 2007). When *pan gu* is mutated, *smaug* mRNA polyadenylation is also reduced (Tadros et al., 2007). Overexpression of a poly-A polymerase can rescue the poly-A tail phenotype of the *pan gu* mutant; however, Smaug protein is still not translated, suggesting that Pan Gu works at multiple levels to control the translation of Smaug (Tadros et al., 2007). Pan Gu is also required for the translation of cyclin B in the early Drosophila embryo (Vardy & Orr-Weaver, 2007). In the case of cyclin B, loss of Pumilio (a translational repressor) can restore cyclin B translation in a *pan gu* mutant background suggesting that at least one additional function of Pan Gu is to inhibit Pumilio (Vardy & Orr-Weaver, 2007). However, Smaug translation is not restored in *pumilio; pan gu* double mutants, indicating that additional mechanisms of Pan Gu translational regulation still remain to be discovered (Tadros et al., 2007).

4. mRNA DEGRADATION

The mature oocyte is stocked full of maternal mRNAs that were transcribed and loaded into the oocyte during oogenesis. In Drosophila, it is estimated that up to 55% of the genes in the genome are represented in maternal transcripts in the oocyte (Tadros et al., 2007). For the transition to a totipotent cell (the embryo), the transcriptional profile of the differentiated cell (the oocyte) must be removed. Thus, while some of the maternal mRNAs are translated in the early embryo, another key facet of egg activation is the targeted degradation of other maternal mRNAs.

Maternal transcript degradation in the early embryo has been observed in a number of species including mouse, zebrafish, and Drosophila (Hamatani, Carter, Sharov, & Ko, 2004; Mathavan et al., 2005; Tadros et al., 2007; Thomsen, Anders, Janga, Huber, & Alonso, 2010). In Drosophila, egg activation is both necessary and sufficient for this maternal mRNA degradation, at least for the specific transcripts examined so far (Bashirullah et al., 1999; Tadros et al., 2003). While the mechanistic connection between the Ca^{2+} signal and maternal mRNA degradation remains elusive, studies in Drosophila provide some insight into the regulation of this process. Here, the maternal protein Smaug is required for two-thirds of the transcript degradation that is observed upon egg activation (Tadros et al., 2007). Studies of two known Smaug targets, *nanos* and *Hsp83*, show that Smaug acts by binding to *cis*-elements, known as Smaug recognition elements, in maternal transcripts and recruiting the CCR4/POP2/NOT-deadenylase complex (Semotok et al., 2005, 2008; Smibert, Wilson, Kerr, & Macdonald, 1996). This leads to the deadenylation of these transcripts and their subsequent degradation. In addition, it has been proposed that Smaug interacts with the piRNA pathway to promote the deadenylation and degradation of a subset of maternal transcripts (Rouget et al., 2010).

Once the zygotic genome is activated, a second wave of maternal transcript degradation is activated which relies on zygotic, as well as maternal products (Bashirullah et al., 1999). We will not discuss this second round of degradation, and instead, refer you to two comprehensive reviews on the maternal and zygotic regulation of mRNA degradation during egg activation and the maternal-to-zygotic transition by Tadros and Lipshitz (2005, 2009), and the role of microRNAs in maternal transcript degradation by Giraldez (2010). These authors also discuss the possible biological consequences of maternal transcript degradation, presenting potential roles in producing localized expression of RNA and proteins or acting as a permissive phenomenon to avoid interfering with zygotic transcripts.

5. SUMMARY

Egg activation represents a cellular transition from a highly differentiated oocyte into a totipotent embryo. There is a single initiating signal at the start of this transition, which ensures that all of the downstream signals and events are properly coordinated within the cell. This is especially important considering each event is regulated at multiple levels. During egg activation, a rise in cytosolic Ca^{2+} independently activates the effector proteins CaMKII and calcineurin, and at least one other pathway, to initiate this cellular transition. [At least one other pathway must be involved as cortical granule exocytosis occurs in the absence of both calcineurin and CaMKII, and CaMKII only partially contributes to the membrane block to polyspermy (Backs et al., 2010; Chang et al., 2009; Gardner, Knott, Jones, & Evans, 2007; Mochida & Hunt, 2007; Nishiyama et al., 2007).] These effectors then have the potential to amplify the initial signal through phosphorylation of multiple target proteins, activating multiple pathways within the cell (Fig. 10.1).

Because the oocyte-to-embryo transition represents a complete change in cellular state, it requires global changes in the composition of both RNAs and proteins within the cell. Indeed, large-scale degradation of maternal mRNAs and proteins, translation of new proteins not present in the oocyte, and protein phosphorylation changes occur during egg activation. To fully understand how the embryo differs from the oocyte will require a complete identification of the molecules that are removed (degraded) and the molecules that take their place. In addition, a consistent trend in the literature is the proposal that finding the targets of the phosphorylation regulators CaMKII and calcineurin will be critical to improving our understanding of how these molecules affect the cellular changes that encompass egg activation. Identifying the proteins that are phosphorylated/dephosphorylated during egg activation will provide new proteins and pathways to explore in connecting Ca^{2+} to all of the egg activation events. Xenopus is the only species in which CaMKII and calcineurin are both known to be critical for egg activation. At present, there is no evidence of a role for calcineurin in mouse egg activation, or for CaMKII in Drosophila egg activation. However, the possibility that roles for both pathways are conserved has not been definitively excluded and remains an important area for future studies.

Our current knowledge of egg activation shows that overlapping regulatory mechanisms ensure these cellular changes are tightly controlled. For example, Emi2 has been shown to be regulated by phosphorylation and subsequent degradation, ubiquitination, and Zn^{2+} levels. Following that, Emi2 is only one regulator of the APC/c, which is also regulated by the activator protein Cdc20 and phosphorylation of other core subunits. All of this works together so that the APC/c can activate the metaphase-to-anaphase transition, while at the same time activation of Wee1B kinase phosphorylates and inhibits Cdc2 (which is necessary for the metaphase arrest). Presumably, in an activated egg, all of these regulatory mechanisms are coordinated by proteins downstream of the Ca^{2+} signal to ensure the efficient transition from meiotic arrest to embryonic mitosis. Understanding how these regulatory mechanisms interrelate, or if they simply represent redundancy within the system, will be critical to our success in understanding this type of cellular transition, as well as manipulating it therapeutically.

ACKNOWLEDGMENTS

We thank Paul Wassarman for the opportunity to write this review, J. Bird, B. LaFlamme, M. Goldberg, A. Clark, and F. Avila for comments on the manuscript, and NIH grant R01-GM044659 (to M. F. W.) for support.

REFERENCES

Backs, J., Stein, P., Backs, T., Duncan, F. E., Grueter, C. E., McAnally, J., et al. (2010). The gamma isoform of CaM kinase II controls mouse egg activation by regulating cell cycle resumption. *Proceedings of the National Academy of Sciences of the United States of America, 107*, 81–86.

Bashirullah, A., Halsell, S. R., Cooperstock, R. L., Kloc, M., Karaiskakis, A., Fisher, W. W., et al. (1999). Joint action of two RNA degradation pathways controls the timing of maternal transcript elimination at the midblastula transition in Drosophila melanogaster. *EMBO Journal, 18*, 2610–2620.

Benoit, P., Papin, C., Kwak, J. E., Wickens, M., & Simonelig, M. (2008). PAP- and GLD-2-type poly(A) polymerases are required sequentially in cytoplasmic polyadenylation and oogenesis in Drosophila. *Development, 135*, 1969–1979.

Bernhardt, M. L., Kong, B. Y., Kim, A. M., O'Halloran, T. V., & Woodruff, T. K. (2012). A zinc-dependent mechanism regulates meiotic progression in mammalian oocytes. *Biology of Reproduction, 86*, 114.

Cargnello, M., & Roux, P. P. (2011). Activation and function of the MAPKs and their substrates, the MAPK-activated protein kinases. *Microbiology and Molecular Biology Reviews, 75*, 50–83.

Casanova, J., & Struhl, G. (1989). Localized surface activity of torso, a receptor tyrosine kinase, specifies terminal body pattern in Drosophila. *Genes & Development, 3*, 2025–2038.

Chang, H. Y., Minahan, K., Merriman, J. A., & Jones, K. T. (2009). Calmodulin-dependent protein kinase gamma 3 (CamKIIgamma3) mediates the cell cycle resumption of metaphase II eggs in mouse. *Development, 136*, 4077–4081.

Cheng, K. C., Klancer, R., Singson, A., & Seydoux, G. (2009). Regulation of MBK-2/DYRK by CDK-1 and the pseudophosphatases EGG-4 and EGG-5 during the oocyte-to-embryo transition. *Cell*, *139*, 560–572.

Chu, T., Henrion, G., Haegeli, V., & Strickland, S. (2001). Cortex, a Drosophila gene required to complete oocyte meiosis, is a member of the Cdc20/fizzy protein family. *Genesis*, *29*, 141–152.

Chung, E., & Chen, R. H. (2003). Phosphorylation of Cdc20 is required for its inhibition by the spindle checkpoint. *Nature Cell Biology*, *5*, 748–753.

Cui, J., Sackton, K. L., Horner, V. L., Kumar, K. E., & Wolfner, M. F. (2008). Wispy, the Drosophila homolog of GLD-2, is required during oogenesis and egg activation. *Genetics*, *178*, 2017–2029.

Dahanukar, A., Walker, J. A., & Wharton, R. P. (1999). Smaug, a novel RNA-binding protein that operates a translational switch in Drosophila. *Molecular Cell*, *4*, 209–218.

Dal Santo, P., Logan, M. A., Chisholm, A. D., & Jorgensen, E. M. (1999). The inositol trisphosphate receptor regulates a 50-second behavioral rhythm in C. elegans. *Cell*, *98*, 757–767.

Driever, W., & Nusslein-Volhard, C. (1988). A gradient of bicoid protein in Drosophila embryos. *Cell*, *54*, 83–93.

Ducibella, T., & Fissore, R. (2008). The roles of Ca2+, downstream protein kinases, and oscillatory signaling in regulating fertilization and the activation of development. *Developmental Biology*, *315*, 257–279.

Ducibella, T., Huneau, D., Angelichio, E., Xu, Z., Schultz, R. M., Kopf, G. S., et al. (2002). Egg-to-embryo transition is driven by differential responses to Ca(2+) oscillation number. *Developmental Biology*, *250*, 280–291.

Ducibella, T., Schultz, R. M., & Ozil, J. P. (2006). Role of calcium signals in early development. *Seminars in Cell & Developmental Biology*, *17*, 324–332.

Gadella, B. M., & Evans, J. P. (2011). Membrane fusions during mammalian fertilization. *Advances in Experimental Medicine and Biology*, *713*, 65–80.

Gardner, A. J., Knott, J. G., Jones, K. T., & Evans, J. P. (2007). CaMKII can participate in but is not sufficient for the establishment of the membrane block to polyspermy in mouse eggs. *Journal of Cellular Physiology*, *212*, 275–280.

Giraldez, A. J. (2010). microRNAs, the cell's Nepenthe: Clearing the past during the maternal-to-zygotic transition and cellular reprogramming. *Current Opinion in Genetics & Development*, *20*, 369–375.

Hamatani, T., Carter, M. G., Sharov, A. A., & Ko, M. S. (2004). Dynamics of global gene expression changes during mouse preimplantation development. *Developmental Cell*, *6*, 117–131.

Hansen, D. V., Tung, J. J., & Jackson, P. K. (2006). CaMKII and polo-like kinase 1 sequentially phosphorylate the cytostatic factor Emi2/XErp1 to trigger its destruction and meiotic exit. *Proceedings of the National Academy of Sciences of the United States of America*, *103*, 608–613.

Hormanseder, E., Tischer, T., Heubes, S., Stemmann, O., & Mayer, T. U. (2011). Non-proteolytic ubiquitylation counteracts the APC/C-inhibitory function of XErp1. *EMBO Reports*, *12*, 436–443.

Horner, V. L., Czank, A., Jang, J. K., Singh, N., Williams, B. C., Puro, J., et al. (2006). The Drosophila calcipressin sarah is required for several aspects of egg activation. *Current Biology*, *16*, 1441–1446.

Horner, V. L., & Wolfner, M. F. (2008a). Mechanical stimulation by osmotic and hydrostatic pressure activates Drosophila oocytes in vitro in a calcium-dependent manner. *Developmental Biology*, *316*, 100–109.

Horner, V. L., & Wolfner, M. F. (2008b). Transitioning from egg to embryo: Triggers and mechanisms of egg activation. *Developmental Dynamics*, *237*, 527–544.

Hudmon, A., & Schulman, H. (2002). Structure-function of the multifunctional Ca^{2+}/calmodulin-dependent protein kinase II. *Biochemical Journal, 364*, 593–611.
Jans, D. A., & Hubner, S. (1996). Regulation of protein transport to the nucleus: Central role of phosphorylation. *Physiological Reviews, 76*, 651–685.
Johnson, J., Bierle, B. M., Gallicano, G. I., & Capco, D. G. (1998). Calcium/calmodulin-dependent protein kinase II and calmodulin: Regulators of the meiotic spindle in mouse eggs. *Developmental Biology, 204*, 464–477.
Juge, F., Zaessinger, S., Temme, C., Wahle, E., & Simonelig, M. (2002). Control of poly(A) polymerase level is essential to cytoplasmic polyadenylation and early development in Drosophila. *EMBO Journal, 21*, 6603–6613.
Kawahara, H., Philipova, R., Yokosawa, H., Patel, R., Tanaka, K., & Whitaker, M. (2000). Inhibiting proteasome activity causes overreplication of DNA and blocks entry into mitosis in sea urchin embryos. *Journal of Cell Science, 113*(Pt 15), 2659–2670.
Kim, A. M., Bernhardt, M. L., Kong, B. Y., Ahn, R. W., Vogt, S., Woodruff, T. K., et al. (2011). Zinc sparks are triggered by fertilization and facilitate cell cycle resumption in mammalian eggs. *ACS Chemical Biology, 6*, 716–723.
Kim, J. H., & Richter, J. D. (2006). Opposing polymerase-deadenylase activities regulate cytoplasmic polyadenylation. *Molecular Cell, 24*, 173–183.
Knott, J. G., Gardner, A. J., Madgwick, S., Jones, K. T., Williams, C. J., & Schultz, R. M. (2006). Calmodulin-dependent protein kinase II triggers mouse egg activation and embryo development in the absence of Ca2+ oscillations. *Developmental Biology, 296*, 388–395.
Krauchunas, A.R., Horner, V.L., Wolfner, M. F., submitted. Identifying protein phosphorylation changes reveal new candidates in the regulation of egg activation and early embryogenesis in D. melanogaster. *Developmental Biology, 370*, 125–134.
Lee, L. A., Van Hoewyk, D., & Orr-Weaver, T. L. (2003). The Drosophila cell cycle kinase PAN GU forms an active complex with PLUTONIUM and GNU to regulate embryonic divisions. *Genes & Development, 17*, 2979–2991.
Lee, K. W., Webb, S. E., & Miller, A. L. (1999). A wave of free cytosolic calcium traverses zebrafish eggs on activation. *Developmental Biology, 214*, 168–180.
Li, L., Zheng, P., & Dean, J. (2010). Maternal control of early mouse development. *Development, 137*, 859–870.
Liu, M. (2011). The biology and dynamics of mammalian cortical granules. *Reproductive Biology and Endocrinology, 9*, 149.
Liu, J., & Maller, J. L. (2005). Calcium elevation at fertilization coordinates phosphorylation of XErp1/Emi2 by Plx1 and CaMK II to release metaphase arrest by cytostatic factor. *Current Biology, 15*, 1458–1468.
Lorca, T., Cruzalegui, F. H., Fesquet, D., Cavadore, J. C., Méry, J., Means, A., et al. (1993). Calmodulin-dependent protein kinase II mediates inactivation of MPF and CSF upon fertilization of Xenopus eggs. *Nature, 366*, 270–273.
Madgwick, S., Levasseur, M., & Jones, K. T. (2005). Calmodulin-dependent protein kinase II, and not protein kinase C, is sufficient for triggering cell-cycle resumption in mammalian eggs. *Journal of Cell Science, 118*, 3849–3859.
Marcello, M. R., & Singson, A. (2010). Fertilization and the oocyte-to-embryo transition in C. elegans. *BMB Reports, 43*, 389–399.
Markoulaki, S., Matson, S., Abbott, A. L., & Ducibella, T. (2003). Oscillatory CaMKII activity in mouse egg activation. *Developmental Biology, 258*, 464–474.
Markoulaki, S., Matson, S., & Ducibella, T. (2004). Fertilization stimulates long-lasting oscillations of CaMKII activity in mouse eggs. *Developmental Biology, 272*, 15–25.
Maruyama, R., Velarde, N. V., Klancer, R., Gordon, S., Kadandale, P., Parry, J. M., et al. (2007). EGG-3 regulates cell-surface and cortex rearrangements during egg activation in Caenorhabditis elegans. *Current Biology, 17*, 1555–1560.

Mathavan, S., Lee, S. G., Mak, A., Miller, L. D., Murthy, K. R., Govindarajan, K. R., et al. (2005). Transcriptome analysis of zebrafish embryogenesis using microarrays. *PLoS Genetics, 1*, 260–276.

McGivern, J. V., Swaney, D. L., Coon, J. J., & Sheets, M. D. (2009). Toward defining the phosphoproteome of Xenopus laevis embryos. *Developmental Dynamics, 238*, 1433–1443.

McLean, J. R., Chaix, D., Ohi, M. D., & Gould, K. L. (2011). State of the APC/C: Organization, function, and structure. *Critical Reviews in Biochemistry and Molecular Biology, 46*, 118–136.

Mei, W., Lee, K. W., Marlow, F. L., Miller, A. L., & Mullins, M. C. (2009). hnRNP I is required to generate the Ca2+ signal that causes egg activation in zebrafish. *Development, 136*, 3007–3017.

Miao, Y. L., Stein, P., Jefferson, W. N., Padilla-Banks, E., & Williams, C. J. (2012). Calcium influx-mediated signaling is required for complete mouse egg activation. *Proceedings of the National Academy of Sciences of the United States of America, 109*, 4169–4174.

Mizushima, N. (2007). Autophagy: Process and function. *Genes & Development, 21*, 2861–2873.

Mochida, S., & Hunt, T. (2007). Calcineurin is required to release Xenopus egg extracts from meiotic M phase. *Nature, 449*, 336–340.

Morin, N., Abrieu, A., Lorca, T., Martin, F., & Dorée, M. (1994). The proteolysis-dependent metaphase to anaphase transition: Calcium/calmodulin-dependent protein kinase II mediates onset of anaphase in extracts prepared from unfertilized Xenopus eggs. *EMBO Journal, 13*, 4343–4352.

Nie, M., Xie, Y., Loo, J. A., & Courey, A. J. (2009). Genetic and proteomic evidence for roles of Drosophila SUMO in cell cycle control, Ras signaling, and early pattern formation. *PLoS One, 4*, e5905.

Nishi, Y., & Lin, R. (2005). DYRK2 and GSK-3 phosphorylate and promote the timely degradation of OMA-1, a key regulator of the oocyte-to-embryo transition in C. elegans. *Developmental Biology, 288*, 139–149.

Nishi, Y., Rogers, E., Robertson, S. M., & Lin, R. (2008). Polo kinases regulate C. elegans embryonic polarity via binding to DYRK2-primed MEX-5 and MEX-6. *Development, 135*, 687–697.

Nishiyama, T., Yoshizaki, N., Kishimoto, T., & Ohsumi, K. (2007). Transient activation of calcineurin is essential to initiate embryonic development in Xenopus laevis. *Nature, 449*, 341–345.

Nomikos, M., Swann, K., & Lai, F. A. (2012). Starting a new life: Sperm PLC-zeta mobilizes the Ca2+ signal that induces egg activation and embryo development: An essential phospholipase C with implications for male infertility. *Bioessays, 34*, 126–134.

Oh, J. S., Susor, A., & Conti, M. (2011). Protein tyrosine kinase Wee1B is essential for metaphase II exit in mouse oocytes. *Science, 332*, 462–465.

Ozil, J. P., Banrezes, B., Tóth, S., Pan, H., & Schultz, R. M. (2006). Ca2+ oscillatory pattern in fertilized mouse eggs affects gene expression and development to term. *Developmental Biology, 300*, 534–544.

Page, A. W., & Orr-Weaver, T. L. (1996). The Drosophila genes grauzone and cortex are necessary for proper female meiosis. *Journal of Cell Science, 109*(Pt 7), 1707–1715.

Parrington, J., Davis, L. C., Galione, A., & Wessel, G. (2007). Flipping the switch: How a sperm activates the egg at fertilization. *Developmental Dynamics, 236*, 2027–2038.

Parry, J. M., & Singson, A. (2011). EGG molecules couple the oocyte-to-embryo transition with cell cycle progression. *Results and Problems in Cell Differentiation, 53*, 135–151.

Parry, J. M., Velarde, N. V., Lefkovith, A. J., Zegarek, M. H., Hang, J. S., Ohm, J., et al. (2009). EGG-4 and EGG-5 Link Events of the Oocyte-to-Embryo Transition with Meiotic Progression in C. elegans. *Current Biology, 19*, 1752–1757.

Pellettieri, J., Reinke, V., Kim, S. K., & Seydoux, G. (2003). Coordinate activation of maternal protein degradation during the egg-to-embryo transition in C. elegans. *Developmental Cell, 5*, 451–462.

Pesin, J. A., & Orr-Weaver, T. L. (2007). Developmental role and regulation of cortex, a meiosis-specific anaphase-promoting complex/cyclosome activator. *PLoS Genetics, 3*, e202.

Pesin, J. A., & Orr-Weaver, T. L. (2008). Regulation of APC/C activators in mitosis and meiosis. *Annual Review of Cell and Developmental Biology, 24*, 475–499.

Peters, J. M., King, R. W., Hoog, C., & Kirschner, M. W. (1996). Identification of BIME as a subunit of the anaphase-promoting complex. *Science, 274*, 1199–1201.

Pfeiffer, M. J., Siatkowski, M., Paudel, Y., Balbach, S. T., Baeumer, N., Crosetto, N., et al. (2011). Proteomic analysis of mouse oocytes reveals 28 candidate factors of the "reprogrammome" *Journal of Proteome Research, 10*, 2140–2153.

Potireddy, S., Vassena, R., Patel, B. G., & Latham, K. E. (2006). Analysis of polysomal mRNA populations of mouse oocytes and zygotes: Dynamic changes in maternal mRNA utilization and function. *Developmental Biology, 298*, 155–166.

Rauh, N. R., Schmidt, A., Bormann, J., Nigg, E. A., & Mayer, T. U. (2005). Calcium triggers exit from meiosis II by targeting the APC/C inhibitor XErp1 for degradation. *Nature, 437*, 1048–1052.

Reber, S., Over, S., Kronja, I., & Gruss, O. J. (2008). CaM kinase II initiates meiotic spindle depolymerization independently of APC/C activation. *The Journal of Cell Biology, 183*, 1007–1017.

Renault, A. D., Zhang, X. H., Alphey, L. S., Frenz, L. M., Glover, D. M., Saunders, R. D., et al. (2003). giant nuclei is essential in the cell cycle transition from meiosis to mitosis. *Development, 130*, 2997–3005.

Rogers, N. T., Halet, G., Piao, Y., Carroll, J., Ko, M. S., & Swann, K. (2006). The absence of a Ca(2+) signal during mouse egg activation can affect parthenogenetic preimplantation development, gene expression patterns, and blastocyst quality. *Reproduction, 132*, 45–57.

Rouget, C., Papin, C., Boureux, A., Meunier, A. C., Franco, B., Robine, N., et al. (2010). Maternal mRNA deadenylation and decay by the piRNA pathway in the early Drosophila embryo. *Nature, 467*, 1128–1132.

Roux, M. M., Townley, I. K., Raisch, M., Reade, A., Bradham, C., Humphreys, G., et al. (2006). A functional genomic and proteomic perspective of sea urchin calcium signaling and egg activation. *Developmental Biology, 300*, 416–433.

Rusnak, F., & Mertz, P. (2000). Calcineurin: Form and function. *Physiological Reviews, 80*, 1483–1521.

Samuel, A. D., Murthy, V. N., & Hengartner, M. O. (2001). Calcium dynamics during fertilization in C. elegans. *BMC Developmental Biology, 1*, 8.

Sato, K., Fukami, Y., & Stith, B. J. (2006). Signal transduction pathways leading to Ca2+ release in a vertebrate model system: Lessons from Xenopus eggs. *Seminars in Cell & Developmental Biology, 17*, 285–292.

Saunders, C. M., Larman, M. G., Parrington, J., Cox, L. J., Royse, J., Blayney, L. M., et al. (2002). PLC zeta: A sperm-specific trigger of Ca(2+) oscillations in eggs and embryo development. *Development, 129*, 3533–3544.

Schmidt, A., Duncan, P. I., Rauh, N. R., Sauer, G., Fry, A. M., Nigg, E. A., et al. (2005). Xenopus polo-like kinase Plx1 regulates XErp1, a novel inhibitor of APC/C activity. *Genes & Development, 19*, 502–513.

Semotok, J. L., Cooperstock, R. L., Pinder, B. D., Vari, H. K., Lipshitz, H. D., & Smibert, C. A. (2005). Smaug recruits the CCR4/POP2/NOT deadenylase complex to trigger maternal transcript localization in the early Drosophila embryo. *Current Biology, 15*, 284–294.

Semotok, J. L., Luo, H., Cooperstock, R. L., Karaiskakis, A., Vari, H. K., Smibert, C. A., et al. (2008). Drosophila maternal Hsp83 mRNA destabilization is directed by multiple

SMAUG recognition elements in the open reading frame. *Molecular and Cellular Biology, 28*, 6757–6772.

Shoji, S., Yoshida, N., Amanai, M., Ohgishi, M., Fukui, T., Fujimoto, S., et al. (2006). Mammalian Emi2 mediates cytostatic arrest and transduces the signal for meiotic exit via Cdc20. *EMBO Journal, 25*, 834–845.

Smibert, C. A., Lie, Y. S., Shillinglaw, W., Henzel, W. J., & Macdonald, P. M. (1999). Smaug, a novel and conserved protein, contributes to repression of nanos mRNA translation in vitro. *RNA, 5*, 1535–1547.

Smibert, C. A., Wilson, J. E., Kerr, K., & Macdonald, P. M. (1996). smaug protein represses translation of unlocalized nanos mRNA in the Drosophila embryo. *Genes & Development, 10*, 2600–2609.

Stricker, S. A. (1999). Comparative biology of calcium signaling during fertilization and egg activation in animals. *Developmental Biology, 211*, 157–176.

Sun, L., Chai, Y., Hannigan, R., Bhogaraju, V. K., & Machaca, K. (2007). Zinc regulates the ability of Cdc25C to activate MPF/cdk1. *Journal of Cellular Physiology, 213*, 98–104.

Suzuki, T., Suzuki, E., Yoshida, N., Kubo, A., Li, H., Okuda, E., et al. (2010). Mouse Emi2 as a distinctive regulatory hub in second meiotic metaphase. *Development, 137*, 3281–3291.

Suzuki, T., Yoshida, N., Suzuki, E., Okuda, E., & Perry, A. C. (2010). Full-term mouse development by abolishing Zn2+-dependent metaphase II arrest without Ca2+ release. *Development, 137*, 2659–2669.

Swan, A., & Schupbach, T. (2007). The Cdc20 (Fzy)/Cdh1-related protein, Cort, cooperates with Fzy in cyclin destruction and anaphase progression in meiosis I and II in Drosophila. *Development, 134*, 891–899.

Tadros, W., Goldman, A. L., Babak, T., Menzies, F., Vardy, L., Orr-Weaver, T., et al. (2007). SMAUG is a major regulator of maternal mRNA destabilization in Drosophila and its translation is activated by the PAN GU kinase. *Developmental Cell, 12*, 143–155.

Tadros, W., Houston, S. A., Bashirullah, A., Cooperstock, R. L., Semotok, J. L., Reed, B. H., et al. (2003). Regulation of maternal transcript destabilization during egg activation in Drosophila. *Genetics, 164*, 989–1001.

Tadros, W., & Lipshitz, H. D. (2005). Setting the stage for development: mRNA translation and stability during oocyte maturation and egg activation in Drosophila. *Developmental Dynamics, 232*, 593–608.

Tadros, W., & Lipshitz, H. D. (2009). The maternal-to-zygotic transition: A play in two acts. *Development, 136*, 3033–3042.

Takeo, S., Hawley, R. S., & Aigaki, T. (2010). Calcineurin and its regulation by Sra/RCAN is required for completion of meiosis in Drosophila. *Developmental Biology, 344*, 957–967.

Takeo, S., Swanson, S. K., Nandanan, K., Nakai, Y., Aigaki, T., Washburn, M. P., et al. (2012). Shaggy/glycogen synthase kinase 3beta and phosphorylation of Sarah/regulator of calcineurin are essential for completion of Drosophila female meiosis. *Proceedings of the National Academy of Sciences of the United States of America, 109*, 6382–6389.

Takeo, S., Tsuda, M., Akahori, S., Matsuo, T., & Aigaki, T. (2006). The calcineurin regulator sra plays an essential role in female meiosis in Drosophila. *Current Biology, 16*, 1435–1440.

Tatone, C., Delle Monache, S., Iorio, R., Caserta, D., Di Cola, M., & Colonna, R. (2002). Possible role for Ca(2+) calmodulin-dependent protein kinase II as an effector of the fertilization Ca(2+) signal in mouse oocyte activation. *Molecular Human Reproduction, 8*, 750–757.

Tatone, C., Iorio, R., Francione, A., Gioia, L., & Colonna, R. (1999). Biochemical and biological effects of KN-93, an inhibitor of calmodulin-dependent protein kinase II, on the initial events of mouse egg activation induced by ethanol. *Journal of Reproduction and Fertility, 115*, 151–157.

Thomsen, S., Anders, S., Janga, S. C., Huber, W., & Alonso, C. R. (2010). Genome-wide analysis of mRNA decay patterns during early Drosophila development. *Genome Biology*, *11*, R93.

Tsaadon, A., Eliyahu, E., Shtraizent, N., & Shalgi, R. (2006). When a sperm meets an egg: Block to polyspermy. *Molecular and Cellular Endocrinology*, *252*, 107–114.

Tsukamoto, S., Kuma, A., & Mizushima, N. (2008). The role of autophagy during the oocyte-to-embryo transition. *Autophagy*, *4*, 1076–1078.

Tsukamoto, S., Kuma, A., Murakami, M., Kishi, C., Yamamoto, A., & Mizushima, N. (2008). Autophagy is essential for preimplantation development of mouse embryos. *Science*, *321*, 117–120.

Vardy, L., & Orr-Weaver, T. L. (2007). The Drosophila PNG kinase complex regulates the translation of cyclin B. *Developmental Cell*, *12*, 157–166.

Verlhac, M. H., Terret, M. E., & Pintard, L. (2010). Control of the oocyte-to-embryo transition by the ubiquitin-proteolytic system in mouse and C. elegans. *Current Opinion in Cell Biology*, *22*, 758–763.

Wakai, T., Vanderheyden, V., & Fissore, R. A. (2011). Ca2+ signaling during mammalian fertilization: Requirements, players, and adaptations. *Cold Spring Harbor Perspectives in Biology*, *3*, a006767.

Wang, S., Kou, Z., Jing, Z., Zhang, Y., Guo, X., Dong, M., et al. (2010). Proteome of mouse oocytes at different developmental stages. *Proceedings of the National Academy of Sciences of the United States of America*, *107*, 17639–17644.

Wu, Q., Guo, Y., Yamada, A., Perry, J. A., Wang, M. Z., Araki, M., et al. (2007). A role for Cdc2- and PP2A-mediated regulation of Emi2 in the maintenance of CSF arrest. *Current Biology*, *17*, 213–224.

Yurttas, P., Morency, E., & Coonrod, S. A. (2010). Use of proteomics to identify highly abundant maternal factors that drive the egg-to-embryo transition. *Reproduction*, *139*, 809–823.

CHAPTER ELEVEN

The Role of the Double Bromodomain-Containing BET Genes During Mammalian Spermatogenesis

Binyamin D. Berkovits*, Debra J. Wolgemuth*,†,‡,§,1

*Department of Genetics and Development, Columbia University Medical Center, New York, USA
†Department of Obstetrics and Gynecology, Columbia University Medical Center, New York, USA
‡The Institute of Human Nutrition, Columbia University Medical Center, New York, USA
§The Herbert Irving Comprehensive Cancer Center, Columbia University Medical Center, New York, USA
[1]Corresponding author: e-mail address: djw3@columbia.edu

Contents

Current Topics in Developmental Biology, Volume 102
ISSN 0070-2153
http://dx.doi.org/10.1016/B978-0-12-416024-8.00011-8

Abstract

The double bromodomain-containing BET (*b*romodomain and *e*xtra *t*erminal) family of proteins is highly conserved from yeast to humans and consists not just of transcriptional regulators but also histone-interacting chromatin remodelers. The four mammalian BET genes are each expressed at unique times during spermatogenesis, and the testis-specific gene *Brdt* is essential for spermatogenesis. Loss of the first bromodomain of BRDT results in improper/incomplete spermatid elongation and severely morphologically defective sperm. The elongation defects observed in mutant spermatids can be directly tied to altered postmeiotic chromatin architecture. BRDT is required for creation/maintenance of the chromocenter of round spermatids, a structure that forms just after completion of meiosis. The chromocenter creates a defined topology in spermatids, and the presence of multiple chromocenters rather than a single intact chromocenter correlates with loss of spermatid polarity, loss of heterochromatin foci at the nuclear envelope, and loss of proper spermatid elongation. BRDT is not only essential for proper chromatin organization but also involved in regulation of transcription and in cotranscriptional processing. That is, transcription and alternative splicing are altered in spermatocytes and spermatids that lack full-length BRDT. Additionally, the transcription of mRNAs with short 3′ UTRs, which is characteristic of round spermatids, is also altered. Examination of the genes regulated by BRDT yields many possible targets that could in part explain the morphologically abnormal sperm produced by the BRDT mutant testes. Thus, BRDT and possibly the other BET genes are required for proper spermatogenesis, which opens up the possibility that the recently discovered small molecule inhibitors of the BET family could be useful as reversible male contraceptives.

1. INTRODUCTION

1.1. The bromodomain and bromodomain-containing proteins

The bromodomain is a conserved protein motif found in all eukaryotes and is the only known protein module that binds acetylated lysine residues (Mujtaba, Zeng, & Zhou, 2007; Sanchez & Zhou, 2009). This motif was first discovered in the *Drosophila* gene *brahma* (*brm*) from which it derives its name (Haynes et al., 1992; Tamkun et al., 1992), and it has subsequently been identified in genes from yeast to humans. There are 42 mammalian bromodomain-containing genes, 12 of which contain multiple bromodomains (Arrowsmith, Bountra, Fish, Lee, & Schapira, 2012; Yang, 2004).

In light of the prevalence of the acetyl posttranslational modification, proteins that can "read" this mark have a wide variety of functions. Not surprisingly, bromodomains are found in proteins with diverse functions, including histone acetyltransferases (Jacobson, Ladurner, King, & Tjian,

2000; Mishima et al., 2011; Ohshima, Suganuma, & Ikeda, 2001; Wang et al., 1997), histone methyltransferases (Gregory et al., 2007; Hess, 2004), histone kinases (Xiao et al., 2009), transcriptional coactivators (Denis, 2010; Ogryzko, Schiltz, Russanova, Howard, & Nakatani, 1996; Tamkun, 1995), transcriptional corepressors (Sankar et al., 2008; Tae et al., 2011; Zeng et al., 2008; Zhou & Grummt, 2005), ubiquitin ligases (Agricola, Randall, Gaarenstroom, Dupont, & Hill, 2011; Allton et al., 2009), sumo ligases (Liang et al., 2011), and chromatin remodelers (Liu, Mulholland, Fu, & Zhao, 2006; Racki et al., 2009; Thompson, 2009; Wilson & Roberts, 2011). The ~110 aa-long bromodomain motif folds into two loops, each containing two alpha helices, thereby creating a hydrophobic pocket between them that interacts with acetylated lysine (Florence & Faller, 2008; Owen et al., 2000). Bromodomains have been shown to bind to acetylated lysines both in histone tails (Jacobson et al., 2000; Jeanmougin, Wurtz, Le Douarin, Chambon, & Losson, 1997) and in non-histone proteins (Barlev et al., 2001; Gamsjaeger et al., 2011).

1.2. The BET subfamily of bromodomain-containing proteins

The BET (bromodomain and extra terminal) family of bromodomain-containing proteins is so-called because each member contains an N-terminal *b*romodomain and a unique C-terminal *e*xtra *t*erminal (ET) domain that is found only in this family (Florence & Faller, 2001). In plant BETs, there is only a single bromodomain, but all other BET members contain a pair of tandem bromodomains near the N-terminus. The ET domain is a protein–protein interacting domain (Matangkasombut, Buratowski, Swilling, & Buratowski, 2000; Platt, Simpson, Mittnacht, & Schulz, 1999) composed of two alpha helices followed by a loop and then a beta sheet and another alpha helix. At the very C-terminus of the ET domain is often found a SEED motif, which comprises serine, aspartic acid, and glutamic acid resides. The SEED domain is found in some but not all plant BET proteins but is conserved from yeast to humans. There are four mammalian BET genes, *Brd2*, *Brd3*, *Brd4*, and *Brdt*, two in yeast, *BDF1* and *BDF2*, and the founding member of the BET family, the *Drosophila female sterile (1) homeotic (fs(1)h)* gene, is the only fly BET gene (Fig. 11.1).

The presence of a single bromodomain in BET proteins in plants, versus the tandem domains in other eukaryotes, has led to the hypothesis that the tandem bromodomains arose from duplication within an ancestral BET gene. Bromodomain 1 (BD1) and bromodomain 2 (BD2) of the BET family

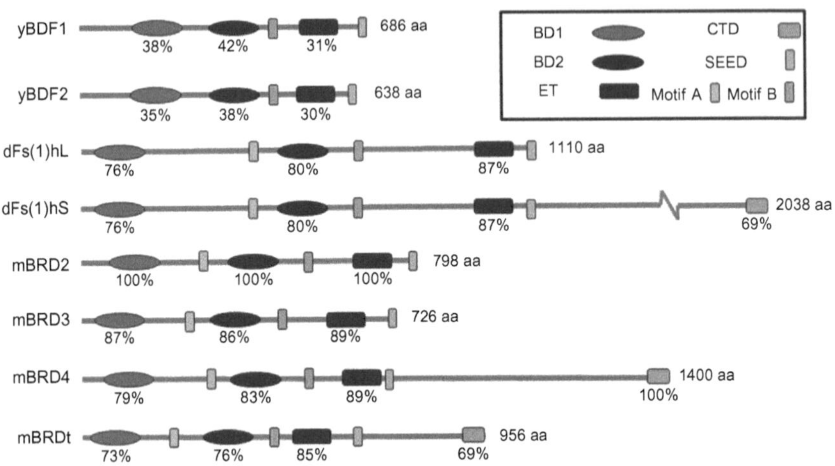

Figure 11.1 Structural comparison of the BET family members in *S. cerevisiae*, *D. melanogaster*, and *M. musculus*. The percentages under each bromodomain and the ET domain indicate the percentage of identical amino acid residues in the domain as compared to the corresponding domain in mBRD2, the founding member of the mouse BET family. The percentages under the CTD are compared to mBRD4. (See Color Insert.)

genes have since diverged sufficiently such that there is greater similarity between BD1 or BD2 from different mammalian BET proteins than there is between BD1 and BD2 in a single BET protein. For instance, BD1 and BD2 of mouse *Brdt* are 73% and 76% identical to BD1 and BD2 of mouse *Brd2*, respectively, but BD1 and BD2 of mouse *Brdt* are only 40% identical to each other (Fig. 11.1). Interestingly, the ET domain of mammalian BET proteins is even more highly conserved than either bromodomain. For example, the ET domains of mouse *Brdt* and mouse *Brd2* are 85% identical (Fig. 11.1). Two other minor motifs, named the A and B motifs, are present in all four mammalian BETs and Drosophila *fs(1)h*; however, motif A is absent from the yeast BET genes (Paillisson et al., 2007).

1.2.1 Drosophila female sterile (1) *homeotic*

Drosophila fs(1)h was the first BET gene identified and was discovered in a screen for genes with a maternal effect phenotype (Gans, Audit, & Masson, 1975; Gans, Forquignon, & Masson, 1980). Although *fs(1)h* is the only *Drosophila* BET gene, it encodes two distinct protein isoforms: fs(1)hS and fs(1)hL (Haynes, Mozer, Bhatia-Dey, & Dawid, 1989). The fs(1)hS protein contains both bromodomains and the ET domain. The fs(1)hL protein contains the entire short isoform (except for 4 aa near the C-terminus) but also has an ~1000 aa-long tail (Fig. 11.1). Numerous mutant alleles of *fs(1)h* have been characterized at the molecular level, revealing modular functions for the fs(1)h protein. The first such mutations were temperature-sensitive mutant alleles; however, it is important to note that even at high non-permissive temperatures, fs(1)h protein was present, albeit at low levels, suggesting that these mutations do not reflect complete loss of function (Florence & Faller, 2008; Gans et al., 1975, 1980). When females were shifted to high (26.5–29° C) temperatures, a reduced number of nonproductive eggs were laid such that the flies were considered to be infertile. Loss of fs(1)h function resulted in alteration of *Ultrabithorax* (*Ubx*) and *trithorax* (*trx*) expression, which in turn caused late segmentation defects (Huang & Dawid, 1990). In a study in which a highly conserved aspartic acid residue in BD1 was mutated to valine, a different mutant phenotype was observed (Florence & Faller, 2008). These mutant flies were also sterile, but the maternal effects resulted in head and gut defects caused by loss of repression of *tailless* (*tll*) and *hückebein* (*hkb*) at the initiation of transcription during embryogenesis. A mutant allele that contained both intact bromodomains but lacked the ET domain showed a phenotype that was similar to the temperature-sensitive mutants, with altered transcription of segment genes, but no head and gut defects (Haynes et al., 1989). The reciprocal nature of the two phenotypes suggests separate functions for the different regions of the fs(1)h protein.

1.2.2 Yeast BDF1 and BDF2

In the yeast *Saccharomyces cerevisiae*, there are two BET genes, *BDF1* and *BDF2*, of which *BDF1* is far more extensively studied. *BDF1* was discovered in a screen for genes affecting snRNA transcription (Lygerou et al., 1994) and along with *BDF2* has been shown to be a general transcription factor (Durant & Pugh, 2007; Matangkasombut et al., 2000). Mutational studies of *BDF1* have also identified a number of other functions beyond its role as a component of the basal transcriptional machinery. Early work on *BDF1* found that although *BDF1* null cells were viable, there were defects in metabolism of certain sugars and meiosis was abnormal (Chua & Roeder, 1995). Interestingly, mutations in BD1 and the SEED domain had no effect on meiosis, suggesting that BDF1, like fs(1)h, has modular functions. *BDF1* has also been shown to affect transcription by delineating heterochromatin–euchromatin boundaries (Ladurner, Inouye, Jain, & Tjian, 2003). That is, BDF1 binds to acetylated H4 found in euchromatin adjacent to constitutive heterochromatin. Its physical presence on chromatin competes with the SIR2 and SIR3 histone deacetylases and stops them from spreading beyond the normally transcriptionally silenced chromatin. Mutation of both bromodomains abolished BDF1's histone-binding and resulted in transcriptional downregulation of genes located near constitutive heterochromatin, reflecting aberrant spreading of heterochromatin.

2. MAMMALIAN BET GENE EXPRESSION DURING SPERMATOGENESIS

All four mammalian BET genes are expressed in the adult testis and the timing of that expression is highly regulated (Shang et al., 2004). BRD4 is strongly expressed and BRD2 more weakly expressed in spermatogonia; however, exactly which types of spermatogonia expresses these two proteins has not yet been determined. BRD2 protein is most strongly expressed in late spermatocyte stages and round spermatids. *Brd3* mRNA is strongly expressed in round spermatids, and interestingly, two distinct mRNA species exist in the testis, the full-length 6.2 kb transcript, which is found in somatic cells, and a testis-specific 3.1 kb transcript (Shang et al., 2004). The significance of this transcript is unknown and nothing is known about BRD3 protein in the testis. There may also be two *Brdt* mRNA transcripts expressed in the murine testis, but the data to date are conflicting. The initial finding was that a short 1.9 kb transcript and a long 3.8 kb transcript were equally abundant in the murine testis (Taniguchi et al., 2001). Another group subsequently reported an abundant 3.5 kb transcript and a less abundantly expressed 4.0 kb transcript (Pivot-Pajot et al., 2003). A third group found only a highly abundant 3.8 kb transcript (Shang et al., 2004). *Brdt* mRNA is most highly expressed in spermatocytes and only to a lesser extent in round spermatids (Shang et al., 2004). BRDT protein expression is present throughout the pachytene stage into diplotene spermatocytes (Shang, Nickerson, Wen, Wang, & Wolgemuth, 2007). However, there are conflicting observations as to whether the protein is present in round spermatids and elongating/compacting spermatids. The first report on BRDT protein expression observed no protein present in round spermatids but strong expression in elongating spermatids that perfectly colocalized with hyperacetylated histone H4 (Govin et al., 2006). This was in direct contradiction to a later report that found BRDT protein highly expressed in round spermatids but absent from elongating spermatids (Shang et al., 2007). Both of these reports examined murine BRDT; a subsequent study reported that rat BRDT was expressed in round spermatids and also in elongating spermatids (Dhar, Thota, & Rao, 2012). However, the expression in elongating spermatids did not extensively overlap with hyperacetylated H4 (only 20%), in contrast to the perfect colocalization that had been observed by Govin et al. (2006). Differences in the specificity and/or sensitivity of the antibodies used in these studies may be responsible for these discrepancies. In this light, it should be noted that the antibody used in Shang et al. (2007) was capable of specifically detecting the truncated BRDT protein as well, thus confirming that it does specifically recognize BRDT.

All three reports do seem to agree that BRDT protein is absent from the nucleus after the removal of the majority of canonical histones is complete. However, this is in contrast to what has been reported in human testis: BRDT protein has been reported in spermatocytes, round and elongated spermatids, and fully mature spermatozoa (Barda et al., 2012; Zheng, Yuan, Zhou, Xu, & Sha, 2005). It should also be noted that there are seven different *BRDT* mRNA transcripts in human testes annotated in the RefSeq database, ranging in length from 3.0 to 3.7 kb. Further, a BRDT splice variant containing seven extra amino acids, termed BRDT-NY, is the isoform found in mature sperm (Zheng et al., 2005). A summary of murine BET testicular expression, including probable BRDT expression, is depicted in Fig. 11.2.

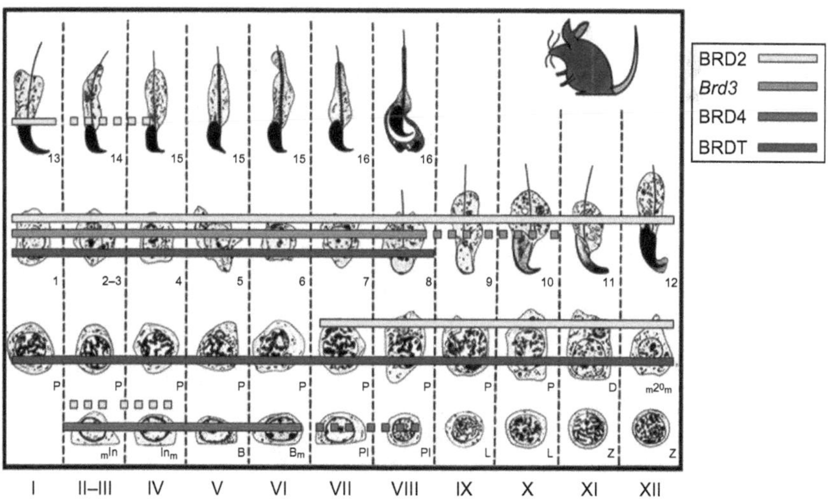

Figure 11.2 Expression patterns of the BET family in the adult mouse testis. Expression depicted for BRD2, BRD4, and BRDT is at the protein level, and for *Brd3*, it is at the mRNA level. Solid bars represent high levels of expression and dashed bars represent less abundant but detectable expression. *The figure is adapted from Shang et al. (2004).* (See Color Insert.)

2.1. Mammalian BRD2

Human *BRD2* (originally known as *RING3*) and mouse *Brd2* (originally known at *Fsrg1*) were the first mammalian BET genes discovered (Beck, Hanson, Kelly, Pappin, & Trowsdale, 1992; Rhee, Brunori, Besset, Trousdale, & Wolgemuth, 1998). Among the murine BETs, *Brd2* is similar to *fs(1)hS* as it has no long C-terminal tail (Fig. 11.1). Recent studies have revealed that the regulation of *BRD2/Brd2* expression is complex (Shang et al., 2011). *BRD2* can be transcribed from multiple promoters which leads to transcripts with both long and short 5′ UTRs, and an alternate splice form exists in which a region of intron 2 is included in transcripts as exon 2A. Only the transcripts with a short 5′ UTR and lacking exon 2A yielded translation of full-length protein *in vitro* and *in vivo* in cultured cells (Shang et al., 2011). Mouse *Brd2* is ubiquitously expressed and long and short transcripts are found in all tissues (Taniguchi et al., 1998), although the ratio of long to short differed. *Brd2* has been reported to be highly expressed in hormonally modulated epithelia, including the mammary gland, ovary, kidney, and uterus (Rhee et al., 1998; Trousdale & Wolgemuth, 2004), as well as in the developing brain and the adult cerebellum and cerebral cortex (Shang et al., 2011).

Like BDF1/2 and fs(1)h, BRD2 has been shown to modulate transcription, in particular, in cell cycle-induced transcriptional activation. For example, BRD2 was found to translocate from cytoplasm to nucleus upon serum stimulation (Denis, Vaziri, Guo, & Faller, 2000; Guo, Faller, & Denis, 2000) and was shown to be nuclear in proliferating cells but cytoplasmic in nondividing, more differentiated cells (Crowley, Brunori, Rhee, Wang, & Wolgemuth, 2004). In the nucleus, BRD2 can act as a scaffold, interacting with TATA-binding protein (TBP) through its BD1 (Peng et al., 2007) and with E2F through the ET domain (Denis et al., 2000; Guo et al., 2000) to activate cell cycle regulatory genes that contain E2F-binding sites in their promoters. BD1 of BRD2 has been shown to bind H4K12ac as a dimer (Kanno et al., 2004) and BD2 binds H4 diacetylated on K5 and K12 (Huang et al., 2007). Although acetylated H4 is often present in the promoters of transcriptionally active genes, BRD2 does not directly interact with the promoters but rather through its association with E2F (Sinha, Faller, & Denis, 2005). It has also been reported that BRD2 binds to the acetylated chromatin in the body of transcriptionally active genes and aids in polymerase progression through the gene (LeRoy, Rickards, & Flint, 2008).

2.2. Mammalian BRD3

Brd3, which also lacks a long C-terminal tail, is thus more closely homologous to *Brd2* than to other mammalian BET genes (Fig. 11.1) and is also ubiquitously expressed (Thorpe et al., 1997) but is the least studied of the mammalian BET genes. BRD3 has been reported to bind via BD1 to GATA1 that was either diacetylated on residues K312 and K315 or tetraacetylated on residues K312, K314, K315, and K316 both *in vitro* and *in vivo* in an erythroid cell line (Gamsjaeger et al., 2011; Lamonica et al., 2011). One study showed that the binding only occurred in the presence of CREB-binding protein, and this binding was required for complete occupancy of GATA1 at its target promoters in erythroid cells (Lamonica et al., 2011). Thus, direct binding of BRD3 to the promoter region of a gene has also yet to be shown, but like BRD2, BRD3 has been reported to bind the acetylated lysines in transcription factors and in chromatin in the body of transcriptionally active genes (LeRoy et al., 2008).

2.3. Mammalian BRD4

Brd4, unlike *Brd2* and *Brd3*, encodes a long C-terminal region after the ET domain and is thus most homologous structurally with *fs(1)h* (Fig. 11.1). Like *fs(1)h*, *Brd4* also produces a short and a long protein variant (Dey et al., 2000; Florence & Faller, 2001). The short variant of BRD4 contains both bromodomains and the ET domain, but due to a unique splicing event, the last 3 aa of the short isoform are different from the amino acid sequence of the long isoform (similar to what occurs in *fs(1)h*). The C-terminal region is over 750 aa long, but the only known motif present is at the very 3′ end and has been named the C-terminal domain (CTD) (Wu & Chiang, 2007). This domain is also present in fs(1)hL.

BRD4 has been well studied as a transcriptional coactivator. BRD4 binds to the positive transcription elongation factor b (P-TEFb) and liberates it from inhibitory binding by HEXIM1. Via BRD4's binding to acetylated histones, it recruits the complex to the promoters of genes poised to be transcribed (Jang et al., 2005; Yang, He, & Zhou, 2008; Yang et al., 2005). However, the mechanism of binding of BRD4 to P-TEFb is somewhat in dispute. The initial report found that BRD4 bound to cyclin T1 (CCNT1), a subunit of P-TEFb, via either of the bromodomains but that an amino acid change in the other subunit of P-TEFb, cyclin-dependent kinase 9 (CDK9), abolished binding (Yang et al., 2005). A later report found that it was the CTD, not the bromodomains, that bound P-TEFb and that the CTDs of BRDT and fs(1)hL were also capable of this interaction (Bisgrove, Mahmoudi, Henklein, & Verdin, 2007). A third report found that it was the CTD that bound P-TEFb and that both subunits, CDK9 and CCNT1, were bound (Rahman et al., 2011). This last report also found that the ET domain of BRD4 (and BRD2 and BRD3 as well) binds to transcriptional activators GLTSCR1, NSD3, and JMJD6 and carries out P-TEFb-independent transcriptional activation (Rahman et al., 2011). Finally, a very recent report found that both observations may be correct, as it was observed that the BD2 in short BRD4 isoforms (lacking the C-terminal tail) bound to a triacetylated form of CCNT1 (Schroder et al., 2012). However, this binding did not alleviate inhibitory binding of HEXIM1, and indeed, binding of BRD4 to P-TEFb through its CTD was required to relieve this inhibition.

BRD4 is also of great interest because it has been reported to be retained on mitotic chromosomes when most chromatin interacting proteins are removed (Dey, Nishiyama, Karpova, McNally, & Ozato, 2009; Dey et al., 2000; Nishiyama, Dey, Miyazaki, & Ozato, 2006). It has been postulated that this "epigenetic memory" would allow for rapid BRD4-dependent transcription of M-G(1) cell cycle genes via P-TEFb recruitment (Ai et al., 2011; Yang et al., 2008). Additionally, it was found that BRD4 plays a role in overall chromatin architecture as, via its CTD, it maintains higher order chromatin structure (Wang, Li, Helfer, Jiao, & You, 2012).

2.4. Mammalian BRDT

Brdt is unique among other BET proteins in that it is nearly exclusively expressed in the testis (Jones, Numata, & Shimane, 1997; Pivot-Pajot et al., 2003; Shang et al., 2004; Taniguchi et al., 2001). Although one report showed mRNA expression in oocytes (Paillisson et al., 2007), this is not confirmed by three other reports (Jones et al., 1997; Pivot-Pajot et al., 2003; Shang et al., 2004). Very low levels of expression in the brain were reported by one group as well (Shang et al., 2004). It is possible that *Brdt* is expressed only in very specific regions of the brain, as there are reports in the databases of its being detected in olfactory neurons [GEO series accession GSE26373 (Clowney et al., 2011)]. Like *Brd4*, *Brdt* has a long C-terminal region after the ET domain which contains a CTD at the very C-terminus, making it most similar structurally to *Brd4* and *fs(1)h* (Florence & Faller, 2001; Wu & Chiang, 2007). Whether a short *Brdt* mRNA, comparable to *fs(1)hS* or short *Brd4*, exists is not clear. No such isoform has been reported in humans (Jones et al., 1997) and no short BRDT protein isoform has been observed to date in mouse or human.

3. BRDT IS REQUIRED FOR PROPER SPERMATOGENESIS AND MALE FERTILITY

The first genetic approach to elucidate BRDT's function generated mice with a 5′ deletion in the endogenous *Brdt* gene that includes the first bromodomain ($Brdt^{\Delta BD1}$) (Shang et al., 2007). Consistent with the expression pattern of *Brdt*, homozygous mutants were viable and females were fully fertile. However, while truncated mutant BRDT protein ($BRDT^{\Delta BD1}$) was produced in this mouse at levels comparable to control, and in the same cell types, the homozygous $Brdt^{\Delta BD1/BD1}$ mice were completely sterile. Additionally, a *BRDT* allele with a point mutation in between BD2 and the B motif that converts an asparagine reside to lysine was found to associate with male infertility in humans (Plaseski, Noveski, Popeska, Efremov, & Plaseska-Karanfilska, 2011). Thus, complete BRDT function, and specifically the first bromodomain, is required for male fertility.

3.1. Chromocenter formation and maintenance

The first visible morphological defect in $Brdt^{\Delta BD1/\Delta BD1}$ mutant testes was observed in step 1 round spermatids (Berkovits & Wolgemuth, 2011). In these cells, the pericentromeric heterochromatin of each spermatid is normally coalesced together in a chromatin structure known as the chromocenter (Meyer-Ficca, Muller-Navia, & Scherthan, 1998; Zalensky et al., 1995). This structure persists throughout spermatid differentiation and can still be found in mature spermatozoa (Hoyer-Fender, Singh, & Motzkus, 2000). In $Brdt^{\Delta BD1/\Delta BD1}$ spermatids, instead of a single focus of heterochromatin in the chromocenter, multiple heterochromatic foci were observed in the majority of spermatids (Berkovits & Wolgemuth, 2011). This multiple chromocenter defect was observed at about the same frequency in all stages of round spermatid differentiation and indeed was observed at the earliest stage of round spermatid differentiation, just after completion of meiosis. Whether the defect in chromatin organization occurred during meiosis in these mutant cells but was only observable postmeiotically or whether the defect originated after the completion of MI remains to be determined.

The mechanisms that result in the coalescence of the centromeric heterochromatin into the chromocenter are not well understood. TATA box-binding protein-like 1 (TBPL1) is required for chromocenter formation but is not itself present in early round spermatids (Martianov et al., 2002). It has been suggested that, in spermatocytes, TBPL1 might regulate the transcription of factors required for chromocenter formation. Loss of high mobility group box 2 (HMGB2) leads to reduced male fertility (Ronfani et al., 2001) and chromocenter defects (Catena, Ronfani, Sassone-Corsi, & Davidson, 2006). This protein associates with euchromatin in spermatocytes and may also regulate transcription of factors involved in chromocenter formation. Interestingly, HMGB family members have been shown to interact with TATA box-binding proteins (Das & Scovell, 2001), suggesting a possible link with TBPL1. Even more interestingly, careful inspection of the data in Ronfani et al. (2001) suggests that HMGB2 appears to be localized within the chromocenter of round spermatids, although the authors do not discuss this. Thus, HMGB2 could play a role in creating and/or maintaining the chromocenter.

BRDT, in contrast, is excluded from the heterochromatin of the chromocenter but exhibits an increased localization directly adjacent to that heterochromatin (Berkovits & Wolgemuth, 2011). As mentioned briefly earlier, yeast BDF1 is known to localize at constitutive heterochromatin–euchromatin boundaries, where it maintains the boundary by halting the spreading of SIR2/3 into euchromatin (Ladurner et al., 2003). The SIR deacetylases in conjunction with the SU(VAR)3-9 methyltransferase are responsible for generating trimethylated H3K9, which is the hallmark of heterochromatin (Schotta, Ebert, & Reuter, 2003; Shankaranarayana, Motamedi, Moazed, & Grewal, 2003). In round spermatids, the mammalian SU(VAR)3-9 homologs SUV39H1/2 are localized in the chromocenter (O'Carroll et al., 2000), but the SIR2 homolog Sirtuin 1 (SIRT1) is found in euchromatin, except in the region of high BRDT expression directly around the chromocenter (Berkovits & Wolgemuth, 2011). In this way, BRDT appears to create a "buffer" region between the deacetylase and the methyltransferases and thus halts ectopic spreading of pericentromeric heterochromatin. Indeed in the $Brdt^{\Delta BD1}$ mutant, SIRT1 is present directly adjacent to the chromocenter (see cartoon in Fig. 11.3), and the level of heterochromatin in the testis is increased by greater than threefold (Berkovits & Wolgemuth, 2011). It may be that an expanded region of heterochromatin keeps the centromeres from coalescing or creates instability in the chromocenter that leads to its fragmentation. It is also possible, and not mutually exclusive, that BRDT modulates gene expression in spermatocytes, as has been hypothesized for TBPL1and HMGB2. Or as previously mentioned, BRDT may alter the

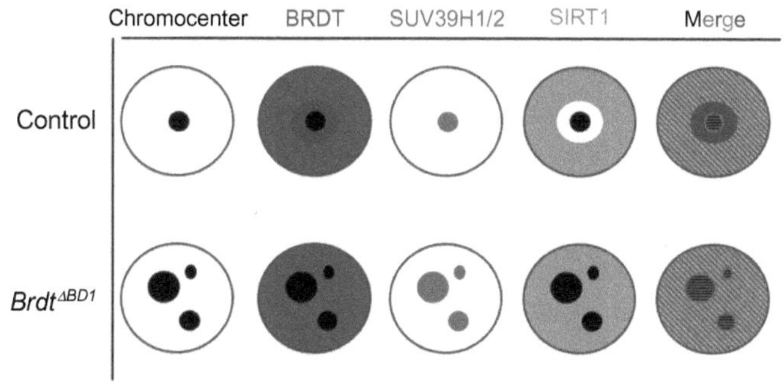

Figure 11.3 Cartoon depicting chromocenter formation and function in control and $Brdt^{\Delta BD1}$ spermatids. BRDT protein is found throughout the nucleus (dark red) but is excluded from the chromocenter (black). In control spermatids, BRDT is more densely localized around the chromocenter (bright red), but this localization is lost in mutant spermatids. SUV39H1/2 (green) is located exclusively in the heterochromatin of the chromocenter both in control and mutant spermatids. SIRT1 (blue) is located throughout the nucleus except in the chromocenter and in the region of high BRDT localization. In control spermatids, the high localization of BRDT around the chromocenter might function to separate SIRT1 and SUV39H1/2. In mutant spermatids, the loss of BRDT localization allows overlap of SIRT1 and SUV39H1/2 which may cause ectopic heterochromatin. (For interpretation of the references to color in this figure legend, the reader is referred to the online version of this chapter.)

chromatin architecture of spermatocytes in such a way that meiosis can be completed, but the chromocenter cannot form or be maintained properly.

3.2. BRDT and spermatid elongation

The failure to form a singular intact chromocenter correlates with many of the later defects observed in $Brdt^{\Delta BD1/\Delta BD1}$ spermatids. At step 9, when spermatid elongation begins in control testes, $Brdt^{\Delta BD1/\Delta BD1}$ spermatids fail to elongate properly, with some spermatids exhibiting stunted elongation and some not elongating at all (Berkovits & Wolgemuth, 2011; Shang et al., 2007). Although chromatin compaction appears to occur in condensing spermatids, without proper elongation condensed but still round spermatid nuclei result. A model proposing a direct connection between the multiple chromocenter defect and loss of nuclear shaping is presented in Fig 11.4. Whole chromosome domains are arranged around the chromocenter, suggesting that the chromocenter is a central organizing structure that helps create a defined nuclear topology in spermatids (Haaf & Ward, 1995; Meyer-Ficca et al., 1998). This hypothesis has been supported by later findings in three different mutant models, which showed that loss of a

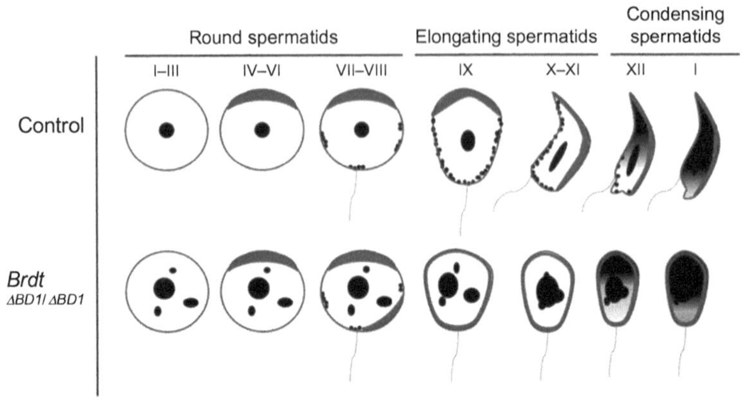

Figure 11.4 Model for how the presence of multiple chromocenters correlates with aberrant localization of H1FNT, loss of heterochromatin foci lining the nuclear membrane, and defective spermatid elongation. In control spermatids, an intact chromocenter (black circle) is present starting in step 1 spermatids. Starting in step 4 spermatids, H1FNT expression (red) begins and is restricted to a subacrosomal chromatin domain. In step 7 spermatids, heterochromatin foci (small black dots) begin to appear at the nuclear membrane in regions where H1FNT is not present. Nuclear elongation occurs in steps 9–11 spermatids, and chromatin condensation begins in step 12 spermatids and is completed in step 13 spermatids. In the *Brdt*$^{\Delta BD1}$ mutant, the majority of spermatids do not have a single intact chromocenter (large black spots). Although H1FNT is initially localized in the subacrosomal region, ectopic localization appears in late mutant round spermatids. In step 9 spermatids, H1FNT is abnormally localized just inside the entire nuclear membrane and blocks the localization of heterochromatin foci in this region. Mutant spermatid nuclei do not fully elongate or take on a normal hook shape. However, chromatin condensation still occurs in steps 12 and 13 spermatids. Heterochromatin is shown in black. (For interpretation of the references to color in this figure legend, the reader is referred to the online version of this chapter.)

singular intact chromocenter can lead to mislocalization of chromatin domains (Berkovits & Wolgemuth, 2011; Catena et al., 2006). Specifically, the presence of multiple chromocenters correlates with a loss of localization of the testis-specific histone variant H1FNT to the nuclear region adjacent to the acrosome (Martianov et al., 2005). In these cells, H1FNT is eventually distributed around the periphery of the nucleus instead of being restricted to the postacrosomal region. H1FNT localization has been implicated in blocking the formation of heterochromatin foci around the nuclear periphery of spermatids (Martianov et al., 2005). Indeed, in *Brdt*$^{\Delta BD1/\Delta BD1}$ spermatids, these foci are not localized to the nuclear periphery (Shang et al., 2007). Heterochromatin foci are hypothesized to connect spermatid chromatin to the actin network in the sperm head and thus coordinate sperm head shaping. In their absence, the sperm nucleus has no tether by which it can be elongated. Although the Sertoli cell's contractile ring may still exert a force on the nucleus to narrow it, the disconnect between the nucleus and the cytoskeleton leads to incomplete or absent elongation.

The link between the chromocenter and proper spermatid nuclear elongation was most evident when the *Brdt*$^{\Delta BD1}$ mutation was examined on two different pure genetic backgrounds (Berkovits & Wolgemuth, 2011). On a pure 129 Sv/Ev background, *Brdt*$^{\Delta BD1/\Delta BD1}$ mice exhibited highly penetrant chromocenter and elongation defects. By contrast, although *Brdt*$^{\Delta BD1/\Delta BD1}$ mutant mice on a C57Bl/6 background were infertile, the chromocenter defect was far less pronounced as was the elongation defect. That these two aspects of the *Brdt*$^{\Delta BD1}$ phenotype should so closely correlate in severity with respect to the specific genetic background strongly supports the link between the chromocenter and spermatid elongation.

3.3. BD1 of BRDT is dispensable for completion of meiosis

As BRDT is expressed from the pachytene through diplotene stages of spermatocytes, we predicted that it would have an important function during meiosis. However, no overt defects were observed in ΔBD1 spermatocytes, and both meiosis I and meiosis II were successfully completed with abundant numbers of round spermatids forming (Shang et al., 2007). Loss of the first bromodomain did lead to an increased expression of the testis-specific histone variant HIST1H1T in spermatocytes and BRDT was shown to bind to a distal region of the *Hist1H1t* promoter (Shang et al., 2007). However, HIST1H1T is dispensable for meiosis as *Hist1h1t* homozygous mutant mice are fertile (Fantz, Hatfield, Horvath, Kistler, & Kistler, 2001; Lin, Sirotkin, & Skoultchi, 2000). Additionally, BRDT was shown to interact with several components of the spliceosome in the testis, and alteration of splicing of at least one target gene was observed in spermatocytes (Berkovits, Wang, Guarnieri, & Wolgemuth, 2012).

We hypothesize that loss of the BD1 does not represent a complete loss of BRDT function, as BET proteins are known to be highly modular. It is therefore possible, even probable, that BRDT is indispensable for meiosis, but that other domains (BD2, the ET domain, the CTD) carryout this function. Such a case is observed in the yeast BET gene *BDF1*. BDF1 is essential for sporulation and *BDF1* null yeast has defects in both meiosis I and meiosis II and fail to complete meiosis (Chua & Roeder, 1995). However, when mutations were made in specific regions of BDF1, only mutations in

BD2 led to failed meiosis. Mutations in BD1 or the ET domain had no overt effect on the completion of meiosis. Thus, although the BD1 of BRDT is not essential for meiotic completion, the BD2 of BRDT may well be involved.

4. BRDT AND REGULATION OF TRANSCRIPTION AND POSTTRANSCRIPTIONAL PROCESSING DURING SPERMATOGENESIS

The BET family of proteins does not directly bind to DNA but rather binds acetylated histone tails and other acetylated proteins and, via these interactions, have been shown to be transcriptional modulators (Florence & Faller, 2001; Lamonica et al., 2011; Peng et al., 2007; Wu & Chiang, 2007). BD1 of BRDT has been shown to bind histone H4 diacetylated on K5 and K8 (Moriniere et al., 2009). Acetylation of H4 tails is well known to be associated with euchromatin (Suka, Suka, Carmen, Wu, & Grunstein, 2001) and active transcription (Grunstein, 1997). It is also known that BRD4, as part of the P-TEFb complex, regulates the active transcription of a large number of genes via BRD4's interaction with acetylated H3 and H4 (Yang et al., 2005). BRD4's binding to the components of the P-TEFb complex (cyclin T and CDK9) is mediated by the CTD of BRD4, and the CTD of BRDT was also found capable of this interaction (Bisgrove et al., 2007). The CTD domain is not present in BRD2 or BRD3; thus BRDT is the only BET protein that would be capable of carrying out P-TEFb-mediated transcriptional regulation in spermatocytes or spermatids (as BRD4 is not expressed in these cells). As previously mentioned, the first documented transcriptional regulation mediated by BRDT was repression of *Hist1h1t* expression and not activation (Shang et al., 2007).

4.1. BRDT and transcriptional regulation

The $Brdt^{\Delta BD1/\Delta BD1}$ phenotype also revealed acrosome detachment, excess cytoplasm, malformed tails and heads, and reduced motility (Berkovits & Wolgemuth, 2011; Shang et al., 2007), defects that cannot all be easily explained by alterations in the chromocenter. The simplest explanation for these defects would be that BRDT is indeed a transcriptional modulator, like other BET family members, and in spermatids, BD1 is required for proper transcription of essential mRNAs, some of which are translated later in spermiogenesis. The phenomenon of translational delay is well characterized in spermatids, as transcription is shutoff after step 8, mRNAs for proteins required for later stages of differentiation must be transcribed prior to step 9 of spermiogenesis and sequestered until later translation (Steger, 2001).

Microarray analysis of control and *Brdt*$^{\Delta BD1/\Delta BD1}$ mutant round spermatids has indeed demonstrated that a significant alteration in transcription occurs in the absence of BD1 of BRDT (Berkovits et al., 2012). Over 1000 genes were upregulated and >400 were repressed in mutant spermatids. Many of these genes were involved in many diverse functions that have putative roles in spermiogenesis (Table 11.1). Some of the more interesting genes with regard to the *Brdt*$^{\Delta BD1/\Delta BD1}$ phenotype include those involved in intracellular protein transport and protein localization, as the highly asymmetric nature of spermatozoa requires extensive cellular restructuring and protein localization

Table 11.1 Biological processes enriched in genes upregulated in *Brdt*$^{\Delta BD1}$ round spermatids

Biological process	No. of genes	Possible genes of interest
mRNA processing	45	*Cnot6l* (RNA supression), *Cstf2* (3′ UTR), *Tardbp* (transcriptional repression), *Srsf2* (alternative splicing)
Intracellular transport	57	*Dynll2* (microtubule motor), *Hook1* (acrosome, manchette, microtubules), *Mtx2* (mitochondrial transport), *Kpna3* (nuclear import), *Cdc42* (acrosome, capacitation)
Macromolecular complex assembly	41	*Capza2* (F-actin cap), *Cdk5* (tail structure), *Tuba1a* and *Tuba1b* (microtubules), *Tubb3* and *Tubb5* (microtubules)
Protein localization	67	*Rab11a* (junction remodeling), *Rab18* (trafficing), *Rab7* (junction remodeling), *Mcfd2* (ER-Golgi transport), *Tmsb4x* (cytoskeletal capping)
RNA localization	14	*Thoc4* (RNA transport, protein interaction), *Fmr1* (translation, tail microtubules), *Srsf13a* (Crem splicing), *Qk* (mRNA stability, translation)
Modification-dependent protein catabolic process	49	*Cdap* (cytoskeletal polarity), *Ranbp2* and *Cbx4* (sumo ligases), 37 genes in the ubiquitination pathway

Continued

Table 11.1 Biological processes enriched in genes upregulated in *Brdt*$^{\Delta BD1}$ round spermatids—cont'd

Biological process	No. of genes	Possible genes of interest
Chromosome organization	41	*Hp1bp3* (heterochromatin structure), *Sgol2* (protects centromeres), Rbl1 (heterochromatin formation), *Cbx1* (Hp1β)
Regulation of cytoskeleton organization	16	*Acpc3* (actin polymerization), *Rac1* (sperm head), *Espn* (ectoplasmic specilization), *Plk4* (centriole, flagella)
Golgi vesicle transport	11	*Sec23a* (acrosome), *Lman1* (ER-Golgi transport), *Golga7* (Golgi to cell surface transport)
Glucose catabolic process	11	*Ldhb* (glycolosis), *Pgk1* (glycolosis first step), *Gapdh* (glycolosis), *Rpia* (carbohydrate metabolism), *Taldo1* (sperm mitochondria)
Negative regulation of gene expression	40	*Ctnnb1* (acrosome, junctions), *Smarca5* (fertilization), *Wt1* (represses Ctnnb1 signaling), *Khdrbs1* (transcription, mRNA sequestering)
Protein complex biogenesis and assembly	26	*Ptk2b* (focal adhesion), *Tmem70* (mitochondrial ATP synthesis), *Gsn* (actin capping)

(Kierszenbaum, Rivkin, & Tres, 2011). For example, defective Golgi vesicle transport could contribute to the observed acrosome defects. The enrichment of genes involved in cytoskeleton organization could also explain some of the sperm structural defects, especially in the tail. As glycolysis is the main source of ATP for flagellar motion (Miki et al., 2004), the inclusion of genes involved in glucose metabolism might also contribute to poor motility. It should be noted that direct binding of BRDT to the promoters of these genes has not been investigated, so it is not known whether BRDT directly regulates them, regulates them as part of a complex (possibly P-TEFb), or whether BRDT may modulate transcription of an intermediate factor. That the transcription of only 400 genes (some of which may not be directly bound by BRDT) are normally activated by BRDT suggests that if BRDT does have a P-TEFb-dependent function, it may not be abrogated in the absence of the first bromodomain. Indeed, studies have shown that the BD2 of BRD4 in conjunction with the CTD is required for P-TEFb function (Schroder et al., 2012).

4.2. BRDT, the spliceosome, and 3′ end formation

Protein–protein interaction studies from our lab suggested that another possible mechanism by which BRDT may regulate gene expression during spermatogenesis is via its interaction with the spliceosome (Berkovits et al., 2012). Mass spectrometry of proteins pulled down in 293T cells ectopically expressing full-length BRDT revealed that BRDT can complex with a variety of spliceosome components. The interaction of BRDT in testicular extracts with four such splicing proteins, SRSF2, DDX5, HNRNPK, and TARDBP, was confirmed, and interestingly, BD1 was not required for this interaction. However, TARDBP, a transcriptional corepressor as well as a splicing modulator, was aberrantly spliced in the mutant testis, demonstrating that while $BRDT^{\Delta BD1}$ may still be capable of complexing with the spliceosome, alternative splicing did not proceed normally. It is interesting to note that alternative splicing is more prevalent in the testis than in any organ other than the brain (Elliott & Grellscheid, 2006; Yeo, Holste, Kreiman, & Burge, 2004).

There was also a striking alteration in the processing of the 3′ UTR of a large number of mRNAs in $Brdt^{\Delta BD1/\Delta BD1}$ round spermatids (Berkovits et al., 2012). A preference for mRNA isoforms that use a more proximal polyadenylation site and thus have a shorter 3′ UTR is a common feature of round spermatids (Liu et al., 2007). The 3′ UTR is a target for RNA-binding proteins (Idler & Yan, 2012) as well as microRNAs (He, Kokkinaki, Pant, Gallicano, & Dym, 2009) and may be involved in the timing and rate of translation of the corresponding mRNA (Yu, Raabe, & Hecht, 2005). In mutant round spermatids, the presence of longer 3′ UTRs correlated with markedly reduced protein levels for the three genes tested. The full extent of alteration in 3′ UTR processing is not known due to the limited number of microarray probes for the majority of genes; however, at least 250 genes, for which there was a microarray probe both in the coding region or proximal 3′ UTR and in the distal 3′ UTR, are predicted to have extended 3′ UTRs and reduced protein levels. Thus, 3′ UTR processing, in addition to transcription and splicing, is another mechanism by which BRDT regulates postmeiotic gene function.

While nothing is known about how BRDT interacts with the spliceosome and the 3′ end formation machinery, there are some observations as to how BRD4, as part of P-TEFb, interacts with these complexes. SRSF2, a major splicing factor and component of the spliceosome, was shown to have an active role in transcriptional elongation mediated by

P-TEFb in cultured mouse embryonic fibroblasts (Lin, Coutinho-Mansfield, Wang, Pandit, & Fu, 2008). Reciprocally, ectopic P-TEFb function led to accumulation of SRSF2 and alterations in splicing in HeLa cells (Barboric et al., 2009). Although P-TEFb and the spliceosome appear to be linked in their control of transcriptional elongation and splicing, the prevailing hypothesis has been that both complexes interact with the CTD of Pol II, but not with each other (Lenasi & Barboric, 2010). It has been demonstrated that some genes with paused Pol II can proceed with low levels of transcription without phosphorylation of the CTD by P-TEFb, but these nascent transcripts lack cotranscriptional processing (Hargreaves, Horng, & Medzhitov, 2009). This suggests that P-TEFb is essential for not just splicing, but mRNA capping and 3′ end maturation. Indeed, the cap-binding protein complex (CBC), which is essential for mRNA 5′ capping, was also found in the complex with P-TEFb, SRSF2, and Pol II in HeLa cells (Lenasi, Peterlin, & Barboric, 2011). There is no direct evidence that P-TEFb interacts with 3′ end maturation proteins, but the CTD of Pol II (Licatalosi et al., 2002) and the spliceosome do (Martinson, 2011). As mentioned previously, we suggest that BRDT might replace BRD4 as part of P-TEFb in spermatocytes and spermatids and thus modulate transcription, splicing, and UTR processing.

5. BRDT AND LATER STAGES OF SPERMIOGENESIS

5.1. BRDT and chromatin compaction

It had been previously reported that BRDT ectopically expressed in cultured cells causes an acetylation-dependent reorganization of chromatin (Pivot-Pajot et al., 2003). This reorganization of chromatin only occurred after the cells were treated with the deacetylase inhibitor trichostatin A (TSA), and both bromodomains and motif B were required. The significance of the ET domain, however, is less clear. The initial report found the ET domain to be dispensable, and in fact, a BRDT fragment lacking the ET domain exhibited enhanced chromatin reorganization (Pivot-Pajot et al., 2003). However, a later study found that a fragment with just the bromodomains and a fragment with the bromodomains and the ET domain had similar reorganization capabilities (Dhar et al., 2012). Whether these different results are due to the different cultured cells used (Cos-7 vs. RAG) is not known, and the general significance of this purported chromatin reorganizing ability is not at all clear.

Hyperacetylation of H4 and H2A beginning in step IX elongating spermatids is a well-documented component of chromatin remodeling during mammalian spermatid differentiation (Meistrich, Trostle-Weige, Lin, Bhatnagar, & Allis, 1992). Species that replace their canonical histones with transition proteins and then protamine all feature this hyperacetylation (Christensen & Dixon, 1982; Gatewood, Cook, Balhorn, Schmid, & Bradbury, 1990; Wolf, Jeppesen, & Mitchell, 1993), while species that do not replace their histones (Kennedy & Davies, 1980). This hyperacetylation is hypothesized to weaken the association of DNA to the histone octamer (Shogren-Knaak et al., 2006) and thus facilitate removal of canonical histones and replacement with first transition proteins and then protamines. Indeed, male mice lacking RNF8, a histone ubiquitin E3 ligase, are infertile due to a failure to deposit the transition proteins and protamines into chromatin (Lu et al., 2010). There is a loss of histone H2A ubiquitination and a loss of histone H4K16 acetylation in the elongating spermatids of these mice. Thus, ubiquitination of H2A is required for complete H4 hyperacetylation and loss of this hyperacetylation was concomitant with retention of canonical histones. The mechanism by which the histone to transition protein exchange is carried out is as yet unknown, and the possibility that BRDT plays a role is enticing. However, for BRDT to carry out such a function *in vivo* it would have to be present in hyperacetylated spermatids and colocalize with hyperacetylated H4. This was indeed reported by one group that examined BRDT and acetylated H4 expression in the mouse testis (Govin et al., 2006). As mentioned previously, while we clearly detected BRDT in murine round spermatids through step 8, we do not detect it in elongating spermatids (Shang et al., 2007) and a third study reported that the expression in elongating rat spermatids was restricted to the anterior region of the spermatid and only minorly overlapped with the nucleus (Dhar et al., 2012). It is difficult to understand how BRDT could be involved in the histone to transition protein exchange if it is not present when this event occurs or if it does not physically colocalize with the majority of nuclear chromatin.

The chromatin remodeling assay that was carried out in non-spermatogenic cells (Dhar et al., 2012; Pivot-Pajot et al., 2003) was extended to cultured round spermatids (Dhar et al., 2012). Round spermatids are normally hypoacetylated due to expression of histone deacetylases, and TSA is capable of inducing hyperacetylation in these cells (Hazzouri et al., 2000). However, when these cultured round spermatids were treated with TSA, there was no observable chromatin remodeling (as had been seen in

TSA-treated BRDT-transfected Cos-7 and RAG cells), despite the fact that round spermatids strongly express BRDT. Only upon overexpression by transfection with a BRDT-expressing plasmid was acetylation-dependent chromatin remodeling observed. Ectopic expression by plasmid transfection, in cells that already express BRDT, likely results in protein levels well beyond physiological relevance. However, there is the possibility that BRDT might normally require another factor for efficient chromatin reorganization and that this factor is only expressed after the round spermatid stage. Further, in the absence of this factor, abnormally high levels of BRDT might be required for reorganization.

It has been suggested that such a factor could be SMARCE1, a Swi/Snf chromatin remodeling factor (Dhar et al., 2012). SMARCE1 and BRDT were found to coimmunoprecipitate and the two proteins partially colocalize in spermatids. Additionally, colocalization was enhanced by H4 hyperacetylation. However, colocalization of BRDT and SMARCE1 was never greater than 20% of their total distribution of expression. While they may indeed interact, neither protein was abundant in elongating spermatid chromatin, and without localization throughout the hyperacetylated spermatid nucleus, it is difficult to envision these proteins facilitating either histone eviction or transition protein depositing.

Last, the phenotype of the *Brdt*$^{\Delta BD1/\Delta BD1}$ mutant is not consistent with a BD1-dependent role in chromatin compaction. The report of BRDT colocalizing with hyperactylated H4 in elongating spermatids resulted in the suggestion that BRDT may itself physically compact chromatin (Govin et al., 2006). In the *Brdt*$^{\Delta BD1/\Delta BD1}$ mutant spermatids, severe morphological abnormalities were initially observed at step IX, just when hyperacetylation occurs (Shang et al., 2007), seemingly supporting this hypothesis. However, the suggestion that BRDT physically compacts acetylated chromatin is difficult to understand, as compaction is not characteristic of elongating hyperacetylated spermatids but rather only begins when transition proteins are integrated into chromatin at step XII (Meistrich, Mohapatra, Shirley, & Zhao, 2003). Thus, a defect at the onset of hyperacetylation temporally correlates with defective elongation and not condensation. Further, highly condensed but nonelongated spermatids are a prominent feature of the mutant testes on all genetic backgrounds (Berkovits & Wolgemuth, 2011; Shang et al., 2007), strongly suggesting that in mutant spermatids, condensation *per se* is not defective. Many other chromatin remodeling events occur during spermatogenesis (Govin, Caron, Lestrat, Rousseaux, & Khochbin, 2004; Lewis, Abbott, & Ausio, 2003), and theoretically may be regulated by

BRDT; however, the canonical histone to transition protein to protamine exchange does not appear to be one of them.

5.2. BRDT in human spermatozoa

Although BRDT protein has not been observed after the removal of 85–95% of the canonical histones in step XII compacting rodent spermatids, BRDT protein has been reported to be present in human spermatozoa (Barda et al., 2012; Zheng et al., 2005). BRDT protein was detected throughout the sperm nucleus, although there did appear to be regions of more intense staining (Barda et al., 2012). The distribution of BRDT did not appear to colocalize with regions containing the retained hyperacetylated canonical histones, suggesting that it did not remain associated with hyperacetylated chromatin after the elongation stage of these cells. A number of transcription factors have been observed in mature spermatozoa and are suggested to have a role in early embryogenesis (Pittoggi et al., 2001). It has also been reported that in sperm nuclei, canonical histones are retained at the promoters of some genes whose transcription is required in the early embryo (Hammoud et al., 2009). Thus, it is possible that during early decompaction of the sperm nucleus, residual BRDT interacts with these retained histones and is involved with modulating transcription of these early paternal genes.

6. BET INHIBITION AS A POSSIBLE CONTRACEPTIVE

6.1. BET family inhibition

There has been a recent explosion in the discovery of small molecules that specifically antagonize the interaction of BET bromodomains with their biological targets. At least five such molecules have been identified, all of which bind in the acetyl-lysine-binding pocket and thus block BET function (Dawson et al., 2011; Filippakopoulos et al., 2010; Hewings et al., 2011; Nicodeme et al., 2010; Zhang et al., 2012). Although BET bromodomains are structurally similar, they do bind very distinct targets and thus may have different binding affinities for these various inhibitory molecules. Compound JQ1 bound to the first bromodomains of the BET proteins with the following order of affinity: BRD4>BRD3>BRD2>BRDT, with very strong binding to the BD1 of BRD4, and very weak binding to the BD1 of BRDT (Filippakopoulos et al., 2010). The second bromodomains all bound JQ1 relatively equally, although the BD2 of BRDT was not investigated.

The I-BET compound was found to bind in a ratio of two molecules to one BET protein, and it bound the tandem bromodomains of BRD2, BRD3, and BRD4 relatively equally (Chung, Coste, et al., 2011). I-BET binding to BRDT was not reported.

Inhibition of BRD4 was shown to downregulate transcription of the oncogene *c-Myc* in a variety of leukemia and lymphoma cell lines (Dawson et al., 2011; Mertz et al., 2011). In mouse models of multiple myeloma, lowering levels of c-MYC resulted in cellular senescence and tumor regression (Delmore et al., 2011). A cell line derived from a rare cancer known as NUT midline carcinoma, in which a translocation of the N-terminus of BRD4 or BRD3 is fused to the testis-specific gene NUT (French, 2010), showed squamous differentiation and loss of proliferation upon treatment with a BET inhibitor (Filippakopoulos et al., 2010). The inflammatory response can also be attenuated by BET inhibitors. One such molecule was shown to repress the transcription of inflammation genes in response to endotoxic shock both in cell lines and in mice (Nicodeme et al., 2010), and another molecule inhibited inflammatory transcription in response to HIV infection in kidney cells (Zhang et al., 2012).

Considering that all four BET proteins are expressed at various stages of male germ cell development, it seems likely that this class of small molecules would have effects on spermatogenesis. A molecule that specifically inhibited BRDT function would be the obvious choice for a male contraceptive. As BRDT bromodomains are the most divergent of any mammalian BET (Fig. 11.1), it would seem likely that a compound could be created that had the highest affinity for BRDT. This would alleviate a possible deleterious affect by a BET inhibitor in any other tissue. Further BRDT is not expressed in mitotically dividing spermatogonia, and thus a BRDT-specific inhibitor would not affect the spermatogonial stem cell population, theoretically making any spermatogenic impairment reversible.

As no BRDT-specific inhibitor has yet been reported, the question arises as to whether inhibition of the other three BET genes might also perturb spermatogenesis. BRD4 and, at lower levels, BRD2 are present in mitotically dividing spermatogonia. Homozygous null mutations of BRD4 and BRD2 are embryonic lethal (Houzelstein et al., 2002; Shang, Wang, Wen, Greenberg, & Wolgemuth, 2009), and as both genes regulate expression of cell cycle genes, it seems likely that their inhibition would disrupt spermatogonial division. Thus, while inhibition of BRD4 and BRD2 function would likely be an effective contraceptive, it is unclear whether such a disruption of spermatogenesis would be reversible. Maintenance of spermatogonial stem cells is required for continued fertility (Singh, Burnicka-Turek, Chauhan, & Hou, 2011) and their number is the main determinant of spermatogenic recovery after insult (Meistrich, 1986, 1993). If the BET genes were inhibited in these cells permanent sterility might result. BRD3, however, did not appear to be expressed in spermatogonia, although its mRNA was abundantly detected in spermatids (Shang et al., 2004). Thus, inhibition of BRD3 might indeed be a useful and reversible method of contraception. However, as BRD3is expressed in many adult tissue (Thorpe et al., 1997), the question of possible toxicity to other organs would need to be investigated.

Lastly, the possibility exists that even though the BET genes are expressed in other tissues, their inhibition may still function as an effective and reversible contraceptive. The testis is an acutely sensitive organ to damage (Bonde, 2010), and there are numerous examples wherein low level doses of a damaging agent resulted in sterility and no other phenotype (Boekelheide & Schoenfeld, 2001; Petrelli & Figa-Talamanca, 2001; Potashnik, Ben-Aderet, Israeli, Yanai-Inbar, & Sober, 1978). Further, inhibition of a widely expressed transcriptional regulator has been shown to have a dramatic and reversible effect on spermatogenesis but no abnormalities in any other tissue/organ (Chung, Wang, et al., 2011). Specifically, low doses of a retinoic acid receptor inhibitor (BMS 189453) over extended dosing periods caused reversible sterility in the mouse model. Thus, low doses of the current inhibitors might have a specific effect on spermatogenesis even if their targets are expressed in other tissues as well. Alternatively, even if the inhibitors only bind weakly to BRDT, this disruption might be sufficient to impair spermatogenesis. An in-depth study of the effect of BET inhibition on spermatogenesis is certainly needed, and a BET inhibitor with higher affinity for BRDT than the other BET proteins could be an ideal reversible male contraceptive.

7. CONCLUSIONS

The members of the BET family are generally thought of as transcriptional regulators, but their role as modulators of chromatin architecture is conserved from yeast to mammals. Both aspects of BET function are at play in the mammalian testis, where full-length BRDT is required both for proper meiotic and postmeiotic transcription and postmeiotic chromosome organization. BRDT is also required in the testis for cotranscriptional processing. This is hypothetically due to BRDT replacing BRD4 in the P-TEFb complex in spermatocytes and spermatids, and this hypothesis must be further investigated to fully understand both BRDT and BRD4 function. Given the distinct patterns of expression of all four BET family members in the testis, it is likely that they each have distinct roles—testing this hypothesis will require the use of conditional knockout models as both *Brd2* and *Brd4* mutant mice are embryonic lethal. The possibility of using BET bromodomain inhibitors as male contraceptives is promising, but much work is still needed to characterize these molecules potential targets, as well as their mechanisms of action, in the testis.

ACKNOWLEDGMENTS

This work was supported in part by a grant from the NIH to D. J. W. (GM081767) and by T32 GM007008 (B. D. B.). We would like to thank Dr. Timothy Bestor for critical comments and suggestions early in preparation of the chapter.

REFERENCES

Agricola, E., Randall, R. A., Gaarenstroom, T., Dupont, S., & Hill, C. S. (2011). Recruitment of TIF1gamma to chromatin via its PHD finger-bromodomain activates its ubiquitin ligase and transcriptional repressor activities. *Molecular Cell*, *43*, 85–96.

Ai, N., Hu, X., Ding, F., Yu, B., Wang, H., Lu, X., et al. (2011). Signal-induced Brd4 release from chromatin is essential for its role transition from chromatin targeting to transcriptional regulation. *Nucleic Acids Research*, *39*, 9592–9604.

Allton, K., Jain, A. K., Herz, H. M., Tsai, W. W., Jung, S. Y., Qin, J., et al. (2009). Trim24 targets endogenous p53 for degradation. *Proceedings of the National Academy of Sciences of the United States of America*, *106*, 11612–11616.

Arrowsmith, C. H., Bountra, C., Fish, P. V., Lee, K., & Schapira, M. (2012). Epigenetic protein families: A new frontier for drug discovery. *Nature Reviews. Drug Discovery*, *11*, 384–400.

Barboric, M., Lenasi, T., Chen, H., Johansen, E. B., Guo, S., & Peterlin, B. M. (2009). 7SK snRNP/P-TEFb couples transcription elongation with alternative splicing and is essential for vertebrate development. *Proceedings of the National Academy of Sciences of the United States of America*, *106*, 7798–7803.

Barda, S., Paz, G., Yogev, L., Yavetz, H., Lehavi, O., Hauser, R., et al. (2012). Expression of BET genes in testis of men with different spermatogenic impairments. *Fertility and Sterility*, *97*(46–52), e5.

Barlev, N. A., Liu, L., Chehab, N. H., Mansfield, K., Harris, K. G., Halazonetis, T. D., et al. (2001). Acetylation of p53 activates transcription through recruitment of coactivators/histone acetyltransferases. *Molecular Cell*, *8*, 1243–1254.

Beck, S., Hanson, I., Kelly, A., Pappin, D. J., & Trowsdale, J. (1992). A homologue of the Drosophila female sterile homeotic (fsh) gene in the class II region of the human MHC. *DNA Sequence*, *2*, 203–210.

Berkovits, B. D., Wang, L., Guarnieri, P., & Wolgemuth, D. J. (2012). The testis-specific double bromodomain-containing protein BRDT forms a complex with multiple spliceosome components and is required for mRNA splicing and 3'-UTR truncation in round spermatids. *Nucleic Acids Research*, *40*, 7162–7175.

Berkovits, B. D., & Wolgemuth, D. J. (2011). The first bromodomain of the testis-specific double bromodomain protein Brdt is required for chromocenter organization that is modulated by genetic background. *Developmental Biology*, *360*, 358–368.

Bisgrove, D. A., Mahmoudi, T., Henklein, P., & Verdin, E. (2007). Conserved P-TEFb-interacting domain of BRD4 inhibits HIV transcription. *Proceedings of the National Academy of Sciences of the United States of America, 104*, 13690–13695.

Boekelheide, K., & Schoenfeld, H. A. (2001). Spermatogenesis by Sisyphus: Proliferating stem germ cells fail to repopulate the testis after 'irreversible' injury. *Advances in Experimental Medicine and Biology, 500*, 421–428.

Bonde, J. P. (2010). Male reproductive organs are at risk from environmental hazards. *Asian Journal of Andrology, 12*, 152–156.

Catena, R., Ronfani, L., Sassone-Corsi, P., & Davidson, I. (2006). Changes in intranuclear chromatin architecture induce bipolar nuclear localization of histone variant H1T2 in male haploid spermatids. *Developmental Biology, 296*, 231–238.

Christensen, M. E., & Dixon, G. H. (1982). Hyperacetylation of histone H4 correlates with the terminal, transcriptionally inactive stages of spermatogenesis in rainbow trout. *Developmental Biology, 93*, 404–415.

Chua, P., & Roeder, G. S. (1995). Bdf1, a yeast chromosomal protein required for sporulation. *Molecular and Cellular Biology, 15*, 3685–3696.

Chung, C. W., Coste, H., White, J. H., Mirguet, O., Wilde, J., Gosmini, R. L., et al. (2011). Discovery and characterization of small molecule inhibitors of the BET family bromodomains. *Journal of Medicinal Chemistry, 54*, 3827–3838.

Chung, S. S., Wang, X., Roberts, S. S., Griffey, S. M., Reczek, P. R., & Wolgemuth, D. J. (2011). Oral administration of a retinoic acid receptor antagonist reversibly inhibits spermatogenesis in mice. *Endocrinology, 152*, 2492–2502.

Clowney, E. J., Magklara, A., Colquitt, B. M., Pathak, N., Lane, R. P., & Lomvardas, S. (2011). High-throughput mapping of the promoters of the mouse olfactory receptor genes reveals a new type of mammalian promoter and provides insight into olfactory receptor gene regulation. *Genome Research, 21*, 1249–1259.

Crowley, T., Brunori, M., Rhee, K., Wang, X., & Wolgemuth, D. J. (2004). Change in nuclear-cytoplasmic localization of a double-bromodomain protein during proliferation and differentiation of mouse spinal cord and dorsal root ganglia. *Brain Research. Developmental Brain Research, 149*, 93–101.

Das, D., & Scovell, W. M. (2001). The binding interaction of HMG-1 with the TATA-binding protein/TATA complex. *The Journal of Biological Chemistry, 276*, 32597–32605.

Dawson, M. A., Prinjha, R. K., Dittman, A., Giotopoulos, G., Bantscheff, M., Chan, W. I., et al. (2011). Inhibition of BET recruitment to chromatin as an effective treatment for MLL-fusion leukaemia. *Nature, 478*, 529–533.

Delmore, J. E., Issa, G. C., Lemieux, M. E., Rahl, P. B., Shi, J., Jacobs, H. M., et al. (2011). BET bromodomain inhibition as a therapeutic strategy to target c-Myc. *Cell, 146*, 904–917.

Denis, G. V. (2010). Bromodomain coactivators in cancer, obesity, type 2 diabetes, and inflammation. *Discovery Medicine, 10*, 489–499.

Denis, G. V., Vaziri, C., Guo, N., & Faller, D. V. (2000). RING3 kinase transactivates promoters of cell cycle regulatory genes through E2F. *Cell Growth & Differentiation, 11*, 417–424.

Dey, A., Ellenberg, J., Farina, A., Coleman, A. E., Maruyama, T., Sciortino, S., et al. (2000). A bromodomain protein, MCAP, associates with mitotic chromosomes and affects G(2)-to-M transition. *Molecular and Cellular Biology, 20*, 6537–6549.

Dey, A., Nishiyama, A., Karpova, T., McNally, J., & Ozato, K. (2009). Brd4 marks select genes on mitotic chromatin and directs postmitotic transcription. *Molecular Biology of the Cell, 20*, 4899–4909.

Dhar, S., Thota, A., & Rao, M. R. (2012). Insights into role of bromodomain, testis-specific (Brdt) in acetylated histone H4-dependent chromatin remodeling in mammalian spermiogenesis. *The Journal of Biological Chemistry, 287*, 6387–6405.

Durant, M., & Pugh, B. F. (2007). NuA4-directed chromatin transactions throughout the Saccharomyces cerevisiae genome. *Molecular and Cellular Biology, 27*, 5327–5335.

Elliott, D. J., & Grellscheid, S. N. (2006). Alternative RNA splicing regulation in the testis. *Reproduction, 132*, 811–819.

Fantz, D. A., Hatfield, W. R., Horvath, G., Kistler, M. K., & Kistler, W. S. (2001). Mice with a targeted disruption of the H1t gene are fertile and undergo normal changes in structural chromosomal proteins during spermiogenesis. *Biology of Reproduction, 64*, 425–431.

Filippakopoulos, P., Qi, J., Picaud, S., Shen, Y., Smith, W. B., Fedorov, O., et al. (2010). Selective inhibition of BET bromodomains. *Nature, 468*, 1067–1073.

Florence, B., & Faller, D. V. (2001). You bet-cha: A novel family of transcriptional regulators. *Frontiers in Bioscience, 6*, D1008–D1018.

Florence, B. L., & Faller, D. V. (2008). Drosophila female sterile (1) homeotic is a multifunctional transcriptional regulator that is modulated by Ras signaling. *Developmental Dynamics, 237*, 554–564.

French, C. A. (2010). NUT midline carcinoma. *Cancer Genetics and Cytogenetics, 203*, 16–20.

Gamsjaeger, R., Webb, S. R., Lamonica, J. M., Billin, A., Blobel, G. A., & Mackay, J. P. (2011). Structural basis and specificity of acetylated transcription factor GATA1 recognition by BET family bromodomain protein Brd3. *Molecular and Cellular Biology, 31*, 2632–2640.

Gans, M., Audit, C., & Masson, M. (1975). Isolation and characterization of sex-linked female-sterile mutants in Drosophila melanogaster. *Genetics, 81*, 683–704.

Gans, M., Forquignon, F., & Masson, M. (1980). The role of dosage of the region 7D1-7D5-6 of the X chromosome in the production of homeotic transformations in Drosophila melanogaster. *Genetics, 96*, 887–902.

Gatewood, J. M., Cook, G. R., Balhorn, R., Schmid, C. W., & Bradbury, E. M. (1990). Isolation of four core histones from human sperm chromatin representing a minor subset of somatic histones. *The Journal of Biological Chemistry, 265*, 20662–20666.

Govin, J., Caron, C., Lestrat, C., Rousseaux, S., & Khochbin, S. (2004). The role of histones in chromatin remodelling during mammalian spermiogenesis. *European Journal of Biochemistry, 271*, 3459–3469.

Govin, J., Lestrat, C., Caron, C., Pivot-Pajot, C., Rousseaux, S., & Khochbin, S. (2006). Histone acetylation-mediated chromatin compaction during mouse spermatogenesis. *Ernst Schering Research Foundation Workshop, 57*, 155–172.

Gregory, G. D., Vakoc, C. R., Rozovskaia, T., Zheng, X., Patel, S., Nakamura, T., et al. (2007). Mammalian ASH1L is a histone methyltransferase that occupies the transcribed region of active genes. *Molecular and Cellular Biology, 27*, 8466–8479.

Grunstein, M. (1997). Histone acetylation in chromatin structure and transcription. *Nature, 389*, 349–352.

Guo, N., Faller, D. V., & Denis, G. V. (2000). Activation-induced nuclear translocation of RING3. *Journal of Cell Science, 113*, 3085–3091.

Haaf, T., & Ward, D. C. (1995). Higher order nuclear structure in mammalian sperm revealed by in situ hybridization and extended chromatin fibers. *Experimental Cell Research, 219*, 604–611.

Hammoud, S. S., Nix, D. A., Zhang, H., Purwar, J., Carrell, D. T., & Cairns, B. R. (2009). Distinctive chromatin in human sperm packages genes for embryo development. *Nature, 460*, 473–478.

Hargreaves, D. C., Horng, T., & Medzhitov, R. (2009). Control of inducible gene expression by signal-dependent transcriptional elongation. *Cell, 138*, 129–145.

Haynes, S. R., Dollard, C., Winston, F., Beck, S., Trowsdale, J., & Dawid, I. B. (1992). The bromodomain: A conserved sequence found in human, Drosophila and yeast proteins. *Nucleic Acids Research, 20*, 2603.

Haynes, S. R., Mozer, B. A., Bhatia-Dey, N., & Dawid, I. B. (1989). The Drosophila fsh locus, a maternal effect homeotic gene, encodes apparent membrane proteins. *Developmental Biology*, *134*, 246–257.
Hazzouri, M., Pivot-Pajot, C., Faure, A. K., Usson, Y., Pelletier, R., Sele, B., et al. (2000). Regulated hyperacetylation of core histones during mouse spermatogenesis: Involvement of histone deacetylases. *European Journal of Cell Biology*, *79*, 950–960.
He, Z., Kokkinaki, M., Pant, D., Gallicano, G. I., & Dym, M. (2009). Small RNA molecules in the regulation of spermatogenesis. *Reproduction*, *137*, 901–911.
Hess, J. L. (2004). MLL: A histone methyltransferase disrupted in leukemia. *Trends in Molecular Medicine*, *10*, 500–507.
Hewings, D. S., Wang, M., Philpott, M., Fedorov, O., Uttarkar, S., Filippakopoulos, P., et al. (2011). 3,5-dimethylisoxazoles act as acetyl-lysine-mimetic bromodomain ligands. *Journal of Medicinal Chemistry*, *54*, 6761–6770.
Houzelstein, D., Bullock, S. L., Lynch, D. E., Grigorieva, E. F., Wilson, V. A., & Beddington, R. S. (2002). Growth and early postimplantation defects in mice deficient for the bromodomain-containing protein Brd4. *Molecular and Cellular Biology*, *22*, 3794–3802.
Hoyer-Fender, S., Singh, P. B., & Motzkus, D. (2000). The murine heterochromatin protein M31 is associated with the chromocenter in round spermatids and Is a component of mature spermatozoa. *Experimental Cell Research*, *254*, 72–79.
Huang, D. H., & Dawid, I. B. (1990). The maternal-effect gene fsh is essential for the specification of the central region of the Drosophila embryo. *The New Biologist*, *2*, 163–170.
Huang, H., Zhang, J., Shen, W., Wang, X., Wu, J., Wu, J., et al. (2007). Solution structure of the second bromodomain of Brd2 and its specific interaction with acetylated histone tails. *BMC Structural Biology*, *7*, 57.
Idler, R. K., & Yan, W. (2012). Control of messenger RNA fate by RNA binding proteins: An emphasis on mammalian spermatogenesis. *Journal of Andrology*, *33*, 309–337.
Jacobson, R. H., Ladurner, A. G., King, D. S., & Tjian, R. (2000). Structure and function of a human TAFII250 double bromodomain module. *Science*, *288*, 1422–1425.
Jang, M. K., Mochizuki, K., Zhou, M., Jeong, H. S., Brady, J. N., & Ozato, K. (2005). The bromodomain protein Brd4 is a positive regulatory component of P-TEFb and stimulates RNA polymerase II-dependent transcription. *Molecular Cell*, *19*, 523–534.
Jeanmougin, F., Wurtz, J. M., Le Douarin, B., Chambon, P., & Losson, R. (1997). The bromodomain revisited. *Trends in Biochemical Sciences*, *22*, 151–153.
Jones, M. H., Numata, M., & Shimane, M. (1997). Identification and characterization of BRDT: A testis-specific gene related to the bromodomain genes RING3 and Drosophila fsh. *Genomics*, *45*, 529–534.
Kanno, T., Kanno, Y., Siegel, R. M., Jang, M. K., Lenardo, M. J., & Ozato, K. (2004). Selective recognition of acetylated histones by bromodomain proteins visualized in living cells. *Molecular Cell*, *13*, 33–43.
Kennedy, B. P., & Davies, P. L. (1980). Acid-soluble nuclear proteins of the testis during spermatogenesis in the winter flounder. Loss of the high mobility group proteins. *The Journal of Biological Chemistry*, *255*, 2533–2539.
Kierszenbaum, A. L., Rivkin, E., & Tres, L. L. (2011). Cytoskeletal track selection during cargo transport in spermatids is relevant to male fertility. *Spermatogenesis*, *1*, 221–230.
Ladurner, A. G., Inouye, C., Jain, R., & Tjian, R. (2003). Bromodomains mediate an acetyl-histone encoded antisilencing function at heterochromatin boundaries. *Molecular Cell*, *11*, 365–376.
Lamonica, J. M., Deng, W., Kadauke, S., Campbell, A. E., Gamsjaeger, R., Wang, H., et al. (2011). Bromodomain protein Brd3 associates with acetylated GATA1 to promote its chromatin occupancy at erythroid target genes. *Proceedings of the National Academy of Sciences of the United States of America*, *108*, E159–E168.

Lenasi, T., & Barboric, M. (2010). P-TEFb stimulates transcription elongation and pre-mRNA splicing through multilateral mechanisms. *RNA Biology*, *7*, 145–150.
Lenasi, T., Peterlin, B. M., & Barboric, M. (2011). Cap-binding protein complex links pre-mRNA capping to transcription elongation and alternative splicing through positive transcription elongation factor b (P-TEFb). *The Journal of Biological Chemistry*, *286*, 22758–22768.
LeRoy, G., Rickards, B., & Flint, S. J. (2008). The double bromodomain proteins Brd2 and Brd3 couple histone acetylation to transcription. *Molecular Cell*, *30*, 51–60.
Lewis, J. D., Abbott, D. W., & Ausio, J. (2003). A haploid affair: Core histone transitions during spermatogenesis. *Biochemistry and Cell Biology*, *81*, 131–140.
Liang, Q., Deng, H., Li, X., Wu, X., Tang, Q., Chang, T. H., et al. (2011). Tripartite motif-containing protein 28 is a small ubiquitin-related modifier E3 ligase and negative regulator of IFN regulatory factor 7. *The Journal of Immunology*, *187*, 4754–4763.
Licatalosi, D. D., Geiger, G., Minet, M., Schroeder, S., Cilli, K., McNeil, J. B., et al. (2002). Functional interaction of yeast pre-mRNA 3' end processing factors with RNA polymerase II. *Molecular Cell*, *9*, 1101–1111.
Lin, S., Coutinho-Mansfield, G., Wang, D., Pandit, S., & Fu, X. D. (2008). The splicing factor SC35 has an active role in transcriptional elongation. *Nature Structural & Molecular Biology*, *15*, 819–826.
Lin, Q., Sirotkin, A., & Skoultchi, A. I. (2000). Normal spermatogenesis in mice lacking the testis-specific linker histone H1t. *Molecular and Cellular Biology*, *20*, 2122–2128.
Liu, D., Brockman, J. M., Dass, B., Hutchins, L. N., Singh, P., McCarrey, J. R., et al. (2007). Systematic variation in mRNA 3'-processing signals during mouse spermatogenesis. *Nucleic Acids Research*, *35*, 234–246.
Liu, H., Mulholland, N., Fu, H., & Zhao, K. (2006). Cooperative activity of BRG1 and Z-DNA formation in chromatin remodeling. *Molecular and Cellular Biology*, *26*, 2550–2559.
Lu, L. Y., Wu, J., Ye, L., Gavrilina, G. B., Saunders, T. L., & Yu, X. (2010). RNF8-dependent histone modifications regulate nucleosome removal during spermatogenesis. *Developmental Cell*, *18*, 371–384.
Lygerou, Z., Conesa, C., Lesage, P., Swanson, R. N., Ruet, A., Carlson, M., et al. (1994). The yeast BDF1 gene encodes a transcription factor involved in the expression of a broad class of genes including snRNAs. *Nucleic Acids Research*, *22*, 5332–5340.
Martianov, I., Brancorsini, S., Catena, R., Gansmuller, A., Kotaja, N., Parvinen, M., et al. (2005). Polar nuclear localization of H1T2, a histone H1 variant, required for spermatid elongation and DNA condensation during spermiogenesis. *Proceedings of the National Academy of Sciences of the United States of America*, *102*, 2808–2813.
Martianov, I., Brancorsini, S., Gansmuller, A., Parvinen, M., Davidson, I., & Sassone-Corsi, P. (2002). Distinct functions of TBP and TLF/TRF2 during spermatogenesis: Requirement of TLF for heterochromatic chromocenter formation in haploid round spermatids. *Development (Cambridge, England)*, *129*, 945–955.
Martinson, H. G. (2011). An active role for splicing in 3'-end formation. *Wiley Interdisciplinary Reviews: RNA*, *2*, 459–470.
Matangkasombut, O., Buratowski, R. M., Swilling, N. W., & Buratowski, S. (2000). Bromodomain factor 1 corresponds to a missing piece of yeast TFIID. *Genes & Development*, *14*, 951–962.
Meistrich, M. L. (1986). Critical components of testicular function and sensitivity to disruption. *Biology of Reproduction*, *34*, 17–28.
Meistrich, M. L. (1993). Effects of chemotherapy and radiotherapy on spermatogenesis. *European Urology*, *23*, 136–141 discussion 142.
Meistrich, M. L., Mohapatra, B., Shirley, C. R., & Zhao, M. (2003). Roles of transition nuclear proteins in spermiogenesis. *Chromosoma*, *111*, 483–488.

Meistrich, M. L., Trostle-Weige, P. K., Lin, R., Bhatnagar, Y. M., & Allis, C. D. (1992). Highly acetylated H4 is associated with histone displacement in rat spermatids. *Molecular Reproduction and Development, 31*, 170–181.

Mertz, J. A., Conery, A. R., Bryant, B. M., Sandy, P., Balasubramanian, S., Mele, D. A., et al. (2011). Targeting MYC dependence in cancer by inhibiting BET bromodomains. *Proceedings of the National Academy of Sciences of the United States of America, 108*, 16669–16674.

Meyer-Ficca, M., Muller-Navia, J., & Scherthan, H. (1998). Clustering of pericentromeres initiates in step 9 of spermiogenesis of the rat (Rattus norvegicus) and contributes to a well defined genome architecture in the sperm nucleus. *Journal of Cell Science, 111*(Pt 10), 1363–1370.

Miki, K., Qu, W., Goulding, E. H., Willis, W. D., Bunch, D. O., Strader, L. F., et al. (2004). Glyceraldehyde 3-phosphate dehydrogenase-S, a sperm-specific glycolytic enzyme, is required for sperm motility and male fertility. *Proceedings of the National Academy of Sciences of the United States of America, 101*, 16501–16506.

Mishima, Y., Miyagi, S., Saraya, A., Negishi, M., Endoh, M., Endo, T. A., et al. (2011). The Hbo1-Brd1/Brpf2 complex is responsible for global acetylation of H3K14 and required for fetal liver erythropoiesis. *Blood, 118*, 2443–2453.

Moriniere, J., Rousseaux, S., Steuerwald, U., Soler-Lopez, M., Curtet, S., Vitte, A. L., et al. (2009). Cooperative binding of two acetylation marks on a histone tail by a single bromodomain. *Nature, 461*, 664–668.

Mujtaba, S., Zeng, L., & Zhou, M. M. (2007). Structure and acetyl-lysine recognition of the bromodomain. *Oncogene, 26*, 5521–5527.

Nicodeme, E., Jeffrey, K. L., Schaefer, U., Beinke, S., Dewell, S., Chung, C. W., et al. (2010). Suppression of inflammation by a synthetic histone mimic. *Nature, 468*, 1119–1123.

Nishiyama, A., Dey, A., Miyazaki, J., & Ozato, K. (2006). Brd4 is required for recovery from antimicrotubule drug-induced mitotic arrest: Preservation of acetylated chromatin. *Molecular Biology of the Cell, 17*, 814–823.

O'Carroll, D., Scherthan, H., Peters, A. H., Opravil, S., Haynes, A. R., Laible, G., et al. (2000). Isolation and characterization of Suv39h2, a second histone H3 methyltransferase gene that displays testis-specific expression. *Molecular and Cellular Biology, 20*, 9423–9433.

Ogryzko, V. V., Schiltz, R. L., Russanova, V., Howard, B. H., & Nakatani, Y. (1996). The transcriptional coactivators p300 and CBP are histone acetyltransferases. *Cell, 87*, 953–959.

Ohshima, T., Suganuma, T., & Ikeda, M. (2001). A novel mutation lacking the bromodomain of the transcriptional coactivator p300 in the SiHa cervical carcinoma cell line. *Biochemical and Biophysical Research Communications, 281*, 569–575.

Owen, D. J., Ornaghi, P., Yang, J. C., Lowe, N., Evans, P. R., Ballario, P., et al. (2000). The structural basis for the recognition of acetylated histone H4 by the bromodomain of histone acetyltransferase gcn5p. *The EMBO Journal, 19*, 6141–6149.

Paillisson, A., Levasseur, A., Gouret, P., Callebaut, I., Bontoux, M., Pontarotti, P., et al. (2007). Bromodomain testis-specific protein is expressed in mouse oocyte and evolves faster than its ubiquitously expressed paralogs BRD2, -3, and -4. *Genomics, 89*, 215–223.

Peng, J., Dong, W., Chen, L., Zou, T., Qi, Y., & Liu, Y. (2007). Brd2 is a TBP-associated protein and recruits TBP into E2F-1 transcriptional complex in response to serum stimulation. *Molecular and Cellular Biochemistry, 294*, 45–54.

Petrelli, G., & Figa-Talamanca, I. (2001). Reduction in fertility in male greenhouse workers exposed to pesticides. *European Journal of Epidemiology, 17*, 675–677.

Pittoggi, C., Magnano, A. R., Sciamanna, I., Giordano, R., Lorenzini, R., & Spadafora, C. (2001). Specific localization of transcription factors in the chromatin of mouse mature spermatozoa. *Molecular Reproduction and Development, 60*, 97–106.

Pivot-Pajot, C., Caron, C., Govin, J., Vion, A., Rousseaux, S., & Khochbin, S. (2003). Acetylation-dependent chromatin reorganization by BRDT, a testis-specific bromodomain-containing protein. *Molecular and Cellular Biology, 23*, 5354–5365.
Plaseski, T., Noveski, P., Popeska, Z., Efremov, G. D., & Plaseska-Karanfilska, D. (2011). Association study of single nucleotide polymorphisms in FASLG, JMJDIA, LOC203413, TEX15, BRDT, OR2W3, INSR and TAS2R38 genes with male infertility. *Journal of Andrology, 33*, 675–683.
Platt, G. M., Simpson, G. R., Mittnacht, S., & Schulz, T. F. (1999). Latent nuclear antigen of Kaposi's sarcoma-associated herpesvirus interacts with RING3, a homolog of the Drosophila female sterile homeotic (fsh) gene. *Journal of Virology, 73*, 9789–9795.
Potashnik, G., Ben-Aderet, N., Israeli, R., Yanai-Inbar, I., & Sober, I. (1978). Suppressive effect of 1,2-dibromo-3-chloropropane on human spermatogenesis. *Fertility and Sterility, 30*, 444–447.
Racki, L. R., Yang, J. G., Naber, N., Partensky, P. D., Acevedo, A., Purcell, T. J., et al. (2009). The chromatin remodeller ACF acts as a dimeric motor to space nucleosomes. *Nature, 462*, 1016–1021.
Rahman, S., Sowa, M. E., Ottinger, M., Smith, J. A., Shi, Y., Harper, J. W., et al. (2011). The Brd4 extraterminal domain confers transcription activation independent of pTEFb by recruiting multiple proteins, including NSD3. *Molecular and Cellular Biology, 31*, 2641–2652.
Rhee, K., Brunori, M., Besset, V., Trousdale, R., & Wolgemuth, D. J. (1998). Expression and potential role of Fsrg1, a murine bromodomain-containing homologue of the Drosophila gene female sterile homeotic. *Journal of Cell Science, 111*, 3541–3550.
Ronfani, L., Ferraguti, M., Croci, L., Ovitt, C. E., Scholer, H. R., Consalez, G. G., et al. (2001). Reduced fertility and spermatogenesis defects in mice lacking chromosomal protein Hmgb2. *Development (Cambridge, England), 128*, 1265–1273.
Sanchez, R., & Zhou, M. M. (2009). The role of human bromodomains in chromatin biology and gene transcription. *Current Opinion in Drug Discovery & Development, 12*, 659–665.
Sankar, N., Baluchamy, S., Kadeppagari, R. K., Singhal, G., Weitzman, S., & Thimmapaya, B. (2008). p300 provides a corepressor function by cooperating with YY1 and HDAC3 to repress c-Myc. *Oncogene, 27*, 5717–5728.
Schotta, G., Ebert, A., & Reuter, G. (2003). SU(VAR)3-9 is a conserved key function in heterochromatic gene silencing. *Genetica, 117*, 149–158.
Schroder, S., Cho, S., Zeng, L., Zhang, Q., Kaehlcke, K., Mak, L., et al. (2012). Two-pronged binding with bromodomain-containing protein 4 liberates positive transcription elongation factor b from inactive ribonucleoprotein complexes. *The Journal of Biological Chemistry, 287*, 1090–1099.
Shang, E., Cui, Q., Wang, X., Beseler, C., Greenberg, D. A., & Wolgemuth, D. J. (2011). The bromodomain-containing gene BRD2 is regulated at transcription, splicing, and translation levels. *Journal of Cellular Biochemistry, 112*, 2784–2793.
Shang, E., Nickerson, H. D., Wen, D., Wang, X., & Wolgemuth, D. J. (2007). The first bromodomain of Brdt, a testis-specific member of the BET sub-family of double-bromodomain-containing proteins, is essential for male germ cell differentiation. *Development (Cambridge, England), 134*, 3507–3515.
Shang, E., Salazar, G., Crowley, T., Wang, X., Lopez, R. A., Wang, X., et al. (2004). Identification of unique, differentiation stage-specific patterns of expression of the bromodomain-containing genes Brd2, Brd3, Brd4, and Brdt in the mouse testis. *Gene Expression Patterns, 4*, 513–519.
Shang, E., Wang, X., Wen, D., Greenberg, D. A., & Wolgemuth, D. J. (2009). Double bromodomain-containing gene Brd2 is essential for embryonic development in mouse. *Developmental Dynamics, 238*, 908–917.

Shankaranarayana, G. D., Motamedi, M. R., Moazed, D., & Grewal, S. I. (2003). Sir2 regulates histone H3 lysine 9 methylation and heterochromatin assembly in fission yeast. *Current Biology, 13*, 1240–1246.

Shogren-Knaak, M., Ishii, H., Sun, J. M., Pazin, M. J., Davie, J. R., & Peterson, C. L. (2006). Histone H4-K16 acetylation controls chromatin structure and protein interactions. *Science, 311*, 844–847.

Singh, S. R., Burnicka-Turek, O., Chauhan, C., & Hou, S. X. (2011). Spermatogonial stem cells, infertility and testicular cancer. *Journal of Cellular and Molecular Medicine, 15*, 468–483.

Sinha, A., Faller, D. V., & Denis, G. V. (2005). Bromodomain analysis of Brd2-dependent transcriptional activation of cyclin A. *The Biochemical Journal, 387*, 257–269.

Steger, K. (2001). Haploid spermatids exhibit translationally repressed mRNAs. *Anatomy and Embryology, 203*, 323–334.

Suka, N., Suka, Y., Carmen, A. A., Wu, J., & Grunstein, M. (2001). Highly specific antibodies determine histone acetylation site usage in yeast heterochromatin and euchromatin. *Molecular Cell, 8*, 473–479.

Tae, S., Karkhanis, V., Velasco, K., Yaneva, M., Erdjument-Bromage, H., Tempst, P., et al. (2011). Bromodomain protein 7 interacts with PRMT5 and PRC2, and is involved in transcriptional repression of their target genes. *Nucleic Acids Research, 39*, 5424–5438.

Tamkun, J. W. (1995). The role of brahma and related proteins in transcription and development. *Current Opinion in Genetics & Development, 5*, 473–477.

Tamkun, J. W., Deuring, R., Scott, M. P., Kissinger, M., Pattatucci, A. M., Kaufman, T. C., et al. (1992). brahma: A regulator of Drosophila homeotic genes structurally related to the yeast transcriptional activator SNF2/SWI2. *Cell, 68*, 561–572.

Taniguchi, Y., Matsuzaka, Y., Fujimoto, H., Miyado, K., Kohda, A., Okumura, K., et al. (1998). Nucleotide sequence of the ring3 gene in the class II region of the mouse MHC and its abundant expression in testicular germ cells. *Genomics, 51*, 114–123.

Taniguchi, Y., Suzuki, H., Ohtsuka, M., Kikuchi, N., Kimura, M., & Inoko, H. (2001). Isolation and characterization of three genes paralogous to mouse Ring3. *Nucleic Acids Research. Supplement, 1*, 247–248.

Thompson, M. (2009). Polybromo-1: The chromatin targeting subunit of the PBAF complex. *Biochimie, 91*, 309–319.

Thorpe, K. L., Gorman, P., Thomas, C., Sheer, D., Trowsdale, J., & Beck, S. (1997). Chromosomal localization, gene structure and transcription pattern of the ORFX gene, a homologue of the MHC-linked RING3 gene. *Gene, 200*, 177–183.

Trousdale, R. K., & Wolgemuth, D. J. (2004). Bromodomain containing 2 (Brd2) is expressed in distinct patterns during ovarian folliculogenesis independent of FSH or GDF9 action. *Molecular Reproduction and Development, 68*, 261–268.

Wang, R., Li, Q., Helfer, C. M., Jiao, J., & You, J. (2012). The bromodomain protein Brd4 associated with acetylated chromatin is important for maintenance of higher-order chromatin structure. *The Journal of Biological Chemistry, 287*, 10738–10752.

Wang, L., Mizzen, C., Ying, C., Candau, R., Barlev, N., Brownell, J., et al. (1997). Histone acetyltransferase activity is conserved between yeast and human GCN5 and is required for complementation of growth and transcriptional activation. *Molecular and Cellular Biology, 17*, 519–527.

Wilson, B. G., & Roberts, C. W. (2011). SWI/SNF nucleosome remodellers and cancer. *Nature Reviews. Cancer, 11*, 481–492.

Wolf, K. W., Jeppesen, P., & Mitchell, A. (1993). Spermatid nucleus of Megaselia scalaris Loew (Insecta, Diptera, Phoridae): A study using anti-histone antibodies, scanning electron microscopy, and a centromere-specific oligonucleotide. *Molecular Reproduction and Development, 35*, 272–276.

Wu, S. Y., & Chiang, C. M. (2007). The double bromodomain-containing chromatin adaptor Brd4 and transcriptional regulation. *The Journal of Biological Chemistry, 282*, 13141–13145.

Xiao, A., Li, H., Shechter, D., Ahn, S. H., Fabrizio, L. A., Erdjument-Bromage, H., et al. (2009). WSTF regulates the H2A.X DNA damage response via a novel tyrosine kinase activity. *Nature, 457*, 57–62.

Yang, X. J. (2004). Lysine acetylation and the bromodomain: A new partnership for signaling. *Bioessays, 26*, 1076–1087.

Yang, Z., He, N., & Zhou, Q. (2008). Brd4 recruits P-TEFb to chromosomes at late mitosis to promote G1 gene expression and cell cycle progression. *Molecular and Cellular Biology, 28*, 967–976.

Yang, Z., Yik, J. H., Chen, R., He, N., Jang, M. K., Ozato, K., et al. (2005). Recruitment of P-TEFb for stimulation of transcriptional elongation by the bromodomain protein Brd4. *Molecular Cell, 19*, 535–545.

Yeo, G., Holste, D., Kreiman, G., & Burge, C. B. (2004). Variation in alternative splicing across human tissues. *Genome Biology, 5*, R74.

Yu, Z., Raabe, T., & Hecht, N. B. (2005). MicroRNA Mirn122a reduces expression of the posttranscriptionally regulated germ cell transition protein 2 (Tnp2) messenger RNA (mRNA) by mRNA cleavage. *Biology of Reproduction, 73*, 427–433.

Zalensky, A. O., Allen, M. J., Kobayashi, A., Zalenskaya, I. A., Balhorn, R., & Bradbury, E. M. (1995). Well-defined genome architecture in the human sperm nucleus. *Chromosoma, 103*, 577–590.

Zeng, L., Yap, K. L., Ivanov, A. V., Wang, X., Mujtaba, S., Plotnikova, O., et al. (2008). Structural insights into human KAP1 PHD finger-bromodomain and its role in gene silencing. *Nature Structural & Molecular Biology, 15*, 626–633.

Zhang, G., Liu, R., Zhong, Y., Plotnikov, A. N., Zhang, W., Zeng, L., et al. (2012). Down-regulation of NF-kappaB transcriptional activity in HIV-associated kidney disease by BRD4 inhibition. *The Journal of Biological Chemistry, 287*, 28840–28851.

Zheng, Y., Yuan, W., Zhou, Z., Xu, M., & Sha, J. H. (2005). Molecular cloning and expression of a novel alternative splice variant of BRDT gene. *International Journal of Molecular Medicine, 15*, 315–321.

Zhou, Y., & Grummt, I. (2005). The PHD finger/bromodomain of NoRC interacts with acetylated histone H4K16 and is sufficient for rDNA silencing. *Current Biology, 15*, 1434–1438.

CHAPTER TWELVE

DMRT Genes in Vertebrate Gametogenesis

David Zarkower[1]

Department of Genetics, Cell Biology, and Development, University of Minnesota, Minneapolis, Minnesota, USA

[1]Corresponding author: e-mail address: zarko001@umn.edu

Contents

Abstract

Genes containing the DM domain DNA-binding motif regulate sex determination and sexual differentiation in a broad variety of metazoans, including nematodes, insects, and vertebrates. They can function in primary sex determination or downstream in sexual differentiation, and they can act either throughout the body or in highly restricted cell types. In vertebrates, several DM domain genes—*DMRT* genes—play critical roles in gonadal differentiation or gametogenesis. *DMRT1* has the most prominent role and likely regulates testicular differentiation in all vertebrates. In the mammalian gonad, *DMRT1* exerts both intrinsic and extrinsic control of gametogenesis; it is required for germ cell differentiation in males and regulates meiosis in both sexes, and it is required in supporting cells for the establishment and maintenance of male fate in the testis. These varied functions of *DMRT1* serve to coordinate gonadal development

Current Topics in Developmental Biology, Volume 102
ISSN 0070-2153
http://dx.doi.org/10.1016/B978-0-12-416024-8.00012-X

and function. In other vertebrates, *DMRT1* regulates gonadal differentiation, and it also appears to have played a central role in the evolution of new sex-determining mechanisms in at least three vertebrate clades. This chapter focuses on the regulation of vertebrate gametogenesis by *DMRT1*.

1. BACKGROUND

DMRT proteins are vertebrate transcription factors containing the DM domain, a DNA-binding motif first identified in two invertebrate sexual regulators, the fly gene *doublesex* (dsx) and the nematode gene *male abnormal 3 (mab-3)*(Baker & Ridge, 1980; Burtis, Coschigano, Baker, & Wensink, 1991; Raymond et al., 1998; Shen & Hodgkin, 1988). Vertebrate *DMRT* genes are expressed in distinct patterns and regulate different developmental processes including somitogenesis and nervous system development (Gennet et al., 2012; Lourenco, Lopes, & Saude, 2011; Sato, Rocancourt, Marques, Thorsteinsdottir, & Buckingham, 2010; Saude, Lourenco, Goncalves, & Palmeirim, 2005; Seo et al., 2006; Yoshizawa et al., 2011). *DMRT1*, one of several *DMRT* genes expressed in the gonad, has been shown to regulate gonadal differentiation, gametogenesis, or sex determination in several vertebrate species (Masuyama et al., 2012; Raymond, Murphy, O'Sullivan, Bardwell, & Zarkower, 2000; Smith et al., 2009). Moreover, *DMRT1* paralogs regulate sex determination in several vertebrate groups with independently evolved sex determination mechanisms (Matson & Zarkower, 2012; Matsuda et al., 2002; Smith et al., 2009; Yoshimoto et al., 2008). It appears, therefore, that the functions of DM domain proteins in sexual regulation are both conserved and dynamic across a broad range of metazoans. This chapter will focus on how *DMRT* genes, primarily *DMRT1*, control vertebrate gonadal development and function, with an emphasis on mammalian gametogenesis.

[*Note on nomenclature*: in this chapter, vertebrate genes will be written generically as *GENE1* and vertebrate proteins as GENE1; in reference to specific species, the appropriate nomenclature will be used—*DMRT1*/DMRT1 in human and *Dmrt1*/DMRT1 in mouse, for example.]

2. DMRT1 IN THE MAMMALIAN FETAL GONAD

2.1. Fetal germ cells

In mammals, germ cells are specified by BMP signaling in the proximal epiblast (an extraembryonic tissue), and then they enter the embryo proper and undertake a long-range migration to populate the primordial gonad or genital ridge (reviewed by Bowles & Koopman, 2010). After germ cells reach the genital ridge, their sex is determined by signals from surrounding somatic supporting cells leading to distinct patterns of gene expression in male and female germ cells by embryonic day 12.5 (E12.5) to E13.5 in the mouse (Jameson et al., 2012; McLaren, 2003). From this point onward, development of male and female germ cells diverges. Female fetal germ cells enter meiosis around E13.5 in the mouse and then arrest in prophase I, completing meiosis only after puberty. By contrast, male fetal germ cells become mitotically quiescent from about E15.5 to birth, when they reinitiate mitosis and rapidly begin to undergo meiosis and spermatogenesis. Figure 12.1 shows an overview of fetal germ cell development in the mouse.

During their development, germ cells tread a narrow path of cell fate: they must remain fully committed to eventual sperm or oocyte differentiation, and yet during several stages of their development, they express potent pluripotency genes, including the "core" reprogramming factors *Sox2*, *Oct3/4*, and *Nanog* (Western, van den Bergen, Miles, & Sinclair, 2010; Yabuta, Kurimoto, Ohinata, Seki, & Saitou, 2006). The pluripotent potential of germ cells becomes evident when they are cultured under appropriate conditions (Guan et al., 2006; Kanatsu-Shinohara et al., 2004; Matsui, Zsebo, & Hogan, 1992). Additionally, when gonads are explanted before E12.5 (Stevens, 1964) or when genes such as *Pten* are mutated (Kimura et al., 2003), germ cells can form teratomas, tumors in which germ cells adopt somatic fates characteristic of all three embryonic germ layers.

2.2. DMRT1 and maintenance of germ cell fate in the fetal testis

Although several *Dmrt* genes are expressed in the fetal mouse gonad in one or both sexes (Kim, Kettlewell, Anderson, Bardwell, & Zarkower, 2003), only *Dmrt1* has been shown to function in the gonad prior to birth. DMRT1

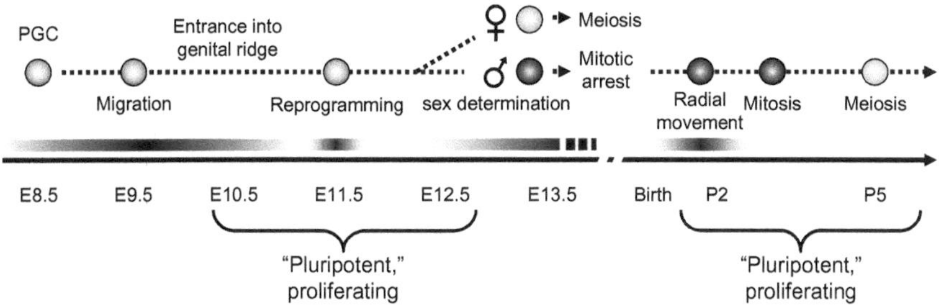

Figure 12.1 Fetal germ cell development in the mouse. Time line indicating some of the major events of germ cell development in mice between mid-gestation and neonatal development.

protein is expressed in the genital ridge of both sexes and becomes male-specific by about E14.5, when testicular and ovarian differentiation are well established (Krentz et al., 2011; Lei et al., 2007; Matson et al., 2010; Raymond et al., 2000). During this period, DMRT1 is expressed in somatic cells (pre-Sertoli cells in the testis and presumptive pregranulosa cells in the ovary) as well as in germ cells. *Dmrt1* is required in both male and female germ cells during the fetal period, but as described below, loss of *Dmrt1* has very different consequences for fetal germ cells in the two sexes.

Testes of C57BL/6J (B6) or mixed genetic background mice mutant for *Dmrt1* exhibit significant misexpression of testicular mRNAs (T. Krentz and D. Zarkower, unpublished) but have apparently normal testicular development prior to birth (Raymond et al., 2000). In mice of the 129Sv strain, however, loss of *Dmrt1* has a much more profound effect on germ cell development, causing a nearly 100% incidence of testicular teratomas (Krentz et al., 2009). Wild-type 129Sv mice are prone to these tumors, which arise in the strain at an incidence of 1–5%, depending on the substrain, whereas other inbred strains, including B6, do not develop teratomas (Stevens, 1967a, 1967b; Stevens & Little, 1954). In *Dmrt1* mutant 129Sv mice, clusters of transformed cells with embryonal carcinoma (EC) morphology form around E15.5 (Krentz et al., 2009). These cells proliferate, escaping the mitotic arrest that normally occurs in male germ cells at this stage, and they undergo differentiation starting around birth, resulting in a large differentiated teratoma within about 3 weeks.

Why do *Dmrt1* mutant germ cells form teratomas? In addition to their failure to exit the cell cycle, the EC-like precursor cells fail to silence expression of pluripotency-related genes such as *Oct3/4*, *Nanog*, and *Sox2*, as well as *E-cadherin*, which has been shown to promote pluripotency in cultured cells (Chou et al., 2008). Chromatin immunoprecipitation (ChIP) showed that DMRT1 binds upstream of the *Sox2* transcriptional start site in E13.5 testes and thus may regulate the pluripotency network via direct repression of *Sox2* transcription. At E15.5, *Dmrt1* mutant 129Sv testes have reduced expression of cell cycle inhibitors including $P18^{INK4c}$ and $P19^{INK4d}$, and DMRT1 binds near the promoter of $\mathrm{P19^{INK4d}}$ in the E13.5 testis, suggesting a potentially direct role in regulation of mitotic proliferation as well (Krentz et al., 2009). mRNA expression profiling in *Dmrt1* mutant testes at E13.5 detected only a small number of changes. Among these was reduced expression of the glial cell-derived neurotrophic factor (GDNF) coreceptor *Ret*, suggesting that reduced GDNF signaling may play a role in teratoma initiation.

Why do *Dmrt1* mutant germ cells form teratomas only in 129Sv mice? Likely candidates to mediate this effect would be genes that are misregulated in *Dmrt1* mutant testes only in the 129Sv strain or genes that are misregulated in *Dmrt1* mutants in both strains but functionally interact with genes that are differentially expressed between the two mouse strains (Fig. 12.2). The former class of genes would be easier to find, and indeed at least one candidate has been identified—*Eras* (ES-expressing Ras), whose expression can drive teratoma formation in ES cells (Takahashi, Mitsui, & Yamanaka, 2003). *Eras* is expressed at similar levels in wild-type testes of B6 and 129Sv mice and in B6 *Dmrt1* mutant testes at E15.5, but its expression is specifically elevated in 129Sv *Dmrt1* mutant testes (Krentz et al., 2009). Genome-wide expression profiling (Krentz and Zarkower, unpublished) suggests that many other genes respond to loss of *Dmrt1* differentially in 129Sv versus B6 testes, and these genes are particularly good candidates to play a role in teratoma progression. It is not yet clear, however, why these genes are affected by *Dmrt1* loss only in 129Sv mice, and this will be an important question to address.

Other genes have been identified whose loss causes teratoma formation in mice, including *Pten*, *Kitl*, and *Dnd1* (Heaney, Lam, Michelson, & Nadeau, 2008; Kimura et al., 2003; Youngren et al., 2005). What is the relationship between *Dmrt1* and other genes that suppress teratomas? Like *Dmrt1*, *Dnd1* mutations normally cause teratomas only in 129Sv mice. However, *Dnd1* mutant germ cells, unlike *Dmrt1* mutant cells, undergo

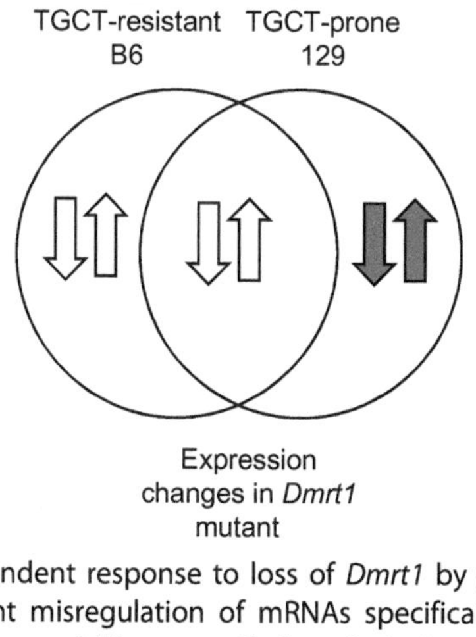

Figure 12.2 Strain-dependent response to loss of *Dmrt1* by genes in the fetal testis. Shaded arrows represent misregulation of mRNAs specifically in *Dmrt1* mutants of the 129Sv genetic background. These are likely to include mRNAs involved in strain-specific germ cell tumor formation.

abnormally high levels of apoptosis, and genetic suppression of this cell death by loss of *Bax* allows *Dnd1* mutant germ cells to form teratomas in B6 testes (Cook, Coveney, Batchvarov, Nadeau, & Capel, 2009). Moreover, *Dnd1* mutant germ cells have normal *Dmrt1* expression, while the *Dnd1* target gene $P27^{Kip1}$ is expressed normally in *Dmrt1* mutants (Krentz et al., 2009). These distinctions suggest that the two genes act in at least partially independent pathways. *Pten* mutant mice also form teratomas irrespective of strain background, suggesting that misregulation of *Dmrt1* is unlikely to be the major factor (Kimura et al., 2003); in addition, *Pten* expression and AKT P-308, which is *Pten*-dependent, appear normal in *Dmrt1* mutant EC cells (Krentz et al., 2009). These results suggest that *Pten* and *Dmrt1* also act in distinct pathways, although it remains possible that *Dmrt1* is one of several genes acting downstream of *Pten*. Although direct links between *Dmrt1* and genes like *Dnd1* and *Pten* have not been established, it seems likely that they regulate common downstream targets or pathways such as those involved in cell cycle and pluripotency regulation.

3. DMRT1 IN HUMAN TESTICULAR GERM CELL CANCER

Human testicular germ cell tumors (TGCTs) are varied and have been divided into three classes mainly on the basis of tumor morphology and presumed progenitor cell (Oosterhuis & Looijenga, 2005). The teratomas resulting from *Dmrt1* loss in the mouse resemble human type I or II tumors in some regards. First, the mouse tumors arise from EC cell progenitors, as do some pediatric type I human TGCTs. Moreover, while type II human tumors arise from carcinoma *in situ* (CIS) cells, these cells are histologically similar to EC cells and express high levels of pluripotency markers such as NANOG and OCT3/4 (Hoei-Hansen et al., 2005; Oosterhuis & Looijenga, 2005). Like the mouse tumors, human TGCTs also have a strong genetic component, including a very strong familial association and a strong influence of ethnic background (Forman et al., 1992; Heimdal et al., 1996). More compellingly, genome-wide association studies have identified *DMRT1* as likely to be involved in human type II TGCTs (Kanetsky et al., 2011; Kratz et al., 2011; Turnbull et al., 2010). Comparison of *DMRT1* expression and expression of genes regulated by *DMRT1* showed similar correlations in human TGCTs to those seen in the mouse, suggesting that *DMRT1* may function similarly in these tumors and hence that *Dmrt1* mutant mice may provide some useful insights into the etiology of human TGCTs (Krentz et al., 2009).

DMRT1 also is associated with the rarer type III TGCTs, also called spermatocytic seminomas. These are tumors of older men that are thought to arise from a later developmental stage of germ cells than the other types (Oosterhuis & Looijenga, 2005). Comparative genome hybridization studies showed that 5/5 spermatocytic seminomas had amplification of the region of chromosome 9p containing *DMRT1* and that within this region, only *DMRT1* was elevated in expression (Looijenga et al., 2006). Canine spermatocytic seminomas also express *DMRT1* (Bush, Gardiner, Palmer, Rajpert-De Meyts, & Veeramachaneni, 2011). These findings suggest that while loss of *DMRT1* can cause proliferation in fetal germ cells, gain of *DMRT1* can lead to proliferation in postnatal, possibly meiotic, cells. There is no mouse model for type III tumors, and it is not yet known whether *DMRT1* overexpression is causative for these tumors or is selected after their initiation.

4. DMRT1 IN THE POSTNATAL MAMMALIAN GONAD

4.1. Establishment of spermatogenesis

Male mammals make a lot of sperm, and they do it continuously for most of adult life. Two factors make possible this sustained high level of gametogenesis (de Rooij & Russell, 2000). First, a stem cell population forms during juvenile testis development and then serves as a source of spermatogonial progenitor cells for the rest of reproductive life. Second, spermatogonia undergo a series of amplifying divisions prior to meiosis, allowing each committed progenitor cell that forms from a stem cell division to give rise to many differentiated spermatogonia. Committed progenitor cells can be produced for decades and each progenitor can give rise to several thousand spermatozoa. To ensure appropriate numbers of gametes without depleting the precursor cell pool, both the rate of amplifying divisions and the number of divisions before the transition from mitosis to meiosis must be tightly controlled.

As described earlier, fetal male germ cells enter mitotic arrest around E15.5. After birth, these cells migrate from a central position in the seminiferous tubule to the periphery, where they resume mitosis and establish the spermatogonial stem cell population (Nagano et al., 2000). In mice, spermatogenesis involves two distinct populations of spermatogonial progenitor cells that can be distinguished by whether or not they express *Ngn3* (Yoshida et al., 2006). One population expresses *Ngn3* and forms the stem cell pool of undifferentiated spermatogonia that will support adult steady-state spermatogenesis. The second population of juvenile spermatogonia enters meiosis without expressing *Ngn3*, contributing to a "first wave" of spermatogenesis that allows earlier reproductive maturity. As described below, DMRT1 regulates both of these cell populations.

4.2. DMRT1 in the juvenile testis

DMRT1 expression in germ cells is silenced around E15.5 in the mouse and reactivated perinatally, coincident with mitotic resumption and migration to the tubule periphery (Krentz et al., 2011; Lei et al., 2007). In *Dmrt1* mutant testes, these events fail to occur and germ cells die by about postnatal day 10 (P10) (Raymond et al., 2000). Conditional gene targeting revealed both intrinsic and extrinsic requirements for DMRT1 in perinatal germ cell development (Kim, Bardwell, & Zarkower, 2007). Germ cell-specific deletion of DMRT1 in fetal germ cells blocked their cell migration, proliferation, and survival, demonstrating that *Dmrt1* is required for the establishment of undifferentiated spermatogonia, including formation of the stem cell pool (Fig. 12.3). By contrast, Sertoli-specific deletion of *Dmrt1* had a later effect on germ cell development, disrupting meiosis in prophase I, presumably due to lack of supporting cell function in the mutant Sertoli cells.

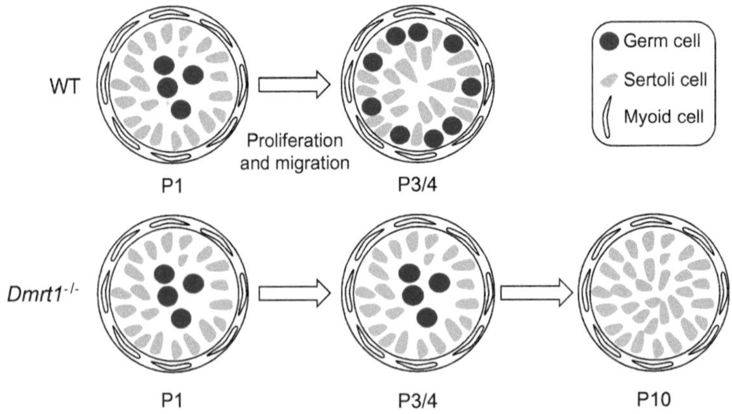

Figure 12.3 Neonatal male germ cell defects in *Dmrt1* mutant mice. In wild-type males (top), between birth and about postnatal day 3 (P3) germ cells proliferate and migrate radially from the interior of the seminferous tubules into close conjunction with the surrounding basement membrane. In *Dmrt1* mutant males (bottom), both proliferation and migration fail, and germ cells are absent by about P10, presumably due to apoptosis. Sertoli cells continue to proliferate but do not differentiate normally, and later undergo transdifferentiation to become feminized granulosa-like cells.

4.3. Adult spermatogenesis and the cycle of the seminiferous epithelium

In addition to its role at the onset of postnatal germ cell development, *Dmrt1* also is required later, during steady-state adult spermatogenesis where it is critical to maintain the balance between mitotic and meiotic germ cells (spermatogonia and spermatocytes). Adult spermatogenesis is a complex process, summarized below, and involves both cell-intrinsic and cell-extrinsic regulation of diverse cell types.

As previously described, abundant spermatogenesis is made possible by a robust stem cell pool together with amplifying divisions of committed progenitor cells. The most primitive germ cell population in the adult testis is undifferentiated type A spermatogonia, which are found in a layer adjacent to the basement membrane that surrounds the seminiferous epithelium. This cell population consists of single cells (A_s) and conjoined chains of two to 16 cells (A_{pr} and $A_{aligned}$, or A_{al}) (de Rooij & Russell, 2000) (diagrammed in Fig. 12.4). The A_s cells comprise the main stem cell pool supporting steady-state spermatogenesis in the adult mouse testis (Nakagawa, Sharma, Nabeshima, Braun, & Yoshida, 2010). A_{al} cells normally proceed to differentiation and meiosis but can be induced, under circumstances such as transplantation or germ cell depletion, to function as stem cells (Nakagawa, Nabeshima, & Yoshida, 2007).

At the onset of spermatogonial differentiation, A_{al} cells become A_1 spermatogonia, which divide five times and differentiate into B spermatogonia. B spermatogonia in turn divide and differentiate into preleptotene spermatocytes, the cells that enter meiotic prophase (de Rooij & Russell, 2000). The transition between A_{al} and A1 spermatogonium commits cells to eventual meiosis and also marks their entry into the cycle of the seminiferous epithelium, and it occurs with a species-specific period (8.6 days in the mouse) (de Rooij, 1998). Prior to the A_{al} to A_1 transition, undifferentiated spermatogonia proliferate independent of the cycle. In the mouse, this transition occurs asynchronously in waves that transit along the seminiferous tubules. The asynchrony in the A_{al} to A_1 transition creates an asynchrony in meiotic progression, which ensures that spermatozoa are produced continuously.

Spermatids require 35 days to differentiate but spermatogonia enter the cycle of the seminiferous epithelium every 8.6 days. As a consequence, differentiating spermatogonia and spermatocytes accumulate in layers above the undifferentiated spermatogonia. The cellular composition of these layers

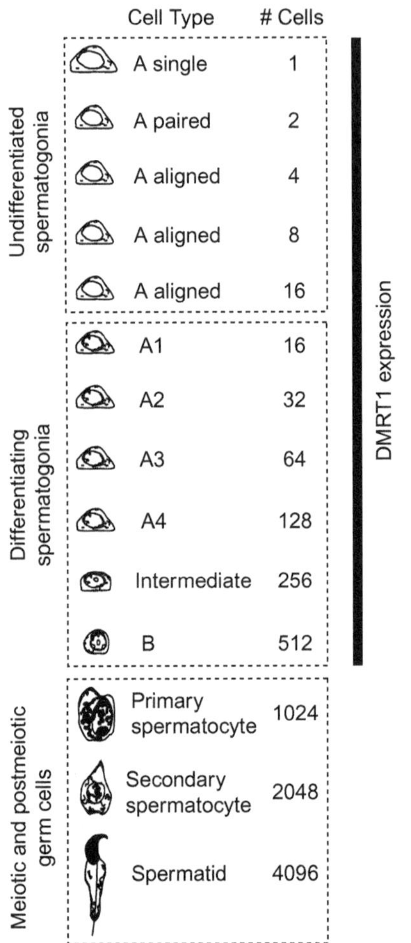

Figure 12.4 Spermatogenesis in the adult mouse. During adult steady-state spermatogenesis, the testis contains two populations of mitotic germ cells—undifferentiated spermatogonia, which include the stem cell pool, and differentiating spermatogonia. The combination of an active stem cell pool and the many mitotic divisions of spermatogonia prior to meiosis (typically nine divisions in the mouse) underlie the vast spermatogenic capacity of the mammalian male. The two spermatogonial populations can be distinguished by expression of markers such as E-cadherin (undifferentiated) and c-KIT (differentiating). Undifferentiated spermatogonia divide independent of the cycle of the seminiferous epithelium, but differentiating spermatogonia are tightly coupled to the cycle. DMRT1 is expressed in both populations, although its level appears higher in undifferentiated spermatogonia. DMRT1 is not expressed in meiotic or postmeiotic germ cells (spermatocytes), and its absence is required for the initiation of meiosis.

differs during the cycle, which allows it to be divided into morphologically distinct stages with unique assortments of cell types (de Rooij & Russell, 2000). Stage VII is of particular importance for the control of spermatogenesis, as it is the stage at which undifferentiated spermatogonia enter the cycle and commit to meiosis.

4.4. DMRT1 regulation of the mitosis/meiosis decision in adult spermatogenesis

In the postnatal gonad, DMRT1 is expressed in mitotic spermatogonia, but not in meiotic or postmeiotic spermatocytes and spermatids (Lei et al., 2007; Matson et al., 2010; Raymond et al., 2000). DMRT1 therefore might potentially function to promote spermatogonial differentiation and proliferation, to inhibit meiosis or possibly both. Conditional deletion of *Dmrt1* in undifferentiated spermatogonia using *Ngn3-cre* suggests that DMRT1 indeed has both functions. Because *Ngn3-cre* is active in A_{al} spermatogonia but not in most spermatogonial stem cells (Nakagawa et al., 2007), the function of *Dmrt1* could be tested in spermatogonia over an extended period by deleting the gene in early progenitor cells as they form but without deleting the gene in the steady-state stem cell pool (Matson et al., 2010).

Deleting *Dmrt1* with *Ngn3-cre* caused a seemingly paradoxical phenotype: germ cell numbers were greatly reduced and yet all stages of spermatogonia and spermatocytes were present, and there was no apparent increase in apoptosis. Moreover, this phenotype remained stable over time, suggesting that there was neither a developmental block nor a loss of stem cell function. The explanation for this unusual set of phenotypes came from examining the abundance of different cell populations and from analysis of markers specific to different stages of germ cell development. In mutant testes, cells expressing the undifferentiated spermatogonial marker E-Cadherin were abundant, but cells expressing the differentiating spermatogonial marker c-KIT were greatly depleted. Strikingly, many E-Cadherin positive cells, including short-chain A_{al} and even apparent A_s cells, also expressed unusually high levels of STRA8, suggesting that they might be inappropriately activating the meiotic program prior to the completion of spermatogonial differentiation. Strongly STRA8-positive cells were present in most tubule sections rather than just in stage VII, suggesting that meiosis might be uncoupled from the cycle of the seminiferous epithelium or that the cycle itself might be disrupted. This latter view was confirmed by conditional deletion of *Dmrt1* in germ cells of adult testes, which caused the rapid disappearance of the basal layer of undifferentiated spermatogonia and the concomitant appearance of prophase spermatocytes at inappropriate stages of the seminiferous epithelial cycle. Together, these results strongly suggest that DMRT1 is required for spermatogonia to continue mitotic proliferation and differentiation and to avoid premature meiotic initiation.

4.5. DMRT1 regulation of RA signaling in adult spermatogonia

Retinoic acid (RA) plays an essential role in mammalian spermatogenesis. Depletion of the RA precursor vitamin A arrests spermatogonia prior to the A_{al} to A_1 transition, and readministration of vitamin A causes a synchronous resumption of meiosis at stage VII (McCarthy & Cerecedo, 1952; Thompson, Howell, & Pitt, 1964; van Pelt & de Rooij, 1990). Thus, RA controls commitment to meiosis and initiation of spermatogonial differentiation, and its activity also plays a role in the asynchrony of the epithelial cycle. Activation of meiosis by RA in the postnatal testis involves transcriptional activation of *Stra8* (Oulad-Abdelghani et al., 1996; Vernet et al., 2006; Zhou et al., 2008). STRA8 is strongly expressed during stage VII in preleptotene spermatocytes, and in *Stra8* mutants, these cells fail to undergo meiosis (Anderson et al., 2008; Mark et al., 2008). The expression of STRA8 at the A_{al} to A_1 transition and at the entry to meiosis implicates RA in controlling both of these processes and points to stage VII as a time of RA signaling activity. Expression of RA metabolic enzymes also strongly suggests that RA activity is high at stage VII and suggests that germ cells play a central role in controlling RA metabolism in the testis (Sugimoto, Nabeshima, & Yoshida, 2011; Vernet et al., 2006). Expression of STRA8 in spermatogonia (low) and preleptotene spermatocytes (high) differs at stage VII even though RA is highly diffusible and the two cell types are often adjacent in the seminiferous epithelium. This expression difference suggests that cell-intrinsic mechanisms must exist to modulate the activity of RA in the two cell-types.

How does DMRT1 restrict meiotic initiation? Vitamin A depletion arrested meiosis and blocked *Stra8* expression in *Dmrt1* mutant germ cells, confirming that an RA-dependent mechanism is involved in the premature meiosis in mutant germ cells. The premature STRA8 expression observed in *Dmrt1* mutant spermatogonia suggests that DMRT1 might somehow suppress RA signaling, might block the transcriptional response to RA signaling, or both. mRNA expression analysis and ChIP suggest that DMRT1 does act specifically to repress *Stra8* transcription and also more generally to limit RA signaling. ChIP analysis found that DMRT1 binds to the *Stra8* proximal promoter between a pair of potential RA response elements and that the binding is germ cell specific. This result suggests that regulation of *Stra8* by DMRT1 is likely to involve direct transcriptional repression, blocking transcriptional activation of *Stra8* by the RA receptor. However, *Stra8* null mutant germ cells can initiate meiosis (Mark et al., 2008), so repressing *Stra8* transcription alone probably is insufficient for DMRT1 to control meiosis. It is likely, therefore, that DMRT1 regulates other meiosis-promoting targets, either directly or indirectly. Reporter gene analysis using an RA-responsive transgene suggested that DMRT1 accomplishes this at least in part by limiting RA signaling activity. In wild-type testes, the reporter was expressed at low levels in spermatogonia and at higher levels in meiotic and postmeiotic germ cells, whereas in mutant testes, the reporter was expressed at high levels in spermatogonia. A significant proportion of RA-responsive mRNAs were elevated in the mutant gonads, further suggesting that RA signaling activity was inappropriately high. Among these overexpressed mRNAs was the mitotic inhibitor *Cdkn1a*, which might provide a mechanistic coupling between RA stimulation of meiotic initiation and inhibition of mitosis.

The reporter and transcriptome data clearly indicate that DMRT1 can limit RA signaling activity. Precisely, how it does so remains unclear, however. Mutant gonads had altered expression of several genes directly involved in RA signaling, including elevated expression of the RA-binding protein gene *Crabp2* and reduced expression of the retinoid signaling inhibitor *Tbx1*, but ChIP did not detect binding of DMRT1 to these genes. ChIP did detect binding of DMRT1 to promoters of the RA synthetic enzymes *Adh4* and *Aldh1a1*, but their mRNA levels were not significantly altered in the mutant gonads and protein expression has not been examined for either gene.

DMRT1 clearly restricts both RA signaling and *Stra8* transcription in spermatogonia. In principle, DMRT1 activity might be limited to preventing meiosis and other regulators might promote spermatogonial differentiation and mitotic proliferation. However, DMRT1 also seems to directly promote spermatogonial development. Comparison of mutant and wild-type vitamin A-deficient (VAD) testes revealed that mutant germ cells were apparently able to initiate (but not complete) meiosis, whereas wild-type germ cells arrested as undifferentiated spermatogonia. In the *Dmrt1* mutant VAD gonads, a number of mRNAs involved in spermatogonial differentiation were reduced, suggesting that the DMRT1 normally activates genes that promote spermatogonial differentiation. Among these was *Sohlh1*, which is coexpressed with *Dmrt1* in spermatogonia and whose promoter was bound by DMRT1 in adult spermatogonia. *Sohlh1* is essential for spermatogonial differentiation (Suzuki et al., 2012) and its activation by DMRT1, together with the activation of a number of other spermatogonial genes, suggests a critical role for DMRT1 in driving spermatogonial development. The dual role of DMRT1 in promoting spermatogonial development and restricting meiosis is diagrammed in Fig. 12.5.

Sertoli cells play a critical supporting role in gametogenesis and thus might be expected to participate in the cycle of the seminiferous epithelium. Transplantation experiments show that germ cells can maintain their species-specific cycle time of entry into differentiation when inserted into the seminiferous epithelium of another species, indicating a germ cell-autonomous component to the epithelial cycle and suggesting that germ cells may have overall control over progression of the cycle (Franca, Ogawa, Avarbock, Brinster, & Russell, 1998). However, Sertoli cells also exhibit cyclical patterns of gene expression (Elftman, 1950; Sugimoto et al., 2011). Although the role of this periodicity is not well understood, they may adapt their metabolism to the changing needs of the local germ cell population or they may help modulate the cyclical development of the germ cells (Sugimoto et al., 2011). Deletion of *Dmrt1* in germ cells caused expression of *Gata1* and androgen receptor (AR) to become noncyclical, with levels typical of stages VII–VIII, possibly indicating that signals from the germ cells help to match the Sertoli cell cycle to that of the germ cells and that these signals are controlled by *Dmrt1* (Matson et al., 2010). The nature of these putative signals has not been explored.

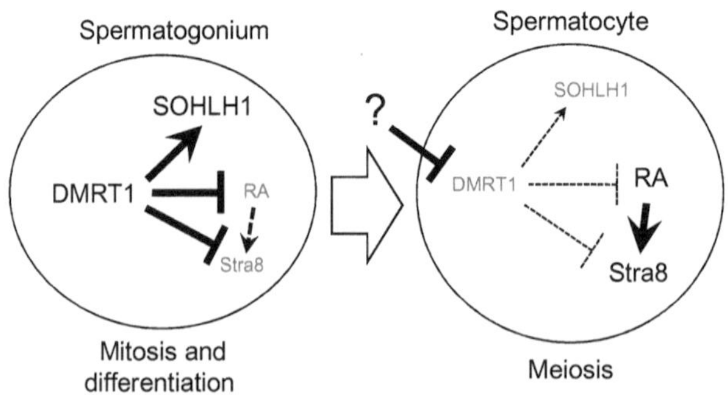

Figure 12.5 DMRT1 and regulation of the mitosis/meosis switch. In spermatogonia, DMRT1 transcriptionally activates SOHLH1 and other genes that promote spermatogonial differentiation and proliferation, inhibits activation of *Stra8* transcription by RA signaling, and generally inhibits RA signaling. In preleptotene spermatocytes, DMRT1 expression is repressed by an unknown mechanism(?), relieving the block on *Stra8* transcriptional and RA signaling and permitting meiotic initiation. *Model is based on data from Matson et al. (2010).*

5. DMRT1 IN SUPPORTING CELLS OF THE MAMMALIAN TESTIS

In the mouse, DMRT1 is expressed in the Sertoli cell lineage from the genital ridge stage onward. Initiation of *Dmrt1* expression in fetal Sertoli cells requires *Gata4* and expression of *Dmrt1* can be stimulated in postnatal primary Sertoli cells by follicle-stimulating hormone (FSH). In their female counterpart, ovarian granulosa cells, *Dmrt1* expression is suppressed by the forkhead transcription factor FOXL2 (Chen & Heckert, 2001; Lei & Heckert, 2004; Lei, Karpova, Hornbaker, Rice, & Heckert, 2009; Manuylov et al., 2011; Ottolenghi et al., 2005; Uhlenhaut et al., 2009). Transgenic analysis suggests that most of the regulatory elements sufficient for *Dmrt1* expression in Sertoli cells are within about 3 kb upstream of the transcriptional start site and are conserved in mammals (Boyer, Dornan, Daneau, Lussier, & Silversides, 2002; Lei et al., 2009).

Loss of *Dmrt1* has no obvious morphological consequence in Sertoli cells prior to birth (Kim, Bardwell, et al., 2007; Raymond et al., 2000). Postnatally, however, mutant Sertoli cells are severely affected and eventually this leads to the loss of most germ cells even when *Dmrt1* is deleted only in the Sertoli cell lineage (Kim, Bardwell, et al., 2007; Matson et al., 2011). Initial analysis of *Dmrt1* mutant Sertoli cells indicated that they fail to complete differentiation and continue to proliferate past the normal time of mitotic arrest (Raymond et al., 2000). More recently, however, it was discovered that the mutant Sertoli cells not only fail to differentiate, but also undergo a profound transdifferentiation of sexual fate. While the focus of this chapter is on the role of *Dmrt* genes in gametogenesis, some of the mechanisms by which *Dmrt1* establishes and maintains somatic cell fate may be relevant to its functions in germ cells and thus they are briefly described here.

At birth, *Dmrt1* mutant Sertoli cells appear normal in number and morphology and express markers such as GATA4 and SOX9 (Raymond et al., 2000). Within about 2 weeks after birth, however, it becomes obvious that they are abnormal: the mutant cells fail to properly upregulate Sertoli cell differentiation markers such as GATA1 and AR (Raymond et al., 2000), they do not become polarized and, remarkably, they begin to express FOXL2 and other genes normally expressed in ovarian granulosa cells (Matson et al., 2011). Over the next 2 weeks, the mutant cells continue to proliferate and switch their gene expression program from male to largely female, extinguishing expression of *Sox9* and other male genes and robustly upregulating *Foxl2*, *Lrh1*, aromatase, and other female genes. Seminiferous tubules break down, the cellular organization of the gonad becomes more ovary-like, and steroidogenesis shifts from androgenic to estrogenic. By adulthood, the mutant gonad has many granulosa-like cells, as well as cells resembling theca cells, the other main somatic component of ovarian follicles. These results show that Sertoli cells cannot complete differentiation without DMRT1 and instead adopt the equivalent female cell fate. Moreover, conditional deletion of *Dmrt1* in the adult testis showed that even fully differentiated Sertoli cells can undergo an apparent direct reprogramming from male to female, changing their morphology and gene expression to become granulosa-like cells. Thus DMRT1 is required not only for the completion of Sertoli cell differentiation but also for maintenance of the Sertoli versus granulosa cell fate decision throughout life.

Similar studies have also shown that ovarian cell fate must be actively maintained into adulthood and that FOXL2 acts with the estrogen receptors ESR1 and ESR2 to suppress expression of SOX9 and DMRT1 as well as other testicular genes (Uhlenhaut et al., 2009; Veitia, 2010). These reciprocal phenotypes indicate that DMRT1 and FOXL2 anchor antagonistic gene regulatory networks that are necessary to maintain sexual fates long after gonadal sex determination is completed. Why such a regulatory system is needed is not obvious, but one possible explanation is that some of the factors required for adult gonadal functions such as gametogenesis (e.g., RA) have the potential to alter somatic cell fate and thus a countering system is needed.

Most of the other essential components of the sex maintenance network remain to be identified. Loss of *Sox9* does not cause sexual transdifferentiation, but this may be due in part to redundancy between *Sox9* and other *Sox* genes including *Sox8*, which also is reduced in *Dmrt1* mutant testes (Barrionuevo et al., 2009; Chang et al., 2008; Matson et al., 2011). Expression studies showed that many of the genes required fetally for sex determination in both sexes are misregulated in *Dmrt1* mutant adult gonads and that DMRT1 can bind near many of these genes in the adult testis (Matson et al., 2011). To elucidate which of these genes are important for sex maintenance and transdifferentiation, it will be necessary to test their roles genetically.

What happens to germ cells in a sexually transdifferentiating testis? As discussed earlier, most germ cells in *Dmrt1* mutant testes die in the first several postnatal weeks. The survivors do not undergo anything resembling normal oogenesis or form normal follicles, but upon gonadotropin stimulation, some adopt an oocyte-like nuclear morphology and express low levels of at least two oocyte-specific proteins, MATER and ZP2, suggesting that they may be partially feminized (Matson et al., 2011). It will be of interest to test whether XX germ cells can undergo more complete feminization in a transdifferentiating testis.

6. DMRT1 IN THE MAMMALIAN OVARY

Dmrt1 is transiently expressed in the genital ridge and early gonad in both sexes before becoming testis-specific around E15.5. Null mutant females are fertile and mutant ovaries are morphologically near-normal (Raymond et al., 2000). However, adult *Dmrt1* mutant ovaries have sharply reduced numbers of follicles (Krentz et al., 2011). mRNA profiling of mutant ovaries at E13.5, just before DMRT1 normally ceases to be expressed, revealed only a small number of expression changes. Among the affected genes, the only one linked to germ cell development was *Stra8*, whose expression was severely reduced in mutant ovaries. ChIP analysis detected DMRT1 binding near *Stra8* at the same location as in male gonads, suggesting that DMRT1 may directly activate transcription of *Stra8* in females, opposite to its role in males. *Stra8* null mutants cannot undergo meiosis, so the residual *Stra8* expression in *Dmrt1* mutant ovaries must be sufficient for meiotic entry and progression. Mutant germ cells have abnormal localization of SYCP3 and γH2AX during meiotic prophase, but these two proteins are not essential for completion of meiosis in females (Celeste et al., 2002; Yuan et al., 2002). Postnatally, mutant ovaries have a deficit in the number of primordial follicles, possibly because the mutant germ cells are deficient in recruitment of granulosa cells.

7. DMRT7 AND SEX CHROMATIN

Mammals have seven *Dmrt* genes and most of them are expressed in the gonads at some stage of development (Kim et al., 2003). In addition to *Dmrt1*, three other paralogs—*Dmrt4*, *Dmrt6*, and *Dmrt7*—have gonadal functions. These genes play very different roles from DMRT1 and from each other, but all three regulate gametogenesis.

Dmrt4 is widely expressed, but mutants of both sexes are fertile and morphologically normal. However, null mutants of both sexes have phenotypes (Balciuniene, Bardwell, & Zarkower, 2006). Mutant females form polyovular follicles (follicles containing two or more oocytes) at an elevated rate, suggesting a role in soma/germ cell interaction during neonatal folliculogenesis. A fraction of mutant males display copulatory behavior toward both sexes rather than just to females, indicating a likely function in the CNS, which has not been investigated in any detail.

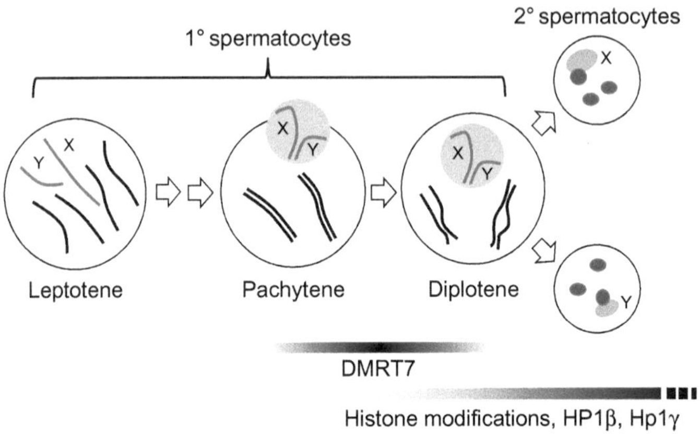

Figure 12.6 The XY body and sex chromatin in mammalian male meiosis. During meiotic prophase I, the sex chromosomes occupy a distinct chromatin domain, the XY body, or sex body (light gray circle), and they undergo transcriptional silencing, referred to as meiotic sex chromosome inactivation (MSCI). The sex chromosomes acquire specific chromatin modifications including heterochromatin proteins and a distinct profile of covalent histone modifications. Silencing of the sex chromosomes persists in secondary spermatocytes and spermatids, where the silenced compartment they occupy is referred to as postmeiotic sex chromatin (PMSC) (light gray ovals). In PMSC, the sex chromosomes have a different suite of chromatin modifications. DMRT7 is expressed in late pachytene and early diplotene spermatocytes and associates with the sex chromosomes. It is not required for MSCI and XY body formation, but is required for the transition from MSCI to PMSC and for germ cell survival beyond the first meiotic division.

Dmrt7 is exclusive to mammals, and in the mouse, DMRT7 protein is expressed only in male germ cells during meiotic prophase, mainly in mid- to late-pachytene spermatocytes (Kawamata, Inoue, & Nishimori, 2007). DMRT7 protein is nuclear and is enriched on the sex chromosomes in the sex body or XY body (Kim, Namekawa, et al., 2007). The XY body (Fig. 12.6) is a condensed chromatin body with distinctive chromatin marks that forms during meiotic prophase (reviewed by Cloutier & Turner, 2010; Heard & Turner, 2011; Ichijima, Sin, & Namekawa, 2012). When the XY body forms, transcription of most sex-linked genes is silenced in a process termed meiotic sex chromosome inactivation, or MSCI. After meiosis, the XY body is no longer visible, but sex chromosome silencing persists and a new set of chromatin marks is recruited to form a chromatin domain termed postmeiotic sex chromatin or PMSC, which persists throughout spermatid differentiation (Namekawa et al., 2006). The significance of MSCI and PMSC is unknown, although mutations that disrupt MSCI cause pachytene arrest (Royo et al., 2010). Among other possibilities, MSCI may prevent illegitimate recombination between the unsynapsed portions of the sex chromosomes or prevent detection of the unpaired chromosomes by a pachytene checkpoint. PMSC has been suggested to prevent aberrant gene expression during the replacement of histones by protamines in round spermatids or set the stage for imprinted inactivation of the paternal X chromosome in female zygotes, although these views are controversial (Huynh & Lee, 2003; Ichijima et al., 2012; McKee & Handel, 1993; Namekawa, Payer, Huynh, Jaenisch, & Lee, 2010; Okamoto et al., 2005).

In *Dmrt7* null mutant testes, spermatocytes progress normally to pachynema and then they arrest and undergo apoptosis, with virtually no cells escaping beyond diplonema (Kawamata & Nishimori, 2006; Kim, Namekawa, et al., 2007). MSCI appears to occur normally in mutant spermatocytes: the XY body forms and the sex chromosomes accumulate the normal chromatin marks and are transcriptionally silenced (Kim, Namekawa, et al., 2007). However, marks of PMSC such as histone H3K9me2 and H3K9me3 and HP1β protein do not accumulate normally on the sex chromosomes. Thus, it appears that DMRT7 is required for the transition between MSCI and PMSC. Based on the enrichment of DMRT7 on the sex chromosomes, it is possible that its transcriptional targets are biased to the sex chromosomes or that DMRT7 plays a relatively direct role in recruitment of chromatin marks to those chromosomes. Unfortunately, because mutant spermatocytes die rapidly by apoptosis after pachynema, it has not been possible to perform expression profiling or extensive biochemical characterization.

Like DMRT7, DMRT6 is expressed exclusively in male germ cells, but earlier in their development, in intermediate and B spermatogonia. *Dmrt6* mutant germ cells arrest at pachynema with defects in sex chromosome pairing, suggesting that DMRT6 activity in spermatogonia is required to establish gene expression necessary for proper handling of the sex chromosomes in meiotic prophase in spermatocytes (T. Zhang and D. Zarkower, unpublished).

8. DMRT1 IN OTHER VERTEBRATES

DMRT1 is expressed in the gonads of all vertebrates examined and has been functionally studied in fish, birds, and amphibians. In birds and amphibians, DMRT1 can determine sex (Smith et al., 2009; Yoshimoto et al., 2008), and thus, it plays a critical if indirect role in controlling gametogenesis. The role of DMRT1 and its close orthologs in birds and amphibians has recently been reviewed elsewhere and will not be considered in detail here (Chue & Smith, 2011; Matson & Zarkower, 2012; Yoshimoto

& Ito, 2011). As described below, in some fish, *Dmrt1* seems to have a more direct role in gametogenesis, either in addition to sex determination or as part of sex determination.

8.1. DMRT1 expression and function in fish

In sex determination an important distinction between mammals and teleost fish is the role of germ cells. Mammalian sex determination occurs in the somatic supporting lineage (bipotential Sertoli/granulosa precursor cells) and although somatic gonad differentiation can be affected by primordial germ cell (PGC) loss, sex determination occurs normally in mammals lacking germ cells. By contrast, experimental manipulation of PCG numbers in medaka has clearly shown that germ cells can play a pivotal role in canalizing sex determination down either a male or female pathway in fish (Saito & Tanaka, 2009). Depletion of PGC numbers can cause full female to male sex reversal (Kurokawa et al., 2007), whereas increased PGC numbers due to the *hotei* mutation in the anti-Mullerian hormone receptor lead to male to female sex reversal (Morinaga et al., 2007). Another, perhaps related, distinction is that in fish such as medaka and zebrafish, dimorphic germ cell development appears to precede dimorphic somatic gonad differentiation; by contrast, in mammals, the order is reversed, and it is the somatic gonadal environment that determines the sex of germ cells.

Dmrt1 in fish became a topic of great interest a decade ago when it was found that a *Dmrt1* homolog, called *Dmy* or *Dmrt1bY*, acts as the Y-linked male sex-determining gene in medaka (Matsuda et al., 2002, 2007; Nanda et al., 2002). *Dmy* was formed by a duplication of the autosomal *Dmrt1* gene (reviewed by Kondo, Nanda, Schmid, & Schartl, 2009). *Dmrt1* retains a function in testicular differentiation in medaka; XY mutants initially develop gonads with male morphology, likely due to the action of *Dmy*, but they later become fertile females with ovaries (Masuyama et al., 2012). It is not yet known whether this transition involves trans-differentiation similar to that in *Dmrt1* mutant mice.

Outside medaka no functional analysis of *Dmrt1* genes has been reported in fish, but a great deal of expression analysis has been performed. While only medaka and its sister species *O. curvinotus* have *Dmy* (Kondo, Nanda, Hornung, Schmid, & Schartl, 2004; Matsuda et al., 2003), *Dmrt1* is present in all fish and its expression has been reported in the gonads of about 20 other fish species (reviewed by Herpin & Schartl, 2011). In every case, *Dmrt1* was expressed only in the testis or in both ovary and testis but with higher testicular expression. The cell types expressing *Dmrt1* appear to vary among fish species, with some species expressing *Dmrt1* just in Sertoli cells, some in both Sertoli and germ cells, and a few apparently only in germ cells. In all cases, *Dmrt1* expression correlated with testicular development (Herpin & Schartl, 2011). Natural sex change after reproductive maturity occurs in many fish species, either male-to-female (protandrous) or female-to-male (protogynous). Dynamic expression of *Dmrt1* in a variety of species during development or regression of testes suggests a role for *Dmrt1* in sex switching (He et al., 2003; Huang et al., 2005; Jeong et al., 2009; Liarte et al., 2007; Shin, An, Park, Jeong, & Choi, 2009; Xia, Zhou, Yao, Li, & Gui, 2007). Indeed, knockdown of *Dmrt1* by RNA interference in the proandrous black porgy reduced the number of germ cells and induced male-to-female sex reversal (Wu et al., 2011). Thus, as in mammals, *Dmrt1* activity is associated with male gonadal differentiation, although with some potentially important differences in the cell types in which it acts.

Gonadal expression has been reported for other *Dmrt* genes, suggesting that this gene family may have multiple roles in gametogenesis in fish (Cao, Cao, & Wu, 2007; Cao, Chen, Wu, Gan, & Luo, 2009; Guan, Kobayashi, & Nagahama, 2000; Guo et al., 2004; Wen et al., 2009). Interestingly, *Dmrt4* is male-specific in medaka and olive flounder (Guan et al., 2000; Wen et al., 2009), but female-specific in tilapia (Cao et al., 2007), perhaps indicating rapid functional change.

8.2. Regulation of *Dmrt1* in fish

Hormonal regulation plays a critical role in gonadal development and function in fish, as in other vertebrates. Although *Dmrt1* functional data are very limited, regulation of *Dmrt1* expression has been investigated in a number of species. Estrogen treatment of males can cause sex reversal in fish and, where it has been examined, this always leads to reduced *Dmrt1* expression (reviewed by Herpin & Schartl, 2011). Conversely, masculinization by treatment with androgens, estrogen receptor antagonists, or aromatase inhibitors (blocking conversion of androgens to estrogens) leads to *Dmrt1* upregulation. It appears therefore that in fish, as in mammals and birds (Matson et al., 2011; Smith, Katz, & Sinclair, 2003; Uhlenhaut et al., 2009), there is antagonism between *Dmrt1* and estrogen signaling. In tilapia, there is evidence that *Dmrt1* can directly repress transcription of *cyp19a1a* (aromatase) in the ovary (Wang et al., 2010). Estrogen treatment does not, however, prevent expression of *Dmy* in sex-reversed XY medaka, suggesting that *Dmy* acts upstream or independent of estrogen signaling, a distinction that may be linked to its primary sex-determining role (Nanda et al., 2002; Suzuki, Nakamoto, Kato, & Shibata, 2005).

cis-regulatory elements controlling *Dmrt1* and *Dmy* expression have been examined in medaka, revealing roles for both transcriptional and posttranscriptional regulation. Early in gonadal development, only *Dmy* is expressed and it determines sex, whereas in the later gonad, *Dmrt1* also is expressed and is present at higher levels than *Dmy*. The regulatory relationship between these genes seems to have been established in part by the fortuitous introduction of *cis* elements by movement of transposons. One such sequence in the *Dmy* promoter mediates negative autoregulation by *Dmy* as well as later repression of *Dmy* by *Dmrt1* (Herpin et al., 2010). The gonad specificity of *Dmy* expression appears to result in part from a short conserved element in the 3′UTR of the *Dmy* transcript that stabilizes the RNA selectively in gonadal cells and can confer gonadal expression on a heterologous mRNA (Herpin, Nakamura, Wagner, Tanaka, & Schartl, 2009). This motif occurs in *Dmrt1* genes in other phyla, suggesting the possibility that this mode of regulation may be conserved, a possibility that has not yet been tested. The ability of meiotic oocytes to overcome male sex determination in XY *hotei* mutants indicates that one function of *Dmy* in medaka must be to limit oogenesis (Morinaga et al., 2007). Indeed, morpholino depletion of *Dmy* in medaka causes an increase in PGC numbers (Herpin et al., 2007). This suggests that *Dmy* may determine sex by two parallel mechanisms: direct determination of Sertoli cell fate combined with an indirect suppression of PCG numbers to reinforce the male sex determination decision.

9. CONCLUSIONS

DM domain genes regulate sexual development in many, probably most, metazoans, and in vertebrates, *DMRT1* appears to be universally required for male gonadogenesis. Closer examination of *DMRT1* expression and function reveals that it plays diverse roles in the male gonad. This is true even in a single species such as the mouse, where it acts multiple times to perform quite distinct roles in the establishment, function, and maintenance of cell fates in the testis, both in the germ line and supporting cell lineage. In other species, less is known, but it seems likely from studies in fish that *Dmrt1* has acquired distinct roles in male gonadogenesis and gametogenesis in different species. The repeated "capture" of vertebrate sex determination mechanisms by *Dmrt1* orthologs underscores that the regulation of gonadal development and gametogenesis are under rapid evolution, presumably reflecting the direct link between gametogenesis and reproductive fitness. Many questions regarding *Dmrt1* function remain to be addressed. These include how *Dmrt1* is inactivated in the mouse to allow meiosis to initiate, how conserved are the downstream targets and functions of *Dmrt1* between mammals and other vertebrates, whether sexual transdifferentiation in *Dmrt1* mutant mice is mechanistically similar to natural sex change in sequentially hermaphroditic fish, and how *Dmrt1* differentially regulates target genes in different cell types and at different developmental stages. With recent advances in genomic technologies and manipulation of gene function in nonmodel vertebrates, answers to these questions should be forthcoming in the relatively near future.

REFERENCES

Anderson, E. L., Baltus, A. E., Roepers-Gajadien, H. L., Hassold, T. J., de Rooij, D. G., van Pelt, A. M., et al. (2008). Stra8 and its inducer, retinoic acid, regulate meiotic initiation in both spermatogenesis and oogenesis in mice. *Proceedings of the National Academy of Sciences of the United States of America*, *105*, 14976–14980.

Baker, B. S., & Ridge, K. A. (1980). Sex and the single cell. I. On the action of major loci affecting sex determination in Drosophila melanogaster. *Genetics*, *94*, 383–423.

Balciuniene, J., Bardwell, V. J., & Zarkower, D. (2006). Mice mutant in the DM domain gene Dmrt4 are viable and fertile but have polyovular follicles. *Molecular and Cellular Biology*, *26*, 8984–8991.

Barrionuevo, F., Georg, I., Scherthan, H., Lecureuil, C., Guillou, F., Wegner, M., et al. (2009). Testis cord differentiation after the sex determination stage is independent of Sox9 but fails in the combined absence of Sox9 and Sox8. *Developmental Biology*, *327*, 301–312.

Bowles, J., & Koopman, P. (2010). Sex determination in mammalian germ cells: Extrinsic versus intrinsic factors. *Reproduction*, *139*, 943–958.

Boyer, A., Dornan, S., Daneau, I., Lussier, J., & Silversides, D. W. (2002). Conservation of the function of DMRT1 regulatory sequences in mammalian sex differentiation. *Genesis*, *34*, 236–243.

Burtis, K. C., Coschigano, K. T., Baker, B. S., & Wensink, P. C. (1991). The doublesex proteins of Drosophila melanogaster bind directly to a sex-specific yolk protein gene enhancer. *The EMBO Journal*, *10*, 2577–2582.

Bush, J. M., Gardiner, D. W., Palmer, J. S., Rajpert-De Meyts, E., & Veeramachaneni, D. N. (2011). Testicular germ cell tumours in dogs are predominantly of spermatocytic seminoma type and are frequently associated with somatic cell tumours. *International Journal of Andrology*, *34*, e288–e295 discussion e295.

Cao, J., Cao, Z., & Wu, T. (2007). Generation of antibodies against DMRT1 and DMRT4 of Oreochromis aurea and analysis of their expression profile in Oreochromis aurea tissues. *Journal of Genetics and Genomics*, *34*, 497–509.

Cao, J., Chen, J., Wu, T., Gan, X., & Luo, Y. (2009). Molecular cloning and sexually dimorphic expression of DMRT4 gene in *Oreochromis aureus*. *Molecular Biology Reports*, *37*, 2781–2788.

Celeste, A., Petersen, S., Romanienko, P. J., Fernandez-Capetillo, O., Chen, H. T., Sedelnikova, O. A., et al. (2002). Genomic instability in mice lacking histone H2AX. *Science*, *296*, 922–927.

Chang, H., Gao, F., Guillou, F., Taketo, M. M., Huff, V., & Behringer, R. R. (2008). Wt1 negatively regulates beta-catenin signaling during testis development. *Development*, *135*, 1875–1885.

Chen, J. K., & Heckert, L. L. (2001). Dmrt1 expression is regulated by follicle-stimulating hormone and phorbol esters in postnatal Sertoli cells. *Endocrinology*, *142*, 1167–1178.

Chou, Y. F., Chen, H. H., Eijpe, M., Yabuuchi, A., Chenoweth, J. G., Tesar, P., Lu, J., McKay, R. D., & Geijsen, N. (2008). The growth factor environment defines distinct pluripotent ground states in novel blastocyst-derived stem cells. *Cell*, *135*, 449–461.

Chue, J., & Smith, C. A. (2011). Sex determination and sexual differentiation in the avian model. *The FEBS Journal*, *278*, 1027–1034.

Cloutier, J. M., & Turner, J. M. (2010). Meiotic sex chromosome inactivation. *Current Biology*, *20*, R962–R963.

Cook, M. S., Coveney, D., Batchvarov, I., Nadeau, J. H., & Capel, B. (2009). BAX-mediated cell death affects early germ cell loss and incidence of testicular teratomas in Dnd1(Ter/Ter) mice. *Developmental Biology*, *328*, 377–383.

de Rooij, D. G. (1998). Stem cells in the testis. *International Journal of Experimental Pathology*, *79*, 67–80.

de Rooij, D. G., & Russell, L. D. (2000). All you wanted to know about spermatogonia but were afraid to ask. *Journal of Andrology*, *21*, 776–798.

Elftman, H. (1950). The Sertoli cell cycle in the mouse. *The Anatomical Record*, *106*, 381–392.

Forman, D., Oliver, R. T., Brett, A. R., Marsh, S. G., Moses, J. H., Bodmer, J. G., et al. (1992). Familial testicular cancer: A report of the UK family register, estimation of risk and an HLA class 1 sib-pair analysis. *British Journal of Cancer*, *65*, 255–262.

Franca, L. R., Ogawa, T., Avarbock, M. R., Brinster, R. L., & Russell, L. D. (1998). Germ cell genotype controls cell cycle during spermatogenesis in the rat. *Biology of Reproduction*, *59*, 1371–1377.

Gennet, N., Gale, E., Nan, X., Farley, E., Takacs, K., Oberwallner, B., et al. (2012). Doublesex and mab-3-related transcription factor 5 promotes midbrain dopaminergic identity in pluripotent stem cells by enforcing a ventral-medial progenitor fate. *Proceedings of the National Academy of Sciences of the United States of America*, *108*, 9131–9136.

Guan, G., Kobayashi, T., & Nagahama, Y. (2000). Sexually dimorphic expression of two types of DM (Doublesex/Mab-3)-domain genes in a teleost fish, the Tilapia (*Oreochromis niloticus*). *Biochemical and Biophysical Research Communications*, *272*, 662–666.

Guan, K., Nayernia, K., Maier, L. S., Wagner, S., Dressel, R., Lee, J. H., et al. (2006). Pluripotency of spermatogonial stem cells from adult mouse testis. *Nature*, *440*, 1199–1203.

Guo, Y., Li, Q., Gao, S., Zhou, X., He, Y., Shang, X., et al. (2004). Molecular cloning, characterization, and expression in brain and gonad of Dmrt5 of zebrafish. *Biochemical and Biophysical Research Communications*, *324*, 569–575.

He, C. L., Du, J. L., Wu, G. C., Lee, Y. H., Sun, L. T., & Chang, C. F. (2003). Differential Dmrt1 transcripts in gonads of the protandrous black porgy, *Acanthopagrus schlegeli*. *Cytogenetic and Genome Research*, *101*, 309–313.

Heaney, J. D., Lam, M. Y., Michelson, M. V., & Nadeau, J. H. (2008). Loss of the transmembrane but not the soluble kit ligand isoform increases testicular germ cell tumor susceptibility in mice. *Cancer Research*, *68*, 5193–5197.

Heard, E., & Turner, J. (2011). Function of the sex chromosomes in mammalian fertility. *Cold Spring Harbor Perspectives in Biology*, *3*, a002675.

Heimdal, K., Olsson, H., Tretli, S., Flodgren, P., Borresen, A. L., & Fossa, S. D. (1996). Familial testicular cancer in Norway and southern Sweden. *British Journal of Cancer*, *73*, 964–969.

Herpin, A., Braasch, I., Kraeussling, M., Schmidt, C., Thoma, E. C., Nakamura, S., et al. (2010). Transcriptional rewiring of the sex determining dmrt1 gene duplicate by transposable elements. *PLoS Genetics*, *6*, e1000844.

Herpin, A., Nakamura, S., Wagner, T. U., Tanaka, M., & Schartl, M. (2009). A highly conserved cis-regulatory motif directs differential gonadal synexpression of Dmrt1 transcripts during gonad development. *Nucleic Acids Research*, *37*, 1510–1520.

Herpin, A., & Schartl, M. (2011). Dmrt1 genes at the crossroads: A widespread and central class of sexual development factors in fish. *The FEBS Journal*, *278*, 1010–1019.

Herpin, A., Schindler, D., Kraiss, A., Hornung, U., Winkler, C., & Schartl, M. (2007). Inhibition of primordial germ cell proliferation by the medaka male determining gene Dmrt I bY. *BMC Developmental Biology*, *7*, 99.

Hoei-Hansen, C. E., Almstrup, K., Nielsen, J. E., Brask Sonne, S., Graem, N., Skakkebaek, N. E., et al. (2005). Stem cell pluripotency factor NANOG is expressed in human fetal gonocytes, testicular carcinoma in situ and germ cell tumours. *Histopathology*, *47*, 48–56.

Huang, X., Guo, Y., Shui, Y., Gao, S., Yu, H., Cheng, H., et al. (2005). Multiple alternative splicing and differential expression of dmrt1 during gonad transformation of the rice field eel. *Biology of Reproduction*, *73*, 1017–1024.

Huynh, K. D., & Lee, J. T. (2003). Inheritance of a pre-inactivated paternal X chromosome in early mouse embryos. *Nature*, *426*, 857–862.

Ichijima, Y., Sin, H. S., & Namekawa, S. H. (2012). Sex chromosome inactivation in germ cells: Emerging roles of DNA damage response pathways. *Cellular and Molecular Life Sciences*, *69*, 2559–2572.

Jameson, S. A., Natarajan, A., Cool, J., DeFalco, T., Maatouk, D. M., Mork, L., et al. (2012). Temporal transcriptional profiling of somatic and germ cells reveals biased lineage priming of sexual fate in the fetal mouse gonad. *PLoS Genetics*, *8*, e1002575.

Jeong, H. B., Park, J. G., Park, Y. J., Takemura, A., Hur, S. P., Lee, Y. D., et al. (2009). Isolation and characterization of DMRT1 and its putative regulatory region in the protogynous wrasse, *Halichoeres tenuispinis*. *Gene*, *438*, 8–16.

Kanatsu-Shinohara, M., Inoue, K., Lee, J., Yoshimoto, M., Ogonuki, N., Miki, H., et al. (2004). Generation of pluripotent stem cells from neonatal mouse testis. *Cell*, *119*, 1001–1012.

Kanetsky, P. A., Mitra, N., Vardhanabhuti, S., Vaughn, D. J., Li, M., Ciosek, S. L., et al. (2011). A second independent locus within DMRT1 is associated with testicular germ cell tumor susceptibility. *Human Molecular Genetics*, *20*, 3109–3117.

Kawamata, M., Inoue, H., & Nishimori, K. (2007). Male-specific function of Dmrt7 by sexually dimorphic translation in mouse testis. *Sexual Development*, *1*, 297–304.

Kawamata, M., & Nishimori, K. (2006). Mice deficient in Dmrt7 show infertility with spermatogenic arrest at pachytene stage. *FEBS Letters*, *580*, 6442–6446.

Kim, S., Bardwell, V. J., & Zarkower, D. (2007). Cell type-autonomous and non-autonomous requirements for Dmrt1 in postnatal testis differentiation. *Developmental Biology*, *307*, 314–327.

Kim, S., Kettlewell, J. R., Anderson, R. C., Bardwell, V. J., & Zarkower, D. (2003). Sexually dimorphic expression of multiple doublesex-related genes in the embryonic mouse gonad. *Gene Expression Patterns*, *3*, 77–82.

Kim, S., Namekawa, S. H., Niswander, L. M., Ward, J. O., Lee, J. T., Bardwell, V. J., et al. (2007). A mammal-specific Doublesex homolog associates with male sex chromatin and is required for male meiosis. *PLoS Genetics*, *3*, e62.

Kimura, T., Suzuki, A., Fujita, Y., Yomogida, K., Lomeli, H., Asada, N., et al. (2003). Conditional loss of PTEN leads to testicular teratoma and enhances embryonic germ cell production. *Development*, *130*, 1691–1700.

Kondo, M., Nanda, I., Hornung, U., Schmid, M., & Schartl, M. (2004). Evolutionary origin of the medaka Y chromosome. *Current Biology, 14*, 1664–1669.

Kondo, M., Nanda, I., Schmid, M., & Schartl, M. (2009). Sex determination and sex chromosome evolution: Insights from medaka. *Sexual Development, 3*, 88–98.

Kratz, C. P., Han, S. S., Rosenberg, P. S., Berndt, S. I., Burdett, L., Yeager, M., et al. (2011). Variants in or near KITLG, BAK1, DMRT1, and TERT-CLPTM1L predispose to familial testicular germ cell tumour. *Journal of Medical Genetics, 48*, 473–476.

Krentz, A. D., Murphy, M. W., Kim, S., Cook, M. S., Capel, B., Zhu, R., et al. (2009). The DM domain protein DMRT1 is a dose-sensitive regulator of fetal germ cell proliferation and pluripotency. *Proceedings of the National Academy of Sciences of the United States of America, 106*, 22323–22328.

Krentz, A. D., Murphy, M. W., Sarver, A. L., Griswold, M. D., Bardwell, V. J., & Zarkower, D. (2011). DMRT1 promotes oogenesis by transcriptional activation of Stra8 in the mammalian fetal ovary. *Developmental Biology, 356*, 63–70.

Kurokawa, H., Saito, D., Nakamura, S., Katoh-Fukui, Y., Ohta, K., Baba, T., et al. (2007). Germ cells are essential for sexual dimorphism in the medaka gonad. *Proceedings of the National Academy of Sciences of the United States of America, 104*, 16958–16963.

Lei, N., & Heckert, L. L. (2004). Gata4 regulates testis expression of Dmrt1. *Molecular and Cellular Biology, 24*, 377–388.

Lei, N., Hornbaker, K. I., Rice, D. A., Karpova, T., Agbor, V. A., & Heckert, L. L. (2007). Sex-specific differences in mouse DMRT1 expression are both cell type- and stage-dependent during gonad development. *Biology of Reproduction, 77*, 466–475.

Lei, N., Karpova, T., Hornbaker, K. I., Rice, D. A., & Heckert, L. L. (2009). Distinct transcriptional mechanisms direct expression of the rat Dmrt1 promoter in sertoli cells and germ cells of transgenic mice. *Biology of Reproduction, 81*, 118–125.

Liarte, S., Chaves-Pozo, E., Garcia-Alcazar, A., Mulero, V., Meseguer, J., & Garcia-Ayala, A. (2007). Testicular involution prior to sex change in gilthead seabream is characterized by a decrease in DMRT1 gene expression and by massive leukocyte infiltration. *Reproductive Biology and Endocrinology, 5*, 20.

Looijenga, L. H., Hersmus, R., Gillis, A. J., Pfundt, R., Stoop, H. J., van Gurp, R. J., et al. (2006). Genomic and expression profiling of human spermatocytic seminomas: Primary spermatocyte as tumorigenic precursor and DMRT1 as candidate chromosome 9 gene. *Cancer Research, 66*, 290–302.

Lourenco, R., Lopes, S. S., & Saude, L. (2011). Left-right function of dmrt2 genes is not conserved between zebrafish and mouse. *PLoS One, 5*, e14438.

Manuylov, N. L., Zhou, B., Ma, Q., Fox, S. C., Pu, W. T., & Tevosian, S. G. (2011). Conditional ablation of Gata4 and Fog2 genes in mice reveals their distinct roles in mammalian sexual differentiation. *Developmental Biology, 353*, 229–241.

Mark, M., Jacobs, H., Oulad-Abdelghani, M., Dennefeld, C., Feret, B., Vernet, N., et al. (2008). STRA8-deficient spermatocytes initiate, but fail to complete, meiosis and undergo premature chromosome condensation. *Journal of Cell Science, 121*, 3233–3242.

Masuyama, H., Yamada, M., Kamei, Y., Fujiwara-Ishikawa, T., Todo, T., Nagahama, Y., et al. (2012). Dmrt1 mutation causes a male-to-female sex reversal after the sex determination by Dmy in the medaka. *Chromosome Research, 20*, 163–176.

Matson, C. K., Murphy, M. W., Griswold, M. D., Yoshida, S., Bardwell, V. J., & Zarkower, D. (2010). The mammalian doublesex homolog DMRT1 is a transcriptional gatekeeper that controls the mitosis versus meiosis decision in male germ cells. *Developmental Cell, 19*, 612–624.

Matson, C. K., Murphy, M. W., Sarver, A. L., Griswold, M. D., Bardwell, V. J., & Zarkower, D. (2011). DMRT1 prevents female reprogramming in the postnatal mammalian testis. *Nature, 476*, 101–104.

Matson, C. K., & Zarkower, D. (2012). Sex and the singular DM domain: Insights into sexual regulation, evolution and plasticity. *Nature Reviews Genetics*, *13*, 163–174.

Matsuda, M., Nagahama, Y., Shinomiya, A., Sato, T., Matsuda, C., Kobayashi, T., et al. (2002). DMY is a Y-specific DM-domain gene required for male development in the medaka fish. *Nature*, *417*, 559–563.

Matsuda, M., Sato, T., Toyazaki, Y., Nagahama, Y., Hamaguchi, S., & Sakaizumi, M. (2003). *Oryzias curvinotus* has DMY, a gene that is required for male development in the medaka, *O. latipes*. *Zoological Science*, *20*, 159–161.

Matsuda, M., Shinomiya, A., Kinoshita, M., Suzuki, A., Kobayashi, T., Paul-Prasanth, B., et al. (2007). DMY gene induces male development in genetically female (XX) medaka fish. *Proceedings of the National Academy of Sciences of the United States of America*, *104*, 3865–3870.

Matsui, Y., Zsebo, K., & Hogan, B. L. (1992). Derivation of pluripotential embryonic stem cells from murine primordial germ cells in culture. *Cell*, *70*, 841–847.

McCarthy, P. T., & Cerecedo, L. R. (1952). Vitamin A deficiency in the mouse. *The Journal of Nutrition*, *46*, 361–376.

McKee, B. D., & Handel, M. A. (1993). Sex chromosomes, recombination, and chromatin conformation. *Chromosoma*, *102*, 71–80.

McLaren, A. (2003). Primordial germ cells in the mouse. *Developmental Biology*, *262*, 1–15.

Morinaga, C., Saito, D., Nakamura, S., Sasaki, T., Asakawa, S., Shimizu, N., et al. (2007). The hotei mutation of medaka in the anti-Mullerian hormone receptor causes the dysregulation of germ cell and sexual development. *Proceedings of the National Academy of Sciences of the United States of America*, *104*, 9691–9696.

Nagano, R., Tabata, S., Nakanishi, Y., Ohsako, S., Kurohmaru, M., & Hayashi, Y. (2000). Reproliferation and relocation of mouse male germ cells (gonocytes) during prespermatogenesis. *The Anatomical Record*, *258*, 210–220.

Nakagawa, T., Nabeshima, Y., & Yoshida, S. (2007). Functional identification of the actual and potential stem cell compartments in mouse spermatogenesis. *Developmental Cell*, *12*, 195–206.

Nakagawa, T., Sharma, M., Nabeshima, Y., Braun, R. E., & Yoshida, S. (2010). Functional hierarchy and reversibility within the murine spermatogenic stem cell compartment. *Science*, *328*, 62–67.

Namekawa, S. H., Park, P. J., Zhang, L. F., Shima, J. E., McCarrey, J. R., Griswold, M. D., et al. (2006). Postmeiotic sex chromatin in the male germline of mice. *Current Biology*, *16*, 660–667.

Namekawa, S. H., Payer, B., Huynh, K. D., Jaenisch, R., & Lee, J. T. (2010). Two-step imprinted X inactivation: Repeat versus genic silencing in the mouse. *Molecular and Cellular Biology*, *30*, 3187–3205.

Nanda, I., Kondo, M., Hornung, U., Asakawa, S., Winkler, C., Shimizu, A., et al. (2002). A duplicated copy of DMRT1 in the sex-determining region of the Y chromosome of the medaka, *Oryzias latipes*. *Proceedings of the National Academy of Sciences of the United States of America*, *99*, 11778–11783.

Okamoto, I., Arnaud, D., Le Baccon, P., Otte, A. P., Disteche, C. M., Avner, P., et al. (2005). Evidence for de novo imprinted X-chromosome inactivation independent of meiotic inactivation in mice. *Nature*, *438*, 369–373.

Oosterhuis, J. W., & Looijenga, L. H. (2005). Testicular germ-cell tumours in a broader perspective. *Nature Reviews. Cancer*, *5*, 210–222.

Ottolenghi, C., Omari, S., Garcia-Ortiz, J. E., Uda, M., Crisponi, L., Forabosco, A., et al. (2005). Foxl2 is required for commitment to ovary differentiation. *Human Molecular Genetics*, *14*, 2053–2062.

Oulad-Abdelghani, M., Bouillet, P., Decimo, D., Gansmuller, A., Heyberger, S., Dolle, P., et al. (1996). Characterization of a premeiotic germ cell-specific cytoplasmic protein

encoded by Stra8, a novel retinoic acid-responsive gene. *The Journal of Cell Biology, 135*, 469–477.
Raymond, C. S., Murphy, M. W., O'Sullivan, M. G., Bardwell, V. J., & Zarkower, D. (2000). Dmrt1, a gene related to worm and fly sexual regulators, is required for mammalian testis differentiation. *Genes and Development, 14*, 2587–2595.
Raymond, C. S., Shamu, C. E., Shen, M. M., Seifert, K. J., Hirsch, B., Hodgkin, J., et al. (1998). Evidence for evolutionary conservation of sex-determining genes. *Nature, 391*, 691–695.
Royo, H., Polikiewicz, G., Mahadevaiah, S. K., Prosser, H., Mitchell, M., Bradley, A., et al. (2010). Evidence that meiotic sex chromosome inactivation is essential for male fertility. *Current Biology, 20*, 2117–2123.
Saito, D., & Tanaka, M. (2009). Comparative aspects of gonadal sex differentiation in medaka: A conserved role of developing oocytes in sexual canalization. *Sexual Development, 3*, 99–107.
Sato, T., Rocancourt, D., Marques, L., Thorsteinsdottir, S., & Buckingham, M. (2010). A Pax3/Dmrt2/Myf5 regulatory cascade functions at the onset of myogenesis. *PLoS Genetics, 6*, e1000897.
Saude, L., Lourenco, R., Goncalves, A., & Palmeirim, I. (2005). terra is a left-right asymmetry gene required for left-right synchronization of the segmentation clock. *Nature Cell Biology, 7*, 918–920.
Seo, K. W., Wang, Y., Kokubo, H., Kettlewell, J. R., Zarkower, D. A., & Johnson, R. L. (2006). Targeted disruption of the DM domain containing transcription factor Dmrt2 reveals an essential role in somite patterning. *Developmental Biology, 290*, 200–210.
Shen, M. M., & Hodgkin, J. (1988). mab-3, a gene required for sex-specific yolk protein expression and a male-specific lineage in C. elegans. *Cell, 54*, 1019–1031.
Shin, H. S., An, K. W., Park, M. S., Jeong, M. H., & Choi, C. Y. (2009). Quantitative mRNA expression of sox3 and DMRT1 during sex reversal, and expression profiles after GnRHa administration in black porgy, Acanthopagrus schlegeli. *Comparative Biochemistry and Physiology. Part B, Biochemistry and Molecular Biology, 154*, 150–156.
Smith, C. A., Katz, M., & Sinclair, A. H. (2003). DMRT1 is upregulated in the gonads during female-to-male sex reversal in ZW chicken embryos. *Biology of Reproduction, 68*, 560–570.
Smith, C. A., Roeszler, K. N., Ohnesorg, T., Cummins, D. M., Farlie, P. G., Doran, T. J., et al. (2009). The avian Z-linked gene DMRT1 is required for male sex determination in the chicken. *Nature, 461*, 267–271.
Stevens, L. C. (1964). Experimental production of testicular teratomas in mice. *Proceedings of the National Academy of Sciences of the United States of America, 52*, 654–661.
Stevens, L. C. (1967a). Origin of testicular teratomas from primordial germ cells in mice. *Journal of the National Cancer Institute, 38*, 549–552.
Stevens, L. C. (1967b). The biology of teratomas. *Advances in Morphogenesis, 6*, 1–31.
Stevens, L. C., & Little, C. C. (1954). Spontaneous testicular teratomas in an inbred strain of mice. *Proceedings of the National Academy of Sciences of the United States of America, 40*, 1080–1087.
Sugimoto, R., Nabeshima, Y., & Yoshida, S. (2011). Retinoic acid metabolism links the periodical differentiation of germ cells with the cycle of Sertoli cells in mouse seminiferous epithelium. *Mechanisms of Development, 128*, 610–624.
Suzuki, H., Ahn, H. W., Chu, T., Bowden, W., Gassei, K., Orwig, K., et al. (2012). SOHLH1 and SOHLH2 coordinate spermatogonial differentiation. *Developmental Biology, 361*, 301–312.
Suzuki, A., Nakamoto, M., Kato, Y., & Shibata, N. (2005). Effects of estradiol-17beta on germ cell proliferation and DMY expression during early sexual differentiation of the medaka *Oryzias latipes*. *Zoological Science, 22*, 791–796.

Takahashi, K., Mitsui, K., & Yamanaka, S. (2003). Role of ERas in promoting tumour-like properties in mouse embryonic stem cells. *Nature, 423*, 541–545.
Thompson, J. N., Howell, J. M., & Pitt, G. A. (1964). Vitamin a and Reproduction in Rats. *Proceedings of the Royal Society of London, Series B: Biological Sciences, 159*, 510–535.
Turnbull, C., Rapley, E. A., Seal, S., Pernet, D., Renwick, A., Hughes, D., et al. (2010). Variants near DMRT1, TERT and ATF7IP are associated with testicular germ cell cancer. *Nature Genetics, 42*, 604–607.
Uhlenhaut, N. H., Jakob, S., Anlag, K., Eisenberger, T., Sekido, R., Kress, J., et al. (2009). Somatic sex reprogramming of adult ovaries to testes by FOXL2 ablation. *Cell, 139*, 1130–1142.
van Pelt, A. M., & de Rooij, D. G. (1990). Synchronization of the seminiferous epithelium after vitamin A replacement in vitamin A-deficient mice. *Biology of Reproduction, 43*, 363–367.
Veitia, R. A. (2010). FOXL2 versus SOX9: A lifelong "battle of the sexes" *BioEssays, 32*, 375–380.
Vernet, N., Dennefeld, C., Guillou, F., Chambon, P., Ghyselinck, N. B., & Mark, M. (2006). Prepubertal testis development relies on retinoic acid but not rexinoid receptors in Sertoli cells. *The EMBO Journal, 25*, 5816–5825.
Vernet, N., Dennefeld, C., Rochette-Egly, C., Oulad-Abdelghani, M., Chambon, P., Ghyselinck, N. B., et al. (2006). Retinoic acid metabolism and signaling pathways in the adult and developing mouse testis. *Endocrinology, 147*, 96–110.
Wang, D. S., Zhou, L. Y., Kobayashi, T., Matsuda, M., Shibata, Y., Sakai, F., et al. (2010). Doublesex- and Mab-3-related transcription factor-1 repression of aromatase transcription, a possible mechanism favoring the male pathway in tilapia. *Endocrinology, 151*, 1331–1340.
Wen, A., You, F., Tan, X., Sun, P., Ni, J., Zhang, Y., et al. (2009). Expression pattern of dmrt4 from olive flounder (Paralichthys olivaceus) in adult gonads and during embryogenesis. *Fish Physiology and Biochemistry, 35*, 421–433.
Western, P. S., van den Bergen, J. A., Miles, D. C., & Sinclair, A. H. (2010). Male fetal germ cell differentiation involves complex repression of the regulatory network controlling pluripotency. *The FASEB Journal, 24*, 3026–3035.
Wu, G. C., Chiu, P. C., Lin, C. J., Lyu, Y. S., Lan, D. S., & Chang, C. F. (2011). Testicular dmrt1 is involved in the sexual fate of the ovotestis in the protandrous Black Porgy. *Biology of Reproduction, 86*, 41–53.
Xia, W., Zhou, L., Yao, B., Li, C. J., & Gui, J. F. (2007). Differential and spermatogenic cell-specific expression of DMRT1 during sex reversal in protogynous hermaphroditic groupers. *Molecular and Cellular Endocrinology, 263*, 156–172.
Yabuta, Y., Kurimoto, K., Ohinata, Y., Seki, Y., & Saitou, M. (2006). Gene expression dynamics during germline specification in mice identified by quantitative single-cell gene expression profiling. *Biology of Reproduction, 75*, 705–716.
Yoshida, S., Sukeno, M., Nakagawa, T., Ohbo, K., Nagamatsu, G., Suda, T., et al. (2006). The first round of mouse spermatogenesis is a distinctive program that lacks the self-renewing spermatogonia stage. *Development, 133*, 1495–1505.
Yoshimoto, S., & Ito, M. (2011). A ZZ/ZW-type sex determination in Xenopus laevis. *The FEBS Journal, 278*, 1020–1026.
Yoshimoto, S., Okada, E., Umemoto, H., Tamura, K., Uno, Y., Nishida-Umehara, C., et al. (2008). A W-linked DM-domain gene, DM-W, participates in primary ovary development in *Xenopus laevis*. *Proceedings of the National Academy of Sciences of the United States of America, 105*, 2469–2474.
Yoshizawa, A., Nakahara, Y., Izawa, T., Ishitani, T., Tsutsumi, M., Kuroiwa, A., et al. (2011). Zebrafish Dmrta2 regulates neurogenesis in the telencephalon. *Genes to Cells, 16*, 1097–1109.

Youngren, K. K., Coveney, D., Peng, X., Bhattacharya, C., Schmidt, L. S., Nickerson, M. L., et al. (2005). The Ter mutation in the dead end gene causes germ cell loss and testicular germ cell tumours. *Nature*, *435*, 360–364.

Yuan, L., Liu, J. G., Hoja, M. R., Wilbertz, J., Nordqvist, K., & Hoog, C. (2002). Female germ cell aneuploidy and embryo death in mice lacking the meiosis-specific protein SCP3. *Science*, *296*, 1115–1118.

Zhou, Q., Li, Y., Nie, R., Friel, P., Mitchell, D., Evanoff, R. M., et al. (2008). Expression of stimulated by retinoic acid gene 8 (Stra8) and maturation of murine gonocytes and spermatogonia induced by retinoic acid in vitro. *Biology of Reproduction*, *78*, 537–545.

CHAPTER THIRTEEN

RanBPM, a Scaffolding Protein for Gametogenesis

Sandrine Puverel[1], Lino Tessarollo[1]
Neural Development Section, Mouse Cancer Genetics Program, Center for Cancer Research, NCI, Frederick, Maryland, USA
[1]Corresponding authors: e-mail address: puverels@gmail.com; tessarol@mail.nih.gov

Contents

Abstract

RanBPM is a multimodular scaffold protein that interacts with a great variety of molecules including nuclear, cytoplasmic, and membrane proteins. By building multiprotein complexes, RanBPM is thought to regulate various signaling pathways, especially in the immune and nervous system. However, the diversity of these interactions does not facilitate the identification of its precise mechanism of action, and therefore the physiological role of RanBPM still remains unclear. Recently, RanBPM has been shown to be critical for the fertility of both genders in mouse. Although mechanistically it is still unclear how RanBPM affects gametogenesis, the data collected so far suggest that it is a key player in this process. Here, we examine the RanBPM sterility phenotype in the context of other genetic mutations affecting mouse gametogenesis to investigate whether this scaffold protein affects the function of other known proteins whose deficiency results in similar sterility phenotypes.

Current Topics in Developmental Biology, Volume 102
ISSN 0070-2153
http://dx.doi.org/10.1016/B978-0-12-416024-8.00013-1

1. INTRODUCTION

Multimodular scaffold proteins are crucial regulators of a great variety of physiological functions. They coordinate the physical assembly of proteins, regulating signal transduction cascades, and shaping signaling responses. RanBPM, also named RanBP9, is a scaffold protein that belongs to the Ran-binding protein family. Proteins of this family were initially identified by yeast two-hybrid as binding partners of the small Ras-like GTPase Ran. However, RanBPM is structurally and functionally unrelated to the other members of this family, as it lacks the consensus Ran-binding domain and seems devoid of any role in nuclear trafficking (Beddow, Richards, Orem, & Macara, 1995). Instead, RanBPM has been described as an adaptor protein. Its structure does not contain any known catalytic domain but includes several protein-binding domains that participate in the formation of large protein complexes. Indeed, RanBPM has been reported to interact so far with more than 45 proteins. These include structural and adhesion proteins, cytosolic kinases, cell surface tyrosine kinase receptors, nuclear receptors, and a number of proteins in the nervous system (Suresh, Ramakrishna, & Baek, 2012). RanBPM has been reported to function in the immune and nervous system (Murrin & Talbot, 2007). Yet, the *in vivo* significance of these activities needs further studies. Moreover, it is still unclear how RanBPM orchestrates the activities of such a broad spectrum of proteins that are functionally unrelated. RanBPM has a multimodular structure with "sticky" properties that makes it possible for it to associate nonspecifically with many other proteins *in vitro*. Indeed, many of its binding partners have been identified by yeast-two-hybrid screening and coimmunoprecipitation experiments in overexpression systems, suggesting that additional functional assays may be needed to validate the existing data. In an attempt to study the *in vivo* function of RanBPM, we recently generated and characterized mice lacking RanBPM (Puverel, Barrick, Dolci, Coppola, & Tessarollo, 2011). This loss-of-function study revealed a crucial role of RanBPM in both spermatogenesis and oogenesis. However, we still do not know how it orchestrates the signal transduction machinery required for the execution of these biological processes. In this chapter, we discuss the RanBPM-deficient phenotypes, the possible mechanism of RanBPM action in relation to its structure, spatiotemporal pattern of expression, and how it relates to other mutations affecting gametogenesis.

2. PROTEIN STRUCTURE

RanBPM comprises five main domains. These include an N-terminal proline-rich domain, a consensus SPRY protein–protein interaction domain originally identified in the SplA kinase and the ryanodine receptor (Ponting, Schultz, & Bork, 1997), a lissencephaly type-1-like homology (LisH) motif described to act as a dimerization domain (Gerlitz, Darhin, Giorgio, Franco, & Reiner, 2005; Kim et al., 2004; Mateja, Cierpicki, Paduch, Derewenda, & Otlewski, 2006), a CTLH (carboxy-terminal to LisH) domain of unknown function, and a C-terminal CRA (CT11-RanBPM) motif, reported to interact with the fragile X mental retardation protein FMRP (Menon, Gibson, & Pastore, 2004). The CRA domain has been reported to contain a nuclear localization signal (Lakshmana et al., 2010). The LisH-CTLH domain is mostly described in proteins involved in the regulation of microtubule dynamics and cell migration and is present in several proteins which associate with RanBPM to form a large protein complex (Emes & Ponting, 2001; Kobayashi et al., 2007; Nishitani et al., 2001). Thus, all these domains provide potential protein-binding sites for a great variety of molecules (Fig. 13.1).

In an initial study, RanBPM was described as a centrosomal 55-kDa protein implicated in the regulation of microtubule nucleation (Nakamura et al., 1998). Later work by the same group revealed that this was instead an N-terminal truncated protein and that the full-length RanBPM cDNA encodes a 90-kDa protein which does not localize to the microtubule-organizing center (Nishitani et al., 2001). However, it is interesting to note that in a recent study, some authors reported the occurrence of a 60-kDa-processed form (C-terminal truncation) with an enhanced stability compared to the full-length RanBPM. While the full-length protein is localized in both the nucleus and cytoplasm, the truncated form is mostly localized in the cytoplasm (Lakshmana et al., 2010). Yet, this fragment retains the capacity to form self-interacting multimeric complexes.

Figure 13.1 Structure of RanBPM protein. RanBPM is composed of five protein–protein interaction domains, including a proline-rich domain (Pro-Rich), a SPRY domain, the functionally related LisH and CTLH domains, and a C-terminal CRA domain. *Adapted from Suresh et al. (2012).* (For color version of this figure, the reader is referred to the online version of this chapter.)

More recently, another protein named RanBP10 has been described as the closest homolog of RanBPM. This protein shares significant sequence similarities and, except for lacking the N-terminal proline-rich domain, it possesses all other domains and these are well conserved. So far, RanBP10 has been reported to play a role in regulating the platelet shape and function in the adult mouse (Kunert et al., 2009). The close structural similarities between RanBPM and RanBP10 suggest the possibility of functional redundancy during the analysis of mutant mice lacking one of these proteins.

3. FUNCTION OUTSIDE THE REPRODUCTIVE SYSTEM

RanBPM is evolutionarily conserved and is ubiquitously expressed; to our knowledge, it has been found in all tissues and cell lines examined so far. It can undergo phosphorylation and ubiquitination, but the precise role of these posttranslational modifications still needs to be evaluated (Denti et al., 2004; Ideguchi et al., 2002). At the subcellular level, RanBPM has been found in the cytoplasm, nucleus, as well as associated with the cell membrane (Denti et al., 2004; Suresh et al., 2012; Valiyaveettil et al., 2008).

Among the best-known partners of RanBPM are the Met tyrosine kinase receptor for hepatocyte growth factor (HGF), whose Ras–Erk pathway is facilitated by RanBPM, and the kelch-repeat protein muskelin which acts on the regulation of cell morphology (Valiyaveettil et al., 2008; Wang, Li, Messing, & Wu, 2002). In addition to its involvement in these general cellular processes, RanBPM is known to play a role in the immune response by interacting with the β2-integrin lymphocyte function-associated antigen 1 (Denti et al., 2004). Moreover, a growing number of studies are reporting a function for RanBPM in the nervous system. For example, it has been involved in the modulation of axonal and neurite outgrowth by cooperating with the Plexin-A receptors and inhibiting the neural cell adhesion molecule L1 signaling (Cheng, Lemmon, & Lemmon, 2005; Togashi, Schmidt, & Strittmatter, 2006). Interestingly, RanBPM binds to both the low- and the high-affinity neurotrophin receptors p75NTR, TrkA, and TrkB (Bai, Chen, & Huang, 2003; Yin et al., 2010; Yuan et al., 2006), interacts with the fragile X mental retardation protein, possibly modulating its RNA-binding properties, and binds to metabotropic glutamate receptors (Menon et al., 2004; Seebahn, Rose, & Enz, 2008). RanBPM has also been found to associate with citron kinase (CITK), regulating the production of pyramidal neurons in the embryonic neocortex (Chang et al., 2010). Finally, RanBPM has been involved in Alzheimer's disease, affecting amyloid precursor protein processing and amyloid β generation (Lakshmana et al., 2009; Woo et al., 2012).

4. PHENOTYPE OF RanBPM-DEFICIENT MICE

While RanBPM has been involved in a number of biological processes, the recent generation of RanBPM-deficient mice has provided a tool that will allow investigators to test their physiological significance *in vivo*. Homozygous mutant mice for RanBPM develop to term, suggesting that the gene is not crucial for embryonic development. However, a little more than half of the mutant animals die postnatally, while the rest survive to adulthood. Surviving $RanBPM^{-/-}$ mice have a normal life span but are significantly smaller than their control littermates. The major organs, including brain, heart, liver, lung, kidney, thymus, spleen, and intestine, are proportional to the body size and do not show any obvious abnormality. By contrast, mutant gonads are severely atrophic leading to sterility of both genders (Puverel et al., 2011) (Fig. 13.2). In the adult mutant testis, seminiferous tubules are decreased in size and completely devoid of germ cells, and the epididymal duct lumen does not contain any sperm. In the mutant ovary, only a few follicles develop in young mice, but disappear in older mice.

5. EXPRESSION IN THE REPRODUCTIVE SYSTEM

RanBPM has been reported to be highly expressed in the mouse testis, especially in the maturating spermatocytes (Shibata et al., 2004). Another study has shown that the RanBPM protein is produced in spermatogonia and primary spermatocytes in the human testis, whereas in the rat testis, it

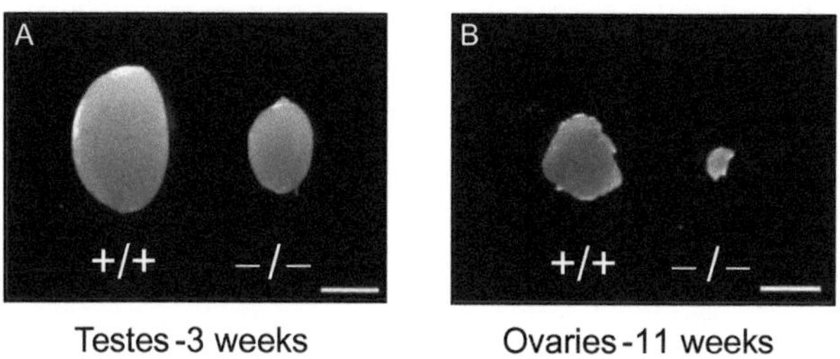

Figure 13.2 Representative pictures showing the gonad atrophy in RanBPM homozygous mutant mice. Mutant and wild-type mouse testes (A) and ovaries (B) from 3-week-old males and 11-week-old females. Scale bars: 2 mm. (For color version of this figure, the reader is referred to the online version of this chapter.)

appears to be expressed only in round and elongated spermatids (Tang et al., 2004). Moreover, gene expression profiling of mouse male meiotic germ cells and northern blot analysis have found RanBPM transcripts in spermatogonia and primary spermatocytes, as well as in spermatids (Rossi et al., 2004).

In order to clarify the RanBPM pattern of expression, we decided to evaluate the activity of this gene during testis development by taking advantage of a *LacZ* reporter gene inserted in the targeted *RanBPM* locus (Puverel et al., 2011). Since the *LacZ* gene product, β-galactosidase, is expressed under the control of the endogenous *RanBPM* promoter, X-gal staining constitutes a sensitive and accurate method to study the expression pattern of the *RanBPM* gene. We found that in males, *RanBPM* mRNA expression starts during embryogenesis; it is first detected in the testis of heterozygous animals at embryonic day 17.5 in clustered mouse vasa homolog (MVH)-positive germ cells. In newborns, expression is still mainly restricted to the gonocytes, while 5 days later, interstitial cells begin expressing *RanBPM*. At 2 weeks, both spermatogonia and Sertoli cells, the supporting cells of the spermatogenic process, are strongly labeled. *RanBPM* transcript is thus sequentially expressed by distinct cell populations during testis postnatal development, yet its expression is always present in the germ cell lineage. Interestingly, *RanBPM* mRNA expression is cyclically regulated during adult steady state spermatogenesis. In the adult testis, expression is detected in spermatogonia residing at the periphery of the seminiferous tubules and maintained in both primary and secondary spermatocytes. Remarkably, the highest expression is observed in primary spermatocytes undergoing the first meiotic division, although it seems to stop at the stage of differentiated tailed spermatids. *RanBPM* expression is thus dynamically regulated during the spermatogenic wave. In addition to *RanBPM* transcriptional activity, we were able to show that the protein is also expressed in the testis. Between birth and 2 weeks, RanBPM protein level is fairly constant but it increases significantly between 2 and 3 weeks, corresponding to the time when the spermatocytes of the first wave of spermatogenesis undergo the first meiotic division. By immunohistochemistry, we found that primary spermatocytes are the cells with the highest RanBPM protein expression.

The first study focusing on RanBPM function in the female reproductive system was conducted in *Drosophila* where there are two proteins (a short and a long) coded by transcripts produced by alternative splicing. The short RanBPM isoform appears to be expressed in all somatic and germline cells of the ovary, while the long isoform is specifically enriched in the germline stem cell niche where it is required for the proper arrangement of niche cells and niche cell size (Dansereau & Lasko, 2008). In mice, the *RanBPM* gene is active in both the developing and adult ovary. Interestingly, *RanBPM* expression starts at the birth and is restricted to the germinal cell lineage, that is, oocytes reaching the dictyate arrest (Puverel et al., 2011). In the adult ovaries, *RanBPM* is present at a high level in oocytes and is maintained during follicle development, from primary to Graafian follicles. As in males, *RanBPM* transcripts are also expressed in the somatic cell lineage in the ovary since expression is detected in the theca cell layers, while granulosa cells are negative. However, whether *RanBPM* transcripts are translated into proteins in every cell type in which they have been detected is still unknown and, as suggested by the results obtained in males, *RanBPM* mRNA expression may be broader than that of the protein (Puverel et al., 2011).

6. RanBPM IS DISPENSABLE FOR PRIMORDIAL GERM CELL MIGRATION

The data available so far suggest that RanBPM is not involved in primordial germ cell (PGC) migration (Puverel et al., 2011). PGCs are derived from the epiblast from where they migrate to colonize the genital ridges by 11.5 days of embryogenesis (Bendel-Stenzel, Anderson, Heasman, & Wylie, 1998; Tres, Rosselot, & Kierszenbaum, 2004). Upon entry into the developing gonads, PGCs coalesce with somatic cells that have initiated sexual differentiation. PGCs proliferate during migration and for a short time after colonization of the gonads. Male germ cells then enter mitotic arrest while female germ cells begin to enter the first meiotic division. Several factors have been involved in this process, including the chemokine stromal cell-derived factor 1 and its receptor CXCR4, as well as stem cell factor (SCF), and the c-Kit tyrosine kinase receptor (Ara et al., 2003; Buehr, McLaren, Bartley, & Darling, 1993; Mahakali Zama, Hudson, & Bedell, 2005; Molyneaux et al., 2003; Runyan et al., 2006). X-gal staining failed to detect any RanBPM expression in E13.5 embryos, as RanBPM expression starts during late embryogenesis in males and at birth in females (Puverel et al., 2011). In addition, the lack of RanBPM does not seem to affect germ cells during embryogenesis, as gonocytes are present in the seminiferous tubules of newborn males, and the number of meiotic oocytes present in the developing ovary of mutants at day E17.5 is comparable with that of WT embryos. These observations suggest that the migration of PGCs occurs normally in mutant mice and that RanBPM does not play a role in the germ cell migration and proliferation events occurring during embryonic development.

7. FUNCTION OF RanBPM DURING POSTNATAL TESTIS DEVELOPMENT

In males, the second week of postnatal life is marked by the initiation of the spermatogenesis process with massive proliferation of spermatogonia. Macroscopically, this transition is reflected by a rapid increase in testis volume as germ cells start filling the seminiferous tubules. RanBPM mutant testes do not show any overt phenotype during the first postnatal week. However, at 2 weeks, contrary to WT testes which undergo a significant growth, they only slightly increase in size. Histological examination at 2 weeks revealed a clear decrease in the diameter of seminiferous tubules, compared to WT (Puverel et al., 2011). These tubules contain a dramatically decreased number of both spermatogonia and spermatocytes. However, the levels of apoptosis were comparable between WT and mutant at this stage, suggesting that the lack of germ cells is not the consequence of cell death events. By contrast, and detectable as early as P8, there was a clear defect in proliferation which was revealed by a dramatic decrease in the number of BrdU-positive cells, as well as an altered organization of the proliferating cells, when compared to WT mouse testes in which these cells are typically forming a continuous ring delineating the seminiferous tubules (Puverel et al., 2011). Altogether, these data demonstrate that RanBPM is a crucial factor for spermatogonia production. Whether RanBPM acts on the differentiation of spermatogonia or on their proliferation remains to be determined.

One of the factors reported to act on spermatogonia proliferation is the tyrosine kinase receptor c-Kit. While the downregulation of its ligand, SCF affects PGCs migration, proliferation, and survival (Mahakali Zama et al., 2005; Runyan et al., 2006), c-Kit also plays a role in spermatogonia proliferation (Blume-Jensen et al., 2000; Kissel et al., 2000). RanBPM has been reported to interact with several tyrosine kinase receptors, including the HGF receptor Met and both the TrkA and TrkB neurotrophin receptors, and to possibly influence their signaling pathways. Considering that c-Kit function is critical for normal spermatogenesis, it is tempting to speculate that RanBPM may interact with this receptor and affect its signaling (Wang et al., 2002; Yin et al., 2010; Yuan et al., 2006). However, c-Kit acts on spermatogonia cell survival and entry into meiosis, and the phenotype of c-Kit mutants with a defect in the PI3-kinase pathway differs from RanBPM mutant mice in that spermatogonia are depleted by apoptosis and meiotic cells are rare or absent (Blume-Jensen et al., 2000; Kissel et al., 2000;

Puverel et al., 2011). Outside the reproductive system, some studies reported a role for RanBPM in cell division. CITK is a protein that localizes to the surface of the lateral ventricles during embryonic cortex development and is essential in neurogenic mitoses (Di Cunto et al., 2000). A biochemical interaction between CITK and RanBPM has been reported recently, and the use of *in utero* RNAi to decrease RanBPM expression leads to an increased number of cells in mitosis, and a concomitant decrease in the number of cells in cytokinesis (Chang et al., 2010). Both RanBPM and RanBP10 have also been described to interact with YPEL5, a protein located at different mitosis-related subcellular structures and involved in cell cycle progression (Hosono et al., 2010). Taken together, these reports suggest that RanBPM could play a role in the proliferation of different cell types. This hypothesis is also supported by the fact that RanBPM is ubiquitously expressed and that its deletion causes partial neonatal lethality and growth retardation for reasons that are still unknown (Puverel et al., 2011).

Androgens are essential hormones for the initiation and maintenance of spermatogenesis. The androgen receptor (AR) is a nuclear receptor acting as a ligand-inducible transcription factor to modulate the expression of a number of target genes. RanBPM has been reported to associate with AR and to enhance its transcriptional activity in a ligand-dependent manner when overexpressed in prostate cancer cell lines (Rao et al., 2002). AR has been reported to be expressed in all cell types of the testis. Knockout mice for AR exhibit a severe phenotype, with an abdominal localization of the testes and an absence of epididymis formation (Wang, Yeh, Tzeng, & Chang, 2009). If RanBPM mediates AR activity in the testis, its effect is thus likely limited to specific cell types. Testicular cell-specific AR knockout mice have been described in the literature and provide a better understanding of the role of this receptor in spermatogenesis (Wang et al., 2009). Interestingly, Sertoli cell-specific and Leydig cell-specific AR mutants exhibit a decrease in testis size and seminiferous tubules development as well as an arrest in germ cell maturation, mainly at the diplotene and round spermatid stages, respectively. By contrast, germ cell-AR-deficient mice are comparable to controls, with a normal range of spermatogenesis. If the defect in postnatal testis development observed in RanBPM-deficient mice was due to an effect of RanBPM on AR activity, one could speculate that this regulation occurs either in Sertoli or Leydig cells. However, such AR deregulation is accompanied by changes in serum hormone levels. In AR-knock-out mice and Sertoli- and Leydig-specific mutants, the levels of testosterone are decreased while LH levels are increased, whereas in RanBPM mutant mice, these hormone levels are not significantly different from those of control littermates (Puverel et al., 2011; Wang et al., 2009). In addition, RanBPM protein expression has been detected in primary spermatocytes, and we have found that RanBPM is acting in germ cells in a cell-autonomous way (see Section 10). Finally, the meiotic arrest observed in AR-deficient mice is unlikely to be due to the same defect occurring in RanBPM mice, since mutant females for AR do not exhibit any arrest in differentiation during the prophase of the first meiotic division, in contrast to RanBPM mutant females (Hu et al., 2004; Shiina et al., 2006) (Section 8).

8. MEIOTIC ARREST IN MICE DEFICIENT FOR RanBPM

Despite the reduction in spermatogonia proliferation, primary spermatocytes are still present, yet in reduced numbers, in the seminiferous tubules of *RanBPM*$^{-/-}$ male mice during the first wave of spermatogenesis. By contrast, postmeiotic cells are totally absent and no sperm is produced at any developmental stage. While somatic Sertoli and Leydig cells appear normal in the young mutant animal, spermatogonia, pachytene, and only a few diplotene spermatocytes are the most mature germ cells present in the tubules. Thus, *RanBPM*$^{-/-}$ males exhibit a stage-specific arrest of spermatogenesis, as generally observed in mutant mice in which male sterility is due to a germinal cell primary defect. The fact that mutant mice are affected at as early as 3 weeks of age shows that the first wave of germ line differentiation does not take place. In mice, the first wave of meiotic division is synchronous throughout the testis and occurs around 3 weeks, when the secondary spermatocytes and round spermatids first appear in the seminiferous tubules (Bellve et al., 1977). At this stage, RanBPM mutant testes display a high level of cell death and primary spermatocytes undergo apoptotic death instead of further developing into spermatids (Puverel et al., 2011). Meiosis is a critical step of gametogenesis. It allows genetic exchange between the paternal and maternal genomes, which occurs by recombination during the prophase of the first meiotic division. Prophase I is tightly regulated and can be divided into different stages (see also Fig. 13.3). The leptotene stage is marked by the compaction of chromatin and the programmed induction of localized double-strand breaks (DSBs) which allows the genetic exchange. Chromosome pairing and synapsis are initiated during zygotene, with full synapsis being achieved at the pachytene stage. During zygotene and pachytene stages, the DSBs are repaired by interaction with the homologs. The subsequent disjunction of homologous chromosomes is initiated at the diplotene stage; chromosomes start to separate but remain attached in the regions where crossing over has occurred (Cohen & Pollard, 2001; Zickler & Kleckner, 1999). The proper alignment and synapsis of homologous chromosomes are achieved by the formation of a large zipper-like tripartite protein complex called the synaptonemal complex, which connects the chromosomes along their entire length (Page & Hawley, 2004; Yang & Wang, 2009).

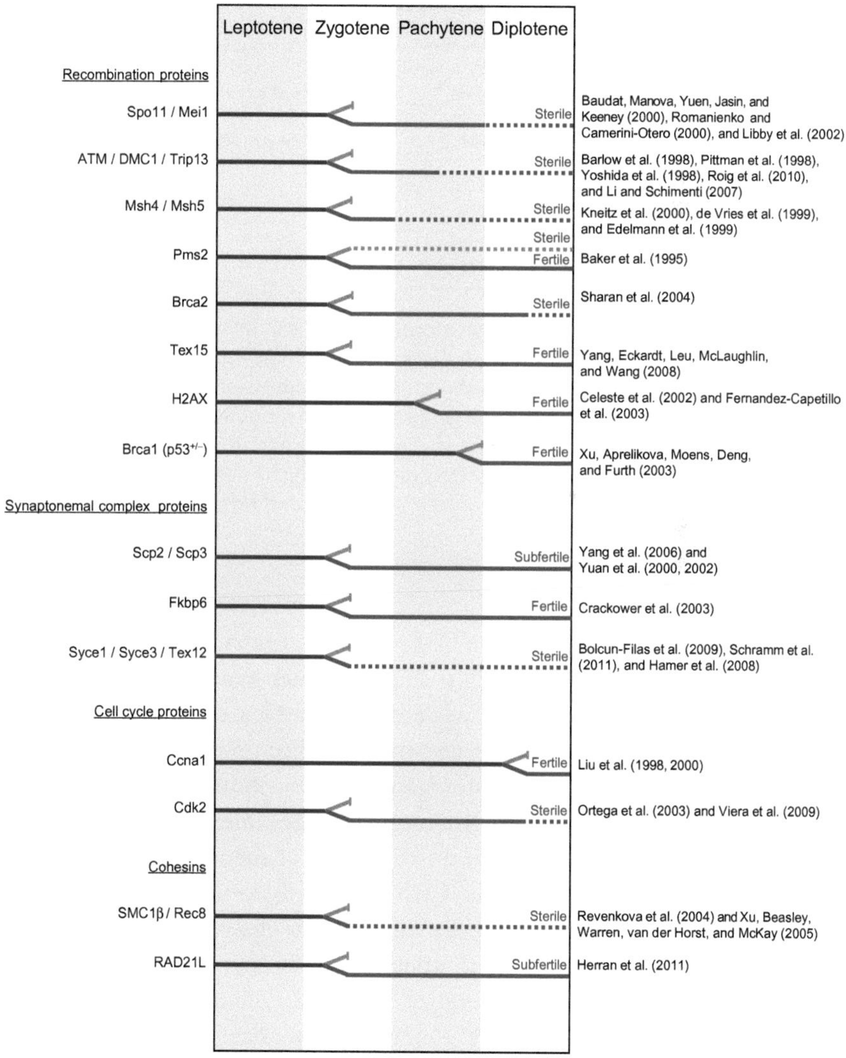

Figure 13.3

(Continued)

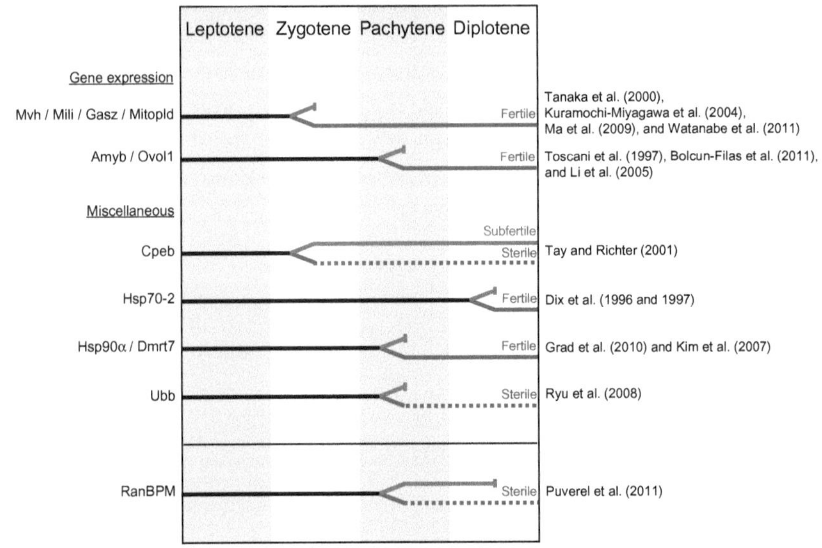

Figure 13.3 Diagram comparing known mouse mutants exhibiting an arrest during the prophase of the first meiotic division. A black solid line indicates stages during which male and female meioses are similar. When meiotic progression is different between the genders, a blue line indicates male progression and a purple line female progression. A meiotic arrest is represented by a solid vertical bar and dotted lines indicate that only a proportion of germ cells persists. Only genes for which information is available for both genders have been included. *Adapted from Hunt and Hassold (2002).* (See Color Insert.)

This protein complex mediates the DSBs resolution events leading to recombination. In males, the X and Y chromosomes synapse only at their pseudoautosomal region forming what is called the sex body, located in the nuclear periphery. Instead, in females, the XX chromosomes synapse along their entire length to undergo homologous recombination. In humans, incorrect assembly of the synaptonemal complex causes infertility by triggering germ cell elimination in males and the formation of aneuploid oocytes in females (Hassold & Hunt, 2001; Judis, Chan, Schwartz, Seftel, & Hassold, 2004; Miyamoto et al., 2003). The high level of apoptosis observed in *RanBPM*$^{-/-}$ seminiferous tubules is correlated to a decrease in the expression of genes that are specifically expressed at the end of Prophase I, such as Calmegin and Hox1.4 (Puverel et al., 2011; Rubin, Toth, Patel, D'Eustachio, & Nguyen-Huu, 1986; Watanabe et al., 1994). Expression of CyclinA1, which has been shown to rise dramatically in late pachytene spermatocytes and to reach a maximum in diplotene cells, is even more affected in RanBPM mutant mice (Sweeney et al., 1996). By contrast, genes expressed earlier in the differentiation process, such as A-myb, Dmc1, and Hsp70.2, are not affected (Mettus et al., 1994; Rosario, Perkins, O'Brien, Allen, & Eddy, 1992; Yoshida et al., 1998; Zakeri, Wolgemuth, & Hunt, 1988). In addition, the genes encoding the synaptonemal complex proteins SCP1 and SCP3 are expressed at normal levels in the mutant, indicative of the presence of pachytene spermatocytes (Klink, Lee, & Cooke, 1997; Meuwissen et al., 1992).

Female germ cells enter meiosis during embryonic development, around E13, and subsequently arrest at the end of Prophase I, around birth (dictyate arrest) (Speed, 1982). Oocytes remain arrested until puberty, when a pool of oocytes is recruited and completes the first meiotic division. The second meiotic round occurs only when fertilization takes place (Edson, Nagaraja, & Matzuk, 2009). Ovaries of adult *RanBPM*$^{-/-}$ females exhibit a severe dysgenesis resulting from an absence of follicles. Premature ovarian failure has many possible causes: for example, it can result from a defect in PGC migration, meiosis, and formation, survival or activation of primordial follicles (Jagarlamudi, Reddy, Adhikari, & Liu, 2010). Analysis of *RanBPM*$^{-/-}$ fetal ovaries at E17.5 revealed that the number of oocytes is comparable to that of controls. However, by postnatal day 0, there is a drastic decrease in their number (Puverel et al., 2011). Interestingly, RanBPM is first expressed at birth in the ovaries suggesting that it is essential from the very earliest stages of its expression since most germ cells are absent immediately after birth. Oocyte loss thus occurs between late embryogenesis and early postnatal development in *RanBPM*$^{-/-}$ females, and the vast majority of the oocytes fail to progress to the dictyate arrest. Early studies of meiosis in mouse fetal ovaries suggest that entry into the pachytene stage starts at E16 (Speed, 1982). Since the majority of the oocytes are between late pachytene and diplotene at the time of birth, the phenotype of RanBPM-deficient mice suggests that oocytes reach the pachytene stage unaffected but display the same meiotic arrest as observed in males. Importantly, this arrest, occurring at the same meiotic stage in both genders, suggests that RanBPM deletion affects mechanisms common to both male and female meiosis. This is even more interesting considering that the timing and synchrony of meiosis show striking differences between the two genders (Hunt & Hassold, 2002; Kolas et al., 2005). In females, the entire germ cell pool synchronously initiates meiosis *in utero* around E13.5 and meiosis transiently arrests at birth. By contrast, male meiosis is not initiated until puberty and spermatocyte development in the adult testis is continuous and proceeds without the arrest periods occurring in females (Nebel, Amarose, & Hacket, 1961). In addition, a growing number of mouse mutants for genes involved in meiotic progression display sexual dimorphism in fertility (Fig. 13.3). These studies revealed clear differences in the stringency of the meiotic events between males and females (Cohen, Pollack, & Pollard, 2006; Handel & Eppig, 1998; Matzuk & Lamb, 2002; Morelli & Cohen, 2005). Female meiosis is more "forgiving" to errors; oocytes with meiotic defects often reach a more advanced stage than spermatocytes and can give rise to aneuploid gametes. Thus, females derived from gene-targeting experiments of genes that affect meiosis display a variety of phenotypes, ranging from subfertility with production of aneuploid embryos to complete infertility. The same meiotic defect in spermatocytes almost always leads to a complete meiotic failure, absence of progression past the zygotene- or pachytene-like stage, and apoptosis of germ cell leading to infertility (Hunt & Hassold, 2002) (see also Fig. 13.3). Although the cellular mechanisms that regulate the Prophase I events, such as homologous recombination and synapsis, appear to be largely conserved between the two genders, these processes are subjects to a different level of control (commonly defined as "checkpoints"), which is more stringent in males than in females.

In order to gain insight into the mechanism of the meiotic defect caused by RanBPM deletion, here, we compare the *RanBPM*$^{-/-}$ mouse meiotic phenotype with those of other mutants, including some of the most recently characterized (Fig. 13.3). Most mouse models lacking proteins involved in the recombination process such as Spo1, known to initiate the DSBs possibly with the cooperation of Mei1, proteins involved in DNA repair such as ATM, Dmc1, the mismatch proteins Msh4 and Msh5, and the recently identified protein Trip13, exhibit a meiotic arrest in spermatogenesis during Prophase I (Baudat et al., 2000; de Vries et al., 1999; Kneitz et al., 2000; Libby et al., 2002; Pittman et al., 1998; Roig et al., 2010; Yoshida et al., 1998). This is caused by a failure of chromosomes to synapse properly, impairing the progression of spermatocytes beyond the zygotene stage. Since, as mentioned above, the mechanisms monitoring meiosis are less stringent in females, meiosis can progress further in females but most oocytes are eliminated soon after birth as the meiotic defects become apparent during the checkpoint process. By contrast, mutant females for Tex15, a gene encoding a protein recently identified as essential for DSBs repair and synapsis, retain normal fertility (Yang et al., 2008). In mice deficient for the synaptonemal complex proteins, spermatocytes also arrest at the zygotene–pachytene transition, with mutant females either retaining fertility (Scp3, Fkbp6) or being sterile (Syce1, Tex12)

(Bolcun-Filas et al., 2009; Crackower et al., 2003; Hamer et al., 2008; Yuan et al., 2002). Throughout the first meiotic prophase, the sister chromatids are held together by cohesin complex proteins. These proteins, such as SMC1β and Rec8, are essential for proper recombination, completion of synapsis, and chiasmata formation, and mutants for these genes exhibit a similar arrest at the zygotene–pachytene transition (Revenkova et al., 2004; Xu et al., 2005). By contrast, in RanBPM mutant mice, meiosis progresses apparently normally up to the pachytene stage. At this stage, there is no obvious defect in chromosome synapsis, since SCP3 immunolabeling in spermatocytes is comparable between mutants and controls (Puverel et al., 2011). The distribution of phosphorylated histone γH2AX, which marks sites of DSBs at the leptotene stage and subsequently converges around the sex chromosomes to form the sex body, also appeared normal in *RanBPM*$^{-/-}$ spermatocytes. Thus, in *RanBPM* homozygous mutant males, meiosis still progresses slightly further than in the mouse mutants mentioned above, and the spermatocytes reach the pachytene stage with no obvious synapsis abnormalities. These data suggest that RanBPM-deficient spermatocytes are able to go through the first checkpoint, which ensures the complete synapsis of homologous chromosomes prior to entry into the pachytene stage, but may be stopped at a later checkpoint (Morelli & Cohen, 2005; Odorisio, Rodriguez, Evans, Clarke, & Burgoyne, 1998). Since recombination is required for the proper synapsis process in mice, we can also suggest that recombination is initiated properly in *RanBPM*$^{-/-}$ male mice. The mechanisms of synapsis and recombination in *RanBPM*$^{-/-}$ oocytes have not yet been elucidated, but the timing of meiotic progression in the embryo suggests that the defect is similar to that of males, making the *RanBPM*$^{-/-}$ mouse phenotype quite unique (see Fig. 13.3). A more detailed analysis of spermatocytes and oocytes using several markers of synapsis and recombination may help determine the exact nature of this defect apparently common to both genders. Nevertheless, we should note that, at this point in time, we still need to investigate whether the phenotype observed in these mutants is uniquely caused by the lack of RanBPM or whether possible compensatory effects by its homolog RanBP10 can influence its penetrance on gametogenesis.

RanBPM has been reported to interact with the MVH protein, a member of the DEAD-box family of genes encoding an ATP-dependent RNA helicase (Fujiwara et al., 1994; Shibata et al., 2004). MVH is part of a group of male-specific regulators that include Mili and Gasz; mutants for these genes display similar phenotypes, marked by a meiotic arrest at the zygotene–pachytene transition (Kuramochi-Miyagawa et al., 2004;

Ma et al., 2009; Tanaka et al., 2000). These genes seem to be involved in posttranscriptional regulation during spermatogenesis via RNA processing. However, contrary to *RanBPM*$^{-/-}$ mice where both males and females are arrested at the pachytene stage, in these mutants, females are completely fertile. It is therefore difficult to link RanBPM and MVH by the phenotypes caused by the deletion of these genes.

One mutation that appears to be the closest to the RanBPM mutant phenotype in terms of timing of meiotic arrest in both genders is the one for the polyubiquitin gene, *Ubb*. *Ubb*$^{-/-}$ germ cells progress to the pachytene stage and form apparently normal synaptonemal complexes in both genders (Ryu et al., 2008). However, in males, spermatocytes do not mature beyond the pachytene stage and undergo cell death, while in females, although some oocytes reach the dictyate arrest, they are not able to complete meiosis I. Therefore, it would be of interest to determine whether the *Ubb*$^{-/-}$ phenotype can be correlated with the defects observed in *RanBPM*$^{-/-}$ mice. However, it is possible that the phenotypes caused by either Ubb or RanBPM loss have in common only the fact that they affect multiple pathways: RanBPM, by interacting with a variety of proteins through its scaffolding activity, and Ubb, in addition to its role in directing protein degradation, by affecting a great variety of other functions including membrane trafficking and transcriptional regulation.

9. ROLE OF RanBPM IN THE ADULT GONAD

As described above, in *RanBPM*$^{-/-}$ pubertal males, germ cell differentiation is arrested at Prophase I. Histological examination of mutant testes at 6 weeks of age revealed totally empty seminiferous tubules (Sertoli cell only), which is followed by tubular degeneration at later stages (Puverel et al., 2011) (Fig. 13.4). This severe phenotype is not typical of mutants with exclusive meiotic arrest. In fact, in mutant mice displaying an arrest in spermatocyte differentiation, the production of germ cells is usually maintained for a certain period despite the elimination of primary spermatocytes. For example, in mice lacking the cyclin-dependent kinase 2 gene, the alteration of chromosome synapsis leads to the elimination of spermatocytes at prophase I, yet some spermatocytes are still produced in 3-month-old mouse testes (Berthet, Aleem, Coppola, Tessarollo, & Kaldis, 2003; Ortega et al., 2003).

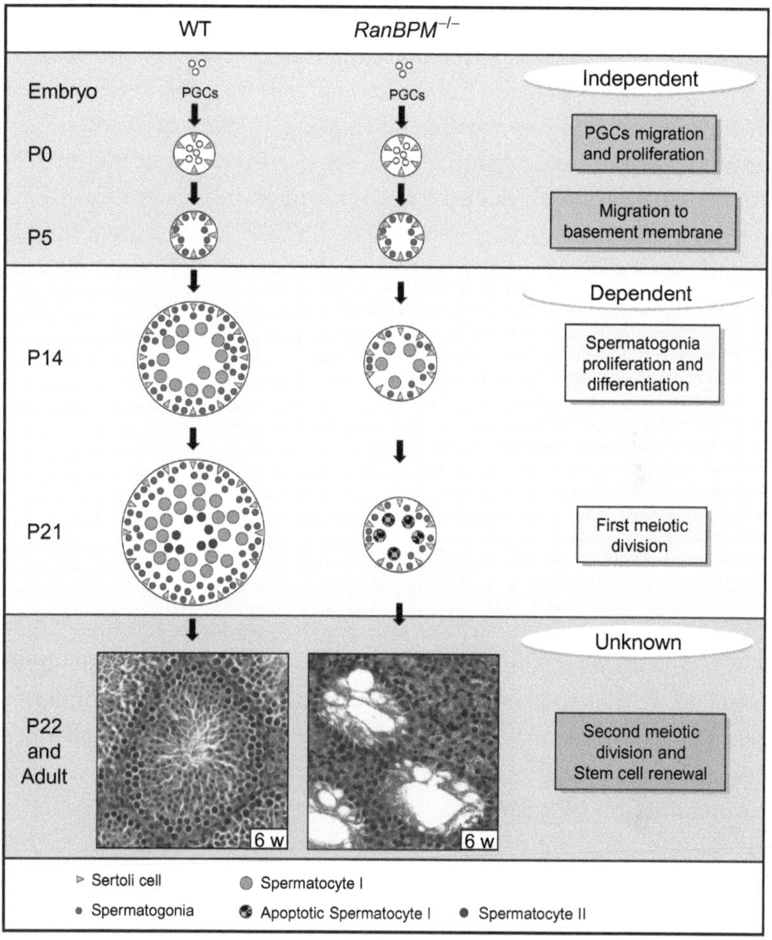

Figure 13.4 Model illustrating the role of RanBPM in spermatogenesis. Schematic representing the stages at which RanBPM plays a role in spermatogenesis. RanBPM is not required for primordial germ cell migration and proliferation and gonocyte migration to the basement membrane of developing seminiferous tubules in the newborn (purple box). However, it is required in the pubertal male at the crucial step of spermatogonia production and for the meiotic progression of primary spermatocytes (yellow box). Whether RanBPM is involved in the second meiotic division and the maintenance of adult spermatogenesis is still unknown, since the elimination of all germ cells in the young mutant does not allow the study of any time point past the meiotic arrest occurring during the prophase of the first meiotic division (gray box). P: postnatal day. (See Color Insert.)

Instead, mice with defects in spermatogonial stem cells (SSCs) display a phenotype that more closely resembles the one observed in the RanBPM-deficient mice. SSCs provide differentiating cells by a self-renewal mechanism, thus keeping the spermatogenesis going (Oatley & Brinster, 2008). Some authors have suggested that SSCs are not involved in the first round of mouse spermatogenesis since the first wave of germline differentiation initiates directly from gonocytes (Yoshida et al., 2006). By using a transgenic system, these authors provided evidence that germ cells originating from gonocytes specifically contribute to the leading edge of pubertal spermatogenesis, generating the first spermatozoa that are released at around P35. Several factors have been shown to be involved in the stem cell renewal process, including the Glial cell line-derived neurotrophic factor Gdnf (Meng et al., 2000), the transcription factor Plzf (Buaas et al., 2004; Costoya et al., 2004), the ERM protein, produced by Sertoli cells (Chen et al., 2005), and the RNA-binding protein NANOS2 (Sada, Suzuki, Suzuki, & Saga, 2009). For example, null mice for GDNF and ERM show an arrest of germ cell production between 4 and 6 weeks (Chen et al., 2005; Meng et al., 2000). Therefore, the similarity of this phenotype with the one displayed by *RanBPM*$^{-/-}$ mutant animals, which are severely depleted of all germ cell types including transit-amplifying spermatogonia as early as 1 month of age, suggests that RanBPM may cause a defect in the spermatogonia self-renewal process as well (Puverel et al., 2011). However, the lack of germ cell production could also be related to the defect in spermatogonia proliferation observed in juvenile animals (Section 7). Taken together, the mutant phenotype displayed by *RanBPM*$^{-/-}$ males shows distinct features, suggesting that this scaffold protein may be critical in more than one mechanism in the control of spermatogenesis.

As outlined in females, RanBPM is crucial for normal oocyte development. Mutant females display a premature ovarian failure and only young adults still retain some follicles in their ovaries (Fig. 13.5). The measure of vaginal impedance in *RanBPM*$^{-/-}$ pubertal females indicates normal estrus cycles, suggesting that follicle and stromal cell functions are intact (Puverel et al., 2011). By contrast, mutant females older than 5 months are not cycling anymore, in accordance with the absence of follicles at this age. Thus, although mutant ovaries contain decreased numbers of follicles, these are sufficient to support normal estrus cycles. The finding that pubertal mutant females are cycling strongly suggests that the sterility of these mice is not caused by hormonal deficiencies and argues in favor of a cell-autonomous role of RanBPM in the oocyte. Oocytes orchestrate and coordinate the development

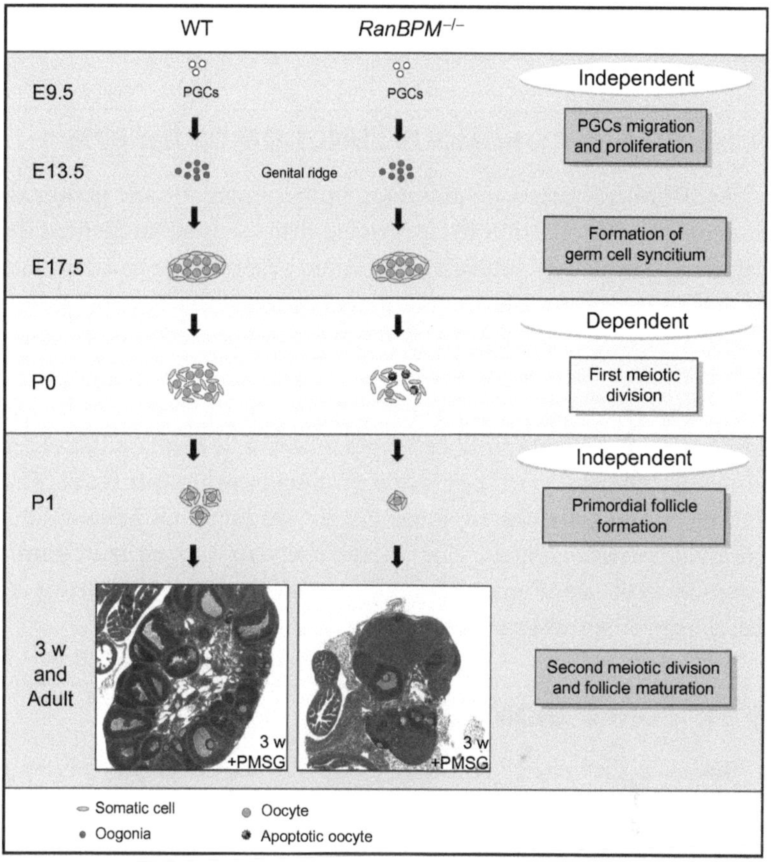

Figure 13.5 Model illustrating the role of RanBPM in oogenesis. Schematic representing the stages at which RanBPM plays a role in oogenesis. As in males, RanBPM is not essential for primordial germ cell proliferation and migration toward the developing gonad (top purple box). However, RanBPM is crucial for the meiotic progression of oocytes, since germ cells exhibit a meiotic arrest during the prophase of the first meiotic division, as in males (yellow box). The observation that some oocytes survive to form primordial follicles and that these follicles can mature upon PMSG stimulation suggests that RanBPM is not required for the maturation of germ cells in pubertal and adult mice, although the ability of these oocytes to produce embryos has not been assessed (bottom purple box). E: embryonic day and P: postnatal day. (See Color Insert.)

of the mammalian ovarian follicles (Eppig, Wigglesworth, & Pendola, 2002). Following the superovulation treatment (PMSG, pregnant mare serum gonadotropin injection), *RanBPM*$^{-/-}$ females developed morphologically normal antral follicles, suggesting that follicles are able to mature in absence of RanBPM (Puverel et al., 2011). Whether *RanBPM*$^{-/-}$ females are able to ovulate and if oocytes are meiotically competent to give rise to normal embryos is still an open question.

10. CELL-AUTONOMOUS FUNCTION OF RanBPM

RanBPM transcripts are present in both the somatic and germinal cell lineages of the testis (Section 5), suggesting that the spermatogenesis defect displayed by *RanBPM*$^{-/-}$ mutant males could be either due to a defect in the Sertoli cell nursing function or to a defect intrinsic to the germ cells. We were able to address this issue by generating chimeric mice with homozygous mutant ES cells to study the extent of the ES cell contribution to the chimeras (Puverel et al., 2011). These experiments showed that while control cells could contribute to all cell types in the male testis, mutant ES cells failed to produce any germ cell while producing other cell types. Importantly, mice with testicular supporting cells produced by mutant ES cells had normal spermatogenesis. Together, these data suggest that RanBPM functions in a cell-autonomous fashion in male germ cells and that it is dispensable from supporting cells for spermatogenesis completion.

11. CONCLUSION

Since its first characterization, RanBPM has been the object of a growing interest from the scientific community. Devoid of any known functional domain, it has been suggested that RanBPM is a scaffolding protein important to assemble high molecular weight protein complexes that shape the activity of various signal transduction pathways. Functionally, RanBPM has been implicated in diverse biological roles, especially in the immune and nervous systems. However, with the recent generation and characterization of a mouse lacking this scaffolding protein, it has become apparent that RanBPM plays a crucial role in spermatogenesis and oogenesis. The main features of this mutant are (1) a defect in postnatal testis development suggesting that RanBPM is essential for spermatogonia production in the young male and (2) a meiotic arrest in both genders during the prophase of the first meiotic division, showing that RanBPM is required for normal progression of meiosis. Even though the exact mechanism of action of RanBPM in the gonad remains to be determined, the results from chimera experiments strongly suggest a cell-autonomous function of RanBPM in germ cells. Importantly, the unique phenotype of RanBPM mutant mice compared to mice lacking other genes critical for normal mouse reproduction (Fig. 13.3) suggests that this scaffold protein may influence the activity of multiple pathways or interact with other, yet unknown, proteins. Regardless, the characterization of RanBPM function in mouse gonad development provides new insight into some key events regulating mammalian gametogenesis.

REFERENCES

Ara, T., Nakamura, Y., Egawa, T., Sugiyama, T., Abe, K., Kishimoto, T., et al. (2003). Impaired colonization of the gonads by primordial germ cells in mice lacking a chemokine, stromal cell-derived factor-1 (SDF-1). *Proceedings of the National Academy of Sciences of the United States of America, 100*, 5319–5323.

Bai, D., Chen, H., & Huang, B. R. (2003). RanBPM is a novel binding protein for p75NTR. *Biochemical and Biophysical Research Communications, 309*, 552–557.

Baker, S. M., Bronner, C. E., Zhang, L., Plug, A. W., Robatzek, M., Warren, G., et al. (1995). Male mice defective in the DNA mismatch repair gene PMS2 exhibit abnormal chromosome synapsis in meiosis. *Cell, 82*, 309–319.

Barlow, C., Liyanage, M., Moens, P. B., Tarsounas, M., Nagashima, K., Brown, K., et al. (1998). Atm deficiency results in severe meiotic disruption as early as leptonema of prophase I. *Development, 125*, 4007–4017.

Baudat, F., Manova, K., Yuen, J. P., Jasin, M., & Keeney, S. (2000). Chromosome synapsis defects and sexually dimorphic meiotic progression in mice lacking Spo11. *Molecular Cell, 6*, 989–998.

Beddow, A. L., Richards, S. A., Orem, N. R., & Macara, I. G. (1995). The Ran/TC4 GTPase-binding domain: Identification by expression cloning and characterization of a conserved sequence motif. *Proceedings of the National Academy of Sciences of the United States of America, 92*, 3328–3332.

Bellve, A. R., Cavicchia, J. C., Millette, C. F., O'Brien, D. A., Bhatnagar, Y. M., & Dym, M. (1977). Spermatogenic cells of the prepuberal mouse. Isolation and morphological characterization. *The Journal of Cell Biology, 74*, 68–85.

Bendel-Stenzel, M., Anderson, R., Heasman, J., & Wylie, C. (1998). The origin and migration of primordial germ cells in the mouse. *Seminars in Cell & Developmental Biology, 9*, 393–400.

Berthet, C., Aleem, E., Coppola, V., Tessarollo, L., & Kaldis, P. (2003). Cdk2 knockout mice are viable. *Current Biology, 13*, 1775–1785.

Blume-Jensen, P., Jiang, G., Hyman, R., Lee, K. F., O'Gorman, S., & Hunter, T. (2000). Kit/stem cell factor receptor-induced activation of phosphatidylinositol 3'-kinase is essential for male fertility. *Nature Genetics, 24*, 157–162.

Bolcun-Filas, E., Hall, E., Speed, R., Taggart, M., Grey, C., de Massy, B., et al. (2009). Mutation of the mouse Syce1 gene disrupts synapsis and suggests a link between synaptonemal complex structural components and DNA repair. *PLoS Genetics, 5*, e1000393.

Bolcun-Filas, E., Bannister, L. A., Barash, A., Schimenti, K. J., Hartford, S. A., Eppig, J. J., et al. (2011). A-MYB (MYBL1) transcription factor is a master regulator of male meiosis. *Development, 138*, 3319–3330.

Buaas, F. W., Kirsh, A. L., Sharma, M., McLean, D. J., Morris, J. L., Griswold, M. D., et al. (2004). Plzf is required in adult male germ cells for stem cell self-renewal. *Nature Genetics, 36*, 647–652.

Buehr, M., McLaren, A., Bartley, A., & Darling, S. (1993). Proliferation and migration of primordial germ cells in We/We mouse embryos. *Developmental Dynamics, 198*, 182–189.

Celeste, A., Petersen, S., Romanienko, P. J., Fernandez-Capetillo, O., Chen, H. T., Sedelnikova, O. A., et al. (2002). Genomic instability in mice lacking histone H2AX. *Science, 296*, 922–927.

Chang, Y., Paramasivam, M., Girgenti, M. J., Walikonis, R. S., Bianchi, E., & LoTurco, J. J. (2010). RanBPM regulates the progression of neuronal precursors through M-phase at the surface of the neocortical ventricular zone. *Developmental Neurobiology, 70*, 1–15.

Chen, C., Ouyang, W., Grigura, V., Zhou, Q., Carnes, K., Lim, H., et al. (2005). ERM is required for transcriptional control of the spermatogonial stem cell niche. *Nature, 436*, 1030–1034.

Cheng, L., Lemmon, S., & Lemmon, V. (2005). RanBPM is an L1-interacting protein that regulates L1-mediated mitogen-activated protein kinase activation. *Journal of Neurochemistry, 94*, 1102–1110.

Cohen, P. E., Pollack, S. E., & Pollard, J. W. (2006). Genetic analysis of chromosome pairing, recombination, and cell cycle control during first meiotic prophase in mammals. *Endocrine Reviews, 27*, 398–426.

Cohen, P. E., & Pollard, J. W. (2001). Regulation of meiotic recombination and prophase I progression in mammals. *BioEssays, 23*, 996–1009.

Costoya, J. A., Hobbs, R. M., Barna, M., Cattoretti, G., Manova, K., Sukhwani, M., et al. (2004). Essential role of Plzf in maintenance of spermatogonial stem cells. *Nature Genetics, 36*, 653–659.

Crackower, M. A., Kolas, N. K., Noguchi, J., Sarao, R., Kikuchi, K., Kaneko, H., et al. (2003). Essential role of Fkbp6 in male fertility and homologous chromosome pairing in meiosis. *Science, 300*, 1291–1295.

Dansereau, D. A., & Lasko, P. (2008). RanBPM regulates cell shape, arrangement, and capacity of the female germline stem cell niche in Drosophila melanogaster. *The Journal of Cell Biology, 182*, 963–977.

de Vries, S. S., Baart, E. B., Dekker, M., Siezen, A., de Rooij, D. G., de Boer, P., et al. (1999). Mouse MutS-like protein Msh5 is required for proper chromosome synapsis in male and female meiosis. *Genes & Development, 13*, 523–531.

Denti, S., Sirri, A., Cheli, A., Rogge, L., Innamorati, G., Putignano, S., et al. (2004). RanBPM is a phosphoprotein that associates with the plasma membrane and interacts with the integrin LFA-1. *The Journal of Biological Chemistry, 279*, 13027–13034.

Di Cunto, F., Imarisio, S., Hirsch, E., Broccoli, V., Bulfone, A., Migheli, A., et al. (2000). Defective neurogenesis in citron kinase knockout mice by altered cytokinesis and massive apoptosis. *Neuron, 28*, 115–127.

Dix, D. J., Allen, J. W., Collins, B. W., Mori, C., Nakamura, N., Poorman-Allen, P., et al. (1996). Targeted gene disruption of Hsp70-2 results in failed meiosis, germ cell apoptosis, and male infertility. *Proceedings of the National Academy of Sciences of the United States of America, 93*, 3264–3268.

Dix, D. J., Allen, J. W., Collins, B. W., Poorman-Allen, P., Mori, C., Blizard, D. R., et al. (1997). HSP70-2 is required for desynapsis of synaptonemal complexes during meiotic prophase in juvenile and adult mouse spermatocytes. *Development, 124*, 4595–4603.

Edelmann, W., Cohen, P. E., Kneitz, B., Winand, N., Lia, M., Heyer, J., et al. (1999). Mammalian MutS homologue 5 is required for chromosome pairing in meiosis. *Nature Genetics, 21*, 123–127.

Edson, M. A., Nagaraja, A. K., & Matzuk, M. M. (2009). The mammalian ovary from genesis to revelation. *Endocrine Reviews, 30*, 624–712.

Emes, R. D., & Ponting, C. P. (2001). A new sequence motif linking lissencephaly, Treacher Collins and oral-facial-digital type 1 syndromes, microtubule dynamics and cell migration. *Human Molecular Genetics, 10*, 2813–2820.

Eppig, J. J., Wigglesworth, K., & Pendola, F. L. (2002). The mammalian oocyte orchestrates the rate of ovarian follicular development. *Proceedings of the National Academy of Sciences of the United States of America, 99*, 2890–2894.

Fernandez-Capetillo, O., Mahadevaiah, S. K., Celeste, A., Romanienko, P. J., Camerini-Otero, R. D., Bonner, W. M., et al. (2003). H2AX is required for chromatin remodeling and inactivation of sex chromosomes in male mouse meiosis. *Developmental Cell, 4*, 497–508.

Fujiwara, Y., Komiya, T., Kawabata, H., Sato, M., Fujimoto, H., Furusawa, M., et al. (1994). Isolation of a DEAD-family protein gene that encodes a murine homolog of Drosophila vasa and its specific expression in germ cell lineage. *Proceedings of the National Academy of Sciences of the United States of America, 91*, 12258–12262.

Gerlitz, G., Darhin, E., Giorgio, G., Franco, B., & Reiner, O. (2005). Novel functional features of the Lis-H domain: Role in protein dimerization, half-life and cellular localization. *Cell Cycle, 4*, 1632–1640.

Grad, I., Cederroth, C. R., Walicki, J., Grey, C., Barluenga, S., Winssinger, N., et al. (2010). The molecular chaperone Hsp90alpha is required for meiotic progression of spermatocytes beyond pachytene in the mouse. *PloS One, 5*, e15770.

Hamer, G., Wang, H., Bolcun-Filas, E., Cooke, H. J., Benavente, R., & Hoog, C. (2008). Progression of meiotic recombination requires structural maturation of the central element of the synaptonemal complex. *Journal of Cell Science, 121*, 2445–2451.

Handel, M. A., & Eppig, J. J. (1998). Sexual dimorphism in the regulation of mammalian meiosis. *Current Topics in Developmental Biology, 37*, 333–358.

Hassold, T., & Hunt, P. (2001). To err (meiotically) is human: The genesis of human aneuploidy. *Nature Reviews. Genetics, 2*, 280–291.

Herran, Y., Gutierrez-Caballero, C., Sanchez-Martin, M., Hernandez, T., Viera, A., Barbero, J. L., et al. (2011). The cohesin subunit RAD21L functions in meiotic synapsis and exhibits sexual dimorphism in fertility. *The EMBO Journal, 30*, 3091–3105.

Hosono, K., Noda, S., Shimizu, A., Nakanishi, N., Ohtsubo, M., Shimizu, N., et al. (2010). YPEL5 protein of the YPEL gene family is involved in the cell cycle progression by interacting with two distinct proteins RanBPM and RanBP10. *Genomics, 96*, 102–111.

Hu, Y. C., Wang, P. H., Yeh, S., Wang, R. S., Xie, C., Xu, Q., et al. (2004). Subfertility and defective folliculogenesis in female mice lacking androgen receptor. *Proceedings of the National Academy of Sciences of the United States of America, 101*, 11209–11214.

Hunt, P. A., & Hassold, T. J. (2002). Sex matters in meiosis. *Science, 296*, 2181–2183.

Ideguchi, H., Ueda, A., Tanaka, M., Yang, J., Tsuji, T., Ohno, S., et al. (2002). Structural and functional characterization of the USP11 deubiquitinating enzyme, which interacts with the RanGTP-associated protein RanBPM. *The Biochemical Journal, 367*, 87–95.

Jagarlamudi, K., Reddy, P., Adhikari, D., & Liu, K. (2010). Genetically modified mouse models for premature ovarian failure (POF). *Molecular and Cellular Endocrinology, 315*, 1–10.

Judis, L., Chan, E. R., Schwartz, S., Seftel, A., & Hassold, T. (2004). Meiosis I arrest and azoospermia in an infertile male explained by failure of formation of a component of the synaptonemal complex. *Fertility and Sterility, 81*, 205–209.

Kim, M. H., Cooper, D. R., Oleksy, A., Devedjiev, Y., Derewenda, U., Reiner, O., et al. (2004). The structure of the N-terminal domain of the product of the lissencephaly gene Lis1 and its functional implications. *Structure, 12*, 987–998.

Kim, S., Namekawa, S. H., Niswander, L. M., Ward, J. O., Lee, J. T., Bardwell, V. J., et al. (2007). A mammal-specific Doublesex homolog associates with male sex chromatin and is required for male meiosis. *PLoS Genetics, 3*, e62.

Kissel, H., Timokhina, I., Hardy, M. P., Rothschild, G., Tajima, Y., Soares, V., et al. (2000). Point mutation in kit receptor tyrosine kinase reveals essential roles for kit signaling in spermatogenesis and oogenesis without affecting other kit responses. *The EMBO Journal, 19*, 1312–1326.

Klink, A., Lee, M., & Cooke, H. J. (1997). The mouse synaptosomal complex protein gene Sycp3 maps to band C of chromosome 10. *Mammalian Genome, 8*, 376–377.

Kneitz, B., Cohen, P. E., Avdievich, E., Zhu, L., Kane, M. F., Hou, H., Jr., et al. (2000). MutS homolog 4 localization to meiotic chromosomes is required for chromosome pairing during meiosis in male and female mice. *Genes & Development, 14*, 1085–1097.

Kobayashi, N., Yang, J., Ueda, A., Suzuki, T., Tomaru, K., Takeno, M., et al. (2007). RanBPM, Muskelin, p48EMLP, p44CTLH, and the armadillo-repeat proteins ARMC8alpha and ARMC8beta are components of the CTLH complex. *Gene, 396*, 236–247.

Kolas, N. K., Marcon, E., Crackower, M. A., Hoog, C., Penninger, J. M., Spyropoulos, B., et al. (2005). Mutant meiotic chromosome core components in mice can cause apparent sexual dimorphic endpoints at prophase or X-Y defective male-specific sterility. *Chromosoma, 114*, 92–102.

Kunert, S., Meyer, I., Fleischhauer, S., Wannack, M., Fiedler, J., Shivdasani, R. A., et al. (2009). The microtubule modulator RanBP10 plays a critical role in regulation of platelet discoid shape and degranulation. *Blood, 114*, 5532–5540.

Kuramochi-Miyagawa, S., Kimura, T., Ijiri, T. W., Isobe, T., Asada, N., Fujita, Y., et al. (2004). Mili, a mammalian member of piwi family gene, is essential for spermatogenesis. *Development, 131*, 839–849.

Lakshmana, M. K., Chung, J. Y., Wickramarachchi, S., Tak, E., Bianchi, E., Koo, E. H., et al. (2010). A fragment of the scaffolding protein RanBP9 is increased in Alzheimer's disease brains and strongly potentiates amyloid-beta peptide generation. *The FASEB Journal, 24*, 119–127.

Lakshmana, M. K., Yoon, I. S., Chen, E., Bianchi, E., Koo, E. H., & Kang, D. E. (2009). Novel role of RanBP9 in BACE1 processing of amyloid precursor protein and amyloid beta peptide generation. *The Journal of Biological Chemistry, 284*, 11863–11872.

Li, B., Nair, M., Mackay, D. R., Bilanchone, V., Hu, M., Fallahi, M., et al. (2005). Ovol1 regulates meiotic pachytene progression during spermatogenesis by repressing Id2 expression. *Development, 132*, 1463–1473.

Li, X. C., & Schimenti, J. C. (2007). Mouse pachytene checkpoint 2 (trip13) is required for completing meiotic recombination but not synapsis. *PLoS Genetics, 3*, e130.

Libby, B. J., De La Fuente, R., O'Brien, M. J., Wigglesworth, K., Cobb, J., Inselman, A., et al. (2002). The mouse meiotic mutation mei1 disrupts chromosome synapsis with sexually dimorphic consequences for meiotic progression. *Developmental Biology, 242*, 174–187.

Liu, D., Liao, C., & Wolgemuth, D. J. (2000). A role for cyclin A1 in the activation of MPF and G2-M transition during meiosis of male germ cells in mice. *Developmental Biology, 224*, 388–400.

Liu, D., Matzuk, M. M., Sung, W. K., Guo, Q., Wang, P., & Wolgemuth, D. J. (1998). Cyclin A1 is required for meiosis in the male mouse. *Nature Genetics, 20*, 377–380.

Ma, L., Buchold, G. M., Greenbaum, M. P., Roy, A., Burns, K. H., Zhu, H., et al. (2009). GASZ is essential for male meiosis and suppression of retrotransposon expression in the male germline. *PLoS Genetics, 5*, e1000635.

Mahakali Zama, A., Hudson, F. P., 3rd, & Bedell, M. A. (2005). Analysis of hypomorphic KitlSl mutants suggests different requirements for KITL in proliferation and migration of mouse primordial germ cells. *Biology of Reproduction, 73*, 639–647.

Mateja, A., Cierpicki, T., Paduch, M., Derewenda, Z. S., & Otlewski, J. (2006). The dimerization mechanism of LIS1 and its implication for proteins containing the LisH motif. *Journal of Molecular Biology, 357*, 621–631.

Matzuk, M. M., & Lamb, D. J. (2002). Genetic dissection of mammalian fertility pathways. *Nature Cell Biology, 4*(Suppl.), s41–s49.

Meng, X., Lindahl, M., Hyvonen, M. E., Parvinen, M., de Rooij, D. G., Hess, M. W., et al. (2000). Regulation of cell fate decision of undifferentiated spermatogonia by GDNF. *Science, 287*, 1489–1493.
Menon, R. P., Gibson, T. J., & Pastore, A. (2004). The C terminus of fragile X mental retardation protein interacts with the multi-domain Ran-binding protein in the microtubule-organising centre. *Journal of Molecular Biology, 343*, 43–53.
Mettus, R. V., Litvin, J., Wali, A., Toscani, A., Latham, K., Hatton, K., et al. (1994). Murine A-myb: Evidence for differential splicing and tissue-specific expression. *Oncogene, 9*, 3077–3086.
Meuwissen, R. L., Offenberg, H. H., Dietrich, A. J., Riesewijk, A., van Iersel, M., & Heyting, C. (1992). A coiled-coil related protein specific for synapsed regions of meiotic prophase chromosomes. *The EMBO Journal, 11*, 5091–5100.
Miyamoto, T., Hasuike, S., Yogev, L., Maduro, M. R., Ishikawa, M., Westphal, H., et al. (2003). Azoospermia in patients heterozygous for a mutation in SYCP3. *Lancet, 362*, 1714–1719.
Molyneaux, K. A., Zinszner, H., Kunwar, P. S., Schaible, K., Stebler, J., Sunshine, M. J., et al. (2003). The chemokine SDF1/CXCL12 and its receptor CXCR4 regulate mouse germ cell migration and survival. *Development, 130*, 4279–4286.
Morelli, M. A., & Cohen, P. E. (2005). Not all germ cells are created equal: Aspects of sexual dimorphism in mammalian meiosis. *Reproduction, 130*, 761–781.
Murrin, L. C., & Talbot, J. N. (2007). RanBPM, a scaffolding protein in the immune and nervous systems. *Journal of Neuroimmune Pharmacology, 2*, 290–295.
Nakamura, M., Masuda, H., Horii, J., Kuma, K., Yokoyama, N., Ohba, T., et al. (1998). When overexpressed, a novel centrosomal protein, RanBPM, causes ectopic microtubule nucleation similar to gamma-tubulin. *The Journal of Cell Biology, 143*, 1041–1052.
Nebel, B. R., Amarose, A. P., & Hacket, E. M. (1961). Calendar of gametogenic development in the prepuberal male mouse. *Science, 134*, 832–833.
Nishitani, H., Hirose, E., Uchimura, Y., Nakamura, M., Umeda, M., Nishii, K., et al. (2001). Full-sized RanBPM cDNA encodes a protein possessing a long stretch of proline and glutamine within the N-terminal region, comprising a large protein complex. *Gene, 272*, 25–33.
Oatley, J. M., & Brinster, R. L. (2008). Regulation of spermatogonial stem cell self-renewal in mammals. *Annual Review of Cell and Developmental Biology, 24*, 263–286.
Odorisio, T., Rodriguez, T. A., Evans, E. P., Clarke, A. R., & Burgoyne, P. S. (1998). The meiotic checkpoint monitoring synapsis eliminates spermatocytes via p53-independent apoptosis. *Nature Genetics, 18*, 257–261.
Ortega, S., Prieto, I., Odajima, J., Martin, A., Dubus, P., Sotillo, R., et al. (2003). Cyclin-dependent kinase 2 is essential for meiosis but not for mitotic cell division in mice. *Nature Genetics, 35*, 25–31.
Page, S. L., & Hawley, R. S. (2004). The genetics and molecular biology of the synaptonemal complex. *Annual Review of Cell and Developmental Biology, 20*, 525–558.
Pittman, D. L., Cobb, J., Schimenti, K. J., Wilson, L. A., Cooper, D. M., Brignull, E., et al. (1998). Meiotic prophase arrest with failure of chromosome synapsis in mice deficient for Dmc1, a germline-specific RecA homolog. *Molecular Cell, 1*, 697–705.
Ponting, C., Schultz, J., & Bork, P. (1997). SPRY domains in ryanodine receptors (Ca(2+)-release channels). *Trends in Biochemical Sciences, 22*, 193–194.
Puverel, S., Barrick, C., Dolci, S., Coppola, V., & Tessarollo, L. (2011). RanBPM is essential for mouse spermatogenesis and oogenesis. *Development, 138*, 2511–2521.
Rao, M. A., Cheng, H., Quayle, A. N., Nishitani, H., Nelson, C. C., & Rennie, P. S. (2002). RanBPM, a nuclear protein that interacts with and regulates transcriptional activity of androgen receptor and glucocorticoid receptor. *The Journal of Biological Chemistry, 277*, 48020–48027.

Revenkova, E., Eijpe, M., Heyting, C., Hodges, C. A., Hunt, P. A., Liebe, B., et al. (2004). Cohesin SMC1 beta is required for meiotic chromosome dynamics, sister chromatid cohesion and DNA recombination. *Nature Cell Biology*, *6*, 555–562.

Roig, I., Dowdle, J. A., Toth, A., de Rooij, D. G., Jasin, M., & Keeney, S. (2010). Mouse TRIP13/PCH2 is required for recombination and normal higher-order chromosome structure during meiosis. *PLoS Genetics*, *6*, e1001062.

Romanienko, P. J., & Camerini-Otero, R. D. (2000). The mouse Spo11 gene is required for meiotic chromosome synapsis. *Molecular Cell*, *6*, 975–987.

Rosario, M. O., Perkins, S. L., O'Brien, D. A., Allen, R. L., & Eddy, E. M. (1992). Identification of the gene for the developmentally expressed 70 kDa heat-shock protein (P70) of mouse spermatogenic cells. *Developmental Biology*, *150*, 1–11.

Rossi, P., Dolci, S., Sette, C., Capolunghi, F., Pellegrini, M., Loiarro, M., et al. (2004). Analysis of the gene expression profile of mouse male meiotic germ cells. *Gene Expression Patterns*, *4*, 267–281.

Rubin, M. R., Toth, L. E., Patel, M. D., D'Eustachio, P., & Nguyen-Huu, M. C. (1986). A mouse homeo box gene is expressed in spermatocytes and embryos. *Science*, *233*, 663–667.

Runyan, C., Schaible, K., Molyneaux, K., Wang, Z., Levin, L., & Wylie, C. (2006). Steel factor controls midline cell death of primordial germ cells and is essential for their normal proliferation and migration. *Development*, *133*, 4861–4869.

Ryu, K. Y., Sinnar, S. A., Reinholdt, L. G., Vaccari, S., Hall, S., Garcia, M. A., et al. (2008). The mouse polyubiquitin gene Ubb is essential for meiotic progression. *Molecular and Cellular Biology*, *28*, 1136–1146.

Sada, A., Suzuki, A., Suzuki, H., & Saga, Y. (2009). The RNA-binding protein NANOS2 is required to maintain murine spermatogonial stem cells. *Science*, *325*, 1394–1398.

Schramm, S., Fraune, J., Naumann, R., Hernandez-Hernandez, A., Hoog, C., Cooke, H. J., et al. (2011). A novel mouse synaptonemal complex protein is essential for loading of central element proteins, recombination, and fertility. *PLoS Genetics*, 7, e1002088.

Seebahn, A., Rose, M., & Enz, R. (2008). RanBPM is expressed in synaptic layers of the mammalian retina and binds to metabotropic glutamate receptors. *FEBS Letters*, *582*, 2453–2457.

Sharan, S. K., Pyle, A., Coppola, V., Babus, J., Swaminathan, S., Benedict, J., et al. (2004). BRCA2 deficiency in mice leads to meiotic impairment and infertility. *Development*, *131*, 131–142.

Shibata, N., Tsunekawa, N., Okamoto-Ito, S., Akasu, R., Tokumasu, A., & Noce, T. (2004). Mouse RanBPM is a partner gene to a germline specific RNA helicase, mouse vasa homolog protein. *Molecular Reproduction and Development*, *67*, 1–7.

Shiina, H., Matsumoto, T., Sato, T., Igarashi, K., Miyamoto, J., Takemasa, S., et al. (2006). Premature ovarian failure in androgen receptor-deficient mice. *Proceedings of the National Academy of Sciences of the United States of America*, *103*, 224–229.

Speed, R. M. (1982). Meiosis in the foetal mouse ovary. I. An analysis at the light microscope level using surface-spreading. *Chromosoma*, *85*, 427–437.

Suresh, B., Ramakrishna, S., & Baek, K. H. (2012). Diverse roles of the scaffolding protein RanBPM. *Drug Discovery Today*, *17*, 379–387.

Sweeney, C., Murphy, M., Kubelka, M., Ravnik, S. E., Hawkins, C. F., Wolgemuth, D. J., et al. (1996). A distinct cyclin A is expressed in germ cells in the mouse. *Development*, *122*, 53–64.

Tanaka, S. S., Toyooka, Y., Akasu, R., Katoh-Fukui, Y., Nakahara, Y., Suzuki, R., et al. (2000). The mouse homolog of Drosophila Vasa is required for the development of male germ cells. *Genes & Development*, *14*, 841–853.

Tang, X., Zhang, J., Cai, Y., Miao, S., Zong, S., Koide, S. S., et al. (2004). Sperm membrane protein (hSMP-1) and RanBPM complex in the microtubule-organizing centre. *Journal of Molecular Medicine*, *82*, 383–388.

Tay, J., & Richter, J. D. (2001). Germ cell differentiation and synaptonemal complex formation are disrupted in CPEB knockout mice. *Developmental Cell, 1*, 201–213.
Togashi, H., Schmidt, E. F., & Strittmatter, S. M. (2006). RanBPM contributes to Semaphorin3A signaling through plexin-A receptors. *The Journal of Neuroscience, 26*, 4961–4969.
Toscani, A., Mettus, R. V., Coupland, R., Simpkins, H., Litvin, J., Orth, J., et al. (1997). Arrest of spermatogenesis and defective breast development in mice lacking A-myb. *Nature, 386*, 713–717.
Tres, L. L., Rosselot, C., & Kierszenbaum, A. L. (2004). Primordial germ cells: What does it take to be alive? *Molecular Reproduction and Development, 68*, 1–4.
Valiyaveettil, M., Bentley, A. A., Gursahaney, P., Hussien, R., Chakravarti, R., Kureishy, N., et al. (2008). Novel role of the muskelin-RanBP9 complex as a nucleocytoplasmic mediator of cell morphology regulation. *The Journal of Cell Biology, 182*, 727–739.
Viera, A., Rufas, J. S., Martinez, I., Barbero, J. L., Ortega, S., & Suja, J. A. (2009). CDK2 is required for proper homologous pairing, recombination and sex-body formation during male mouse meiosis. *Journal of Cell Science, 122*, 2149–2159.
Wang, D., Li, Z., Messing, E. M., & Wu, G. (2002). Activation of Ras/Erk pathway by a novel MET-interacting protein RanBPM. *The Journal of Biological Chemistry, 277*, 36216–36222.
Wang, R. S., Yeh, S., Tzeng, C. R., & Chang, C. (2009). Androgen receptor roles in spermatogenesis and fertility: Lessons from testicular cell-specific androgen receptor knockout mice. *Endocrine Reviews, 30*, 119–132.
Watanabe, D., Yamada, K., Nishina, Y., Tajima, Y., Koshimizu, U., Nagata, A., et al. (1994). Molecular cloning of a novel Ca(2+)-binding protein (calmegin) specifically expressed during male meiotic germ cell development. *The Journal of Biological Chemistry, 269*, 7744–7749.
Watanabe, T., Chuma, S., Yamamoto, Y., Kuramochi-Miyagawa, S., Totoki, Y., Toyoda, A., et al. (2011). MITOPLD is a mitochondrial protein essential for nuage formation and piRNA biogenesis in the mouse germline. *Developmental Cell, 20*, 364–375.
Woo, J. A., Jung, A. R., Lakshmana, M. K., Bedrossian, A., Lim, Y., Bu, J. H., et al. (2012). Pivotal role of the RanBP9-cofilin pathway in Abeta-induced apoptosis and neurodegeneration. *Cell Death and Differentiation, 19*, 1413–1423.
Xu, X., Aprelikova, O., Moens, P., Deng, C. X., & Furth, P. A. (2003). Impaired meiotic DNA-damage repair and lack of crossing-over during spermatogenesis in BRCA1 full-length isoform deficient mice. *Development, 130*, 2001–2012.
Xu, H., Beasley, M. D., Warren, W. D., van der Horst, G. T., & McKay, M. J. (2005). Absence of mouse REC8 cohesin promotes synapsis of sister chromatids in meiosis. *Developmental Cell, 8*, 949–961.
Yang, F., De La Fuente, R., Leu, N. A., Baumann, C., McLaughlin, K. J., & Wang, P. J. (2006). Mouse SYCP2 is required for synaptonemal complex assembly and chromosomal synapsis during male meiosis. *The Journal of Cell Biology, 173*, 497–507.
Yang, F., Eckardt, S., Leu, N. A., McLaughlin, K. J., & Wang, P. J. (2008). Mouse TEX15 is essential for DNA double-strand break repair and chromosomal synapsis during male meiosis. *The Journal of Cell Biology, 180*, 673–679.
Yang, F., & Wang, P. J. (2009). The Mammalian synaptonemal complex: A scaffold and beyond. *Genome Dynamics, 5*, 69–80.
Yin, Y. X., Sun, Z. P., Huang, S. H., Zhao, L., Geng, Z., & Chen, Z. Y. (2010). RanBPM contributes to TrkB signaling and regulates brain-derived neurotrophic factor-induced neuronal morphogenesis and survival. *Journal of Neurochemistry, 114*, 110–121.
Yoshida, K., Kondoh, G., Matsuda, Y., Habu, T., Nishimune, Y., & Morita, T. (1998). The mouse RecA-like gene Dmc1 is required for homologous chromosome synapsis during meiosis. *Molecular Cell, 1*, 707–718.

Yoshida, S., Sukeno, M., Nakagawa, T., Ohbo, K., Nagamatsu, G., Suda, T., et al. (2006). The first round of mouse spermatogenesis is a distinctive program that lacks the self-renewing spermatogonia stage. *Development, 133*, 1495–1505.

Yuan, L., Liu, J. G., Zhao, J., Brundell, E., Daneholt, B., & Hoog, C. (2000). The murine SCP3 gene is required for synaptonemal complex assembly, chromosome synapsis, and male fertility. *Molecular Cell, 5*, 73–83.

Yuan, Y., Fu, C., Chen, H., Wang, X., Deng, W., & Huang, B. R. (2006). The Ran binding protein RanBPM interacts with TrkA receptor. *Neuroscience Letters, 407*, 26–31.

Yuan, L., Liu, J. G., Hoja, M. R., Wilbertz, J., Nordqvist, K., & Hoog, C. (2002). Female germ cell aneuploidy and embryo death in mice lacking the meiosis-specific protein SCP3. *Science, 296*, 1115–1118.

Zakeri, Z. F., Wolgemuth, D. J., & Hunt, C. R. (1988). Identification and sequence analysis of a new member of the mouse HSP70 gene family and characterization of its unique cellular and developmental pattern of expression in the male germ line. *Molecular and Cellular Biology, 8*, 2925–2932.

Zickler, D., & Kleckner, N. (1999). Meiotic chromosomes: Integrating structure and function. *Annual Review of Genetics, 33*, 603–754.

CHAPTER FOURTEEN

K^+ and Cl^- Channels and Transporters in Sperm Function

C.M. Santi*, G. Orta†, L. Salkoff*, P.E. Visconti‡,1, A. Darszon†,1, C.L. Treviño†,1

*Department of Anatomy and Neurobiology, Washington University School of Medicine, St Louis, Missouri, USA
†Departamento de Genética del Desarrollo y Fisiología Molecular, Instituto de Biotecnología, Universidad Nacional Autónoma de México, Cuernavaca, Morelos, Mexico
‡Department of Veterinary and Animal Science, University of Massachusetts, Amherst Massachusetts, USA
[1]Corresponding authors: e-mail address: pvisconti@vasci.umass.edu; darszon@ibt.unam.mx; ctrevino@ibt.unam.mx

Contents

Abstract

To succeed in fertilization, spermatozoa must decode environmental cues which require a set of ion channels. Recent findings have revealed that K^+ and Cl^- channels participate in some of the main sperm functions. This work reviews the evidence indicating the involvement of K^+ and Cl^- channels in motility, maturation, and the acrosome reaction, and the advancement in identifying their molecular identity and modes of regulation. Improving our insight on how these channels operate will strengthen our ability to surmount some infertility problems, improve animal breeding, preserve biodiversity, and develop selective and secure male contraceptives.

Current Topics in Developmental Biology, Volume 102
ISSN 0070-2153
http://dx.doi.org/10.1016/B978-0-12-416024-8.00014-3

1. INTRODUCTION

Generating a unique individual requires that spermatozoa reach and fertilize the egg to deliver their genetic information. Cells build up and maintain ion concentration gradients across their membranes using ion pumps and transporters as a means of coding information related to their state which will then be used to respond to their changing environment. Ion channels are efficient transporters moving millions of ions per second which allow to rapidly modify the cell electric potential and the concentrations of internal second messengers within a wide time-range, depending on how they are regulated (Hille, 2001).

Spermatozoa are quite small, morphologically complex, and differentiated cells (Yanagimachi, 1994). Achieving their main goal of fertilizing the female gamete requires that sperm are equipped to decode multiple signals during their maturation and along their journey to reach the egg. Mammalian sperm travel through convoluted paths while maturing in the epididymis and then face many environmental changes while traveling in female reproductive tract (Dacheux et al., 2012; Hung & Suarez, 2010; Visconti, Krapf, de la Vega-Beltran, Acevedo, & Darszon, 2011). The minimal set of ion channels and transporters needed for sperm to succeed in their fundamental task is still not fully known (Darszon, Nishigaki, Beltran, & Trevino, 2011; Lishko et al., 2012; Publicover & Barratt, 2012). In any case, ion channels play a major role in sperm maturation and in the regulation of motility and the acrosome reaction (AR) (Darszon et al., 2011; Lishko et al., 2012). Since mature sperm seem essentially unable to perform gene transcription or protein synthesis, the proteins needed for their function are generated during their differentiation (Baker, 2011).

Establishing the functional presence of an ion channel in a specific cell is not easy as these proteins are minor membrane components. Inklings of the characteristics of some sperm ion channels were initially derived from planar bilayers with incorporated sperm plasma membranes (reviewed in Darszon, Labarca, Nishigaki, & Espinosa, 1999). In fact, some of the first recordings of sperm K^+ and Cl^- single-channel activity were obtained in planar bilayers with incorporated sea urchin sperm plasma membranes (Labarca et al., 1996; Lievano, Sanchez, & Darszon, 1985; Morales, de la Torre, Moy, Vacquier, & Darszon, 1993) and of Ca^{2+} permeable channels from boar sperm plasma membranes (Cox & Peterson, 1989). For many years, achieving electrophysiological recordings in sperm to study their ion channels was exceedingly difficult (Darszon et al., 1999; Guerrero, Sanchez, & Darszon, 1987; Jimenez-Gonzalez, Michelangeli, Harper, Barratt, & Publicover, 2006; Kirichok & Lishko, 2011; Ren & Xia, 2010; Weyand et al., 1994). Fortunately, this experimental bottleneck was surmounted by obtaining whole-cell patch-clamp recordings sealing on the cytoplasmic droplet of mouse epididymal sperm (Kirichok, Navarro, & Clapham, 2006). This strategy was extended to human spermatozoa (Kirichok & Lishko, 2011) that have a cytoplasmic droplet in the head–flagellar junction (Cooper, 2011). This approach has allowed the characterization of CatSper (mouse and human) (Kirichok et al., 2006) and SLO3 (mouse) (Navarro, Kirichok, & Clapham, 2007; Santi et al., 2010; Schreiber et al., 1998; Zeng, Yang, Kim, Lingle, & Xia, 2011), so far, the only sperm-specific channels described whose elimination leads to infertility. CatSper is a tetrameric, Ca^{2+}-permeable cation channel potently regulated by intracellular pH (pHi) (Lishko et al., 2012), and SLO3 is a pHi regulated K^+ channel which will be described in detail later. In addition, whole-cell recordings have revealed the presence of a voltage-sensitive H^+ channel involved in the pHi regulation in human sperm and much less in mouse sperm (Kirichok & Lishko, 2011), and of ATP-gated channels of the purinergic family, P2X2, in mouse epididymal sperm (Navarro, Miki, & Clapham, 2011). The presence of ion currents consistent with the properties of TRPM8 and absent in sperm from TRPM8 null mice has been recorded in testicular sperm (Gibbs et al., 2011; Martinez-Lopez et al., 2011). Functional evidence indicates that this channel is also present in human sperm (De Blas et al., 2009). An alternative to obtaining whole-cell recordings in sperm without using the cytoplasmic droplet has been recently described. This modified perforated patch-clamp strategy allows sealing directly onto the head of mature human spermatozoa where CatSper and the first human sperm Cl^- currents have been documented displaying characteristics associated with Ca^{2+}-dependent Cl^- channels (Orta et al., 2012).

Demonstrating the presence of a specific ion channel in a cell now requires controlled immunological or proteomic detection combined with electrophysiological, ion-sensitive fluorescent functional assays and pharmacology. In species that allow it, knocking out a specific ion channel from spermatozoa, might reveal its function and if good antibodies are available, establish its presence (Kirichok et al., 2006; Santi et al., 2010). This review will discuss the K^+ and Cl^- channels for which there is evidence indicating their presence in sperm and their participation in sperm's main functions: epididymal maturation, capacitation, motility, and/or the AR. Due to the particular relevance of these ions in the regulation of the membrane potential (Em), special emphasis will be given in Sections 3.4 and 3.5 on the discussion of their role in the control of sperm-resting Em and the changes in Em that accompany the capacitation process.

2. EPIDIDYMAL MATURATION

After differentiation in the testis, sperm travel along the epididymis, a specialized duct of the male reproductive system that performs four important functions related to spermatozoa: transport, concentration, maturation, and storage (Turner, 2008). The function of various K^+ channels has been implicated at least during the transport and maturation processes. Transport of sperm through the epididymis is achieved by hydrostatic pressure and by smooth muscle contractions of the epididymis (Bellentani et al., 2011). It has been reported that voltage-gated K^+ channels play a role in the control of smooth muscle contraction. For example, Erg (*ether-a-go-go* related gene) K^+ channels participate in the contractibility of smooth muscle cells in addition to their contribution to membrane repolarization during the cardiac action potential. The former function is evident as pharmacological inhibition of Erg K^+ channels increases contraction activity of different smooth muscle cells. Mewe et al. (2008) demonstrated through RT-PCR and Western blot experiments the presence of *erg1a and erg1b* isoforms in the bovine epididymal duct. They also reported that Erg channel inhibitors increase the contractile activity of the duct, likely by depolarizing the resting potential and increasing its excitability. Additionally, Bellentani et al. (2011) and Nojimoto et al. (2009) showed that sibutramide (a K^+-channel blocker) increases the mechanical activity of the epididymis and of the vas deferens in rats, respectively. In both cases, the effect was attributed to the blockage of voltage-dependent K^+ channels implicated in smooth muscle contraction (Bellentani et al., 2011; Nojimoto et al., 2009).

The fact that sperm transport requires a relatively long time in many species—ranging 10–13 days (except for human sperm in which the transport time is between 2 and 6 days)—supports the notion that epididymal passage entails an indispensable maturation step rather than simply acting as a sperm conduit (Turner, 2008). Sperm from the caput epididymis are mostly immotile and are unable to undergo capacitation and fertilize the egg. In addition, such maturation process is evident by the greater fertilization ability of sperm obtained from cauda compared to that of sperm obtained from corpus epididymis.

The epididymal maturational process is complex and involves a series of modifications in the sperm, such as changes in the plasma membrane composition, modification, and/or remodeling which occur in the absence of transcription and *de novo* protein synthesis (Dun, Aitken, & Nixon, 2012). Although the complete process has not yet been fully elucidated, one important aspect is that cauda spermatozoa exhibit an increased volume regulation capacity. As spermatozoa leave the testis to transit into the epididymis, they encounter an increasing osmolarity ranging from 280 (rete testis fluid) to up to 400 mmol/kg (cauda epididymis fluid) (Yeung, Barfield, & Cooper, 2006). Upon ejaculation into the female reproductive tract, spermatozoa experience hypo-osmotic stress, which is counterbalanced through the process known as regulatory volume decrease (RVD) involving influx and efflux of water and osmolytes (Yeung et al., 2006).

2.1. Transporters involved in epididymal maturation

The role of K^+ channels during RVD is inferred by the observation that quinine, a general K^+-channel blocker, produces cell swelling upon a hypo-osmotic challenge; in other words, RVD is impaired when the channels are blocked. This notion is further supported by the fact that valinomycin (a K^+ ionophore) can reverse the quinine effect (Yeung et al., 2006). Cooper and Yeung (2007) summarized the pharmacological approaches that have been used by several laboratories to dissect the possible roles of various K^+, Cl^-, and K^+/Cl^- transporters in sperm RVD. Although an unequivocal identification is not possible due to a lack of specificity among blockers, the survey suggested the participation of the following K^+ channels in sperm RVD: $K_V1.5$ and $K_V7.1$, mink, and TASK2. The presence of $K_V1.5$ (human and mouse), mink (mouse), and TASK2 (human and mouse) has been confirmed by Western blot analyses (Cooper & Yeung, 2007). Immunocytochemistry studies localized all these channels to the flagellum (Cooper & Yeung, 2007). Although sperm are believed by most researchers to be translationally and transcriptionally inactive after leaving the testis, transcripts for $K_V1.5$, *mink*, and *TAKS2* were detected in human sperm (Cooper & Yeung, 2007) suggesting that their protein products are synthesized in spermatids and remain in posttesticular sperm. There is also data supporting the presence of a variety of K^+ channels in epididymis from several species using RT-PCR and immunodetection techniques. For example, evidence for the presence of K_{ATP} channels derived from RT-PCR and Western blot has been reported for rat and mouse epididymis, and in mature sperm of bovine, feline, canine, mouse, and human origin (Acevedo et al., 2006; Lybaert et al., 2008).

As in somatic cells, the aforementioned evidence for a role of K^+ channels in sperm volume regulation during epididymal maturation suggests a parallel involvement of Cl^- channels in compensating the positive charges and maintaining electroneutrality. The identity of Cl^- channels involved in volume regulation is not well understood. It has been proposed that ClC-2 (CLCN2) and ClC-3 (CLCN3) play a role in somatic cells (Furst et al., 2002; Nilius & Droogmans, 2003); however, their function is still controversial (Sardini et al., 2003). In sperm, CLCN3 was detected by Western blot and localized to the sperm tail by immunofluorescence (Yeung, Barfield, & Cooper, 2005). While the function of K^+ and Cl^- channels in the regulation of sperm volume is still under study, their presence in sperm from several species suggests that they may play an important role during epididymal maturation and warrants further research.

3. CAPACITATION

Mammalian sperm acquire fertilization capacity only after residing in the female genital tract for a finite period of time (Austin, 1952; Chang, 1951). This maturation process is called capacitation and results in two major changes in sperm physiology: (1) they develop a distinctive motility pattern known as hyperactivation and (2) they become competent to undergo the AR, an exocytotic event that allows the sperm to fertilize the egg. Among physiological changes which take place during capacitation are: (a) activation of PKA (Harrison, 2004); (b) intracellular alkalinization (Zeng, Clark, & Florman, 1995); (c) increase in intracellular Ca^{2+} concentration ($[Ca^{2+}]i$) (Baldi et al., 1991; Breitbart, 2003; DasGupta, Mills, & Fraser, 1993; Suarez, Varosi, & Dai, 1993; Xia & Ren, 2009); (d) changes in the plasma membrane composition (Cross, 1998; Davis, 1981; Gadella & Harrison, 2000; Go & Wolf, 1983; Travis & Kopf, 2002; Visconti et al., 1999); (e) activation of PKA, upstream of an increase in tyrosine phosphorylation (Visconti, Bailey, et al., 1995; Visconti, Moore, et al., 1995); and (f) hyperpolarization of the sperm plasma membrane (Arnoult et al., 1999; Demarco et al., 2003; Munoz-Garay et al., 2001; Zeng et al., 1995).

3.1. Changes in Em during sperm capacitation

The role of Em in regulating any of the multiple processes that take place during capacitation at the molecular level is an important avenue of research that has many questions yet to be answered. Sperm membrane hyperpolarization is a key event in mouse and bovine sperm capacitation (Arnoult, Cardullo, Lemos, & Florman, 1996; Arnoult et al., 1999; Demarco et al.,

2003; Munoz-Garay et al., 2001; Zeng et al., 1995), but as yet has not been demonstrated in human sperm. This may be due to the fact that a much smaller fraction of human spermatozoa is capacitated at any given time compared to mouse sperm (Cohen-Dayag, Tur-Kaspa, Dor, Mashiach, & Eisenbach, 1995). Thus, capturing voltage changes occurring in a small fraction of sperm may be difficult. Sperm Em was described in mouse and bovine sperm populations using the voltage-sensitive dye DisC3 by several groups showing that the sperm-resting potential is relatively depolarized (between −30 and −40 mV) but hyperpolarizes during capacitation (Demarco et al., 2003; Espinosa & Darszon, 1995; Hernandez-Gonzalez et al., 2006; Munoz-Garay et al., 2001; Santi et al., 2010; Zeng et al., 1995). However, the absolute level of membrane hyperpolarization has not been well established. Since this type of experiment uses a nonhomogeneous population of sperm where only a portion of the population may be capacitated *in vitro*, the Em absolute values have to be taken cautiously. Therefore, the ∼−60 mV value obtained in sperm populations in capacitated conditions could be an average value in a mixed population of sperm (and thus an underestimation of the real Em of a subfraction of capacitated sperm).

In an effort to circumvent this problem, Arnoult et al. (1999) measured Em in individual spermatozoa using the voltage-sensitive dye di8-ANEPPS. Their results corroborate earlier experiments showing the Em of noncapacitated sperm to be relatively depolarized as reported earlier, and that two populations of sperm are present under capacitated conditions: one that is hyperpolarized (∼−80 mV) and may represent capacitated sperm able to undergo the AR and another population that appears to be noncapacitated with resting potentials averaging ∼−43 mV. Regarding the role of Em changes in capacitation, some experiments suggest that hyperpolarization is essential for the sperm to acquire the ability to undergo a physiological AR. For example, if capacitation is inhibited by the application of high external KCl during this process, the AR induced by zona pellucida (ZP) is significantly reduced (Arnoult et al., 1999; Zeng et al., 1995). The authors of those experiments reached the conclusion that a significant hyperpolarization is an essential part of the capacitation process and is required for the AR to take place.

3.2. What is the functional role of hyperpolarization during capacitation?

3.2.1 Hyperpolarization and sperm motility

As spermatozoa must find and deliver their genetic material to the homologous female gamete, highly sophisticated strategies to regulate their motility and succeed in their search for the egg or oocyte have evolved. Motility is therefore a fundamental function of the mammalian male gamete. It is not surprising that a multitude of gene-targeted transgenic mice, without obvious sperm morphological defects, nevertheless have motility alterations and are sterile (e.g., *sAC, PKA, sNHE, GAPDHs, CatSper, PMCA4, SLO3*) (Esposito et al., 2004; Miki et al., 2004; Nolan et al., 2004; Okunade et al., 2004; Quill, Ren, Clapham, & Garbers, 2001; Ren et al., 2001; Santi et al., 2010; Wang et al., 2007; Zeng et al., 2011). The flagellum generates the driving force for motility in most male gametes, including some plant kingdom species. The propelling machinery of cilia and flagella is the axoneme (Lindemann & Goltz, 1988) which consists of nine microtubule doublets around a central pair of microtubules (commonly denoted the 9+2 structure). The axonemal motor proteins that generate the force that slides the microtubules producing flagellar beating are the dynein ATPases whose activity is modulated by pH, ATP, ADP, Ca^{2+}, and phosphorylation (Christen, Schackmann, & Shapiro, 1983; Lindemann & Goltz, 1988). Ion transport which supports and controls flagellar beating plays key roles in sperm motility regulation (Guerrero et al., 2011; Kaupp, Kashikar, & Weyand, 2008).

Mammalian spermatozoa display two main modes of motility, activated and hyperactivated. Mature sperm stored in the caudal epididymis and vas deferens before ejaculation are immotile or weakly motile. Sperm activate upon release by ejaculation into media containing bicarbonate and Ca^{2+} and swim propelled by a relatively low-amplitude flagellar beat corresponding to the activated mode (Wennemuth, Carlson, Harper, & Babcock, 2003). The activation of the sperm AdCyc10 (a.k.a. SACY (soluble adenylate cyclase) by HCO_3^- and the consequent cAMP/PKA activation is the main factor responsible for the activated motility (Carlson, Hille, & Babcock, 2007; Esposito et al., 2004; Hess et al., 2005; Nolan et al., 2004; Xie et al., 2006). Subsequently, the initiation of hyperactivated sperm motility, characterized by high amplitude and asymmetrical flagellar beating, allows the sperm to detach from temporary binding sites along the female genital tract and penetrate the extracellular matrix of cumulus cells and the ZP surrounding the oocyte (Suarez, 2008). Though it is believed that hyperactivation is encompassed by the capacitation process, there is a relative independence between them. Exactly how hyperactivation is triggered remains not well understood; however, it has been shown to involve a rise in $[Ca^{2+}]i$ carried mainly by CatSper channels in the flagellar plasma membrane. This Ca^{2+} channel, only present in the sperm flagella, is weakly voltage dependent and activated by an increase in pHi (Kirichok et al., 2006; Ren et al., 2001). CatSper null male mice are infertile mainly because of failure to hyperactivate (Carlson et al., 2005, 2003; Ren et al., 2001; Quill et al., 2001). Hyperpolarization of the sperm plasma could play an important role in hyperactivation by facilitating Ca^{2+} influx through CatSper channels during cytosolic alkalinization due to an increase in the driving force on Ca^{2+} (Navarro et al., 2007). Since CatSper is only weakly voltage dependent, it is likely to retain a significant conductance during capacitation-induced hyperpolarization (Kirichok et al., 2006).

3.2.2 Hyperpolarization and the AR

Originally it was proposed that hyperpolarization occurring during capacitation might be necessary for the AR to occur by enabling sperm to generate transient $[Ca^{2+}]$i elevations; the proposed mechanism involved removing the inactivation of T-type voltage-dependent Ca^{2+} channels (Ca_V3), which could then be subject to induction by physiological agonists (e.g., ZP) (Arnoult et al., 1996, 1999; Santi, Darszon, & Hernandez-Cruz, 1996; Zeng et al., 1995). However, $Ca_V3.1$ and 3.2 knockout mice are fertile (Stamboulian et al., 2004) and Ca_V channels, although recorded in testicular sperm, were not found in epididymal sperm (Martinez-Lopez et al., 2009; Ren & Xia, 2010). These findings bring into question the participation of Ca_V3 channels in the mouse sperm AR, although they do not completely rule it out. Alternatively, hyperpolarization of the sperm plasma membrane could play an important role in sustaining Ca^{2+} influx, essential for the AR to take place, through other Ca^{2+} entry pathways like TRP or SOCs channels that have been described in sperm plasma membrane (Darszon et al., 2011). Several lines of evidence suggest that hyperpolarization of the sperm plasma membrane is necessary for the AR to take place, although the actual role of hyperpolarization is not yet understood.

3.3. What is the molecular mechanism that underlies the sperm hyperpolarization associated with capacitation?

The mechanism of hyperpolarization is not very well understood yet, but evidence about the molecular players is beginning to emerge. In the absence of changes in the bulk ionic concentration, hyperpolarization could, for instance, be due to (1) an increase in K^+ permeability caused by the activation of K^+ selective channels and (2) a reduction of Na^+ permeability, by decreasing the activity of Na^+ channels. Although not directly affecting membrane voltage, the regulation and activity of Cl^- permeability through Cl^- channels and transporters could also play a direct or indirect role in the regulation of the sperm plasma Em (see below).

3.4. Role of K^+ channels in sperm capacitation

The resting potential of sperm plasma membrane prior to capacitation is around −40 mV (Arnoult et al., 1999; Demarco et al., 2003; Espinosa & Darszon, 1995; Munoz-Garay et al., 2001; Zeng et al., 1995), significantly less negative than the equilibrium potential for K^+ ions ($E_K \sim -90$ mV). This indicates that K^+ permeability is not the sole determinant of sperm-resting Em at least in noncapacitated sperm. The opening of K^+ channels in these conditions will displace the Em toward E_K resulting in membrane hyperpolarization, while the closure of these channels will cause depolarization. Since the hyperpolarization that accompanies mouse sperm capacitation is influenced by the external K^+ concentration and K^+-channel blockers, it is reasonable to assume that this hyperpolarization could be due to the opening of these channels (Arnoult et al., 1999; Zeng et al., 1995). The types of K^+ channels present in spermatogenic cells and sperm and their possible roles in sperm capacitation and other sperm functions will be discussed in the following sections.

K^+ channels can be classified in four major classes: (1) voltage-gated K^+ channels (K_V channels) that open or close in response to changes in Em; (2) ion-activated K^+ channels that open in response to the presence of Ca^{2+} or other intracellular signaling molecules; (3) inwardly rectifying K^+ channels that conduct K^+ more easily in the inward direction (into the cell) than outward; (4) tandem pore domain K^+ channels which may be constitutively open or possess high basal activity. The first two above mentioned classes of K^+ channels, voltage gated and ion activated, may have overlapping properties as in the case of SLO3, a unique sperm-specific K^+ channel which is modulated by intracellular H^+ but is also voltage gated. More will be said about SLO3 later in this review. Biochemical, molecular biology, and electrophysiological data support the presence of several of the above mentioned K^+ channels in spermatogenic cells and sperm (Acevedo et al., 2006; Chan et al., 1998; Felix et al., 2002; Hagiwara & Kawa, 1984; Jacob, Hurley, Goodwin, Cooper, & Benoff, 2000; Martinez-Lopez et al., 2009; Munoz-Garay et al., 2001; Navarro et al., 2007; Salvatore, D'Adamo, Polishchuk, Salmona, & Pessia, 1999; Santi et al., 2010; Schreiber et al., 1998; Wu et al., 1998).

3.4.1 Delayed outward voltage-dependent K^+ currents

Initial whole-cell patch-clamp studies in rodent spermatogenic cells showed the presence of only one functional type of K^+ channel, a delayed non-inactivating tetraethyl ammonium (TEA)-sensitive K^+ channel (Hagiwara

& Kawa, 1984). The molecular identity of these currents was not known at that time. Subsequent studies done by Felix et al. (2002) revealed the presence of at least two types of delayed rectifier currents in spermatogenic cells, one sensitive and one resistant to external TEA. It was speculated that the TEA-sensitive channel was probably $K_V3.1$ that was identified by RT-PCR in spermatogenic cells and by immunocytochemistry in mature sperm. On the other hand, the delayed rectifier component less sensitive to TEA was hypothesized to correspond to the SLO3 K^+ channels (see Section 3.4.3). Other channels that were identified by immunocytochemistry and RT-PCR in mouse sperm and spermatogenic cells, respectively, were $K_V1.1$, $K_V1.2$, and GIRK1 channels (Felix et al., 2002). Also, Jacob et al. (2000), using RT-PCR and Northern blot analysis showed the presence of $K_V1.3$ mRNA in rat spermatogenic cells. Ca^{2+}-dependent K^+ currents resembling delayed rectifier currents have also been recorded in *Xenopus* oocytes injected with RNAs from rat spermatogenic cells. These currents showed the typical properties of the SLO1 family of Ca^{2+}-activated K^+ channels such as block by charybdotoxin and low concentrations of TEA (Chan et al., 1998). Of all the delayed rectifier currents mentioned above, only currents having the properties of SLO3 have been observed in mature corpus epididymal sperm.

3.4.2 Inward rectifier K^+ currents

Inward rectifiers are a class of K^+ channels that conduct larger inward currents at membrane voltages negative to the K^+ equilibrium potential than outward currents at positive voltages. This property, called inward rectification, enables these channels to function at negative voltages. Inward rectifier K^+ channels differ in their degree of rectification. Rectification is not an inherent property of the channel protein itself, but reflects strong voltage dependence of channel block by intracellular cations such as Mg^{2+} and polyamines (Hibino et al., 2010). Among the various classes of inward rectifiers are K_{ATP} channels which are heteromeric complexes of two types of protein subunits, the Kir 6 subfamily and the sulfonylurea receptors (SURs). SUR is a member of the ATP-binding cassette (ABC) family. These channels are blocked by elevated levels of intracellular ATP, and several therapeutic agents including sulfonylureas, and respond to K^+ channel openers like pinacil and diazoxide (Aguilar-Bryan & Bryan, 1999; Aguilar-Bryan et al., 1998; Mannhold, 2004; Seino, 1999). Different types of SUR subunits endow the channels with differential sensitivity to sulfonylureas: SUR1 possesses a high-affinity binding site for tolbutamide and glibenclamide, whereas SUR2A binds glibenclamide but not tolbutamide with high affinity (Dorschner, Brekardin, Uhde, Schwanstecher, & Schwanstecher, 1999).

Two types of inward rectifier K^+ channels were electrophysiologically characterized in spermatogenic cells using the whole-cell patch-clamp technique: (1) K^+ channels with strong inward rectification were reported in mouse spermatids and primary spermatocytes by Munoz-Garay et al. (2001). These currents were highly K^+ selective, showed strong inward rectification, and were inhibited by intracellular acidification. Additionally, they were blocked by 0.3–1 mM external Ba^{2+}, which also inhibits the hyperpolarization associated with capacitation and partially inhibits the AR induced by ZP (Munoz-Garay et al., 2001). (2) Weakly rectifying K^+ channels sensitive to ATP (K_{ATP} channels) were also reported by Darszon's group and others in spermatogenic cells (Acevedo et al., 2006; Zhou et al., 2011). These currents were sensitive to micromolar concentrations of the K_{ATP}-channel blockers tolbutamide and glibeclamide, and channel activity also increased when glucose was removed from the external media, a maneuver to reduce internal ATP (Acevedo et al., 2006). It has also been shown that tolbutamide produces both an inhibition of the hyperpolarization at micromolar concentrations and an inhibition of the AR induced by ZP in a dose-dependent manner, suggesting a plausible participation of K_{ATP} channels in the hyperpolarization associated with capacitation (Acevedo et al., 2006).

The molecular identities of the K_{ATP} channels were determined by isolating *Kir 6.1, Kir 6.2*, and *SUR1* and *SUR2* transcripts in mouse and rat spermatogenic cells. In addition, immunocytochemistry experiments showed that Kir 6.1 and Kir 6.2 as well as SUR1 and SUR2B subunits are expressed in both spermatogenic cells and mature sperm (Acevedo et al., 2006; Lybaert, Leleux, Meuris, & Lebrun, 2010; Lybaert et al., 2008). In mouse, while Kir 6.1 was detected in the flagellum midpiece, the Kir 6.2 was localized both in the midpiece and in the postacrosomal region of the sperm head. The SUR1 subunit displays the same distribution pattern as Kir 6.2, suggesting that Kir 6.2 and SUR1 subunits probably are colocalized and form K_{ATP} channels similar to those seen in pancreatic beta cells. The SUR 2 subunit that was detected in the flagellum principal piece and to lesser extent in the midpiece (Acevedo et al., 2006; Zhou et al., 2011) showed that in rat, Kir 6.2 colocalized with SUR2B in acrosome of spermatids.

3.4.3 SLO3 K^+ channels

Although several types of K^+ currents have been detected in spermatogenic cells by electrophysiological methods, and voltage-dependent K^+ channels were identified by immunocytochemistry in spermatogenic cells and sperm, these types of currents have not been identified by patch-clamp experiments in mature spermatozoa. The only K^+ current detected to date in corpus epididymal sperm is a pH-sensitive K^+ current, originally named IK_{Sper} (Navarro et al., 2007) that was later identified as the SLO3 current (Santi et al., 2010; Zeng et al., 2011). IK_{Sper} was first recorded in 2007 by Navarro et al. using the whole-cell patch-clamp technique (Navarro et al., 2007). The current activates at ~ -70 mV (in 5 mM external K^+, 130 mM internal K^+, and in the absence of external Na^+) and shows weak voltage and time-dependent activation. Consistent with the properties of SLO3, the current is potentiated by intracellular alkalinization increasing approximately eight-fold when pHi changed from 6.0 to 8.0. Pharmacological studies on IK_{Sper} revealed that this current is sensitive to quinine, clofilium, EIPA, mibefradil, and Ba^{2+} and is insensitive to 10 mM external TEA and $CdCl_2$. These authors also showed that intracellular alkalinization under current clamp conditions produced a rapid hyperpolarization of the sperm plasma membrane due to the activation of a K^+ current which has the same pharmacology of IK_{Sper}. They concluded that the hyperpolarization of the sperm plasma membrane produced by intracellular alkalinization is due to the activation of IK_{Sper} channels. In 2009, Martínez-López et al. described a pH-sensitive K^+ current present in testicular sperm (Martinez-Lopez et al., 2009). Like IK_{Sper} recorded in more mature sperm, this current is also activated by membrane depolarization and intracellular alkalinization; it is strongly blocked by 1 mM Ba^{2+} and weakly blocked by external TEA. The current was also activated by intracellular cAMP. Although the molecular identity of IK_{Sper} and the pH-sensitive current present in testicular sperm was not proven at the time, some of the electrophysiological characteristics of these currents, such as outward rectification, pH modulation, and modulation by PIP_2 (Tang, Zhang, Xia, Ren, & Logothetis, 2010), suggested that the channel carrying those currents could be the high-conductance K^+ channel SLO3. However, validation of this hypothesis had to wait until 2010–2011 when Santi et al. (2010) and Zeng et al. (2011) using SLO3 knockout mice strains corroborated that SLO3 channels were responsible for these currents.

Slo3 is one of four mammalian genes encoding high-conductance potassium (K^+) channels of the SLO K^+ channel family. *Slo3* channels from mouse were first cloned in the Salkoff lab in 1998, from a testis cDNA library based on its homology to the BK (mSLO1 channel). Expression of these channels in *Xenopus* oocytes showed that they are high-conductance K^+ channels activated by both voltage and intracellular alkalinization (Schreiber et al., 1998). These channels are highly homologous to SLO1 (BK), channels especially in the transmembrane regions with most of the differences located in the C-terminus. Unlike SLO1 which is conserved in *Drosophila* (Atkinson, Robertson, & Ganetzky, 1991), *C. elegans*, and mammals, SLO3 channels are only present in mammals and are only expressed in testis (Martinez-Lopez et al., 2009; Santi et al., 2010; Schreiber et al., 1998; Zeng et al., 2011). An interesting property of SLO3 channels is their low sequence conservation among different mammalian species. While SLO1 channels are extremely conserved in evolution, we found that SLO3 is highly divergent. This is not surprising for genes that mediate sexual reproduction (Swanson & Vacquier, 2002; Torgerson, Kulathinal, & Singh, 2002; Wyckoff, Wang, & Wu, 2000). Positive Darwinian selection has been reported in many genes involved in the immunity of reproduction and also in ion channels like CatSper (Podlaha, Webb, Tucker, & Zhang, 2005). The mouse, bovine and very recently human SLO3 channels (mSLO3, bSLO3) are the only species of SLO3 channels that have been cloned and expressed in *Xenopus* oocytes (Leonetti, Yuan, Hsiung, & Mackinnon, 2012; Santi, Butler, Kuhn, Wei, & Salkoff, 2009; Schreiber et al., 1998).

The first evidence about the molecular identity of the pH-sensitive K^+ current in sperm and its physiological role was obtained by Santi et al. (2010). Using a SLO3 knockout mouse in which the first two coding exons of the *Kcnu1* gene (*slo3* gene) were deleted, they showed that mutant testicular sperm lack the pH-sensitive K^+ current. Male SLO3 knockout mice are infertile; homozygous mutant males mated to wild-type females do not produce offspring in extended mating periods. On the other hand, homozygous mutant females and heterozygous males have normal fertility consistent with the male-specific expression of SLO3. This result was corroborated in 2011 by Zeng et al. (2011), who showed the absence of IK_{Sper} current in corpus epididymal sperm in another SLO3 knockout mouse which lacks the last coding exon (exon 27) of the *Kcnu1* gene (Zeng et al., 2011).

Experiments measuring Em before and after capacitation with voltage-sensitive dyes in SLO3 knockout sperm populations confirmed that SLO3 channels, directly or indirectly, are the main channels responsible for the hyperpolarization that occurs during *in vitro* capacitation; apparently no other channels can compensate for the loss of SLO3. Mutant sperm show a small but significant depolarization after capacitation (Santi et al., 2010). Consistent with these results, current clamp experiments demonstrated that the application of NH_4Cl failed to hyperpolarize mutant sperm, resulting in a small depolarization instead (Zeng et al., 2011).

The SLO3 knockout mouse has helped to unravel the physiological role of this channel in mouse sperm. Since hyperpolarization is removed, the

SLO3 knockout mouse would be very helpful in understanding the impact of membrane voltage in different sperm functions. Although SLO3 mutant sperm can, to some degree, undergo the spontaneous AR, they fail to undergo this exocytotic event when exposed to solubilized ZP. This phenotype is rescued by incubation of the mutant sperm with valinomycin, a K^+ ionophore that hyperpolarizes the sperm Em bringing it to the E_K. This result supports the hypothesis that membrane hyperpolarization during capacitation is a key factor required for the induction of the AR (Zeng et al., 1995). Interestingly, although A23187 is capable of inducing a significant AR in the mutant sperm, the efficiency is significantly lower than in wild-type sperm (Santi et al., 2010). This intriguing result suggests that, in addition to Ca^{2+} entry, other voltage-sensitive processes might be required for the AR to take place, and are thus deficient in SLO3 mutant sperm.

In addition to impaired Ca^{2+} entry, fertility depends in part, on the ability of sperm to respond to osmotic challenges encountered in their journey to meet the egg. Therefore, volume regulation may rely on the movement of K^+ (Barfield, Yeung, & Cooper, 2005; Yeung et al., 2006). Perhaps not coincidentally, volume regulation has a distinctive pharmacology that overlaps with the pharmacology of SLO3 channels, an area that remains to be explored (Yeung et al., 2006). Thus, SLO3 channels could well participate in volume control. When volume regulation fails, sperm swell and undergo characteristic morphological changes. Angulated sperm fail to migrate from the uterus to the oviduct, a deficiency resulting in infertility (Yeung & Cooper, 2001). Both Santi et al. (2010) and Zeng et al. (2011) reported that 60–70% of SLO3 mutant sperm are angulated when they are isolated in 330 mOsm/kg medium. Zeng et al. (2011) also found that this SLO3 mutant phenotype is rescued by isolating sperm in a higher osmolarity medium of 430 mOsm/kg.

3.5. Role of Cl^- in sperm capacitation

Work from our group has recently shown that when sperm are incubated in media lacking Cl^- anions, most of the capacitation-associated processes are blocked (Hernandez-Gonzalez et al., 2007; Wertheimer et al., 2008). In particular, Cl^--free media support neither the increase in tyrosine phosphorylation nor the hyperpolarization of the sperm Em. Not surprisingly, sperm incubated in these conditions do not hyperactivate, do not acquire the ability to undergo the AR, and do not fertilize a metaphase II-arrested egg *in vitro*. Interestingly, although in the absence of Cl^-, cAMP agonists induced phosphorylation events, the activation of the cAMP pathway is not sufficient to allow the sperm to fertilize *in vitro*. Also, using a Cl^- indicator MQAE, we have shown that intracellular Cl^- concentration ($[Cl^-]i$) increased during capacitation (Hernandez-Gonzalez et al., 2007; Wertheimer et al., 2008). These results indicate that the regulation of Cl^- homeostasis in sperm is needed for capacitation and suggest that one or more Cl^- transport systems are present in sperm. Despite these findings, little is known regarding specific sperm Cl^- transporters and how Cl^- affects capacitation. Below, we summarize some of the Cl^- translocation systems that have been proposed to be expressed in sperm with special emphasis on those that might play a role in the capacitation process.

In all cells, $[Cl^-]i$ levels are established by the relative contributions of all Cl^- transporters present in their plasma membrane. Those Cl^- transporters can be divided in two categories: Cl^- channels and specialized Cl^- carriers (Jentsch, Neagoe, & Scheel, 2005; Nilius & Droogmans, 2003). Among the Cl^- channels, four structural families have been identified: (1) CFTR channels; (2) the γ-aminobutyric (GABA)-gated and related glycine-gated neurotransmitter receptors; (3) Ca^{2+}-activated Cl^- channels (CaCCs); and (4) CLC channels. On the other hand, Cl^- can also cross the plasma membrane through secondary active transporters. In these molecules, Cl^- translocation is coupled to the movement of another ion in either opposite direction (antiporter) or in the same direction (cotransporter or symporter). Therefore, the energy to transport one of the ions against its electrochemical gradient is obtained by coupling it to the translocation of a second ion down its electrochemical gradient. Therefore, the process does not require the direct use of ATP. The Cl^- carrier proteins include (1) the electroneutral cation-Cl^- cotransporter family. This later family is composed of seven members, one Na^+/Cl^- cotransporter (NCC), two $Na^+/K^+/2Cl^-$ cotransporters (NKCC), and four Na^+-independent K^+/Cl^- cotransporters (KCC); and (2) the electroneutral Cl^-/HCO_3^- exchanger family.

3.5.1 CFTR channels

The cystic fibrosis transmembrane conductance regulator (CFTR) is a unique member of the ABC transporter family that forms an anion channel modulated by cAMP/PKA and ATP. The anion selectivity sequence of cAMP-regulated channels in cells containing either endogenous or recombinant CFTR is $Br^- > Cl^- > I^- > F^-$ (Anderson et al., 1991). CFTR is composed of five domains: two membrane-spanning domains (MSDs), two nucleotide-binding domains (NBDs), and a regulatory (R) domain. While the MSDs form the channel pore, phosphorylation of the R domain determines channel activity; and ATP hydrolysis by the NBDs controls the channel-gating properties (Sheppard & Welsh, 1999). It is well established that mutations in CFTR cause cystic fibrosis (CF), a disease characterized by defective Cl^- and HCO_3^- transport. About the role of CFTR in reproduction, although more than 95% of all CF male patients are infertile because of congenital bilateral absence of the vas deferens, it is still unknown if CFTR mutations are involved in other forms of male infertility. Using specific antibodies, our group and others have shown that CFTR is present in both human and mouse sperm (Chan et al., 2006; Hernandez-Gonzalez et al., 2007; Li et al., 2010; Xu et al., 2007). It has also been shown that the fertilizing capacity of sperm obtained from heterozygous CFTR mutant mice is also significantly lower than that of wild type (Xu et al., 2007). More recently, whole-cell patch-clamp recordings from testicular and epididymal mouse sperm revealed membrane currents containing a Cl^- selective component that is ATP dependent, stimulated by cAMP, cGMP, and genistein, and inhibited by DPC and $CFTR_{inh}$-172 (Fierro et al., 2012). Furthermore, the Cl^- current component activated by cAMP and inhibited by $CFTR_{inh}$-172 is absent in recordings on testicular sperm from mice in which CFTR was replaced by a loss-of-function mutation of the *CFTR* gene (ΔF508). Altogether these findings indicate that CFTR is present in mature mouse sperm and support the hypothesis that this Cl^- channel is involved in the regulation of capacitation.

The mechanism by which Cl^- and other anions are involved in the regulation of the sperm Em is not well understood. When Cl^- is replaced by nonpermeable anions (e.g., gluconate or methanesulfonate), there is no change in the sperm-resting membrane potential (Hernandez-Gonzalez et al., 2007). However, as mentioned above, in conditions that support capacitation, the associated hyperpolarization is inhibited in Cl^- free medium. Because CFTR is mainly a Cl^- transporter, one possibility is that this channel mediates the role of Cl^- in the regulation of both the resting sperm Em and the capacitation-associated hyperpolarization. Three lines of evidence support this hypothesis in mouse sperm: (1) the CFTR inhibitor diphenylamine-2-carboxylic acid (DPC 250 μM) inhibits the capacitation-associated hyperpolarization and decreases the ZP-induced AR without affecting the increase in tyrosine phosphorylation; (2) a CFTR agonist (genistein; 5–10 μM) promotes hyperpolarization in noncapacitated mouse sperm; and (3) addition of permeable analogs of cAMP to noncapacitated mouse sperm elevates $[Cl^-]i$ (Hernandez-Gonzalez et al., 2007). In addition to its role as a Cl^- channel, CFTR is also known to interact with and regulate other ion channels including epithelial Na^+ channels (ENaC) (Berdiev, Qadri, & Benos, 2009; Konig, Schreiber, Voelcker, Mall, & Kunzelmann, 2001; Kunzelmann & Schreiber, 1999; Perez-Cornejo & Arreola, 2004). As mentioned in the previous section, the sperm-resting Em is relatively depolarized and cannot be explained only by active K^+ channels. An approximate 10% contribution of Na^+ permeability would predict an Em of −35 mV which is close to experimental observations. Consistent with a Na^+ contribution: (1) when sperm are incubated in media in which Na^+ is replaced by choline$^+$ or glutamine$^+$, the sperm Em is hyperpolarized to an Em approaching the K^+ equilibrium (Hernandez-Gonzalez et al., 2006); (2) addition of pulses of Na^+ to sperm incubated in Na^+-free media induces sperm depolarization, suggesting the presence of an open Na^+ channel in these conditions; and (3) using the Na^+ indicator, CoroNa Red in combination with flow cytometry analysis, we have recently shown that the intracellular Na^+ ($[Na^+]i$) decreases when the sperm are incubated under capacitating conditions (Escoffier, Krapf, Navarrete, Darszon, & Visconti, 2012). All these results are consistent with ENaC channels being present in the membrane. In this regard, ENaC subunits have been detected in mature mouse sperm by Western blot analysis and immunofluorescence (Hernandez-Gonzalez et al., 2006). Furthermore, amiloride, a compound known to block these channels, induces sperm hyperpolarization in noncapacitated sperm to levels similar to those observed in capacitated conditions and blocks the capacitation-induced decrease in $[Na^+]i$. Altogether these results indicate that Na^+ influx is involved in establishing the sperm-resting Em and support the hypothesis that ENaC downregulation plays a role in the capacitation-associated sperm hyperpolarization. Although it has been demonstrated in other systems (Konig et al., 2001) that CFTR inhibits ENaC through an increase in the $[Cl^-]i$ (Stutts et al., 1995), there is no direct evidence of CFTR and ENaC interaction in sperm. However, the hypothesis that activation of CFTR downregulates ENaC is supported by the findings that activation of CFTR by genistein hyperpolarizes the sperm Em (Hernandez-Gonzalez et al., 2007) and decreases $[Na^+]i$ (Escoffier et al., 2012). Interestingly, immunofluorescence experiments show midpiece localization for both ENaC alpha subunit and CFTR (Hernandez-Gonzalez et al., 2006, 2007).

3.5.2 GABA and glycine channels

$GABA_A$ receptors are Cl^- channels that mediate inhibitory neurotransmission in the central nervous system (CNS). They were first identified pharmacologically as being activated by GABA and the selective agonist muscimol, blocked by bicuculline and picrotoxin, and modulated by benzodiazepines, barbiturates, and certain other CNS depressants (Macdonald & Olsen, 1994; Sieghart, 1995). In sperm, the presence of $GABA_A$ receptors has been studied by Meizel's group (Meizel, 1997; Wistrom & Meizel, 1993). It has been shown that GABA can induce the AR (Shi, Yuan, & Roldan, 1997) and also that $GABA_A$ receptors can modulate the response to progesterone in these cells (Hu, He, Wu, Yan, & Koide, 2002; Ritta, Calamera, & Bas, 1998; Shi & Roldan, 1995; Turner, Garcia, & Meizel, 1994). The role of GABA in the regulation of the AR will be discussed in Section 4.2. Regarding sperm capacitation, it is not clear whether GABA has an effect on this process. Ritta et al. have studied the role of this neurotransmitter in both human and bovine sperm capacitation, and their experiments suggested that GABA has a role in the regulation of sperm motility (Ritta, Bas, & Tartaglione, 2004; Ritta et al., 1998). Also in rat sperm, GABA and progesterone have been proposed to accelerate capacitation and hyperactivated motility, followed by an increase in the AR. Bicuculline and picrotoxin, antagonists of $GABA_A$ receptor/Cl^- channels, inhibit the effects of both GABA and progesterone (Jin et al., 2009). This evidence suggests that activation of $GABA_A$ receptor/Cl^- channels may contribute to sperm capacitation and hyperactivated motility.

3.5.3 Ca^{2+}-activated Cl^- channels

CaCCs are activated by increases in cytosolic free Ca^{2+} concentrations due to release from intracellular stores or influx through plasma membrane channels. The molecular identity of some types of Cl^- channels is still unknown. Recently, three research groups have arrived independently at the identification of TMEM16A (also known as anoctamin-1) as a membrane protein strongly related to the activity of CaCCs (Caputo et al., 2008; Ferrera, Caputo, & Galietta, 2010; Schroeder, Cheng, Jan, & Jan, 2008; Yang et al., 2008). TMEM16A is part of a family of proteins that includes nine other members named as TMEM16B-K (Galindo & Vacquier, 2005). TMEM16B coexpression with CaCCs, exhibits biophysical characteristics (voltage dependence, unitary conductance) different from those associated with TMEM16A coexpression (Pifferi, Dibattista, & Menini, 2009; Scudieri, Sondo, Ferrera, & Galietta, 2012). All TMEM16 proteins have a similar putative topology, consisting of eight transmembrane segments and cytosolic N- and C-termini (Galietta, 2009).

Patch-clamp studies in cell-attached mode "mapping" ion channel activity in human sperm head reveals a Cl^--permeable channel showing long stable openings. Different parts of the sperm head possess different channels, and there is remarkable clustering which may have a functional significance

(Jimenez-Gonzalez, Gu, Kirkman-Brown, Barratt, & Publicover, 2007). Recently, a CaCC in human sperm was characterized (Orta et al., 2012). The use of the perforated patch-clamp technique to obtain whole-cell recordings from the head of mature human spermatozoa revealed the presence of CaCC currents which could be carried by TMEM16A proteins. CaCCs play an important role in the physiology of human spermatozoa and are likely to participate in the solubilized ZP-induced AR, but their role in sperm motility is not yet determined.

3.5.4 Secondary active Cl^- transporters

As mentioned in the previous section, there is evidence supporting a role of CFTR in the changes in sperm Em that occur during capacitation. However, CFTR inhibitors do not inhibit other aspects of capacitation, such as the increase in tyrosine phosphorylation, which may suggest that, in addition to CFTR, other Cl^- transport systems are present in sperm. Cl^- can also be translocated into sperm through electroneutral carriers (e.g., NCC, NKCC, and KCC). Although these three cation/Cl^- cotransporters can work in both ways (e.g., Cl^- influx or efflux), under physiological conditions, the NCC and the NKCC carriers normally transport Cl^- into the cell, while the KCC family transport Cl^- out of the cell (Russell, 2000). During capacitation, it has been shown that $[Cl^-]i$ is increased (Hernandez-Gonzalez et al., 2007; Meizel & Turner, 1996); therefore, if present, NCC and NKCC might play a role in the regulation of Cl^- homeostasis during this process. This hypothesis was tested by observing the effect of Cl^- transport inhibitors on capacitation (Wertheimer et al., 2008). General Cl^- transport blockers such as stilbenes (e.g., DIDS and SITS) reduced sperm capacitation parameters to similar levels as those observed in the absence of Cl^-. However, most of the other Cl^- transport inhibitors tested, including the NCC inhibitor thiazide, failed to block capacitation-associated processes. Exceptions were bumetanide and furosemide, two NKCC inhibitors, which blocked the increase in tyrosine phosphorylation, hyperactivation, and the ability of the sperm to fertilize *in vitro*. However, the concentration necessary for these inhibitory effects was higher than that reported to be effective in inhibiting NKCC (Garg et al., 2007; Russell, 2000). Moreover, although the presence of Cl^- and Na^+ is essential for the increase in tyrosine phosphorylation, K^+, another ion needed for the function of NKCCs is not required. These results suggest that high concentrations of bumetanide may act on a mechanism different from the inhibition of NKCC. Conversely, the ZP-induced AR was inhibited at a much lower concentration of bumetanide and was dependent on the presence of Cl^-, K^+, and Na^+ suggesting that NKCC might have a role in the preparation of the sperm for the physiologically induced AR. Interestingly, *NKCC1* transcripts are present in spermatids, and null mutants of this protein have defects in spermatogenesis and are infertile (Pace et al., 2000). Although likely to be present in sperm, more research will be needed to understand the role of these cation/Cl^- exchangers in capacitation.

3.5.5 Cl^-/HCO_3^- exchangers

Cl^- carriers also include those molecules that exchange Cl^- for HCO_3^- in either direction. In sperm, HCO_3^- has been shown to activate cAMP synthesis through the atypical soluble adenylyl cyclase (SACY) (Hess et al., 2005; Okamura, Tajima, Soejima, Masuda, & Sugita, 1985). The specific carriers responsible for HCO_3^- transport have not yet been fully defined. Our group has provided evidence that Na^+/HCO_3^- cotransporters are present in mouse sperm and are responsible for initial HCO_3^- influxes (Demarco et al., 2003). However, it is not clear whether other HCO_3^- transport systems can also play a role in the control of HCO_3^- levels in sperm. In this respect, Cl^-/HCO_3^- exchangers have been proposed to play a role in the regulation of HCO_3^- homeostasis. These exchangers are expressed in multiple cell types and are relevant to the regulation of pHi, cell volume, and the regulation of Em through their contribution to the Cl^- gradient. The molecules responsible for the exchange of HCO_3^- and Cl^- belong to two evolutionary independent gene superfamilies, *SLC4* and *SLC26*, which exhibit unique patterns of anion selectivity and tissue distribution. The *SLC4* superfamily is composed by three genes (*AE1*, *AE2*, and *AE3*), each of them, represented by more than one alternative spliced sequences. The *SLC26* gene superfamily, comprising the SLC26 transporter superfamily, is represented by 11 genes. Of those, only *SLC26A3*, *SLC26A4*, and *SLC26A6* have Cl^-/HCO_3^- exchange activity.

Although little is known about the contribution of the *SLC4* and *SLC26* genes to sperm function, their presence has been proposed for many years (Ruknudin & Silver, 1990; Visconti, Muschietti, Flawia, & Tezon, 1990). From the SLC4 superfamily, only AE2 is prominent in testicular germ cells. Its expression pattern suggests a role for this gene product in either spermatogenesis or later on in sperm function. The *AE2* gene is represented by five splice variants (*AE2a, AE2b1, AE2b2, AE2c1*, and *AE2c2*). Mice lacking expression of all of them die before weaning due to severely retarded development (Gawenis et al., 2004); however, those mice retaining

AE2c but lacking AE2a, AE2b1, and AE2b2 have a milder phenotype. Consistent with the observation that AE2 is highly expressed in the testis, these mice are infertile and exhibit testicular dysplasia (Medina et al., 2003).

Regarding the SLC26 superfamily, recent work from our group and others using Western blot and immunofluorescence approaches show the presence of SLC26A3 and SLC26A6 in the sperm midpiece (Chan et al., 2009; Chavez et al., 2011; Chen et al., 2009). In this chapter, we provided evidence that these transporters coprecipitate with CFTR and that tenidap, a SLC26A3-specific inhibitor, blocked the capacitation-associated hyperpolarization and the ZP-induced AR. However, tenidap did not block the activation of a cAMP pathway and the increase in tyrosine phosphorylation, suggesting that these transporters are not directly involved in the regulation of AdCyc10.

4. ACROSOME REACTION

The acrosome is a large secretory vesicle located at the posterior end of the sperm head (Yanagimachi, 1998). The AR is a unique, single-vesicle exocytotic event required for sperm of many species to achieve fusion with the female gamete. During the AR, now considered as a multistep process, many fusion points occur between the sperm head plasma membrane and the outer acrosomal membrane. Plasma membrane–outer acrosome hybrid vesicles are liberated as a result of the multiple fenestrations that occur in this irreversible reaction, which leads to acrosomal content release of hydrolytic enzymes. Not surprisingly, the fusion machinery conserved in many neuroendocrinal secretory cells and regulated by Ca^{2+} is present in sperm and involved in the AR (Bello, Zanetti, Mayorga, & Michaut, 2012; Castillo Bennett, Roggero, Mancifesta, & Mayorga, 2010). At present, where the AR takes place in the female tract and the identity of the molecules inducing this exocytotic reaction is controversial (Jin et al., 2011; Visconti & Florman, 2010; Yanagimachi, 1998). However, among the many inducers of the AR described, the ZP and progesterone are the ones considered more physiologically relevant (Litscher, Williams, & Wassarman, 2009; Mayorga, Tomes, & Belmonte, 2007). In addition to its participation in sperm–egg signaling, ZP is also a protective layer (Litscher et al., 2009). The ZP-induced AR requires the convergence of several transduction pathways (for review, see Mayorga et al., 2007) which result in a complex cascade of $[Ca^{2+}]i$ changes. The physiologically relevant AR requires changes in $[Ca^{2+}]i$; external and internal sources contribute to its modulation

(Breitbart, Rotman, Rubinstein, & Etkovitz, 2010; Costello et al., 2009; Darszon et al., 2011; Florman, Jungnickel, & Sutton, 2008).

4.1. Ca^{2+} channels and the AR

Three distinct Ca^{2+} channels have been proposed to contribute to $[Ca^{2+}]i$ responses associated with the AR. Though not fully understood, these channels are functionally linked (reviewed in Darszon et al., 2011; Florman et al., 2008; Publicover, Harper, & Barratt, 2007). The first is a voltage-dependent Ca^{2+} (Ca_V) channel whose involvement in the AR was deduced from functional and pharmacological observations. Supporting this possibility, sperm from many mammalian species undergo $[Ca^{2+}]i$ increases in response to K^+ depolarization and ZP that are sensitive to Ca_V-channel blockers (reviewed in Darszon et al., 2011; Florman et al., 2008; Lishko et al., 2012; Ren & Xia, 2010). Ca_V channels have been proposed to participate in the transient elevation induced by ZP which last ~1 s and has been best characterized in mouse sperm (Florman et al., 2008). However, even though $Ca_V3.2$ (T-type Ca^{2+} channel) has been considered the most likely Ca_V candidate to participate in the mouse AR (Arnoult et al., 1996; Escoffier et al., 2007; Lievano et al., 1996; Trevino et al., 2004), knockout mice lacking this channel are fertile and undergo the ZP-induced AR. In addition, recent patch-clamp whole-cell recordings on the cytoplasmic droplet of epididymal mouse sperm have failed to detect Ca_Vs (Xia, Reigada, Mitchell, & Ren, 2007). These findings have questioned the presence of functional Ca_Vs in mature sperm and their involvement in the AR, in spite of solid immunological data demonstrating their presence (Escoffier et al., 2007; Trevino et al., 2004). Alternatively, the fertile phenotype of $Ca_V3.2$ null male mice can be explained by compensation from other Ca_V channels, by the possible participation of high-voltage-activated Ca^{2+} channels (HVA) in the AR (Escoffier et al., 2007), or by activation of CatSper channels as recently proposed (Xia et al., 2007).

Following the fast transient $[Ca^{2+}]i$ elevation induced by ZP, a sustained $[Ca^{2+}]i$ increase lasting up to minutes occurs. This $[Ca^{2+}]i$ change results from the release of Ca^{2+} from internal stores (i.e., the acrosome). As the IP3 receptor, the second type of Ca^{2+} channel involved in the AR, is activated due to IP3 production (reviewed in Publicover et al., 2007; Florman et al., 2008; Darszon et al., 2011). Ca^{2+} store emptying leads to the opening of plasma membrane Ca^{2+} channels (SOCs) which in turn, as in somatic cells, cause a sustained $[Ca^{2+}]i$ increase. Since the SOC machinery is more complex than originally thought (Moreno & Vaca, 2011), several components such as STIM, ORAI, and TRPCs may participate in the sustained Ca^{2+} uptake which is the third type of Ca^{2+} channels involved in the AR. Preliminary reports of the presence of STIM and ORAI in human and mouse sperm are consistent with this proposal (Costello et al., 2009; Darszon et al., 2012).

4.2. Cl^- channels and the AR

While there is no doubt that Ca^{2+} fluxes are key to the AR, little is known about the participation of Cl^- movement during this event. As mentioned in Section 3, using different approaches, several Cl^- channels and exchangers have been detected in sperm from various species, and it has been previously established that Cl^- is essential for capacitation. Regarding the participation of Cl^- channels in the mammalian sperm AR, niflumic acid (NFA) has been shown to inhibit Ca^{2+}-induced hyperpolarization partially driven by Cl^- (Espinosa et al., 1998). Furthermore, this stilbene inhibited the first Cl^- single-channel activity recorded in mammalian sperm as well as the AR induced by solubilized ZP, progesterone, and GABA in mouse sperm (Espinosa et al., 1998). Recently, our group has provided evidence that anion-channel blockers like NFA, DIDS, and others inhibit the mouse and human sperm AR as well as Cl^- channels detected in these cells (Espinosa & Darszon, 1995; Espinosa et al., 1998; Orta et al., 2012).

In addition, neurotransmitter receptors have been identified and implicated during the AR of sperm from several species, based primarily on pharmacological and genetic evidence. $GABA_A$ and glycine receptors are of particular interest in this review, since they are also associated with Cl^- fluxes. Although the best studied inductor of the AR is ZP3, GABA and glycine are present in the female reproductive tract and have been shown to also induce the AR, possibly through Cl^- fluxes. For example, Burrello et al. (2004) reported AR induction by GABA in human sperm. They also observed that the AR induced by progesterone is blocked in the presence of picrotoxin (a GABA Cl^- channel inhibitor) and that GABA and Pg together exert a stronger AR induction compared to each component alone. These researchers postulated that GABA present in the follicular fluid, together with Pg, activates the same receptor (probably a $GABA_A$-like receptor). However, when they used specific inhibitors of $GABA_A$ and $GABA_B$ receptors (bicuculline and saclofen, respectively), the AR induced by follicular fluid was unaffected, unless these drugs where used simultaneously. Burrello et al. (2004) concluded that both $GABA_A$ and $GABA_B$ receptors participate in the human sperm AR. However, it has been shown that the Ca^{2+} channel CatSper can be activated by progesterone in human sperm (Lishko, Botchkina, & Kirichok, 2011; Strunker et al., 2011), and due to the promiscuous activation of this channel (Brenker et al., 2012), it is now important to establish whether GABA is acting via CatSper or via a GABA receptor. In any case, the induction of the AR by GABA has been corroborated by several researchers in different species such as mouse, rat, human, porcine, guinea pig, and bull (Espinosa et al., 1998; Hu et al., 2002; Melendrez & Meizel, 1995; Roldan, Murase, & Shi, 1994; Shi & Roldan, 1995; Shi et al., 1997). The participation of CatSper in this induction is yet to be established.

Pharmacological evidence has also implicated the glycine receptor in the ZP- or glycine-induced AR in human, mouse, and hamster sperm (Bray, Son, Kumar, Harris, & Meizel, 2002; Llanos, Ronco, Aguirre, & Meizel, 2001; Sato, Son, & Meizel, 2000). Sperm from glycine receptor null mice are unable to respond to ZP, although fertilization *in vitro* can still proceed, albeit at a lower rate and attributable to a spontaneous AR (Meizel & Son, 2005). It has been proposed that a Cl^- efflux takes place via a glycine receptor causing a depolarization, which in turn may trigger the voltage-dependent Ca^{2+} channel opening required for the ZP-induced AR (Llanos et al., 2001). This hypothesis is also under scrutiny, as both the functional presence of Ca_vs and the relevance of ZP-induced AR is questioned in the light of new evidence regarding this process (reviewed in Darszon et al., 2011).

As discussed earlier, there is also electrophysiological evidence for the presence of Cl^- channels in sperm. The first recordings were performed by directly patching mouse sperm. Using this technique, Espinosa et al. (1998) recorded an anion channel in epididymal sperm with biophysical properties and sensitivity to NFA, similar to the Ca^{2+}-dependent Cl^- channels (Hogg, Wang, & Large, 1994). Recently, Orta et al. (2012) reported the presence of CaCCs, possibly TMEM16A, in human sperm. Interestingly, the pharmacology profile of the recorded currents was consistent with inhibition of the ZP3-induced AR. $TMEM16A_{inh}$ (20 μM), so far the most specific antagonist of TMEM16A Cl^- channels, inhibited nearly 80% of the AR, supporting the participation of TMEM channels during this process. In this regard, the AR involves a Ca^{2+}-dependent swelling of the acrosome (Zanetti & Mayorga, 2009), a process in which Cl^- currents may participate.

Interestingly, evidence gathered in recent years has uncovered an unexpected pharmacological overlap between CaCCs and the large conductance, Ca^{2+}-gated K^+ channels (BK_{Ca} or $K_{Ca}1.1$) (Greenwood & Leblanc, 2007; Sones, Leblanc, & Greenwood, 2009). Various compounds with structurally different characteristics considered to be Cl^--channel blockers, such as NFA, anthracene-9-carboxylate, and ethacrynic acid, enhance $K_{Ca}1.1$ currents (Greenwood & Large, 1995; Ottolia & Toro, 1994; Toma, Greenwood, Helliwell, & Large, 1996). Since there are evidences of the presence of $K_{Ca}1.1$ channels in mammalian sperm (Rossato, Di Virgilio, Rizzuto, Galeazzi, & Foresta, 2001; Wu et al., 1998), the combined effects of anion-channel blockers such as NFA on CaCCs and $K_{Ca}1.1$s could account for their potent ability to inhibit the AR. The large $[Ca^{2+}]i$ changes that occur during the AR result in significant morphological sperm head alterations which seem to involve acrosome swelling and an RVD in which CaCCs may participate. By blocking CaCCs, NFA, DIDS, and $TMEM16A_{inh}$ could alter the regulatory volume decrease that appears to be important to regulate the distance between the outer acrosomal membrane and the plasma membrane, which is critical for acrosome exocytosis (Zanetti & Mayorga, 2009).

5. FINAL REMARKS

As mentioned in the introduction, the role of ionic fluxes in sperm has been hindered due to many factors which include: (1) the small sperm size which has made the use of electrophysiological techniques very difficult; and (2) the lack of transcription and translation, making it very challenging to perform knock down or to express exogenous proteins. In the past years, the use of genetically modified mice has accelerated the process of identifying molecules which are essential for different aspects of sperm function. Among them, knockout mice for two sperm-specific ion channels, CatSper and SLO3, were key to reveal that these channels are necessary for fertilization. Also, the increasing sensitivity of mass spectrometry has helped to identify some of the proteins associated with these ion channel-associated proteins. Finally, new patch-clamp techniques have allowed the study of ion movements in sperm. Overall, the chapters discussed in this review support the idea that the activity of sperm ion channels is involved in the regulation of sperm maturation, capacitation, and the AR. However, currently little is known about how these sperm ion channels are regulated during these processes. Several lines of experimentation now show promise in uncovering these mechanisms including identifying posttranslational modifications in those proteins that modulate ion fluxes and making direct electrophysiological observations in sperm which has only recently been possible. These as well as other new techniques will almost certainly accelerate the pace of our understanding of the physiology of sperm. Finally, from the applied point of view, a better understanding of sperm ion channels exposed to the extracellular milieu offers opportunities for intervention in fertility treatments as well as for their use as possible contraceptive targets.

ACKNOWLEDGMENTS

We would like to thank Alice Butler and Ana Laura González-Cota for editing and correcting this chapter. This work was supported by DGAPA: IN202312 (to A. D.) and IN202212 (to C. T.), CONACyT: 49113 and 128566 (to A. D.) and 99333 (to C. T.), and NIH: R01 HD44044 and HD038082 (to P. E. V.) and HD069631 (to C. S.).

REFERENCES

Acevedo, J. J., Mendoza-Lujambio, I., de la Vega-Beltran, J. L., Trevino, C. L., Felix, R., & Darszon, A. (2006). KATP channels in mouse spermatogenic cells and sperm, and their role in capacitation. *Developmental Biology, 289*, 395–405.

Aguilar-Bryan, L., & Bryan, J. (1999). Molecular biology of adenosine triphosphate-sensitive potassium channels. *Endocrine Reviews, 20*, 101–135.

Aguilar-Bryan, L., Clement, J. P., 4th, Gonzalez, G., Kunjilwar, K., Babenko, A., & Bryan, J. (1998). Toward understanding the assembly and structure of KATP channels. *Physiological Reviews, 78*, 227–245.

Anderson, M. P., Gregory, R. J., Thompson, S., Souza, D. W., Paul, S., Mulligan, R. C., et al. (1991). Demonstration that CFTR is a chloride channel by alteration of its anion selectivity. *Science, 253*, 202–205.

Arnoult, C., Cardullo, R. A., Lemos, J. R., & Florman, H. M. (1996). Activation of mouse sperm T-type Ca2+ channels by adhesion to the egg zona pellucida. *Proceedings of the National Academy of Sciences of the United States of America, 93*, 13004–13009.

Arnoult, C., Kazam, I. G., Visconti, P. E., Kopf, G. S., Villaz, M., & Florman, H. M. (1999). Control of the low voltage-activated calcium channel of mouse sperm by egg ZP3 and by membrane hyperpolarization during capacitation. *Proceedings of the National Academy of Sciences of the United States of America, 96*, 6757–6762.

Atkinson, N. S., Robertson, G. A., & Ganetzky, B. (1991). A component of calcium-activated potassium channels encoded by the Drosophila slo locus. *Science, 253*, 551–555.

Austin, C. R. (1952). The capacitation of the mammalian sperm. *Nature, 170*, 326.

Baker, M. A. (2011). The 'omics revolution and our understanding of sperm cell biology. *Asian Journal of Andrology, 13*, 6–10.

Baldi, E., Casano, R., Falsetti, C., Krausz, C., Maggi, M., & Forti, G. (1991). Intracellular calcium accumulation and responsiveness to progesterone in capacitating human spermatozoa. *Journal of Andrology, 12*, 323–330.

Barfield, J. P., Yeung, C. H., & Cooper, T. G. (2005). Characterization of potassium channels involved in volume regulation of human spermatozoa. *Molecular Human Reproduction, 11*, 891–897.

Bellentani, F. F., Fernandes, G. S., Perobelli, J. E., Pacini, E. S., Kiguti, L. R., Pupo, A. S., et al. (2011). Acceleration of sperm transit time and reduction of sperm reserves in the epididymis of rats exposed to sibutramine. *Journal of Andrology, 32*, 718–724.

Bello, O. D., Zanetti, M. N., Mayorga, L. S., & Michaut, M. A. (2012). RIM, Munc13, and Rab3A interplay in acrosomal exocytosis. *Experimental Cell Research, 318*, 478–488.

Berdiev, B. K., Qadri, Y. J., & Benos, D. J. (2009). Assessment of the CFTR and ENaC association. *Molecular BioSystems, 5*, 123–127.

Bray, C., Son, J. H., Kumar, P., Harris, J. D., & Meizel, S. (2002). A role for the human sperm glycine receptor/Cl(−) channel in the acrosome reaction initiated by recombinant ZP3. *Biology of Reproduction, 66*, 91–97.

Breitbart, H. (2003). Signaling pathways in sperm capacitation and acrosome reaction. *Cellular and Molecular Biology (Noisy-le-Grand, France), 49*, 321–327.

Breitbart, H., Rotman, T., Rubinstein, S., & Etkovitz, N. (2010). Role and regulation of PI3K in sperm capacitation and the acrosome reaction. *Molecular and Cellular Endocrinology, 314*, 234–238.

Brenker, C., Goodwin, N., Weyand, I., Kashikar, N. D., Naruse, M., Krahling, M., et al. (2012). The CatSper channel: A polymodal chemosensor in human sperm. *The EMBO Journal, 31*, 1654–1665.

Burrello, N., Vicari, E., D'Amico, L., Satta, A., D'Agata, R., & Calogero, A. E. (2004). Human follicular fluid stimulates the sperm acrosome reaction by interacting with the gamma-aminobutyric acid receptors. *Fertility and Sterility, 82*(Suppl. 3), 1086–1090.

Caputo, A., Caci, E., Ferrera, L., Pedemonte, N., Barsanti, C., Sondo, E., et al. (2008). TMEM16A, a membrane protein associated with calcium-dependent chloride channel activity. *Science, 322*, 590–594.

Carlson, A. E., Hille, B., & Babcock, D. F. (2007). External Ca2+ acts upstream of adenylyl cyclase SACY in the bicarbonate signaled activation of sperm motility. *Developmental Biology, 312*, 183–192.

Carlson, A. E., Quill, T. A., Westenbroek, R. E., Schuh, S. M., Hille, B., & Babcock, D. F. (2005). Identical phenotypes of CatSper1 and CatSper2 null sperm. *The Journal of Biological Chemistry, 280*, 32238–32244.

Carlson, A. E., Westenbroek, R. E., Quill, T., Ren, D., Clapham, D. E., Hille, B., et al. (2003). CatSper1 required for evoked Ca2+ entry and control of flagellar function in sperm. *Proceedings of the National Academy of Sciences of the United States of America, 100*, 14864–14868.

Castillo Bennett, J., Roggero, C. M., Mancifesta, F. E., & Mayorga, L. S. (2010). Calcineurin-mediated dephosphorylation of synaptotagmin VI is necessary for acrosomal exocytosis. *The Journal of Biological Chemistry, 285*, 26269–26278.

Chan, H. C., Ruan, Y. C., He, Q., Chen, M. H., Chen, H., Xu, W. M., et al. (2009). The cystic fibrosis transmembrane conductance regulator in reproductive health and disease. *The Journal of Physiology, 587*, 2187–2195.

Chan, H. C., Shi, Q. X., Zhou, C. X., Wang, X. F., Xu, W. M., Chen, W. Y., et al. (2006). Critical role of CFTR in uterine bicarbonate secretion and the fertilizing capacity of sperm. *Molecular and Cellular Endocrinology, 250*, 106–113.

Chan, H. C., Wu, W. L., Sun, Y. P., Leung, P. S., Wong, T. P., Chung, Y. W., et al. (1998). Expression of sperm Ca2+-activated K+ channels in Xenopus oocytes and their modulation by extracellular ATP. *FEBS Letters, 438*, 177–182.

Chang, M. C. (1951). Fertilizing capacity of spermatozoa deposited into the fallopian tubes. *Nature, 168*, 697–698.

Chavez, J. C., Hernandez-Gonzalez, E. O., Wertheimer, E., Visconti, P. E., Darszon, A., & Trevino, C. L. (2012). Participation of the Cl−/HCO3− exchangers SLC26A3 and SLC26A6, the Cl− channel CFTR and the regulatory factor SLC9A3R1 in mouse sperm capacitation. *Biology of Reproduction, 86*, 1–14.

Chen, W. Y., Xu, W. M., Chen, Z. H., Ni, Y., Yuan, Y. Y., Zhou, S. C., et al. (2009). Cl− is required for HCO3− entry necessary for sperm capacitation in guinea pig: Involvement of a Cl−/HCO3− exchanger (SLC26A3) and CFTR. *Biology of Reproduction, 80*, 115–123.
Christen, R., Schackmann, R. W., & Shapiro, B. M. (1983). Metabolism of sea urchin sperm. Interrelationships between intracellular pH, ATPase activity, and mitochondrial respiration. *The Journal of Biological Chemistry, 258*, 5392–5399.
Cohen-Dayag, A., Tur-Kaspa, I., Dor, J., Mashiach, S., & Eisenbach, M. (1995). Sperm capacitation in humans is transient and correlates with chemotactic responsiveness to follicular factors. *Proceedings of the National Academy of Sciences of the United States of America, 92*, 11039–11043.
Cooper, T. G. (2011). The epididymis, cytoplasmic droplets and male fertility. *Asian Journal of Andrology, 13*, 130–138.
Cooper, T. G., & Yeung, C. H. (2007). Involvement of potassium and chloride channels and other transporters in volume regulation by spermatozoa. *Current Pharmaceutical Design, 13*, 3222–3230.
Costello, S., Michelangeli, F., Nash, K., Lefievre, L., Morris, J., Machado-Oliveira, G., et al. (2009). Ca2+-stores in sperm: Their identities and functions. *Reproduction, 138*, 425–437.
Cox, T., & Peterson, R. N. (1989). Identification of calcium conducting channels in isolated boar sperm plasma membranes. *Biochemical and Biophysical Research Communications, 161*, 162–168.
Cross, N. L. (1998). Role of cholesterol in sperm capacitation. *Biology of Reproduction, 59*, 7–11.
Dacheux, J. L., Belleannee, C., Guyonnet, B., Labas, V., Teixeira-Gomes, A. P., Ecroyd, H., et al. (2012). The contribution of proteomics to understanding epididymal maturation of mammalian spermatozoa. *Systems Biology in Reproductive Medicine, 58*, 197–210.
Darszon, A., Labarca, P., Nishigaki, T., & Espinosa, F. (1999). Ion channels in sperm physiology. *Physiological Reviews, 79*, 481–510.
Darszon, A., Nishigaki, T., Beltran, C., & Trevino, C. L. (2011). Calcium channels in the development, maturation, and function of spermatozoa. *Physiological Reviews, 91*, 1305–1355.
Darszon, A., Sanchez-Cardenas, C., Orta, G., Sanchez-Tusie, A. A., Beltran, C., Lopez-Gonzalez, I., et al. (2012). Are TRP channels involved in sperm development and function? *Cell and Tissue Research, 349*, 749–764.
DasGupta, S., Mills, C. L., & Fraser, L. R. (1993). Ca(2+)-related changes in the capacitation state of human spermatozoa assessed by a chlortetracycline fluorescence assay. *Journal of Reproduction and Fertility, 99*, 135–143.
Davis, B. K. (1981). Timing of fertilization in mammals: Sperm cholesterol/phospholipid ratio as a determinant of the capacitation interval. *Proceedings of the National Academy of Sciences of the United States of America, 78*, 7560–7564.
De Blas, G. A., Darszon, A., Ocampo, A. Y., Serrano, C. J., Castellano, L. E., Hernandez-Gonzalez, E. O., et al. (2009). TRPM8, a versatile channel in human sperm. *PLoS One, 4*, e6095.
Demarco, I. A., Espinosa, F., Edwards, J., Sosnik, J., De La Vega-Beltran, J. L., Hockensmith, J. W., et al. (2003). Involvement of a Na+/HCO-3 cotransporter in mouse sperm capacitation. *The Journal of Biological Chemistry, 278*, 7001–7009.
Dorschner, H., Brekardin, E., Uhde, I., Schwanstecher, C., & Schwanstecher, M. (1999). Stoichiometry of sulfonylurea-induced ATP-sensitive potassium channel closure. *Molecular Pharmacology, 55*, 1060–1066.
Dun, M. D., Aitken, R. J., & Nixon, B. (2012). The role of molecular chaperones in spermatogenesis and the post-testicular maturation of mammalian spermatozoa. *Human Reproduction Update, 18*, 420–435.

Escoffier, J., Boisseau, S., Serres, C., Chen, C. C., Kim, D., Stamboulian, S., et al. (2007). Expression, localization and functions in acrosome reaction and sperm motility of Ca(V) 3.1 and Ca(V)3.2 channels in sperm cells: An evaluation from Ca(V)3.1 and Ca(V)3.2 deficient mice. *Journal of Cellular Physiology, 212*, 753–763.

Escoffier, J., Krapf, D., Navarrete, F., Darszon, A., & Visconti, P. E. (2012). Flow cytometry analysis reveals a decrease in intracellular sodium during sperm capacitation. *Journal of Cell Science, 125*, 473–485.

Espinosa, F., & Darszon, A. (1995). Mouse sperm membrane potential: Changes induced by Ca2+. *FEBS Letters, 372*, 119–125.

Espinosa, F., de la Vega-Beltran, J. L., Lopez-Gonzalez, I., Delgado, R., Labarca, P., & Darszon, A. (1998). Mouse sperm patch-clamp recordings reveal single Cl- channels sensitive to niflumic acid, a blocker of the sperm acrosome reaction. *FEBS Letters, 426*, 47–51.

Esposito, G., Jaiswal, B. S., Xie, F., Krajnc-Franken, M. A., Robben, T. J., Strik, A. M., et al. (2004). Mice deficient for soluble adenylyl cyclase are infertile because of a severe sperm-motility defect. *Proceedings of the National Academy of Sciences of the United States of America, 101*, 2993–2998.

Felix, R., Serrano, C. J., Trevino, C. L., Munoz-Garay, C., Bravo, A., Navarro, A., et al. (2002). Identification of distinct K+ channels in mouse spermatogenic cells and sperm. *Zygote, 10*, 183–188.

Ferrera, L., Caputo, A., & Galietta, L. J. (2010). TMEM16A protein: A new identity for Ca (2+)-dependent Cl(−) channels. *Physiology (Bethesda, Md.), 25*, 357–363.

Fierro, D. F., Acevedo, J. J., Martinez, P., Escoffier, J., Sepulveda, F. V., Balderas, E., et al. (2012). Electrophysiological evidence for the presence of cystic fibrosis transmembrane conductance regulator (CFTR) in mouse sperm. *Journal of Cellular Physiology*. http://dx.doi.org/10.1002/jcp. 24166 [Epub ahead of print].

Florman, H. M., Jungnickel, M. K., & Sutton, K. A. (2008). Regulating the acrosome reaction. *The International Journal of Developmental Biology, 52*, 503–510.

Furst, J., Gschwentner, M., Ritter, M., Botta, G., Jakab, M., Mayer, M., et al. (2002). Molecular and functional aspects of anionic channels activated during regulatory volume decrease in mammalian cells. *Pflügers Archiv, 444*, 1–25.

Gadella, B. M., & Harrison, R. A. (2000). The capacitating agent bicarbonate induces protein kinase A-dependent changes in phospholipid transbilayer behavior in the sperm plasma membrane. *Development, 127*, 2407–2420.

Galietta, L. J. (2009). The TMEM16 protein family: A new class of chloride channels? *Biophysical Journal, 97*, 3047–3053.

Galindo, B. E., & Vacquier, V. D. (2005). Phylogeny of the TMEM16 protein family: Some members are overexpressed in cancer. *International Journal of Molecular Medicine, 16*, 919–924.

Garg, P., Martin, C. F., Elms, S. C., Gordon, F. J., Wall, S. M., Garland, C. J., et al. (2007). Effect of the Na-K-2Cl cotransporter NKCC1 on systemic blood pressure and smooth muscle tone. *American Journal of Physiology. Heart and Circulatory Physiology, 292*, H2100–H2105.

Gawenis, L. R., Ledoussal, C., Judd, L. M., Prasad, V., Alper, S. L., Stuart-Tilley, A., et al. (2004). Mice with a targeted disruption of the AE2 Cl−/HCO3− exchanger are achlorhydric. *The Journal of Biological Chemistry, 279*, 30531–30539.

Gibbs, G. M., Orta, G., Reddy, T., Koppers, A. J., Martinez-Lopez, P., Luis de la Vega-Beltran, J., et al. (2011). Cysteine-rich secretory protein 4 is an inhibitor of transient receptor potential M8 with a role in establishing sperm function. *Proceedings of the National Academy of Sciences of the United States of America, 108*, 7034–7039.

Go, K. J., & Wolf, D. P. (1983). The role of sterols in sperm capacitation. *Advances in Lipid Research, 20*, 317–330.

Greenwood, I. A., & Large, W. A. (1995). Comparison of the effects of fenamates on Ca-activated chloride and potassium currents in rabbit portal vein smooth muscle cells. *British Journal of Pharmacology, 116*, 2939–2948.

Greenwood, I. A., & Leblanc, N. (2007). Overlapping pharmacology of Ca2+-activated Cl- and K+ channels. *Trends in Pharmacological Sciences, 28*, 1–5.

Guerrero, A., Carneiro, J., Pimentel, A., Wood, C. D., Corkidi, G., & Darszon, A. (2011). Strategies for locating the female gamete: The importance of measuring sperm trajectories in three spatial dimensions. *Molecular Human Reproduction, 17*, 511–523.

Guerrero, A., Sanchez, J. A., & Darszon, A. (1987). Single-channel activity in sea urchin sperm revealed by the patch-clamp technique. *FEBS Letters, 220*, 295–298.

Hagiwara, S., & Kawa, K. (1984). Calcium and potassium currents in spermatogenic cells dissociated from rat seminiferous tubules. *The Journal of Physiology, 356*, 135–149.

Harrison, R. A. (2004). Rapid PKA-catalysed phosphorylation of boar sperm proteins induced by the capacitating agent bicarbonate. *Molecular Reproduction and Development, 67*, 337–352.

Hernandez-Gonzalez, E. O., Sosnik, J., Edwards, J., Acevedo, J. J., Mendoza-Lujambio, I., Lopez-Gonzalez, I., et al. (2006). Sodium and epithelial sodium channels participate in the regulation of the capacitation-associated hyperpolarization in mouse sperm. *The Journal of Biological Chemistry, 281*, 5623–5633.

Hernandez-Gonzalez, E. O., Trevino, C. L., Castellano, L. E., de la Vega-Beltran, J. L., Ocampo, A. Y., Wertheimer, E., et al. (2007). Involvement of cystic fibrosis transmembrane conductance regulator in mouse sperm capacitation. *The Journal of Biological Chemistry, 282*, 24397–24406.

Hess, K. C., Jones, B. H., Marquez, B., Chen, Y., Ord, T. S., Kamenetsky, M., et al. (2005). The "soluble" adenylyl cyclase in sperm mediates multiple signaling events required for fertilization. *Developmental Cell, 9*, 249–259.

Hibino, H., Inanobe, A., Furutani, K., Murakami, S., Findlay, I., & Kurachi, Y. (2010). Inwardly rectifying potassium channels: Their structure, function, and physiological roles. *Physiological Reviews, 90*, 291–366.

Hille, B. (2001). *Ion channels of excitable membranes*. Sunderland, MA: Sinauer Associates Inc.

Hogg, R. C., Wang, Q., & Large, W. A. (1994). Action of niflumic acid on evoked and spontaneous calcium-activated chloride and potassium currents in smooth muscle cells from rabbit portal vein. *British Journal of Pharmacology, 112*, 977–984.

Hu, J. H., He, X. B., Wu, Q., Yan, Y. C., & Koide, S. S. (2002). Biphasic effect of GABA on rat sperm acrosome reaction: Involvement of GABA(A) and GABA(B) receptors. *Archives of Andrology, 48*, 369–378.

Hung, P. H., & Suarez, S. S. (2010). Regulation of sperm storage and movement in the ruminant oviduct. *Society of Reproduction and Fertility Supplement, 67*, 257–266.

Jacob, A., Hurley, I. R., Goodwin, L. O., Cooper, G. W., & Benoff, S. (2000). Molecular characterization of a voltage-gated potassium channel expressed in rat testis. *Molecular Human Reproduction, 6*, 303–313.

Jentsch, T. J., Neagoe, I., & Scheel, O. (2005). CLC chloride channels and transporters. *Current Opinion in Neurobiology, 15*, 319–325.

Jimenez-Gonzalez, M. C., Gu, Y., Kirkman-Brown, J., Barratt, C. L., & Publicover, S. (2007). Patch-clamp 'mapping' of ion channel activity in human sperm reveals regionalisation and co-localisation into mixed clusters. *Journal of Cellular Physiology, 213*, 801–808.

Jimenez-Gonzalez, C., Michelangeli, F., Harper, C. V., Barratt, C. L., & Publicover, S. J. (2006). Calcium signalling in human spermatozoa: A specialized 'toolkit' of channels, transporters and stores. *Human Reproduction Update, 12*, 253–267.

Jin, J. Y., Chen, W. Y., Zhou, C. X., Chen, Z. H., Yu-Ying, Y., Ni, Y., et al. (2009). Activation of GABAA receptor/Cl- channel and capacitation in rat spermatozoa: HCO3- and Cl- are essential. *Systems Biology in Reproductive Medicine, 55*, 97–108.

Jin, M., Fujiwara, E., Kakiuchi, Y., Okabe, M., Satouh, Y., Baba, S. A., et al. (2011). Most fertilizing mouse spermatozoa begin their acrosome reaction before contact with the zona pellucida during in vitro fertilization. *Proceedings of the National Academy of Sciences of the United States of America, 108*, 4892–4896.
Kaupp, U. B., Kashikar, N. D., & Weyand, I. (2008). Mechanisms of sperm chemotaxis. *Annual Review of Physiology, 70*, 93–117.
Kirichok, Y., & Lishko, P. V. (2011). Rediscovering sperm ion channels with the patch-clamp technique. *Molecular Human Reproduction, 17*, 478–499.
Kirichok, Y., Navarro, B., & Clapham, D. E. (2006). Whole-cell patch-clamp measurements of spermatozoa reveal an alkaline-activated Ca2+ channel. *Nature, 439*, 737–740.
Konig, J., Schreiber, R., Voelcker, T., Mall, M., & Kunzelmann, K. (2001). The cystic fibrosis transmembrane conductance regulator (CFTR) inhibits ENaC through an increase in the intracellular Cl- concentration. *EMBO Reports, 2*, 1047–1051.
Kunzelmann, K., & Schreiber, R. (1999). CFTR, a regulator of channels. *The Journal of Membrane Biology, 168*, 1–8.
Labarca, P., Santi, C., Zapata, O., Morales, E., Beltr'an, C., Lievano, A., et al. (1996). A cAMP regulated K+-selective channel from the sea urchin sperm plasma membrane. *Developmental Biology, 174*, 271–280.
Leonetti, M. D., Yuan, P., Hsiung, Y., & Mackinnon, R. (2012). Functional and structural analysis of the human SLO3 pH- and voltage-gated K+ channel. *Proceedings of the National Academy of Sciences of the United States of America.* In press, PMID: 23129643.
Li, C. Y., Jiang, L. Y., Chen, W. Y., Li, K., Sheng, H. Q., Ni, Y., et al. (2010). CFTR is essential for sperm fertilizing capacity and is correlated with sperm quality in humans. *Human Reproduction, 25*, 317–327.
Lievano, A., Sanchez, J. A., & Darszon, A. (1985). Single-channel activity of bilayers derived from sea urchin sperm plasma membranes at the tip of a patch-clamp electrode. *Developmental Biology, 112*, 253–257.
Lievano, A., Santi, C. M., Serrano, C. J., Trevino, C. L., Bellve, A. R., Hernandez-Cruz, A., et al. (1996). T-type Ca2+ channels and alpha1E expression in spermatogenic cells, and their possible relevance to the sperm acrosome reaction. *FEBS Letters, 388*, 150–154.
Lindemann, C. B., & Goltz, J. S. (1988). Calcium regulation of flagellar curvature and swimming pattern in triton X-100—Extracted rat sperm. *Cell Motility and the Cytoskeleton, 10*, 420–431.
Lishko, P. V., Botchkina, I. L., & Kirichok, Y. (2011). Progesterone activates the principal Ca2+ channel of human sperm. *Nature, 471*, 387–391.
Lishko, P. V., Kirichok, Y., Ren, D., Navarro, B., Chung, J. J., & Clapham, D. E. (2012). The control of male fertility by spermatozoan ion channels. *Annual Review of Physiology, 74*, 453–475.
Litscher, E. S., Williams, Z., & Wassarman, P. M. (2009). Zona pellucida glycoprotein ZP3 and fertilization in mammals. *Molecular Reproduction and Development, 76*, 933–941.
Llanos, M. N., Ronco, A. M., Aguirre, M. C., & Meizel, S. (2001). Hamster sperm glycine receptor: Evidence for its presence and involvement in the acrosome reaction. *Molecular Reproduction and Development, 58*, 205–215.
Lybaert, P., Leleux, F., Meuris, S., & Lebrun, P. (2010). Evidence for KATP channel involvement in calcium influx in murine spermatozoa. In *11th international symposium on spermatology*, 95 pp.
Lybaert, P., Vanbellinghen, A. M., Quertinmont, E., Petein, M., Meuris, S., & Lebrun, P. (2008). KATP channel subunits are expressed in the epididymal epithelium in several mammalian species. *Biology of Reproduction, 79*, 253–261.
Macdonald, R. L., & Olsen, R. W. (1994). GABAA receptor channels. *Annual Review of Neuroscience, 17*, 569–602.

Mannhold, R. (2004). KATP channel openers: Structure-activity relationships and therapeutic potential. *Medicinal Research Reviews*, *24*, 213–266.

Martinez-Lopez, P., Santi, C. M., Trevino, C. L., Ocampo-Gutierrez, A. Y., Acevedo, J. J., Alisio, A., et al. (2009). Mouse sperm K+ currents stimulated by pH and cAMP possibly coded by Slo3 channels. *Biochemical and Biophysical Research Communications*, *381*, 204–209.

Martinez-Lopez, P., Trevino, C. L., de la Vega-Beltran, J. L., Blas, G. D., Monroy, E., Beltran, C., et al. (2011). TRPM8 in mouse sperm detects temperature changes and may influence the acrosome reaction. *Journal of Cellular Physiology*, *226*, 1620–1631.

Mayorga, L. S., Tomes, C. N., & Belmonte, S. A. (2007). Acrosomal exocytosis, a special type of regulated secretion. *IUBMB Life*, *59*, 286–292.

Medina, J. F., Recalde, S., Prieto, J., Lecanda, J., Saez, E., Funk, C. D., et al. (2003). Anion exchanger 2 is essential for spermiogenesis in mice. *Proceedings of the National Academy of Sciences of the United States of America*, *100*, 15847–15852.

Meizel, S. (1997). Amino acid neurotransmitter receptor/chloride channels of mammalian sperm and the acrosome reaction. *Biology of Reproduction*, *56*, 569–574.

Meizel, S., & Son, J. H. (2005). Studies of sperm from mutant mice suggesting that two neurotransmitter receptors are important to the zona pellucida-initiated acrosome reaction. *Molecular Reproduction and Development*, *72*, 250–258.

Meizel, S., & Turner, K. O. (1996). Chloride efflux during the progesterone-initiated human sperm acrosome reaction is inhibited by lavendustin A, a tyrosine kinase inhibitor. *Journal of Andrology*, *17*, 327–330.

Melendrez, C. S., & Meizel, S. (1995). Studies of porcine and human sperm suggesting a role for a sperm glycine receptor/Cl- channel in the zona pellucida-initiated acrosome reaction. *Biology of Reproduction*, *53*, 676–683.

Mewe, M., Wulfsen, I., Schuster, A. M., Middendorff, R., Glassmeier, G., Schwarz, J. R., et al. (2008). Erg K+ channels modulate contractile activity in the bovine epididymal duct. *American Journal of Physiology. Regulatory, Integrative and Comparative Physiology*, *294*, R895–R904.

Miki, K., Qu, W., Goulding, E. H., Willis, W. D., Bunch, D. O., Strader, L. F., et al. (2004). Glyceraldehyde 3-phosphate dehydrogenase-S, a sperm-specific glycolytic enzyme, is required for sperm motility and male fertility. *Proceedings of the National Academy of Sciences of the United States of America*, *101*, 16501–16506.

Morales, E., de la Torre, L., Moy, G. W., Vacquier, V. D., & Darszon, A. (1993). Anion channels in the sea urchin sperm plasma membrane. *Molecular Reproduction and Development*, *36*, 174–182.

Moreno, C., & Vaca, L. (2011). SOC and now also SIC: Store-operated and store-inhibited channels. *IUBMB Life*, *63*, 856–863.

Munoz-Garay, C., De la Vega-Beltran, J. L., Delgado, R., Labarca, P., Felix, R., & Darszon, A. (2001). Inwardly rectifying K(+) channels in spermatogenic cells: Functional expression and implication in sperm capacitation. *Developmental Biology*, *234*, 261–274.

Navarro, B., Kirichok, Y., & Clapham, D. E. (2007). KSper, a pH-sensitive K+ current that controls sperm membrane potential. *Proceedings of the National Academy of Sciences of the United States of America*, *104*, 7688–7692.

Navarro, B., Miki, K., & Clapham, D. E. (2011). ATP-activated P2X2 current in mouse spermatozoa. *Proceedings of the National Academy of Sciences of the United States of America*, *108*, 14342–14347.

Nilius, B., & Droogmans, G. (2003). Amazing chloride channels: An overview. *Acta Physiologica Scandinavica*, *177*, 119–147.

Nojimoto, F. D., Piffer, R. C., Kiguti, L. R., Lameu, C., de Camargo, A. C., Pereira, O. C., et al. (2009). Multiple effects of sibutramine on ejaculation and on vas deferens and seminal vesicle contractility. *Toxicology and Applied Pharmacology*, *239*, 233–240.

Nolan, M. A., Babcock, D. F., Wennemuth, G., Brown, W., Burton, K. A., & McKnight, G. S. (2004). Sperm-specific protein kinase A catalytic subunit Calpha2 orchestrates cAMP signaling for male fertility. *Proceedings of the National Academy of Sciences of the United States of America*, *101*, 13483–13488.

Okamura, N., Tajima, Y., Soejima, A., Masuda, H., & Sugita, Y. (1985). Sodium bicarbonate in seminal plasma stimulates the motility of mammalian spermatozoa through direct activation of adenylate cyclase. *The Journal of Biological Chemistry*, *260*, 9699–9705.

Okunade, G. W., Miller, M. L., Pyne, G. J., Sutliff, R. L., O'Connor, K. T., Neumann, J. C., et al. (2004). Targeted ablation of plasma membrane Ca2+-ATPase (PMCA) 1 and 4 indicates a major housekeeping function for PMCA1 and a critical role in hyperactivated sperm motility and male fertility for PMCA4. *The Journal of Biological Chemistry*, *279*, 33742–33750.

Orta, G., Ferreira, G., Jose, O., Treviño, C. L., Beltran, C., & Darszon, A. (2012). Human spermatozoa possess a calcium-dependent chloride channel that may participate in the acrosomal reaction. *The Journal of Physiology*, *590*, 2659–2675.

Ottolia, M., & Toro, L. (1994). Potentiation of large conductance KCa channels by niflumic, flufenamic, and mefenamic acids. *Biophysical Journal*, *67*, 2272–2279.

Pace, A. J., Lee, E., Athirakul, K., Coffman, T. M., O'Brien, D. A., & Koller, B. H. (2000). Failure of spermatogenesis in mouse lines deficient in the Na(+)-K(+)-2Cl(-) cotransporter. *The Journal of Clinical Investigation*, *105*, 441–450.

Perez-Cornejo, P., & Arreola, J. (2004). Regulation of Ca(2+)-activated chloride channels by cAMP and CFTR in parotid acinar cells. *Biochemical and Biophysical Research Communications*, *316*, 612–617.

Pifferi, S., Dibattista, M., & Menini, A. (2009). TMEM16B induces chloride currents activated by calcium in mammalian cells. *Pflügers Archiv*, *458*, 1023–1038.

Podlaha, O., Webb, D. M., Tucker, P. K., & Zhang, J. (2005). Positive selection for indel substitutions in the rodent sperm protein catsper1. *Molecular Biology and Evolution*, *22*, 1845–1852.

Publicover, S. J., & Barratt, C. L. (2012). Chloride channels join the sperm 'channelome'. *The Journal of Physiology*, *590*, 2553–2554.

Publicover, S., Harper, C. V., & Barratt, C. (2007). [Ca2+]i signalling in sperm—Making the most of what you've got. *Nature Cell Biology*, *9*, 235–242.

Quill, T. A., Ren, D., Clapham, D. E., & Garbers, D. L. (2001). A voltage-gated ion channel expressed specifically in spermatozoa. *Proceedings of the National Academy of Sciences of the United States of America*, *98*, 12527–12531.

Ren, D., Navarro, B., Perez, G., Jackson, A. C., Hsu, S., Shi, Q., et al. (2001). A sperm ion channel required for sperm motility and male fertility. *Nature*, *413*, 603–609.

Ren, D., & Xia, J. (2010). Calcium signaling through CatSper channels in mammalian fertilization. *Physiology (Bethesda, Md.)*, *25*, 165–175.

Ritta, M. N., Bas, D. E., & Tartaglione, C. M. (2004). In vitro effect of gamma-aminobutyric acid on bovine spermatozoa capacitation. *Molecular Reproduction and Development*, *67*, 478–486.

Ritta, M. N., Calamera, J. C., & Bas, D. E. (1998). Occurrence of GABA and GABA receptors in human spermatozoa. *Molecular Human Reproduction*, *4*, 769–773.

Roldan, E. R., Murase, T., & Shi, Q. X. (1994). Exocytosis in spermatozoa in response to progesterone and zona pellucida. *Science*, *266*, 1578–1581.

Rossato, M., Di Virgilio, F., Rizzuto, R., Galeazzi, C., & Foresta, C. (2001). Intracellular calcium store depletion and acrosome reaction in human spermatozoa: Role of calcium and plasma membrane potential. *Molecular Human Reproduction*, *7*, 119–128.

Ruknudin, A., & Silver, I. A. (1990). Ca2+ uptake during capacitation of mouse spermatozoa and the effect of an anion transport inhibitor on Ca2+ uptake. *Molecular Reproduction and Development*, *26*, 63–68.

Russell, J. M. (2000). Sodium-potassium-chloride cotransport. *Physiological Reviews, 80*, 211–276.

Salvatore, L., D'Adamo, M. C., Polishchuk, R., Salmona, M., & Pessia, M. (1999). Localization and age-dependent expression of the inward rectifier K+ channel subunit Kir 5.1 in a mammalian reproductive system. *FEBS Letters, 449*, 146–152.

Santi, C. M., Butler, A., Kuhn, J., Wei, A., & Salkoff, L. (2009). Bovine and mouse SLO3 K+ channels: Evolutionary divergence points to an RCK1 region of critical function. *The Journal of Biological Chemistry, 284*, 21589–21598.

Santi, C. M., Darszon, A., & Hernandez-Cruz, A. (1996). A dihydropyridine-sensitive T-type Ca2+ current is the main Ca2+ current carrier in mouse primary spermatocytes. *The American Journal of Physiology, 271*, C1583–C1593.

Santi, C. M., Martinez-Lopez, P., de la Vega-Beltran, J. L., Butler, A., Alisio, A., Darszon, A., et al. (2010). The SLO3 sperm-specific potassium channel plays a vital role in male fertility. *FEBS Letters, 584*, 1041–1046.

Sardini, A., Amey, J. S., Weylandt, K. H., Nobles, M., Valverde, M. A., & Higgins, C. F. (2003). Cell volume regulation and swelling-activated chloride channels. *Biochimica et Biophysica Acta, 1618*, 153–162.

Sato, Y., Son, J. H., & Meizel, S. (2000). The mouse sperm glycine receptor/chloride channel: Cellular localization and involvement in the acrosome reaction initiated by glycine. *Journal of Andrology, 21*, 99–106.

Schreiber, M., Wei, A., Yuan, A., Gaut, J., Saito, M., & Salkoff, L. (1998). Slo3, a novel pH-sensitive K+ channel from mammalian spermatocytes. *The Journal of Biological Chemistry, 273*, 3509–3516.

Schroeder, B. C., Cheng, T., Jan, Y. N., & Jan, L. Y. (2008). Expression cloning of TMEM16A as a calcium-activated chloride channel subunit. *Cell, 134*, 1019–1029.

Scudieri, P., Sondo, E., Ferrera, L., & Galietta, L. J. (2012). The anoctamin family: TMEM16A and TMEM16B as calcium-activated chloride channels. *Experimental Physiology, 97*, 177–183.

Seino, S. (1999). ATP-sensitive potassium channels: A model of heteromultimeric potassium channel/receptor assemblies. *Annual Review of Physiology, 61*, 337–362.

Sheppard, D. N., & Welsh, M. J. (1999). Structure and function of the CFTR chloride channel. *Physiological Reviews, 79*, S23–S45.

Shi, Q. X., & Roldan, E. R. (1995). Evidence that a GABAA-like receptor is involved in progesterone-induced acrosomal exocytosis in mouse spermatozoa. *Biology of Reproduction, 52*, 373–381.

Shi, Q. X., Yuan, Y. Y., & Roldan, E. R. (1997). gamma-Aminobutyric acid (GABA) induces the acrosome reaction in human spermatozoa. *Molecular Human Reproduction, 3*, 677–683.

Sieghart, W. (1995). Structure and pharmacology of gamma-aminobutyric acidA receptor subtypes. *Pharmacological Reviews, 47*, 181–234.

Sones, W. R., Leblanc, N., & Greenwood, I. A. (2009). Inhibition of vascular calcium-gated chloride currents by blockers of KCa1.1, but not by modulators of KCa2.1 or KCa2.3 channels. *British Journal of Pharmacology, 158*, 521–531.

Stamboulian, S., Kim, D., Shin, H. S., Ronjat, M., De Waard, M., & Arnoult, C. (2004). Biophysical and pharmacological characterization of spermatogenic T-type calcium current in mice lacking the CaV3.1 (alpha1G) calcium channel: CaV3.2 (alpha1H) is the main functional calcium channel in wild-type spermatogenic cells. *Journal of Cellular Physiology, 200*, 116–124.

Strunker, T., Goodwin, N., Brenker, C., Kashikar, N. D., Weyand, I., Seifert, R., et al. (2011). The CatSper channel mediates progesterone-induced Ca2+ influx in human sperm. *Nature, 471*, 382–386.

Stutts, M. J., Canessa, C. M., Olsen, J. C., Hamrick, M., Cohn, J. A., Rossier, B. C., et al. (1995). CFTR as a cAMP-dependent regulator of sodium channels. *Science, 269*, 847–850.

Suarez, S. S. (2008). Regulation of sperm storage and movement in the mammalian oviduct. *The International Journal of Developmental Biology, 52*, 455–462.

Suarez, S. S., Varosi, S. M., & Dai, X. (1993). Intracellular calcium increases with hyperactivation in intact, moving hamster sperm and oscillates with the flagellar beat cycle. *Proceedings of the National Academy of Sciences of the United States of America, 90*, 4660–4664.

Swanson, W. J., & Vacquier, V. D. (2002). The rapid evolution of reproductive proteins. *Nature Reviews. Genetics, 3*, 137–144.

Tang, Q. Y., Zhang, Z., Xia, J., Ren, D., & Logothetis, D. E. (2010). Phosphatidylinositol 4,5-bisphosphate activates Slo3 currents and its hydrolysis underlies the epidermal growth factor-induced current inhibition. *The Journal of Biological Chemistry, 285*, 19259–19266.

Toma, C., Greenwood, I. A., Helliwell, R. M., & Large, W. A. (1996). Activation of potassium currents by inhibitors of calcium-activated chloride conductance in rabbit portal vein smooth muscle cells. *British Journal of Pharmacology, 118*, 513–520.

Torgerson, D. G., Kulathinal, R. J., & Singh, R. S. (2002). Mammalian sperm proteins are rapidly evolving: Evidence of positive selection in functionally diverse genes. *Molecular Biology and Evolution, 19*, 1973–1980.

Travis, A. J., & Kopf, G. S. (2002). The role of cholesterol efflux in regulating the fertilization potential of mammalian spermatozoa. *The Journal of Clinical Investigation, 110*, 731–736.

Trevino, C. L., Felix, R., Castellano, L. E., Gutierrez, C., Rodriguez, D., Pacheco, J., et al. (2004). Expression and differential cell distribution of low-threshold Ca(2+) channels in mammalian male germ cells and sperm. *FEBS Letters, 563*, 87–92.

Turner, T. T. (2008). De Graaf's thread: The human epididymis. *Journal of Andrology, 29*, 237–250.

Turner, K. O., Garcia, M. A., & Meizel, S. (1994). Progesterone initiation of the human sperm acrosome reaction: The obligatory increase in intracellular calcium is independent of the chloride requirement. *Molecular and Cellular Endocrinology, 101*, 221–225.

Visconti, P. E., Bailey, J. L., Moore, G. D., Pan, D., Olds-Clarke, P., & Kopf, G. S. (1995). Capacitation of mouse spermatozoa. I. Correlation between the capacitation state and protein tyrosine phosphorylation. *Development, 121*, 1129–1137.

Visconti, P. E., & Florman, H. M. (2010). Mechanisms of sperm-egg interactions: Between sugars and broken bonds. *Science Signaling, 3*, pe35.

Visconti, P. E., Galantino-Homer, H., Ning, X., Moore, G. D., Valenzuela, J. P., Jorgez, C. J., et al. (1999). Cholesterol efflux-mediated signal transduction in mammalian sperm. beta-cyclodextrins initiate transmembrane signaling leading to an increase in protein tyrosine phosphorylation and capacitation. *The Journal of Biological Chemistry, 274*, 3235–3242.

Visconti, P. E., Krapf, D., de la Vega-Beltran, J. L., Acevedo, J. J., & Darszon, A. (2011). Ion channels, phosphorylation and mammalian sperm capacitation. *Asian Journal of Andrology, 13*, 395–405.

Visconti, P. E., Moore, G. D., Bailey, J. L., Leclerc, P., Connors, S. A., Pan, D., et al. (1995). Capacitation of mouse spermatozoa. II. Protein tyrosine phosphorylation and capacitation are regulated by a cAMP-dependent pathway. *Development, 121*, 1139–1150.

Visconti, P. E., Muschietti, J. P., Flawia, M. M., & Tezon, J. G. (1990). Bicarbonate dependence of cAMP accumulation induced by phorbol esters in hamster spermatozoa. *Biochimica et Biophysica Acta, 1054*, 231–236.

Wang, D., Hu, J., Bobulescu, I. A., Quill, T. A., McLeroy, P., Moe, O. W., et al. (2007). A sperm-specific Na+/H+ exchanger (sNHE) is critical for expression and in vivo bicarbonate regulation of the soluble adenylyl cyclase (sAC). *Proceedings of the National Academy of Sciences of the United States of America, 104*, 9325–9330.

Wennemuth, G., Carlson, A. E., Harper, A. J., & Babcock, D. F. (2003). Bicarbonate actions on flagellar and Ca2+ -channel responses: Initial events in sperm activation. *Development, 130*, 1317–1326.

Wertheimer, E. V., Salicioni, A. M., Liu, W., Trevino, C. L., Chavez, J., Hernandez-Gonzalez, E. O., et al. (2008). Chloride is essential for capacitation and for the

capacitation-associated increase in tyrosine phosphorylation. *The Journal of Biological Chemistry*, *283*, 35539–35550.
Weyand, I., Godde, M., Frings, S., Weiner, J., Muller, F., Altenhofen, W., et al. (1994). Cloning and functional expression of a cyclic-nucleotide-gated channel from mammalian sperm. *Nature*, *368*, 859–863.
Wistrom, C. A., & Meizel, S. (1993). Evidence suggesting involvement of a unique human sperm steroid receptor/Cl- channel complex in the progesterone-initiated acrosome reaction. *Developmental Biology*, *159*, 679–690.
Wu, W. L., So, S. C., Sun, Y. P., Chung, Y. W., Grima, J., Wong, P. Y., et al. (1998). Functional expression of P2U receptors in rat spermatogenic cells: Dual modulation of a Ca (2+)-activated K+ channel. *Biochemical and Biophysical Research Communications*, *248*, 728–732.
Wyckoff, G. J., Wang, W., & Wu, C. I. (2000). Rapid evolution of male reproductive genes in the descent of man. *Nature*, *403*, 304–309.
Xia, J., Reigada, D., Mitchell, C. H., & Ren, D. (2007). CATSPER channel-mediated Ca2+ entry into mouse sperm triggers a tail-to-head propagation. *Biology of Reproduction*, *77*, 551–559.
Xia, J., & Ren, D. (2009). The BSA-induced Ca2+ influx during sperm capacitation is CATSPER channel-dependent. *Reproductive Biology and Endocrinology*, 7, 119.
Xie, F., Garcia, M. A., Carlson, A. E., Schuh, S. M., Babcock, D. F., Jaiswal, B. S., et al. (2006). Soluble adenylyl cyclase (sAC) is indispensable for sperm function and fertilization. *Developmental Biology*, *296*, 353–362.
Xu, W. M., Shi, Q. X., Chen, W. Y., Zhou, C. X., Ni, Y., Rowlands, D. K., et al. (2007). Cystic fibrosis transmembrane conductance regulator is vital to sperm fertilizing capacity and male fertility. *Proceedings of the National Academy of Sciences of the United States of America*, *104*, 9816–9821.
Yanagimachi, R. (1994). Mammalian fertilization. In E. Knobile & J. D. Neill (Eds.), *The physiology of reproduction* (pp. 189–317). New York: Raven Press.
Yanagimachi, R. (1998). Intracytoplasmic sperm injection experiments using the mouse as a model. *Human Reproduction*, *13*(Suppl. 1), 87–98.
Yang, Y. D., Cho, H., Koo, J. Y., Tak, M. H., Cho, Y., Shim, W. S., et al. (2008). TMEM16A confers receptor-activated calcium-dependent chloride conductance. *Nature*, *455*, 1210–1215.
Yeung, C. H., Barfield, J. P., & Cooper, T. G. (2005). Chloride channels in physiological volume regulation of human spermatozoa. *Biology of Reproduction*, *73*, 1057–1063.
Yeung, C. H., Barfield, J. P., & Cooper, T. G. (2006). Physiological volume regulation by spermatozoa. *Molecular and Cellular Endocrinology*, *250*, 98–105.
Yeung, C. H., & Cooper, T. G. (2001). Effects of the ion-channel blocker quinine on human sperm volume, kinematics and mucus penetration, and the involvement of potassium channels. *Molecular Human Reproduction*, 7, 819–828.
Zanetti, N., & Mayorga, L. S. (2009). Acrosomal swelling and membrane docking are required for hybrid vesicle formation during the human sperm acrosome reaction. *Biology of Reproduction*, *81*, 396–405.
Zeng, Y., Clark, E. N., & Florman, H. M. (1995). Sperm membrane potential: Hyperpolarization during capacitation regulates zona pellucida-dependent acrosomal secretion. *Developmental Biology*, *171*, 554–563.
Zeng, X. H., Yang, C., Kim, S. T., Lingle, C. J., & Xia, X. M. (2011). Deletion of the Slo3 gene abolishes alkalization-activated K+ current in mouse spermatozoa. *Proceedings of the National Academy of Sciences of the United States of America*, *108*, 5879–5884.
Zhou, M., He, H. J., Tanaka, O., Sekiguchi, M., Kawahara, K., & Abe, H. (2011). Different localization of ATP sensitive K+ channel subunits in rat testis. *Anatomical Record (Hoboken)*, *294*, 729–737.

INDEX

Note: Page numbers followed by "*f*" indicate figures, and "*t*" indicate tables.

D

H

I

J

K

N

R

S

T

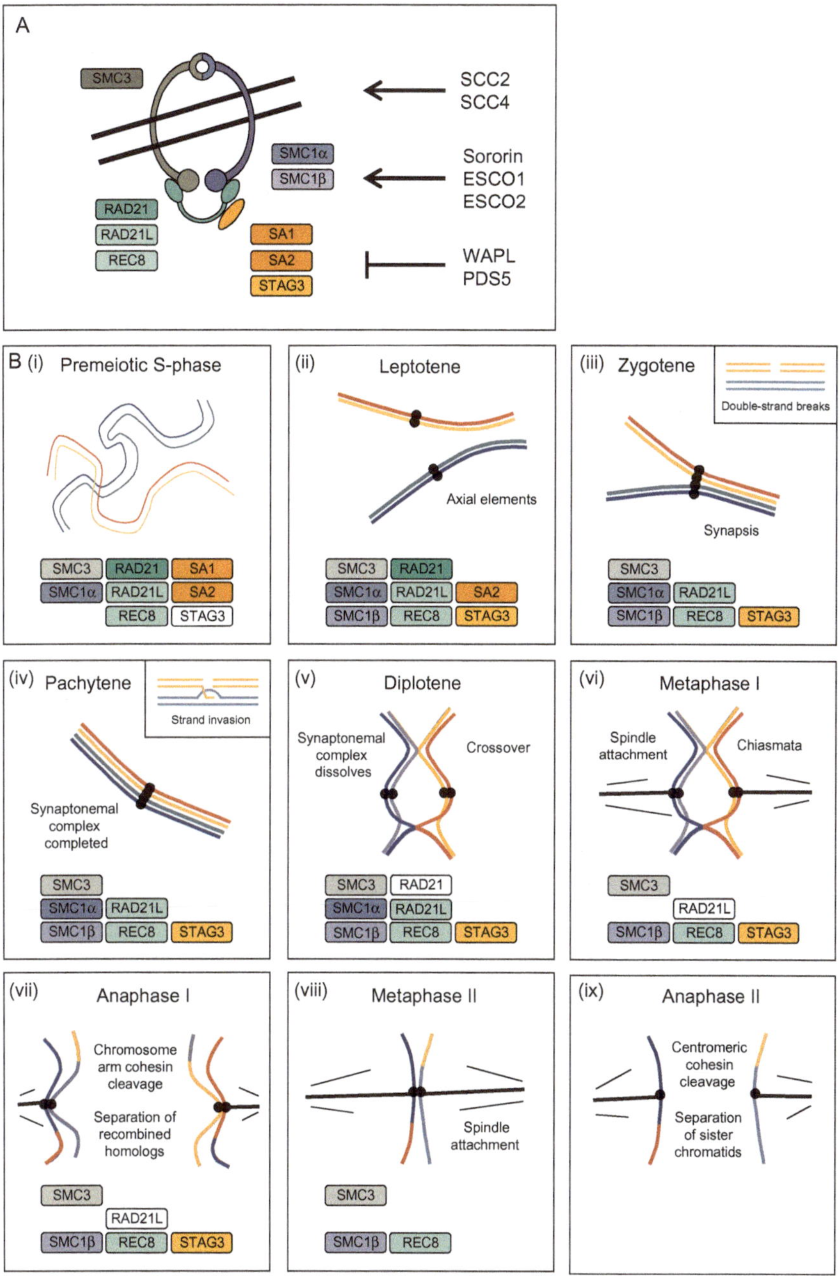

François McNicoll *et al.*, Figure 1.1 Cohesin in meiosis. (A) Model of the cohesin ring encircling two chromatids. SMC3 (gray) is present in all cohesin complexes. There are two SMC1 genes and proteins: SMC1α (dark blue) and a meiosis-specific SMC1β (light blue). The tripartite ring closes via the association of an α-kleisin subunit, of which three variants exist: the ubiquitous RAD21 (dark turquoise), and two meiosis-specific forms, REC8 and RAD21L (light turquoise). A third component, of which there are three variants, associates with the complex via binding to the α-kleisin: canonical SA1 or SA2 (dark orange) or the meiosis-specific STAG3 (SA3) (light orange). Loading of the cohesin complex onto chromosomes and its maintenance on chromosomes are controlled by loading factors, establishment and antiestablishment factors. Loading of the cohesin complex onto mitotic chromosomes is performed by a complex of SCC2–SCC4 (kollerin) and cohesin dissociation by PDS5-WAPL (releasin). Cohesin acetyltransferases (ESCO1 and ESCO2) are necessary to establish cohesion in mitotic S-phase through acetylation of SMC3, which recruits Sororin, a maintenance factor that counteracts releasin activity during mitotic S and G2 phases. (B) Scheme of the meiotic stages from (i) to (ix), showing the progression of one pair of homologous chromosomes (one red and the other blue, each drawn as two single lines representing sister chromatids without chromatin loops for illustration purposes) through the different stages. In reality, the progression is continuous. The presence of specific cohesin proteins as far as known is presented in the lower sections of the panels (neither the relative amounts nor the potential interactions between the subunits are taken into consideration). There are conflicting reports on the presence of cohesin proteins at some of the stages. In such cases, the cohesins are depicted in noncolored boxes. There is very little information about SA1. The data illustrated here are based on analyses of spermatocytes. (i) During premeiotic S-phase, newly formed sister chromatids (red or blue) are held together by cohesin complexes (not shown). The mitotic cohesin subunits are present. The meiosis-specific SMC1β is not yet present; however, REC8, RAD21L as well as perhaps STAG3 in a few cells are already beginning to be expressed; (ii) during leptotene, the chromosomes begin to condense and axial elements form, STAG3 and SMC1β are now present on the chromosomes; (iii) synapsis of homologous chromosomes begins during zygotene, facilitated by frequent DNA DSBs, of which one is represented in the inset, DSBs are initiated in leptotene (iv) formation of the SC is complete in pachytene with all homologs fully synapsed, meiotic recombination proceeds as indicated in the inset; (v) crossovers (two examples are shown) which have been formed between homologs during pachytene, physically link the homologs together in diplotene. At this stage, the SC has largely dissolved; however, sister chromatid cohesion is maintained. Oocytes will arrest shortly after this stage, at a stage called dictyate arrest (not shown)—for many years in humans—and cohesion must be maintained during this time; (vi) at metaphase I, spindle attachments are formed at mono-oriented centromeres of homologs and chiasmata still resist microtubule pulling forces; (vii) cleavage of the α-kleisin subunit of cohesin by separase results in separation of homologs as chiasmata resolve in absence of arm cohesion. Centromeric cohesion is protected by Shugoshin/PP2A (not shown) and a pool of dephosphorylated cohesin subunits; (viii) sister chromatids align on the metaphase plate during metaphase II and spindle microtubules attach to bioriented kinetochores; (ix) cohesion is lost and sister chromatids are pulled apart in anaphase II, creating haploid gametes.

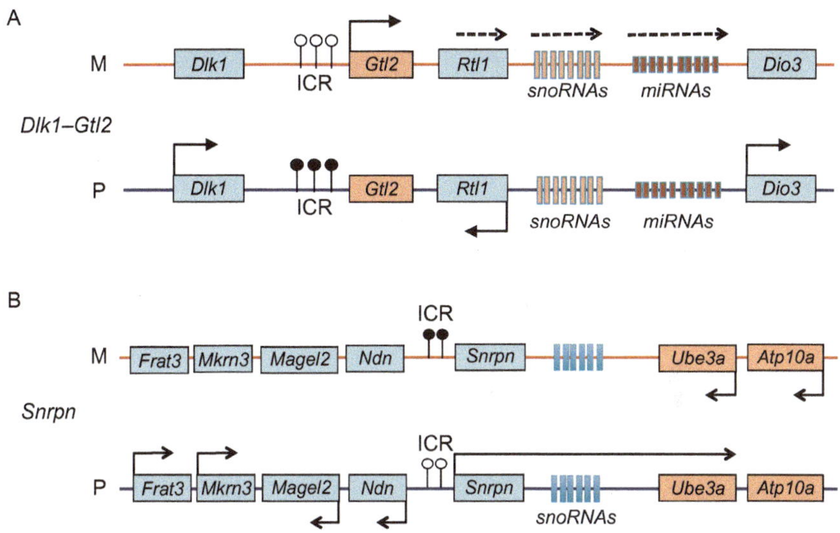

Xiajun Li, Figure 2.4 Diagrams are for the *Dlk1–Gtl2* and *Snrpn* imprinted regions. The relative distances for these genes are not drawn to scale. (A) The imprinted *Dlk1–Gtl2* region. There is an ICR located between the imprinted *Dlk1* gene and the imprinted *Gtl2* gene. This ICR is methylated (filled lollipops) on the paternal (P) chromosome but not methylated (unfilled lollipops) on the maternal (M) chromosome. When this ICR is methylated, the imprinted *Dlk1*, *Rtl1*, and *Dio3* genes are expressed from the paternal chromosome. These are all protein-coding genes. In contrast, the imprinted noncoding RNA *Gtl2* gene, many clustered *snoRNA* genes and *miRNA* genes, as well as the antisense transcript for *Rtl1* are expressed from the maternal chromosome that contains an unmethylated ICR. The line and the arrow above the genes indicate the transcriptional start site and direction of the transcriptional elongation. The dashed arrow indicates the transcriptional direction for the noncoding RNA transcript with uncertain transcriptional start sites. (B) The imprinted *Snrpn* region. There is an ICR near the *Snrpn* imprinted gene. When this ICR is methylated (filled lollipops) on the maternal (M) chromosome, *Ube3a* and *Atp10* are only expressed from the maternal chromosome. In contrast, the ICR on the paternal (P) chromosome is not methylated (unfilled lollipops). A long transcript is initiated at the beginning of the imprinted *Snrpn* gene and elongated on this imprinted region to cover a cluster of *snoRNA* genes. This transcript is extended to the imprinted *Ube3a* region on the paternal chromosome that is in the antisense orientation to the *Ube3a* transcript that is normally expressed from the maternal chromosome in brain. In addition, the imprinted genes including *Frat3*, *Mkrn3*, *Magel2*, and *Ndn* will also be expressed from the paternal chromosome but not from the maternal chromosome. The line above the genes indicates the transcriptional start site. The arrow points to the direction of the transcriptional elongation.

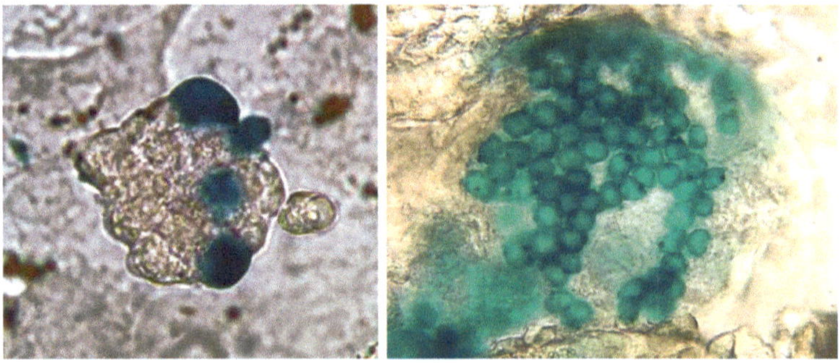

Makoto C. Nagano and Jonathan R. Yeh, Figure 3.4 The germ cell community *in vitro* (A) and *in vivo* (B). (A) A spermatogonial aggregate that emerges in the presence of GDNF and FGF2. SSCs represent a small subpopulation of these cells. Blue staining (LacZ) indicates activation of Wnt/β-catenin signaling. (B) Aggregating germ cells observed in recipient testes 2 weeks after transplantation. Cells are stained for β-galactosidase activity.

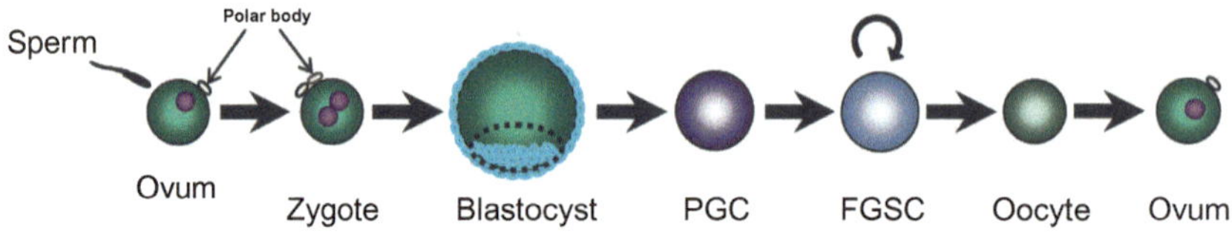

Ji Wu *et al.*, Figure 4.2 Origin and differentiation of FGSCs. PGC, primordial germ cell.

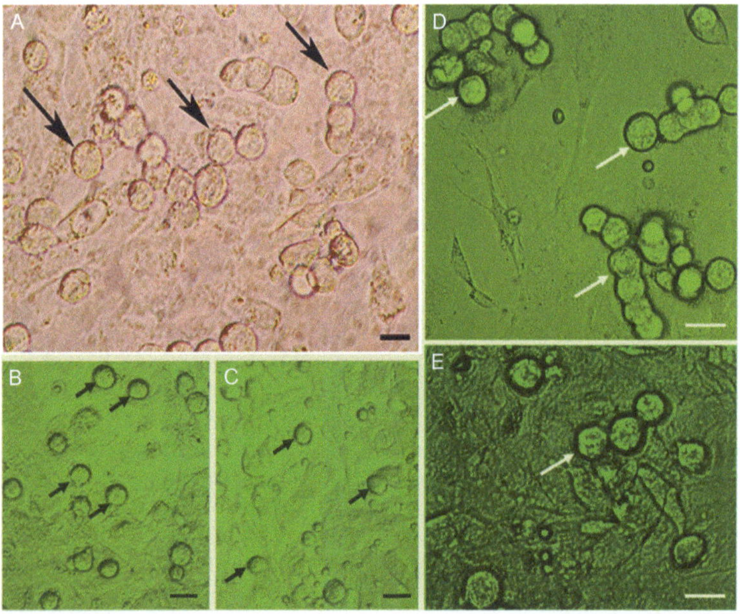

Ji Wu *et al.*, Figure 4.3 Characteristic of FGSC is similar to that of SSCs. (A) Morphology of adult mouse SSCs (arrows) cultured for 24 h after isolation (Yuan et al., 2009). (B, C) Morphology of FGSCs (arrows) from neonatal (B) and adult (C) ovaries cultured for 24 h after isolation. (D, E) The FGSCs isolated from neonatal (E) and adult (F) mice expanded and were undergoing mitosis (arrows) (Zhang, Yang, et al., 2011). Scale bars, 10 μm (A), 20 μm (B–E).

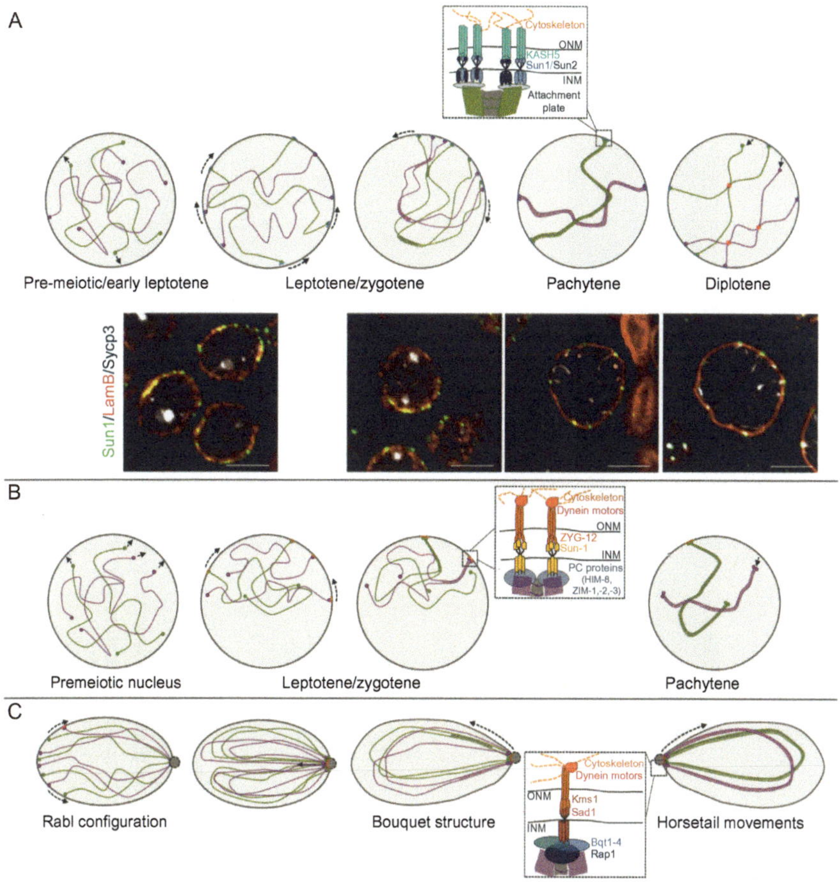

Martin P. Kracklauer *et al.*, Figure 5.2 Schematic representation of meiotic telomere dynamics in different model organisms. (A) The meiotic LINC complex (see inset) responsible for telomere attachment in the mouse is composed of Sun1 and Sun2 interacting with KASH5 in the outer nuclear membrane (ONM). Corresponding immunofluorescence images of mouse spermatocytes stained with Sun1, Lamin B, and Sycp3 show meiotic telomere dynamics *in vivo*. Sun1 signals are located at telomere ends of synaptonemal complexes (labeled by Sycp3) and lie within the NE labeled by lamin B. Scale bars 5 μm. (B) Meiotic telomere attachment in *C. elegans* (see inset) is mediated by the INM protein SUN-1 interacting with the KASH domain protein ZYG-12. (C) Attachment and clustering of telomeres at the spindle pole body in *S. pombe* (see inset) is mediated by the SUN domain protein Sad1 binding to the KASH domain protein Kms1.

A

B

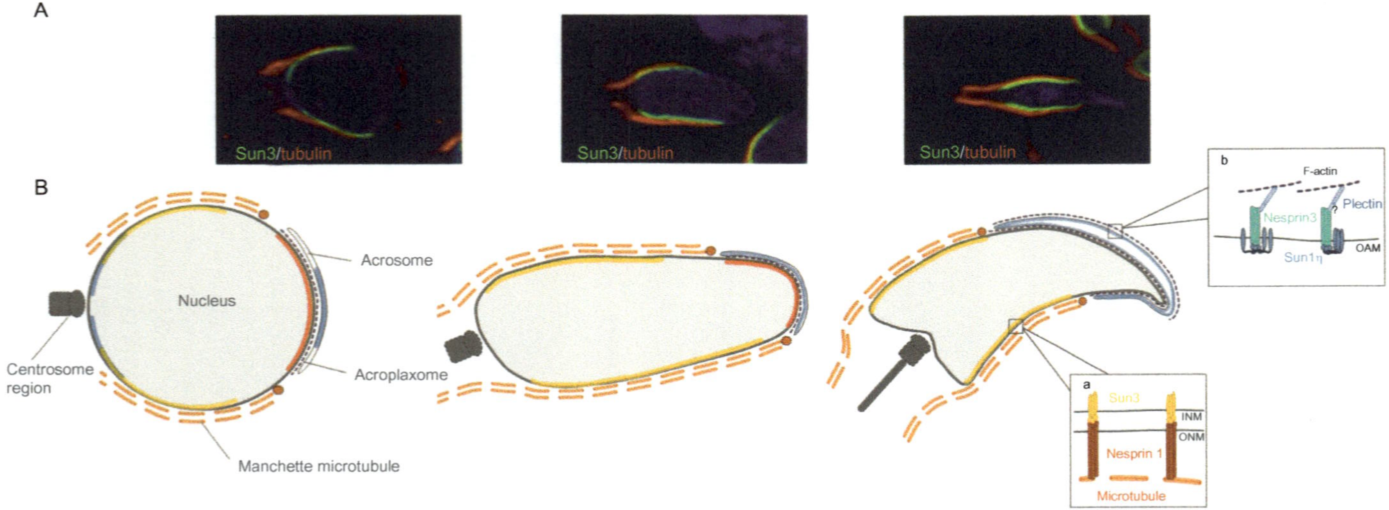

Martin P. Kracklauer *et al.*, Figure 5.3 Distribution of LINC complex proteins during mouse spermiogenesis. (A) Sun3 (in green) is localized within the NE only at sites, where the cytoplasmic manchette (in red) is in close contact with the NE. (B) Schematic representation of the redistribution of spermiogenic SUN domain proteins during nuclear elongation. Yellow: Sun3, blue: Sun1 and Sun1η, and red: Spag4L. Posterior and anterior spermiogenesis-specific LINC complexes are shown in insets a and b.

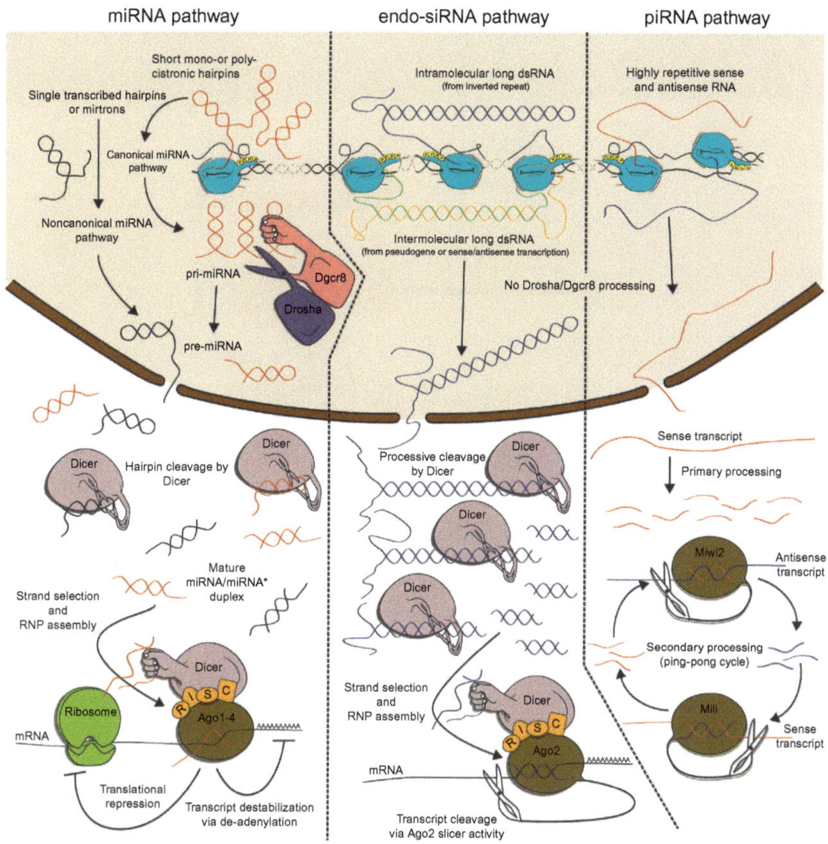

Matthew S. Cook and Robert Blelloch, Figure 6.2 Schematic representation of small RNA biogenesis and function in mammals. (*miRNA pathway*) Canonical miRNAs are transcribed by RNA polymerase II as mono- or poly-cistronic hairpins known as primary miRNA (pri-miRNA) transcripts, which are subsequently cleaved by the microprocessor (Drosha/Dgcr8 complex) into pre-miRNAs. Noncanonical miRNAs are transcribed directly as single short hairpins, or they can be derived from intronic regions of a host mRNA, by-passing Microprocessor cleavage or other unknown endonuclease activity. After export to the cytoplasm, the RNase III-containing enzyme Dicer cleaves the hairpin, producing the mature miRNA/miRNA* duplex. The Dicer/miRNA complex then associates with Argonaute, where it guides one strand of the miRNA/miRNA* duplex into RISC with the complementary mRNA based on its seed sequence match. Target repression is generally achieved by inhibition of the translational machinery and/or destabilization of the transcript via deadenylation of the mRNA. (*endo-siRNA pathway*) Long double-stranded RNAs (dsRNAs) are the precursors to siRNAs. Whether long dsRNAs form from intramolecular or intermolecular complementary binding, they are exported from the nucleus without processing. In the cytoplasm, Dicer binds to and cleaves the dsRNA in a processive fashion, generating short dsRNA duplexes 21 nucleotides (21-nt) in length. The Dicer/siRNA complex guides one strand of the siRNA duplex into the Ago2-containing RISC based on perfect complementarity with the appropriate mRNA. Target repression is thought to be achieved by cleavage of the target transcript by Ago2, but little is known about endo-siRNA/target interactions in mammals. (*piRNA pathway*) Highly repetitive

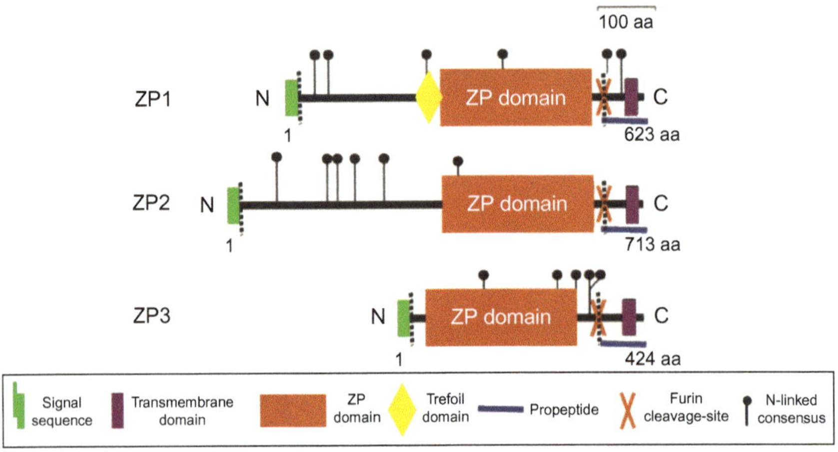

Paul M. Wassarman and Eveline S. Litscher, Figure 9.2 Schematic representation of the overall architecture of mouse ZP glycoproteins mZP1, mZP2, and mZP3. The polypeptide of each ZP glycoprotein is drawn to scale with the N- and C-termini indicated. Shown are the signal peptide (green), trefoil domain (yellow), ZP domain (red), consensus furin cleavage-site (red X), transmembrane domain (purple), and C-terminal propeptide region (blue bar). Putative N-linked glycosylation sites are shown (black lollipops). The number of amino acids in each polypeptide is indicated.

---

regions of the genome are transcribed to produce long sense and antisense RNAs. The sense transcript is exported, and a poorly understood primary processing event produces ~27-nt fragments that are then incorporated into a secondary processing event. Miwi2 forms a complex with the short sense RNAs bound to a long complementary antisense transcript, which is then cleaved into short RNAs that then associate with Mili to cleave more sense transcripts, producing a positive feedback loop known as the ping-pong cycle. The ultimate functional mechanism of piRNAs is poorly understood in mammals, but it is possible that PIWI-related proteins associate with piRNAs and recruit chromatin-modifying machinery to specific regions of the genome to promote epigenetic changes that repress expression of mobile genetic elements.

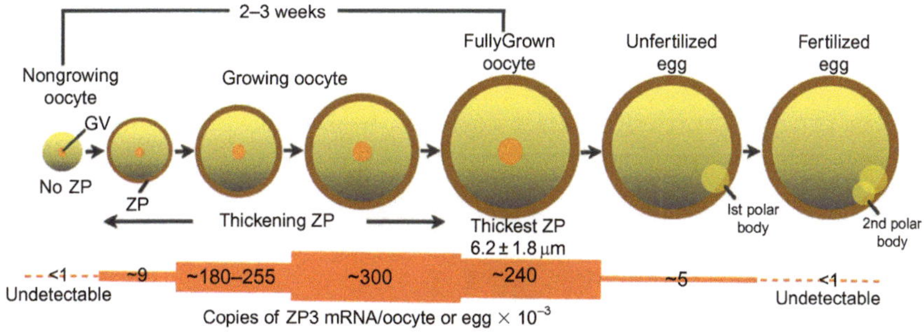

Paul M. Wassarman and Eveline S. Litscher, Figure 9.5 Schematic representation of ZP3 mRNA accumulation during oocyte growth, ovulation of eggs, and fertilization of eggs in the mouse. ZP3 mRNA is undetectable in non-growing oocytes, appears just as oocytes begin to grow, accumulates to high levels during the final stages of oocyte growth, decreases dramatically during ovulation of eggs, and is undetectable in fertilized eggs.

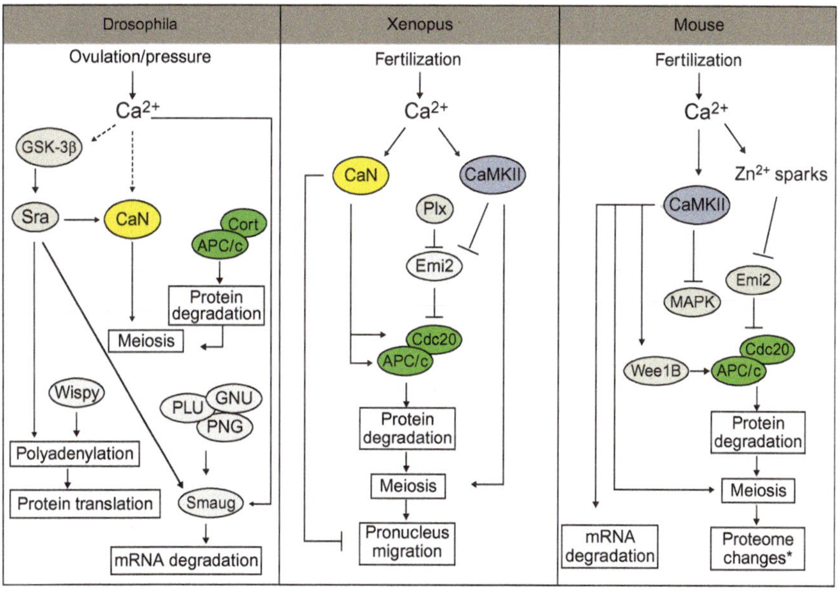

Amber R. Krauchunas and Mariana F. Wolfner, Figure 10.1 Egg activation pathways in Drosophila, Xenopus, and mouse showing only those aspects for which experimental evidence is reported in the literature. See text for references. *Knott et al. (2006) and Backs et al. (2010) show that the protein patterns observed by 2D- and 1D-gel electrophoresis in activated eggs are regulated by CaMKII and attribute these changes to maternal mRNA recruitment and protein synthesis. However, it cannot be ruled out that some of the observed changes are due to posttranslational modifications. As such, we have placed "proteome changes" in the pathway to encompass both of these possibilities.

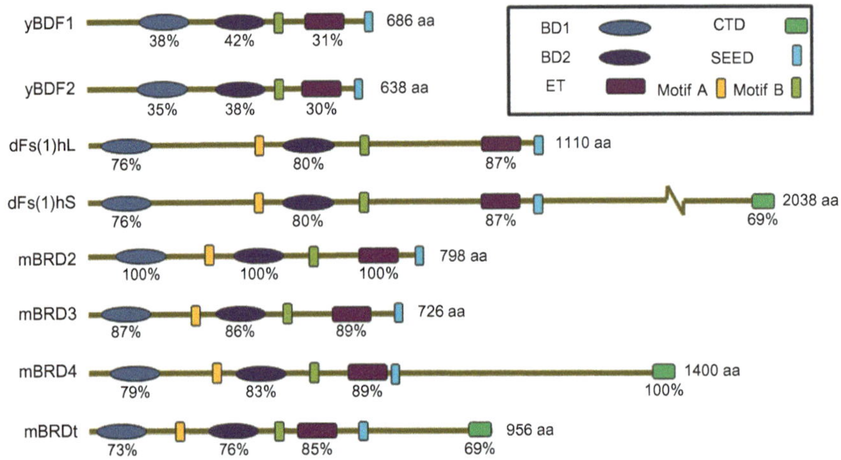

Binyamin D. Berkovits and Debra J. Wolgemuth, Figure 11.1 Structural comparison of the BET family members in *S. cerevisiae*, *D. melanogaster*, and *M. musculus*. The percentages under each bromodomain and the ET domain indicate the percentage of identical amino acid residues in the domain as compared to the corresponding domain in mBRD2, the founding member of the mouse BET family. The percentages under the CTD are compared to mBRD4.

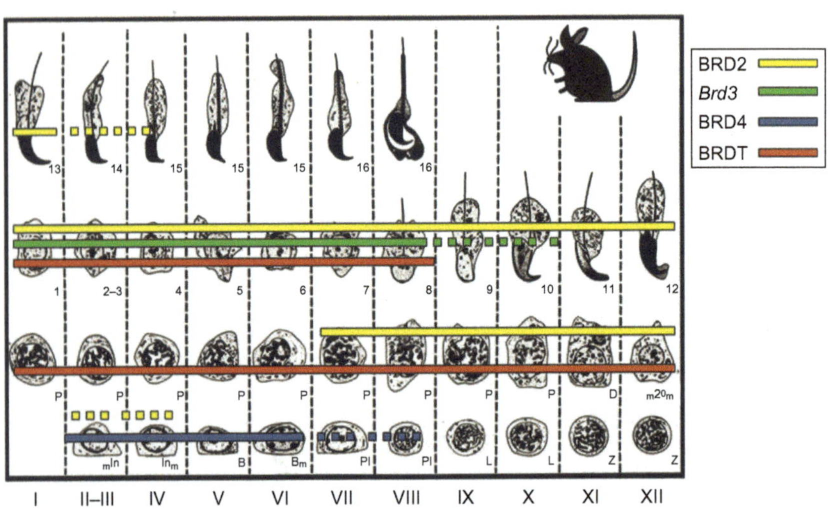

Binyamin D. Berkovits and Debra J. Wolgemuth, Figure 11.2 Expression patterns of the BET family in the adult mouse testis. Expression depicted for BRD2, BRD4, and BRDT is at the protein level, and for *Brd3*, it is at the mRNA level. Solid bars represent high levels of expression and dashed bars represent less abundant but detectable expression. *The figure is adapted from Shang et al. (2004).*

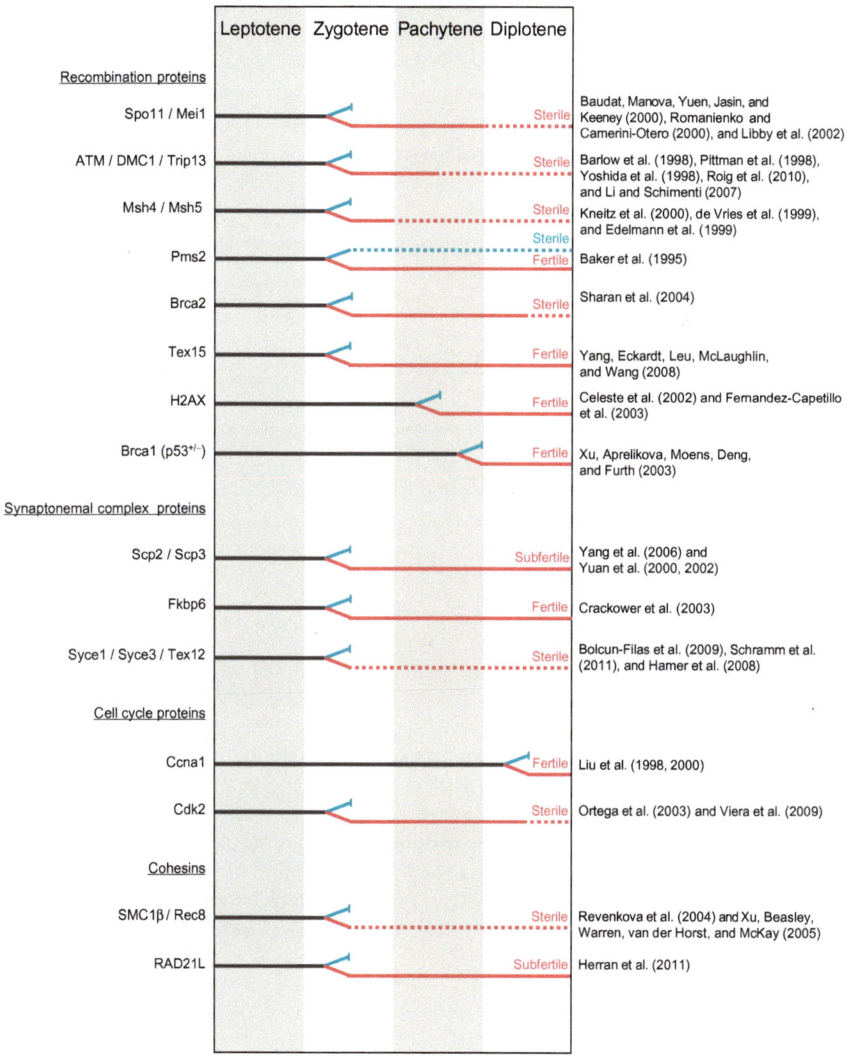

Leptotene
Zygotene
Pachytene
Diplotene
Recombination proteins
Spo11 / Mei1
Sterile
Baudat, Manova, Yuen, Jasin, and Keeney (2000), Romanienko and Camerini-Otero (2000), and Libby et al. (2002)
ATM / DMC1 / Trip13
Sterile
Barlow et al. (1998), Pittman et al. (1998), Yoshida et al. (1998), Roig et al. (2010), and Li and Schimenti (2007)
Msh4 / Msh5
Sterile
Kneitz et al. (2000), de Vries et al. (1999), and Edelmann et al. (1999)
Pms2
Sterile
Fertile
Baker et al. (1995)
Brca2
Sterile
Sharan et al. (2004)
Tex15
Fertile
Yang, Eckardt, Leu, McLaughlin, and Wang (2008)
H2AX
Fertile
Celeste et al. (2002) and Fernandez-Capetillo et al. (2003)
Brca1 (p53+/−)
Fertile
Xu, Aprelikova, Moens, Deng, and Furth (2003)
Synaptonemal complex proteins
Scp2 / Scp3
Subfertile
Yang et al. (2006) and Yuan et al. (2000, 2002)
Fkbp6
Fertile
Crackower et al. (2003)
Syce1 / Syce3 / Tex12
Sterile
Bolcun-Filas et al. (2009), Schramm et al. (2011), and Hamer et al. (2008)
Cell cycle proteins
Ccna1
Fertile
Liu et al. (1998, 2000)
Cdk2
Sterile
Ortega et al. (2003) and Viera et al. (2009)
Cohesins
SMC1β / Rec8
Sterile
Revenkova et al. (2004) and Xu, Beasley, Warren, van der Horst, and McKay (2005)
RAD21L
Subfertile
Herran et al. (2011)

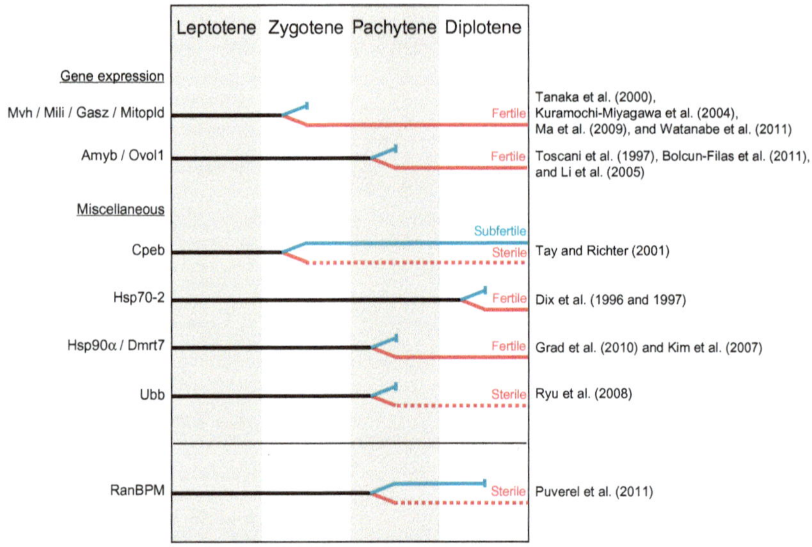

Figure 13.3 Diagram comparing known mouse mutants exhibiting an arrest during the prophase of the first meiotic division. A black solid line indicates stages during which male and female meioses are similar. When meiotic progression is different between the genders, a blue line indicates male progression and a purple line female progression. A meiotic arrest is represented by a solid vertical bar and dotted lines indicate that only a proportion of germ cells persists. Only genes for which information is available for both genders have been included. *Adapted from Hunt and Hassold (2002).*

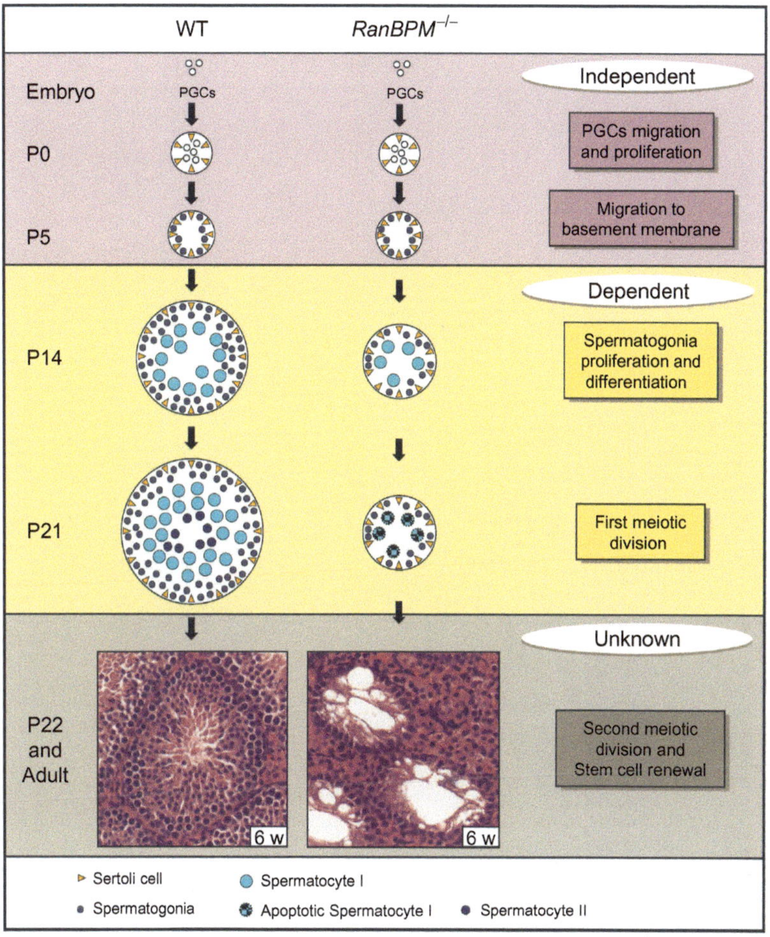

Sandrine Puverel and Lino Tessarollo, Figure 13.4 Model illustrating the role of RanBPM in spermatogenesis. Schematic representing the stages at which RanBPM plays a role in spermatogenesis. RanBPM is not required for primordial germ cell migration and proliferation and gonocyte migration to the basement membrane of developing seminiferous tubules in the newborn (purple box). However, it is required in the pubertal male at the crucial step of spermatogonia production and for the meiotic progression of primary spermatocytes (yellow box). Whether RanBPM is involved in the second meiotic division and the maintenance of adult spermatogenesis is still unknown, since the elimination of all germ cells in the young mutant does not allow the study of any time point past the meiotic arrest occurring during the prophase of the first meiotic division (gray box). P: postnatal day.

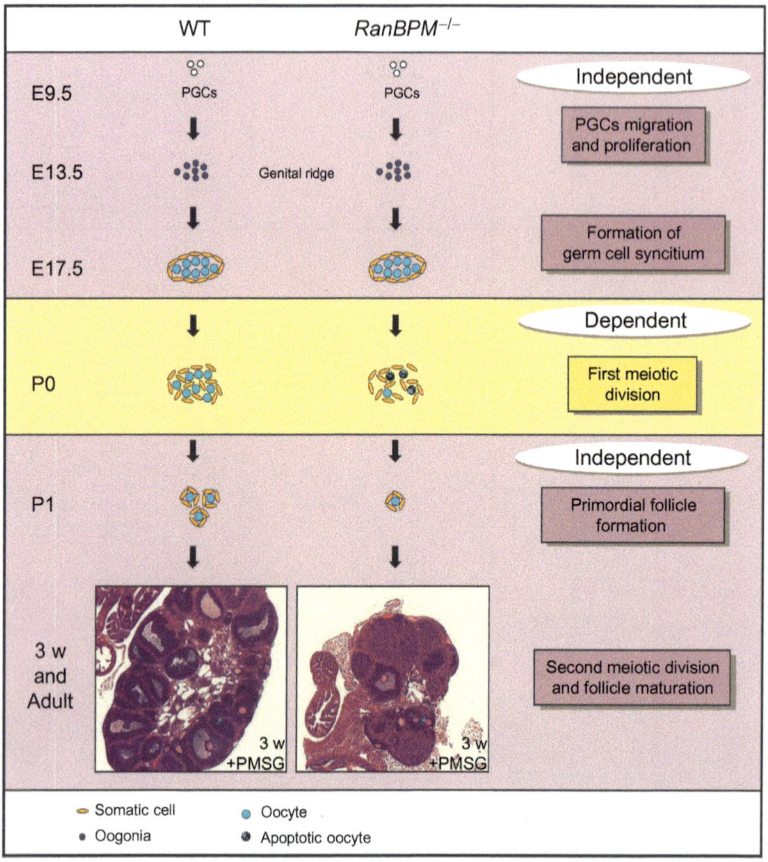

Sandrine Puverel and Lino Tessarollo, Figure 13.5 Model illustrating the role of RanBPM in oogenesis. Schematic representing the stages at which RanBPM plays a role in oogenesis. As in males, RanBPM is not essential for primordial germ cell proliferation and migration toward the developing gonad (top purple box). However, RanBPM is crucial for the meiotic progression of oocytes, since germ cells exhibit a meiotic arrest during the prophase of the first meiotic division, as in males (yellow box). The observation that some oocytes survive to form primordial follicles and that these follicles can mature upon PMSG stimulation suggests that RanBPM is not required for the maturation of germ cells in pubertal and adult mice, although the ability of these oocytes to produce embryos has not been assessed (bottom purple box). E: embryonic day and P: postnatal day.

FSC
www.fsc.org
MIX
Papier | Fördert gute Waldnutzung
FSC® C083411

Kolibri 360 GmbH
Ferdinand-Jühlke-Straße 7
99095 Erfurt, Deutschland
produktsicherheit@kolibri360.de